RECENT DEVELOPMENTS IN PLASMON-SUPPORTED RAMAN SPECTROSCOPY

45 Years of Enhanced Raman Signals

RECENT DEVELOPMENTS IN PLASMON-SUPPORTED RAMAN SPECTROSCOPY

45 Years of Enhanced Raman Signals

Editors

Katrin Kneipp
Berlin, Germany

Yukihiro Ozaki
Kwansei Gakuin University, Japan

Zhong-Qun Tian
Xiamen University, China

World Scientific

NEW JERSEY • LONDON • SINGAPORE • BEIJING • SHANGHAI • HONG KONG • TAIPEI • CHENNAI • TOKYO

Published by

World Scientific Publishing Europe Ltd.

57 Shelton Street, Covent Garden, London WC2H 9HE

Head office: 5 Toh Tuck Link, Singapore 596224

USA office: 27 Warren Street, Suite 401-402, Hackensack, NJ 07601

Library of Congress Cataloging-in-Publication Data
Names: Kneipp, Katrin, editor. | Ozaki, Y. (Yukihiro), editor. | Tian, Zhong-Qun, editor.
Title: Recent developments in plasmon-supported Raman spectroscopy : 45 years of enhanced Raman signals / edited by Katrin Kneipp (Berlin, Germany), Yukihiro Ozaki (Kwansei Gakuin University, Japan), Zhong-Qun Tian (Xiamen University, China).
Description: New Jersey : World Scientific, 2017. | Includes bibliographical references.
Identifiers: LCCN 2017030465 | ISBN 9781786344236 (hc : alk. paper)
Subjects: LCSH: Raman spectroscopy.
Classification: LCC QC454.R36 R43 2017 | DDC 543/.57--dc23
LC record available at https://lccn.loc.gov/2017030465

British Library Cataloguing-in-Publication Data
A catalogue record for this book is available from the British Library.

For any available supplementary material, please visit
http://www.worldscientific.com/worldscibooks/10.1142/Q0124#t=suppl

Desk Editors: Ram Mohan/Jennifer Brough/Shi Ying Koe

Typeset by Stallion Press
Email: enquiries@stallionpress.com

Printed in Singapore

About the Editors

Katrin Kneipp received her Diploma in Physics, Ph.D. degree and Dr.Sc. from the Friedrich-Schiller University in Jena, and facultas docendi in experimental physics from the Humboldt University Berlin. She has held various posts at German universities in Jena and Berlin and worked in industries and at the Academy of Sciences of the former German Democratic Republic. Katrin came as a Heisenberg fellow to MIT in 1993 and continued to work there as Rockefeller-Mauze Visiting Professor. Between 2003 and 2009, she was appointed as Associate Professor at Harvard University Medical School and at the Harvard-MIT Division of Health Sciences and Technology. After returning to Europe, she was Professor at the Department of Physics of the Danish Technical University until her retirement. Katrin is known for her pioneering work in the fields of single-molecule SERS spectroscopy, plasmonic spectroscopy, and bio-nanophotonics. She is a fellow of the American Physical Society and the Society for Applied Spectroscopy.

Yukihiro Ozaki obtained his Ph.D. (1978) in chemistry from Osaka University. He is currently a Professor in the Department of Chemistry, Kwansei Gakuin University, Sanda, Japan. He has been active in the research of a variety of molecular spectroscopy techniques, covering IR, Raman, NIR and far-ultraviolet (FUV) spectroscopy. His contributions to SERS/TERS spectroscopy involve: (i) mechanism of SERS, (ii) development of SERS/TERS spectrometers and experimental techniques, (iii) application of TERS to nanomaterials and liquid TERS, and (iv) three-dimensional SERS imaging. Prof. Ozaki has received many awards including the 1998 Tomas Hirschfeld Award, the 2005 Science and Technology Award of Japanese Government (Ministry of Education, Culture, Sports, Science and Technology), and Gerald Birth Award, Bomen-Michelson Award, and Chemical Society of Japan Award. Also, he is a fellow of the Royal Society of Chemistry, the Chemical Society of Japan and the Society for Applied Spectroscopy.

Zhong-Qun Tian received his B.S. in Chemistry from Xiamen University (1982) and his Ph.D., on electrochemical surface-enhanced Raman spectroscopy, supervised by Professor Martin Fleischmann (FRS), from the University of Southampton (1987). He then returned to China and has worked at Xiamen University ever since. He was appointed as a Full Professor in 1992. His main research interests are surface-enhanced Raman spectroscopy, spectro-electrochemistry, plasmonics and nanochemistry. Tian was elected as a Member of the Chinese Academy of Sciences, Fellow of Royal Society of Chemistry (2005), and Fellow of The World Academy of Sciences (2014). He is currently Director of Collaborative Innovation Center of Chemistry for Energy and Materials, a member of the advisory board for over 10 international journals including *J. Am. Chem. Soc., Chem. Sci., Chem. Comm. and Phys. Chem. Chem. Phys.*, and an Associate Editor for *Chem. Soc. Rev., J. Raman Spectrosc. and Sci. China Chem.*

Contents

Introduction

Recent Developments in Plasmon-Supported Raman Spectroscopy: 45 Years of Enhanced Raman Signals

During the early 1970s, Fleischman, Hendra, and McQuillan at the University of Southampton found surprisingly strong and potential-dependent Raman signals from pyridine adsorbed on a rough silver electrode. In 1977, Van Duyne and Jeanmaire and, independently, Creighton and Albrecht confirmed this observation and concluded that the enormously strong Raman signal measured from pyridine on a rough silver electrode cannot be explained by a larger number of molecules contributing to the scattering signal and must be caused by a true enhancement of the Raman scattering efficiency for the molecules adsorbed on the electrode surface. This might have lead to the term "surface-enhanced Raman scattering (SERS)". In 1978, it was proposed that the enhancement of the Raman signal can be understood in terms of surface plasmon resonances related to the bumps on the silver electrodes and that similar enhancement effects should exist for molecules adsorbed on metal colloids. On the other hand, since the early days of SERS, experiments suggested a second enhancement mechanism which might occur due to so-called "chemical or electronic effects". Charge transfer between molecule and metal is considered as most likely basic process for those "chemical contributions" to SERS.

Within the next years, a theory of "electromagnetic SERS enhancement" based on enhanced fields in the close vicinity of metal nanostructures which occur due to resonances between surface plasmons and the excitation and scattered optical fields was developed. It started with computing field enhancement for simple plasmonic structures, such as silver and gold spheres, nanorods, and dimers formed by nanospheres, followed by the discussion and computation of more complex plasmonic structures such as fractal aggregates and sophisticated nanoantennas. Experiments conducted in laboratories around the world demonstrated agreement with those computations and pointed on the electromagnetic enhancement mechanism as the key effect to explain SERS. So, from today's insight, plasmon-enhanced Raman scattering might be a more appropriate name for the phenomenon. New experimental methods for exploring plasmonic nanostructures, such as electron energy loss spectroscopy (EELS) as well as deeper theoretical insight into plasmonic spectra including understanding hybridization of plasmon modes, the role of dark modes and Fano resonances and considering non-local and quantum effects, have advanced the understanding of electromagnetic SERS enhancement.

Giant SERS enhancement has been obtained for composites of nanostructures such as fractal aggregates or regular top-down arrangement of nanoparticles with small interparticle gaps. Enhancement factors of about $10^{10} - 10^{11}$ were inferred for dye molecules on silver nanoaggregates in surface-enhanced resonance Raman experiments about 10 years after the discovery of the effect. At that time, it was already stated that SERS might have the capability to surpass fluorescence as tool for ultrasensitive detection and that single-molecule detection using SERS should be possible. Experiments such as vibrational pumping in non-resonant SERS identified unexpectedly large effective SERS cross-sections and showed that only very few molecules in highly confined areas, so-called hot spots, essentially contribute to the obtained SERS signal. Reports of single-molecule SERS attracted much attention and renewed the interest in SERS in the middle of the 1990s. One approach exploits extremely high local fields which exist for silver and gold nanoaggregates at near infrared excitation and measures single molecule Raman spectra under molecular non-resonant conditions. Another way to accomplish single molecule SERS is based on surface-enhanced resonance Raman scattering.

A combination of highly confined probed volumes produced by the local field of a plasmonic metal tip apex with scanning probe capabilities is employed in tip-enhanced Raman scattering (TERS). The idea was first put forward in 1985, and was realized in 2000 independently by several groups. Today, optimized TERS experiments can selectively probe the local chemical and electronic structure at nanometer spatial resolution. First observations of surface-enhanced femtosecond stimulated Raman scattering open up exciting new ways for probing ultrafast processes that might occur in plasmon-mediated interaction between molecules, surfaces and light.

Extremely high detection sensitivity along with molecular structural selectivity as well as capabilities of nanoscale resolution of SERS and TERS open up many fields of practical application in sensing and imaging. SERS imaging allows low-concentration molecules to be visualized. Nowadays one can develop even dynamic SERS imaging with 3D mapping in a living cell. TERS provides a label-free ultrahigh-resolution imaging technique for a variety of samples from 1D and 2D nanomaterials to DNA. Overall, Raman scattering performed in high local fields of plasmonic nanostructures allows us to detect and structurally identify single molecules and to collect vibrational information at subnanometer resolution. Vice versa, SERS signatures of single molecules can provide unique information on the plasmonic nearfield.

Almost 45 years after the first measurements of enhanced Raman signals, this book provides 16 chapters on recent developments in SERS and related techniques. In the first chapter, Mark Stockman points out some basic physics behind SERS. Tamitake Itoh provides an overview about the electromagnetic SERS enhancement and recent studies in this field. In particular, nonlinear or two-photon excited Raman processes benefit from plasmonic support since they scale with enhanced local fields to higher powers. Janina Kneipp and

Katrin Kneipp introduce surface-enhanced nonlinear Raman processes and discuss two-photon excited hyper Raman scattering for vibrational probing.

Most exciting current practical applications of SERS seem to occur for advanced sensing and imaging in the biomedical field. SERS nanosensors inside a biological object such as a living cell can provide information on structures and processes from molecular perspectives. Mostafa A. El-Sayed and Sajanlal R. Panikkanvalappil report elastic and inelastic light scattering of Au nanoparticles to detect and characterize molecular cell functions. Atsushi Taguchi discusses deep ultraviolet (DUV) light as an excitation source of SERS and TERS, particularly also for bio-analysis and imaging.

The development of nanotechnology enables progress in controlled preparation of plasmonic nanostructures and smart SERS-active structures for analytical applications and sensing. Moreover, well defined structures enable new basic experimental studies in SERS. Kenneth Crozier and Wenqi Zhu report lithographic fabrication of gold dimers with Angstrom-scale gaps. Such structures with controlled atomic scale gaps allow to study the impact of electron tunneling upon plasmonic near-field enhancement via SERS. The chapter by Dang Yuan Lei provides a comprehensive review on recent developments in the design and fabrication of hierarchically arranged porous nanostructures (and their complementary structures) based SERS substrates that exhibit excellent detection sensitivity and reproducibility as well as other new functionalities. Zhong-Qun Tian and Jian-Feng Li and their co-worker Rajapandiyan Panneerselvam report on shell-isolated nanoparticle-enhanced Raman spectroscopy (SHINERS). This new technique expands the generality of SERS and provides vast practical applications to various materials and can be used to investigate many chemical, electrochemical, biological and catalytic processes.

The new branch of TERS has brought about a surge recently, by demonstrating both very high spatial resolution and high sensitivity. Richard P. Van Duyne and George C. Schatz together with Naihao Chiang, Guillaume Goubert, Eric A. Pozzi, Michael

O. McAnally, Craig Chapman and Nan Jiang provide an introduction on TERS and discuss ultrahigh vacuum TERS (UHV-TERS). In particular, they discuss a combination of stimulated Raman time-resolved techniques with TERS which will enable investigations of transient behavior in molecular systems with nanoscale spatial resolution. Bin Ren together with Jin-Hui Zhong, Xiang Wang, Teng-Xiang Huang and Sheng-Chao Huang discuss the electromagnetic enhancement mechanism in TERS, sensitivity and spatial resolution of the technique. Application of TERS for surface and interface analysis in ambient air, ultrahigh vacuum, aqueous solution and electrochemical environment are discussed. Yukihiro Ozaki and Prompong Pienpinijtham focus on TERS in solution including a theoretical point of view as well as some promising potential applications of TERS in liquids and solutions. Yukihiro Ozaki, Sanpon Vantasin, Yoshito Okuno and Yuika Saito discuss TERS as a unique technique for probing very small portions of nanomaterials such as buckminister fullerene, graphenes, and carbon nanotubes. Zhenchao Dong together with Yao Zhang, Yang Zhang and Jianguo Hou describe their recent experiments on TERS in the UHV and low-temperature scanning tunneling microscope (STM). Using such STM-controlled TERS technique enables to access the structure and conformation of a single molecule with both chemical recognition and subnanometer resolution.

Despite that the key effect in SERS is caused by plasmonic field enhancement, understanding and modeling small chemical contributions to SERS and spectral changes related to chemisorption could be the basis for sensitive probes to explore surface processes. George C. Schatz, Lindsey R. Madison and Mark A. Ratner study the chemical effect of binding R6G to a silver surface and a silver nanoparticle model and calculate Raman spectra for R6G interacting with a 20-atom silver cluster, and with a flat Ag(111) surface.

Recently, graphene has attracted also some interest as a substrate for the enhancement of Raman signals of adsorbed molecules. The effect of graphene-enhanced Raman scattering (GERS) can be considered as a kind of model system for chemical SERS enhancement. Mildred Dresselhaus, Jing Kong, Xi Ling and Shengxi Huang provide an overview on the use of graphene and also other 2D materials as substrates for chemical enhancement.

Bing Zhao, Wei Ji and Xiao Xia Han report on charge-transfer-induced enhancement of Raman scattering based on nanostructured semiconductors including also some promising applications.

We cordially thank all our colleagues who contributed to this book for their time and effort for writing such excellent chapters on their recent work. Additionally, we are grateful to Dr. Ren Hu, Dr. Xiang Wang, Dr. Sanpon Vantasin and Dr. Tamitake Itoh who provided technical editorial support.

Plasmon-supported Raman spectroscopy is a rapidly emerging field and this book can only provide a small glimpse. Nevertheless, we hope that it will foster and stimulate future work in this exciting area and, maybe, attract young researchers and students to enter the field.

Katrin Kneipp
Yukihiro Ozaki
Zhong-Qun Tian

Chapter 1

Nanoplasmonics Fundamentals and Surface-Enhanced Raman Scattering as a Physical Phenomenon

Mark I. Stockman

Center for Nano-Optics (CeNO) and
Department of Physics and Astronomy
Georgia State University,
Atlanta, GA 30303, USA
mstockman@gsu.edu

1. Introduction

Giant surface-enhanced Raman scattering (SERS) discovered for organic molecules adsorbed on electrochemically roughened silver surfaces has served as a power impetus to the entire field of metal nano-optics called today as nanoplasmonics. It is an important fundamental phenomenon that exhibits the highest nanoplasmonic enhancement and allows for unprecedented access to the most localized and enhanced nanoplasmonic fields. On the other hand, over the years, it has become a powerful applied tool in diverse physical, chemical, and biomedical fields. In the following, we present a brief introduction to the nanoplasmonic fundamentals and the major electromagnetic pathways to SERS: resonant enhancement, structural enhancement, and adiabatic concentration.

SERS was discovered almost 45 years ago by Fleischman, Hendra, and McQuillan [1], followed by observations of Jeanmaire and

Van Duyne [2], and of Albrecht and Creighton [3]. As time passed, it has become abundantly clear that the major contribution to SERS of electromagnetic nature is due to the highly localized and enhanced nanoplasmonic fields [4]. Even before the term "nanoplasmonics" was introduced [5], theory was able to correctly reproduce the magnitude of the SERS enhancement, and even more importantly, to predict its spectral dependence with the enhancement increasing as the frequency is reduced toward near-infrared spectral region [6] where average enhancements much greater than 10^6 have been predicted.

A major breakthrough was the discovery of single-molecule Raman scattering by Kneipp *et al.* [7] and by Nie and Emery [8]. Though the Raman enhancements observed in these two publications were similar, their interpretations were quite contrasting. While in Ref. [7], the enhancement was attributed to molecules adsorbed on a colloidal clusters of silver, in contrast, in Ref. [8], it was interpreted as coming from single silver nanoparticles. In the course of further investigations, it was reliably established that the enhancement of SERS in silver colloids is due to the formation of fractal clusters of silver.

A fundamental question regarding the single-molecule SERS was whether it was originating from a single molecule or a few molecules. This was answered in Ref. [9] by using chemical spectral shift originating from vibrations of ordinary or heavy hydrogen atoms. It was shown that the observed shift was due to either hydrogen or deuterium atoms, but not an average value. This proves that a single active hot spot contributes to the SERS that can bind not more than one molecule of the analyte.

Thus, the existence of the single-molecule SERS from complex nanoplasmonic systems (roughened silver surfaces or fractal clusters of silver) is now established beyond a reasonable doubt. The major part of the SERS enhancement is of electromagnetic nature as was predicted two decades ago [6]. However, there is some evidence of the so-called chemical (or electronic) enhancement [10]; this is due to the modification of electronic states upon binding to metal and can also be considered as a result of the chemical modification of the molecules caused by the binding [4].

In this chapter, we will concentrate on the main electromagnetic mechanism of the SERS. There are three non-mutually exclusive pathways to such enhancement:

(i) **Resonant enhancement:** This is due to a general physical phenomenon of resonance where the field amplitude is transferred to and accumulated by a resonant object (plasmonic metal hot spot in this case) during many coherent cycles of the oscillations. In this case, the enhancement is determined by the quality factor of the resonance, $Q = \omega/(2\gamma)$, where ω is the resonant frequency and γ is the spectral width of the resonance. The Raman scattering is a linear two-photon process where one photon is absorbed and one photon spontaneously emitted. The probability of the absorption of the first photon is enhanced proportionally to the field intensity, i.e. by a factor of $\sim Q^2$. Due to the reciprocity (time reversal symmetry), the virtual electronic state excited by the first photon decays into a real photon with the rate that is likewise enhanced by a factor $\sim Q^2$. Thus, the SERS is enhanced by a factor

$$G^{\text{SERS}} \sim Q^4. \tag{1}$$

For a good plasmonic metal like silver, $Q \gtrsim 10^2$, so $G_{\text{res}}^{\text{SERS}} \gtrsim 10^8$.

(ii) **Shape (non-resonant) enhancement, also known as the "lightning rode effect":** As an example, consider a metal nanorod of length $l \lesssim \lambda$, where λ is wavelength. Electric field F of the optical wave polarized parallel to the nanorod produces potential difference with amplitude $\Delta U \sim Fl$. This potential difference is applied to a small-scale feature of the rod (a sharp nanotip with curvature radius ρ or to a narrow nanogap with a width ρ, where $\rho \ll l$) creating a field enhanced by a factor of $\sim l/\rho \gg 1$. Correspondingly, the rates in either absorption or emission channels are enhanced by a factor $\sim (l/\rho)^2$, which brings about a SERS enhancement by a factor

$$G_{\text{shape}}^{\text{SERS}} \sim (l/\rho)^4 \gg 1. \tag{2}$$

Note that this enhancement is multiplicative with the resonant enhancement (described above in point 1). Combined, they can render very high enhancement factors as in the case of a self-similar nanolens [11, 12]. Such nanolenses do indeed exhibit very high values of the SERS enhancement in the experiment [13]. A similar mechanism also causes enhancement in near-field scanning nanoscopy [14]. Note that the common type of optical nanoscopy, i.e. tip-enhanced Raman scattering (TERS) aperturless nanoscopy, is based mostly on the shape enhancement.

(iii) **Adiabatic concentration (nanofocusing) in plasmonic tapers [15]:** In this case, surface plasmon polaritons are excited at the surface of an adiabatically tapered nanotip. Due to the adiabaticity of the tapering, they propagate toward the sharp tip without experiencing back reflection or radiation into the free space. This leads to a three-dimensional nanofocusing, creating a hot spot at the tip of the taper with the intensity enhancement factor of $\sim (r/\lambda)^3 \gg 1$, where r is the radius of curvature of the tip. A molecule at the tip will experience SERS enhancement on the order of the square of this factor, i.e. $G^{\text{SERS}}_{\text{adiabatic}} \sim (r/\lambda)^6$. An advantage of the adiabatic nanofocusing is that it gives a very stable and robust hot spot at the tip. This approach has shown itself as a powerful tool in nanoscopy [16–21], especially in SERS nanoscopy and nanospectroscopy [22–33].

In the following sections, we will consider fundamentals and examples of SERS for the three basic mechanisms mentioned above.

2. Resonant Nanoplasmonic Mechanism of SERS

Nanoplasmonics is a branch of optical condensed matter science devoted to optical phenomena on the nanoscale in nanostructured metal systems. A remarkable property of such systems is their ability to keep the optical energy concentrated on the nanoscale due to modes called surface plasmons (SPs). It is well known [34] that the existence of SPs depends entirely on the fact that dielectric function ε_m has a negative real part, $\text{Re}\varepsilon_m < 0$. The SPs are well pronounced as resonances when the losses are small enough, i.e. $\text{Im}\varepsilon_m \ll -\text{Re}\varepsilon_m$. This is a known property of a good plasmonic metal, valid, e.g. for silver in most of the visible region. We will call a substance a good

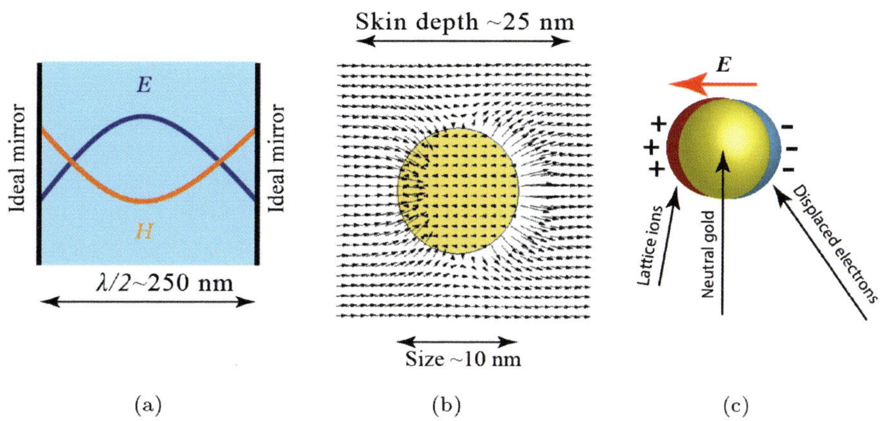

Figure 1. (a) Localization of optical fields by ideal mirrors and (b) by a gold nanoparticle. (c) Schematic representation of charge separation is shown in panel. (Reprinted from Ref. [35] with permission from The Optical Society of America.)

plasmonic metal if these two properties are satisfied simultaneously:

$$\mathrm{Re}\varepsilon_m < 0, \quad \mathrm{Im}\varepsilon_m \ll -\mathrm{Re}\varepsilon_m. \tag{3}$$

There is a limit to which an electromagnetic wave can be concentrated. We immediately note that, as we explain below, nanoplasmonics is about concentration of *electromechanical* energy at optical frequencies (in contrast to electromagnetic energy) on the nanoscale.

The scale of the concentration of electromagnetic energy is determined by the wavelength and can be understood from Figure 1(a). Naively, let us try to achieve the strongest light localization using two parallel perfect mirrors forming an ideal Fabry–Perot resonator. A confined wave (resonator mode) should propagate normally to the surface of the mirrors. In this case, its electric field **E** is parallel to the surface of the mirror. The ideal mirror can be thought of as a metal with a zero skin depth that does not allow the electric field of the wave to penetrate inside. Therefore, the field is zero inside the mirror and, due to the Maxwell boundary conditions, must be zero on the surface of the mirror. The same condition should be satisfied at the surface of the second mirror. Thus, the length L of this Fabry–Perot cavity should be equal to an integer number n of the half-wavelengths of light in the inner dielectric, $L = n\lambda/2$. The minimum length of this resonator is obviously $\lambda/2$. This implies that light cannot be confined tighter than to a length of $\lambda/2$ in each direction, with the minimum modal volume of $\lambda^3/8$.

One may think that it is impossible to achieve a localization of the optical energy to smaller volume than $\lambda^3/8$ by any means because the ideal mirrors provide the best confinement of electromagnetic waves. There are two implied assumptions: (i) the optical energy is electromagnetic energy, and (ii) the best confinement is provided by ideal mirrors. Both these assumptions must be abandoned to achieve nanolocalization of optical energy.

Consider a nanoplasmonic system whose size is less than or comparable to the skin depth,

$$l_s = \lambdabar \left[\mathrm{Re} \left(\frac{-\varepsilon_m^2}{\varepsilon_m + \varepsilon_d} \right)^{1/2} \right]^{-1}, \tag{4}$$

where $\lambdabar = \lambda/(2\pi) = \omega/c$ is reduced vacuum wavelength, and ε_d is the permittivity of the embedding dielectric. For single-valence plasmonic metals (silver, gold, copper, alkaline metals) $l_s \approx 25$ nm in the entire optical region.

For such a plasmonic nanosystem with a characteristic size $R \lesssim l_s$, the optical electric field penetrates the entire system and drives oscillations of the metal electrons. The total energy of the system in this case is a sum of the potential energy of the electrons in the electric field and their mechanical kinetic energy. While the magnetic field is present, non-relativistic electron interaction with it is weak, proportional to a small parameter $v_F/c \sim \alpha \sim 10^{-2}$, where v_F is the electron speed at the Fermi surface, c is speed of light, and $\alpha = e^2/\hbar c$ is the fine structure constant. Thus, in this limit, which is conventionally called quasi-static, the effects of the magnetic component of the total energy are relatively small. Hence, this total energy is mostly *electromechanical* (and not electromagnetic) energy. This is why the wavelength, which determines the length scale of the energy exchange between the electric and magnetic components of an electromagnetic wave, does not define the limit of the spatial localization of energy. Because the size of the system, R, is smaller than any electromagnetic length scale, of which the smallest is l_s, it is R that defines the spatial scale of the optical energy localization. Thus, the optical fields are confined on the nanoscale, and their spatial distribution scales with the system's size. This physical picture is at the heart of the nanoplasmonics.

Consider as an example a gold nanosphere of radius $R < l_s$, e.g. $R \sim 10$ nm, subjected to a plane electromagnetic wave, as shown in Figure 1(b). The field penetrates the metal and causes displacement of electrons with respect to the lattice resulting in the opposite charges appearing at the opposing surfaces, as illustrated in Figure 1(c). The attraction of these charges causes a restoring force that along with the (effective) mass of the electrons defines an electromechanical oscillator called an SP. When the frequency ω_{sp} of this SP is close to the frequency of the excitation light wave, a resonance occurs leading to the enhanced local field at the surface, as illustrated in Figure 1(b).

This resonant enhancement also has an adverse side: loss of energy is always associated with a resonance. The rate of this loss is proportional to $\text{Im}\varepsilon_m$ [36]. This leads to a finite lifetime of SPs. The decay rate of the plasmonic field γ is $\gamma \propto \text{Im}\varepsilon_m$. The corresponding expression has originally been obtained in Ref. [37] and is also reproduced as follows for convenience:

$$\gamma = \frac{\text{Im}s(\omega)}{\frac{\partial \text{Re}s(\omega)}{\partial \omega}} \approx \frac{\text{Im}\varepsilon_m(\omega)}{\frac{\partial \text{Re}\varepsilon_m(\omega)}{\partial \omega}}, \tag{5}$$

where $s(\omega) = \frac{\varepsilon_d}{\varepsilon_d - \varepsilon_m(\omega)}$ is Bergman's spectral parameter [34]. Note that γ does not explicitly depend on the system geometry but only on the optical frequency ω and the permittivities. However, the system's geometry determines the SP frequency, ω, and, thus, implicitly enters these equations.

Apart from γ, an important parameter is the quality factor,

$$Q = \frac{\omega}{2\gamma} = \frac{\omega \frac{\partial \text{Re}\varepsilon_m(\omega)}{\partial \omega}}{2\text{Im}\varepsilon_m(\omega)}. \tag{6}$$

Another, approximate expression for Q is

$$Q \approx \frac{-\text{Re}\varepsilon_m(\omega)}{\text{Im}\varepsilon_m(\omega)}. \tag{7}$$

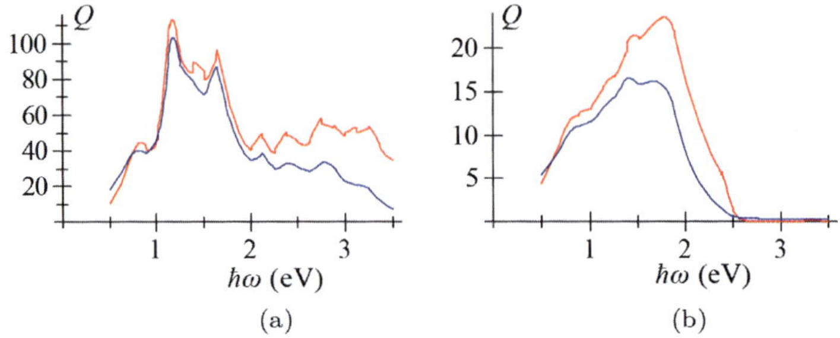

Figure 2. (a) Quality factor Q for silver and (b) for gold calculated according to Equations (6) (red) and (7) (blue) as a function of frequency ω. (Reprinted from Ref. [35] with permission from The Optical Society of America.)

The quality factor determines how many optical cycles free SP oscillations undergo before field decays. It also shows how many times the local optical field at the surface of a plasmonic nanoparticle exceeds the external field.

The SP quality factors, Q, calculated according to Equations (6) and (7) for gold and silver using the permittivity data of [38], are shown in Figure 2. The Q-factors found from these two definitions agree reasonably well in the red to near-infrared (near-ir) region, but not in the yellow to blue region of the visible spectrum. The reason is that these two definitions would be equivalent if metals' permittivity were precisely described by a Drude-type formula $\mathrm{Re}\varepsilon_m(\omega) = -\omega_p^2/\omega^2$, where ω_p is the bulk plasma frequency, and $\hbar\omega_p \approx 9$ eV for one-electron metals such as silver, copper, gold, and alkaline metals. This formula is reasonably well applicable in the red and longer wavelength part of the spectrum, but not in the yellow to blue part where the D-band transitions are important. Note that silver is a much better plasmonic metal than gold: its Q-factor is several-fold of that of gold.

As we have already mentioned, the SERS enhancement $G_{\mathrm{res}}^{\mathrm{SERS}} \sim Q^4$. More precisely, taking into account the dependence of Q on frequency, the enhancement in the excitation channel is $\sim Q^2(\omega_p)$, and that in the radiation channel is $\sim Q^2(\omega_R)$, resulting in

$$G_{\mathrm{res}}^{\mathrm{SERS}} \sim Q^2(\omega_p) Q^2(\omega_R), \tag{8}$$

where ω_p is the excitation (pumping) frequency, and ω_R is the radiation (Raman) frequency. In reality, the Raman shift is small enough, so a difference between $Q(\omega_p)$ and $Q(\omega_R)$ can be neglected, which brings about the simplest estimate for the resonant SERS enhancement

$$G_{\text{res}}^{\text{SERS}} \sim Q^4(\omega_p). \tag{9}$$

Equations (8) and (9) give qualitative description of the SERS electromagnetic enhancement coefficient, $G_{\text{res}}^{\text{SERS}}$. Quantitative expression for G^{SERS} in the quasi-static approximation was obtained in Ref. [39] using Green's function method. Note that it is mostly of resonant nature, but also includes shape enhancement as reproduced by the quasi-static approximation. In the case of strong enhancement, $G^{\text{SERS}} \gg 1$, assuming the excitation field of unit amplitude polarized along the z-direction, the result is expressed as

$$G^{\text{SERS}} = \frac{1}{|s(\omega_R)|^2} \sum_{\alpha=x,y,z} \left| \sum_{\beta=x,y,z} g_{\beta\alpha}(\mathbf{r}_0;\omega_R) g_{\beta z}(\mathbf{r}_0,\omega) \right|^2, \tag{10}$$

where $g_{\beta\alpha}(\mathbf{r}_0;\omega)$ is enhancement factor of the local field polarized along a β direction excited by the field polarized along the α direction ($\alpha, \beta = x, y, z$) at a frequency ω at a point $\mathbf{r}_0$. Strictly speaking, in contrast to Equation (8), SERS enhancement coefficient G^{SERS} (10) does not factorize into a product of the local field enhancements at the excitation and emission frequency. In fact, Equation (8) can be written in terms of the amplitude enhancement factors such as

$$G_{\text{res}}^{\text{SERS}} = \sum_{\alpha=x,y,z} |g_{\alpha z}(\mathbf{r}_0;\omega_R)|^2 \sum_{\alpha=x,y,z} |g_{\alpha z}(\mathbf{r}_0;\omega)|^2. \tag{11}$$

It agrees with Equation (10) only on the order of magnitude.

It is known experimentally that a large SERS enhancement is observed in fractal clusters of colloidal silver [4]. The colloidal clusters are large enough (thousands or more of metal–nanosphere monomers), so the numerical solution of the eigenmode problem is still above the existing computer capabilities for reasonable resources used. Therefore, we used dipolar spectral theory where only dipole fields of the monomers were taken into account [6].

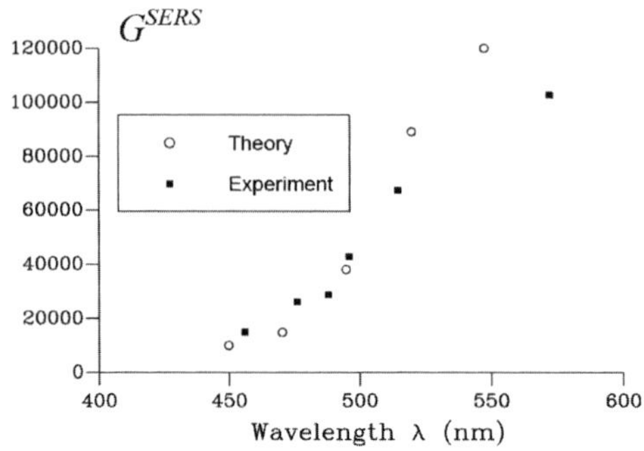

Figure 3. SERS enhancement factor G^{SERS} averaged over ensemble of fractal clusters of silver as a function of excitation wavelength λ. (Adapted from Ref. [6] with permission from the American Physical Society.)

The dependence of enhancement factor G^{SERS} averaged over binding sites in individual clusters and over the clusters is shown in Figure 3. We use the dielectric permittivity for silver from Ref. [38]. Clusters are simulated in the cluster–cluster aggregation model [40, 41]. As one can see from Figure 3, the SERS enhancement dramatically increases from the blue region where $G^{\text{SERS}} \lesssim 10^4$ to the red spectral region where it grows to $G^{\text{SERS}} \gtrsim 10^5$. This is an expected range and behavior of G^{SERS} that is due to the increase of the Q-factor toward the red region.

To understand the SERS enhancement in Figure 3, we note that Equations (8) and (9) give enhancement at an individual hot spot at the frequency of a resonant SP mode. If one averages over different hot spots, it would reduce the enhancement by a factor of $\gamma/\omega \sim Q^{-1}$ and result in the SERS resonant enhancement estimate

$$G_{\text{res}}^{\text{SERS}} \sim Q^3(\omega_p). \tag{12}$$

For a typical value of $Q \lesssim 10^2$, we obtain from Equation (12) $G_{\text{res}}^{\text{SERS}} \lesssim 10^6$ in a qualitative agreement with Figure 3.

From these data, it is also clear that the phenomenon of the single-molecule SERS cannot be explained by the averaged behavior of G^{SERS}. Therefore, we explore a cause of the SERS enhancement associated with the giant fluctuations of local fields existing in random, especially fractal, systems [42]. These fluctuations play an important role in single-molecule SERS because G^{SERS} is proportional to the square of the local field intensity enhancement, Q^4 [cf. Equation (9)] despite the fact that Raman scattering is an optically linear effect. As a result, the strong fluctuations cause higher average of and a very high maximum value of G^{SERS}. One can say that the single molecule SERS is a result of a giant positive fluctuation at the site where the Raman molecule is bound. This peculiarity has been well been understood starting from an early article Ref. [6].

Unfortunately, accurate many-multipole computations for large fractal clusters are still more complex than is possible to carry out with realistic volumes of computations at the present time. Therefore, we concentrate here on another model of random nanosystems, namely random planar composites (RPC). Such a composite is generated on a planar rectangular grid (say, in the xz plane) by randomly filling grid cells with metal monomers with some probability until a given fill factor f is achieved. The monomers are supposed to be small metal cubes in the xz plane with small but finite thickness in the y-direction.

To improve the numerical convergence and avoid singularities at the metal surface, the RPC is smoothed out with a Gaussian filter over a fraction of nm length. The eigenmodes and Green's function are found numerically using a finite difference method of the third-order accuracy in the grid step, as described in Ref. [43]. Note that in the quasi-static approximation, there is no dependence on the total size of the system, only on its form, since the wavelength λ is much larger than the system's size. Therefore, we give all dimensions arbitrarily on the nanometer scale with correct relative magnitudes.

We have calculated the enhancement factor of SERS from the exact formula (11). The corresponding results for G^{SERS} are presented in Figure 4 for a single molecule located at different planes above the metal surface. The frequency chosen of $\omega_p = 1.55$ eV corresponds to $\lambda = 800$ nm, i.e. to the radiation of a Ti:sapphire laser as one of the most common optical generators. As we see from panel (a), the enhancement is very large, up to $G^{\text{SERS}} \approx 6 \times 10^9$ for a molecule at the metal surface. This is a large value, but still much less than $G^{\text{SERS}} \sim 10^{12}$ believed to be needed for the observable

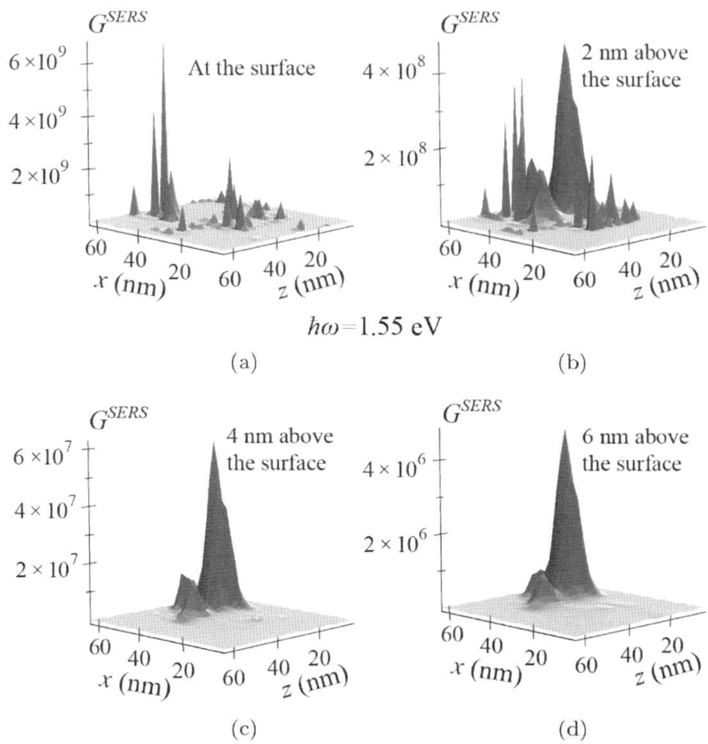

Figure 4. SERS enhancement factor G^{SERS} for single molecule situated at different planes above metal surface as a function of the position in the plane. (a) Molecule position is at the metal surface, and (b)–(d) for 2, 4 and 6 nm, respectively, above this surface. The excitation frequency is $\hbar\omega_p = 1.55$ eV, the Raman radiation frequency is $\hbar\omega_R = 1.5$ eV. (Adapted from Ref. [39] with permission from Springer.)

single-molecule SERS. Note that this value is still significantly greater than the simplified estimate $G^{\mathrm{SERS}}_{\mathrm{res}} \sim 10^8$ obtained from Equation (9).

With the distance from the metal surface increased by just a few grid steps nanometers, the SERS enhancement is drastically, by orders of magnitude, suppressed. This reduction in G^{SERS} is stronger than exponential and is certainly due to very high multipoles of the surface plasmon eigenmodes involved. Similar results have been obtained for other excitation frequencies (data not shown). Note that not only the maximum enhancement changes with the distance to the

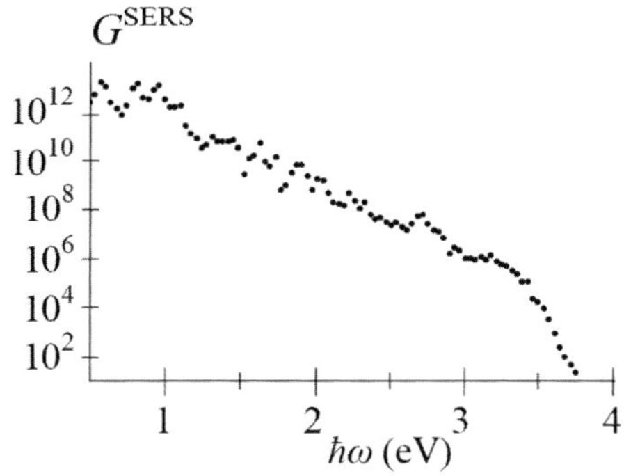

Figure 5. SERS enhancement factor for single molecule situated at the "hottest spot" at metal surface as a function of excitation frequency $\hbar\omega_p$. Raman frequency shift $\Delta\omega_R = \hbar\omega_p - \hbar\omega_R = 0.05$ eV. (Adapted from Ref. [39] with permission from Springer.)

metal RPC surface but also the spatial dependence of $G^{\rm SERS}$ experiences dramatic redistribution, i.e. the eigenmodes contributing at various distances are to a significant degree different.

For the same system, we show in Figure 5 the frequency dependence of the maximum enhancement factor of the single-molecule SERS. For each frequency, the Raman molecule is put at the position of the maximum enhancement (the Raman "hottest spot"). The Raman frequency shift $\Delta\omega^R = \hbar\omega_p - \hbar\omega_R = 0.05$ eV is constant for all these spectral points. As we see, there is a dramatic increase of the single-molecule SERS for the frequency changing from the near-ultraviolet to near-infrared, where it reaches the values $G^{\rm SERS} \sim 10^{12}$ typical for the single-molecule SERS.

3. Shape Enhancement of SERS in Nanolenses

The oldest, most established and employed method based on SERS for nanoscopy with chemical resolution uses a pointed plasmonic nanoprobe (nanocone), which is excited by a laser radiation focused on the tip and linearly polarized along the axis of the nanocone [44–48]. Sometimes, this method is also called the tip-enhanced Raman scattering (TERS) [49–51]. This is one of the most developed areas in the near-field optics and nanoplasmonics where many excellent reviews and monographs exist, see, e.g. Refs. [44, 49, 52].

Conventionally, it was believed that such pointed probes are not resonant and their field enhancement is due to the factor of shape — see Equation (2). From this estimate, it follows that the sharpest tip, which can only be fabricated when the surface of the metal is very smooth, will possess the highest enhancements. However, it was found both experimentally and computationally that this was not the case in the visible range [49]. In fact, the tips that have very rough surfaces, which remind by their structure fractal clusters, give the highest enhancements. Thus, the mechanisms of TERS and SERS appear to be similar and, to a significant degree, also combination of resonance and shape enhancements.

It appears to be very attractive to have a fully controllable and reproducible SERS system that provides both the resonant and shape enhancement factors. This was introduced in Ref. [11] as a self-similar nanolens and will be described below.

The fact is that the fractals are the most efficient enhancers of the SERS [4, 53]. This is likely to be related to two reasons: (i) self-similarity of fractals and (ii) giant fluctuations of local optical fields. As we argued, the self-similarity in random systems leads to giant fluctuations [42]. However, it has been important to realize that fractality by itself can lead to the giant enhancement of local fields [11].

The enhancing effect of the self-similar geometry can be qualitatively understood from the following arguments. First, as soon as the size of the system is much less than λ, the plasmonic mode frequencies stop depending on size; they only depend on the form factor. Consider a fractal system on the largest scale within the self-similarity range. There are plasmonic eigenmodes localized at this maximum scale. Such eigenmodes resonant to the external radiation will generate enhanced local fields whose characteristic extension is also this maximum scale. Therefore, these large-scale local fields are almost uniform on some much smaller scale that is also present due to the self-similarity of the fractal. These fields, whose magnitude is enhanced by the Q-factor, play a role of the external excitation for the smaller-scale eigenmodes. These smaller-scale eigenmodes will respond in kind creating on their scale a local field enhanced by a

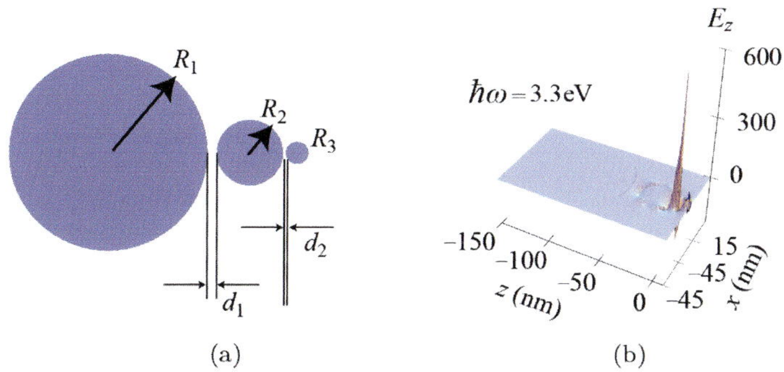

Figure 6. (a) Geometry of a silver nanolens consisting of three silver nanospheres is shown in the cross-section through its plane of symmetry. Nanosphere radii are: $R_1 = 45$ nm, $R_2 = 15$ nm, and $R_3 = 5$ nm. The separations between the nanosphere surfaces are $d_1 = 4.5$ nm and $d_2 = 1.5$ nm. (b) Local field enhancement E_z is computed from Eq. (10) and is shown as a function of the position in the cross-section through the xz plane of symmetry. (Adapted from Ref. [11] with permission from the American Physical Society.)

factor of $\sim Q^2$. This transfer of excitation toward the progressively smaller scales enhances the local field by a factor of Q each time it takes place. It is, however, limited by the minimum scale of the fractal.

The above-described idea of the self-similar transfer of excitation down the spatial scales can be implemented in a simple, fully controllable, and reproducible system shown in Figure 6(a). This system comprises a linear aggregate of three silver nanospheres whose radii decrease by a factor of 1/3 from one sphere to another (from bottom to top). This specific reduction factor should be significantly less than 1, but its precise value is not of a principal importance. The larger nanosphere's radius should be much smaller than wavelength, and the minimum radius should be not too small to use the continuous electrodynamics. Our specific choice of the three nanosphere radii is $R_1 = 45$ nm, $R_2 = 15$ nm, and $R_3 = 5$ nm. The gaps between surfaces of the nanosphere are chosen to be ≈ 0.3 of the radius of the smaller one of the adjacent nanospheres, i.e. $d_1 = 0.3R_2 = 4.5$ nm, and $d_2 = 0.3R_3 = 1.5$ nm. This is a finite, deterministic, self-similar system that possesses the above-discussed mechanism of enhancement based on the excitation transfer down spatial scales.

The external field with frequency close to the nanosphere surface plasmon resonance excites the local field around the largest nanosphere enhanced by a factor of $\sim Q$. The next nanosphere is small enough not to perturb this resonance significantly. This local field is nearly uniform on the scale of the next, smaller nanosphere and plays a role of the excitation field for it. This in turn creates the local enhancement by a factor of $\sim Q^2$. Similarly, the local fields around the smallest nanosphere are enhanced by a factor of $\sim Q^3 \sim 10^3$ for the realistic $Q \sim 10$ in the blue spectral region where the nanospheres have their surface plasmon resonances. This process cannot be continued further because the next nanosphere would be too small for the continuous electrodynamics to apply.

The nanolens problem solution has been found in Ref. [11] in the quasi-static approximation using a multipolar expansion for Green's function. The local fields shown in Figure 6(b). As we see, there is the hottest spot of local fields (nanofocus of this nanolens) present in the smallest gap at the surface of the intermediate nanosphere. At this spot, the local field is enhanced by a factor of $E_z \approx 500$ in an excellent qualitative agreement with our order-of-magnitude estimate. From Eq. (9), we can estimate the SERS enhancement factor as $G^{\mathrm{SERS}} \gtrsim 10^{10}$, which approaches a magnitude needed for single-molecule SERS.

A full electrodynamic solution for the three-nanosphere self-similar nanolens has been obtained in Ref. [12]. It is in a semi-quantitative agreement with the above-discussed quasi-static theory. For small nanolenses, the results are consistent. As the size of the nanolens increases, the full electrodynamic solution shows resonances. In a resonance, the enhancement may exceed that predicted quasi-statically; out of resonance, it is smaller, e.g. for a $R_3 = 5$ nm smallest nanosphere, by a factor of ≈ 3. For $R_3 \leq 2.5$ nm, the full electrodynamic and quasi-static theories are in quantitative agreement.

Experimentally, the enhanced SERS from nanolenses was observed in Ref. [13] where gold nanolenses obtained by ablation were shown to provide enhancement $G^{\mathrm{SERS}} \sim 10^9$. Yet another experimental progress was top-down nanofabrication of various nanolenses with the number of the individual nanospheres ranging between three and eight [54, 55]. These nanospheres provided single-molecule or a few-molecule SERS sensitivity and were used, in particular, to detect point-like mutations in peptides related to the protein encoded by the BRCA1 (breast cancer suppressor) gene [54].

Note that recently there has been much active work done on other approaches to synthesis, SERS, and fluorescence-enhancement applications of the nanolens substrates [14, 56–62]. Currently, it is clear that such fully controllable and purposely designed nanolens optically enhancing substrates are highly promising for biomedical, environmental, defense, and other applications.

4. Adiabatic Nanofocusing and SERS

The central problem of the nano-optics is the delivery and concentration (nanofocusing) of the optical radiation energy on the nanoscale. This is a formidable problem because the wavelength of light is on the microscale, many orders of magnitude larger than the size of a nanosystem. Coupling laser radiation to the nanoscale through, e.g. tapered optical fibers or by focusing on metal tips leads to an enormous loss: only a miniscule part of the excitation energy is transferred to the nanoscale. For instance, for a nanosphere, the optical absorption cross-section at the maximum of the plasmonic dipolar eigenmode is

$$\sigma_a = 3\pi\lambdabar^2 \left(\frac{R}{\lambdabar}\right)^3 Q, \tag{13}$$

where $\lambdabar = \lambda/2\pi$ is reduced wavelength. For a small nanoparticle, $R \ll \lambdabar$, then the absorption cross-section is small, $\sigma_a \ll \lambda^2$. Thus, for a typical focus-spot radius of ~ 1 micron, only a small fraction of optical energy will be absorbed. Increasing the size of the nanoparticle to $\sim\lambdabar$ will not be useful because it will lead to a greatly increased radiative loss.

We have proposed [15] an approach to nanofocus and concentrate in three dimensions the optical radiation energy on the nanoscale without major losses. This can be done by exciting the surface plasmon polaritons (SPPs) propagating toward a tip of a tapered surface-plasmonic waveguide. This propagation of SPPs causes their rapid adiabatic slowing down and asymptotic stopping. This phenomenon leads to giant concentration of energy on the nanoscale. The SPPs are adiabatically transformed into localized surface plasmons (SPs) that are purely electric oscillations that can and do nanolocalize [63], leading to the 3D nanofocusing. This leads to the appearance of a hotspot of local optical fields at the nanotip, which is an efficient source of energy on the nanoscale. A great advantage of this source is that there is no background of the excitation radiation that illuminates the tip and the substrate in a direct excitation used in nanoscopy.

To illustrate this phenomenon of the rapid adiabatic nanofocusing in 3D, we discuss the following example. Consider a tapered nanoplasmonic waveguide, which is a silver cone in vacuum; its angle of opening is 0.04 radian, as shown in Figure 7(a). The vacuum reduced wavelength of the excitation radiation is $\bar{\lambda} = 100$ nm, which corresponds to red light of $\lambda = 630$ nm.

The SPPs are efficiently excited at the wide end of the waveguide nanowire by using, e.g. grating or Kretschmann geometry and propagating to the tip as indicated. This propagation causes accumulation of the SPP energy at the tip and the corresponding increase of the local fields by more than three orders of magnitude. As shown in

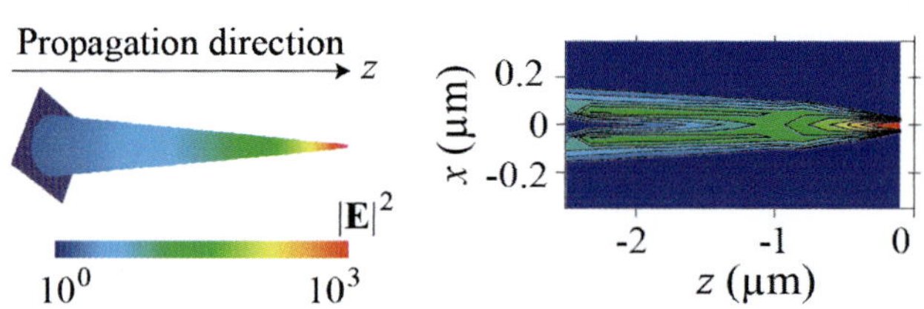

(a) (b)

Figure 7. (a) Geometry of the nanoplasmonic waveguide. The propagation direction of the SPPs is indicated by the arrow. Intensity $I(\mathbf{r}) = |\mathbf{E}(\mathbf{r})|^2$ of the local fields relative to the excitation field is shown by color. The scale of the intensities is indicated by the color bar in the center. (b) Local electric field intensity $|\mathbf{E}(\mathbf{r})|^2$ is shown in the longitudinal cross-section of the system, normalized to the initial intensity. The radius of the waveguide gradually decreases from 50 to 2 nm. (Adapted from Ref. [15] with permission from the American Physical Society.)

Figure 7(b), the intensity of the local optical field is sharply concentrated in 3D in a nanolayer at the surface of the metal, which is a signature of SPPs. The hot spot of local fields is created in a nanosize region at the very tip.

If it was not a plasmonic nanowire waveguide, but the conventional tapered optical fiber supporting guided photonic modes, then there would be a cut-off at some waveguide radius beyond which the propagation is not possible [64], the wave would be reflected back with only a short evanescent tail in the forward direction. Then no field enhancement would occur, and only an exponentially small part of the incident energy would reach the tip.

Note that it was suggested earlier [65] that the propagation of SPPs toward the tip can produce energy concentration. Similarly, the nanofocusing of optical energy due to propagation toward the tip of a plasmonic wedge or a cone has also been predicted [66, 67]. However, in contrast to Ref. [15], these articles have not introduced the adiabatic concentration as a general approach to connecting, reversibly — in both the directions — optical far field and near field.

The physical reason that the nanoplasmonic waveguide is an efficient energy concentrator can be inferred from Figure 8(a). Both the phase and group velocities of SPPs asymptotically tend to zero toward the nanotip. Consequently, the SPPs are slowed down and adiabatically stopped at $z \to 0$, which leads to their accumulation

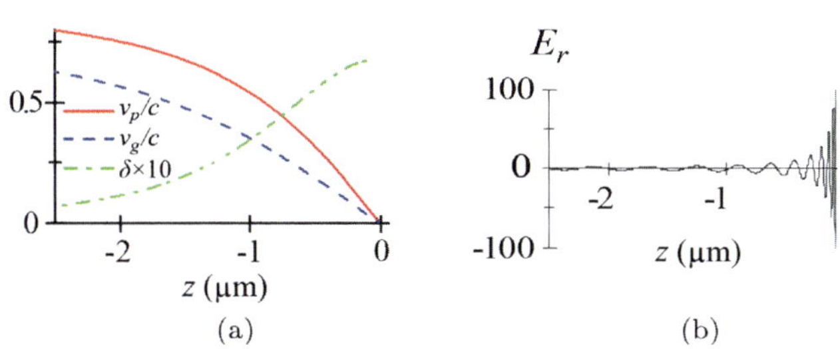

Figure 8. (a) Phase velocity v_p, group velocity v_g, and adiabatic parameter δ (scaled by a factor of 10) are shown as functions of the plasmonic waveguide radius. (b) Radial optical electric field E_r at the surface of the metal nanocone waveguide in the units of the excitation field against the coordinate z in the propagation direction. (Adapted from Refs. [15, 68] with permission from the American Physical Society.)

at the tip. Correspondingly, in Figure 8(b), the local optical field is oscillating in space with wavelength progressively decreasing and its amplitude increasing by more than an order of magnitude. The highest enhancement is in fact limited only by the minimum tip size that can be considered on the basis of continuous electrodynamics. Importantly, being adiabatic to prevent the back reflection and 3D scattering, this process should be as rapid as possible to prevent losses in the metal.

Theoretically, we consider a nanoplasmonic waveguide that consists of a metal nanocone whose axis coincides with the coordinate z-axis and whose dielectric function $\varepsilon_m(\omega)$ is uniform in space, where ω is the optical excitation frequency. The local radius, $R(z)$, of this nanocone is a smooth function of z and is assumed to decrease adiabatically from microscale for z large negative to a nanoscale size at $z \to 0$, as discussed above, see Figure 7. This nanocone is surrounded by a dielectric medium with dielectric constant ε_d. Using the adiabatic nature of the tapering, we will employ the eikonal approximation [36], also called the Wentzel–Kramers–Brillouin (WKB) or quasi-classical approximation in quantum mechanics [69].

We consider an axially uniform SPP mode that is a TM_{00} wave whose magnetic field has the ϕ polarization, and electric field $\mathbf{E}$ has both transverse (radial) component E_r and longitudinal component E_z. In the eikonal (WKB) approximation, this field has the form

$$\mathbf{E}(\mathbf{r}, z, t) = \mathbf{E}_0(\mathbf{r})A(z)\exp[ik_0\varphi(\mathbf{r}) - i\omega t], \tag{14}$$

where $\mathbf{r}$ is a 2D vector in the xy plane, $k_0 = 1/\lambda\!\!\!^{-}$ is the wave vector in vacuum, and $A(z)$ is a slow-varying pre-exponential factor with boundary value $A(0) = 1$. From the Maxwell equations, using the corresponding boundary conditions at the interface, for the SPP guided mode, we find the phase (eikonal) as

$$\varphi = k_0 \int n(z)dz, \tag{15}$$

where $n(z)$ is the effective SPP index of the plasmonic waveguide at a point z, which is determined by an equation

$$\frac{\varepsilon_m}{\kappa_m}\frac{I_1(k_0\kappa_m R)}{I_0(k_0\kappa_m R)} + \frac{\varepsilon_d}{\kappa_d}\frac{K_1(k_0\kappa_d R)}{K_0(k_0\kappa_d R)} = 0, \tag{16}$$

where I_p and K_p ($p = 0, 1$) are the modified Bessel functions, and the complex decrements of the field in the metal and dielectric are $\kappa_m = \sqrt{n^2 - \varepsilon_m}$ and $\kappa_d = \sqrt{n^2 - \varepsilon_d}$. This equation determines n as a function of the local wire radius R, which together with the grading dependence $R = R(z)$ defines the required effective index $n(z)$. Under the conventional plasmonic condition $\text{Re}\varepsilon_m < -\varepsilon_d$, Equation (16) has nearly real solutions corresponding to the propagating SPPs. For a thick wire ($k_0 R \gg 1$), the solution is, understandably, the same as for the flat surface, $n = \sqrt{\varepsilon_m \varepsilon_d / (\varepsilon_m + \varepsilon_d)}$. For a thin, nanoscale-radius wire ($k_0 R \ll 1$) with logarithmic precision, we have

$$n(R) \approx \frac{1}{k_0 R} \left[-\frac{\varepsilon_m}{2\varepsilon_d} \left(\ln \sqrt{-\frac{4\varepsilon_m}{\varepsilon_d}} - \gamma \right) \right]^{-1/2}, \tag{17}$$

where $\gamma \approx 0.57721$ is the Euler constant. Note that at the tip $n \to \infty$, SPPs do asymptotically stop, i.e. both the phase velocity $v_p = c/n$ and group velocity $v_g = c[d(n\omega)/d\omega]^{-1}$ tend to zero $\propto k_0 R$ for $k_0 R \to 0$. The point $R = 0$ (or, equivalently, $z = 0$) is an essential singularity. The time, t_0, to reach this point diverges logarithmically.

$$t_0 \propto \int n(R) dR \propto -\ln(k_0 R) \to \infty. \tag{18}$$

The eikonal parameter (also called WKB or adiabatic parameter) is defined as $\delta = |R' d(k_0 n)^{-1}/dR|$, where $R' = dR/dz$ is the wire grading. For the applicability of the eikonal (WKB) approximation, it is necessary and sufficient that $\delta \ll 1$. At the nanoscale tip of the wire, which is the critical site for the adiabaticity (eikonal approximation applicability), from we obtain

$$\delta \approx \left| R' \left[-\frac{\varepsilon_m}{2\varepsilon_d} \left(\ln \sqrt{-\frac{4\varepsilon_m}{\varepsilon_d}} - \gamma \right) \right]^{1/2} \right|. \tag{19}$$

Thus, δ stays finite at the tip and can always be made small enough by choosing sufficiently small grading R', so the eikonal (WKB) approximation is valid for the entire wire waveguide, including the stopping point at the tip. This conclusion does not rely or significantly depend on $\text{Im}\varepsilon_m$ (the optical losses in the system).

We consider a conic nanowire of silver with $R' = -0.02$, see Figure 7(a). In Figure 8(a), along with the phase velocity v_p and group velocity v_g, we also show the adiabatic (WKB) parameter, δ. It is of principal importance that this adiabatic parameter does stay finite and small, $\delta \le 0.07$, throughout the entire length of the system, ensuring the global applicability of the WKB (eikonal) approximation, including the essentially singular point at $z = 0$.

The SPP electric fields are found from the Maxwell equations in the WKB approximation in the form:

$$
\begin{aligned}
E_z(r,z) &= \theta(R-r)I_0(k_0\kappa_m r) + \theta(r-R)BK_0(k_0\kappa_d r), \\
E_z(r,z) &= \theta(R-r)i\frac{n}{\kappa_m}I_1(k_0\kappa_m r) + \theta(r-R)i\frac{n}{\kappa_d}BK_1(k_0\kappa_d r),
\end{aligned} \tag{20}
$$

where $B = I_0(k_0\kappa_m R)/K_0(k_0\kappa_m R)$, and $\theta(\ldots)$ denotes the Heaviside unit-step θ-function.

To determine the pre-exponential $A(z)$ in Eq. (20), we use the energy flux conservation in terms of the Pointing vector flux through the normal (xy) plane, obtaining

$$
A \propto \mathrm{Re}\left[\frac{n^*\varepsilon_m^*}{|\kappa_m|^2}|K_0(k_0\kappa_d R)|^2\int_0^R |I_1(k_0\kappa_m r)|^2 r dr \right.
$$
$$
\left. + \frac{n^*\varepsilon_d^*}{|\kappa_d|^2}|I_0(k_0\kappa_m R)|^2\int_R^\infty |K_1(k_0\kappa_d r)|^2 r dr\right]^{-\frac{1}{2}}, \tag{21}
$$

where all the spatially varying quantities, n, κ_m, and κ_d, are functions of local radius R of the wire, as originally given by Equation (16). The required dependence, $A = A(z)$, is obtained by substituting the grading relation $R = R(z)$. We only indicate the proportionality of A: the total scale of A is undetermined by the equations and is defined by the total power of the propagating SPP wave. This completes the eikonal (WKB) solution.

The intensity of the optical electric fields has already been discussed in conjunction with Figure 7 where they are shown on the logarithmic scale. For this example, we set the minimum radius of the wire to be $R_{\min} = 0.02\lambda = 2$ nm to avoid the effects of the spatial dispersion of the dielectric response that are important at shorter distances, cf. Refs. [70, 71], and the maximum radius (at $z = -2.5\mu$m)

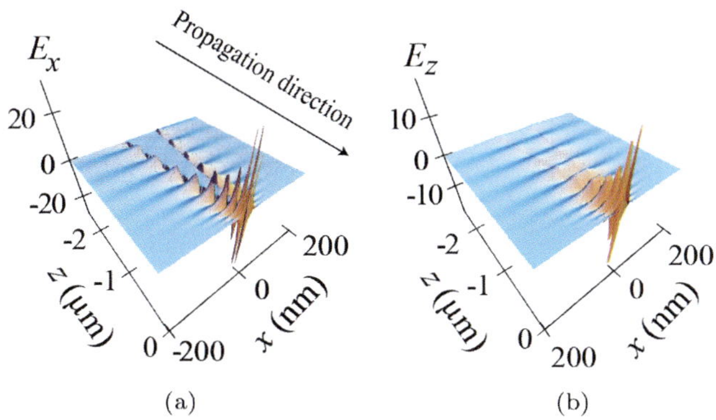

Figure 9. Snapshot of instantaneous fields (at some arbitrary moment of time): (a) Normal component E_x of the local optical electric field is shown in the longitudinal cross-section (xz) plane of the system. (b) The same as (a) but for longitudinal component E_z. The fields are in the units of the far-zone (excitation) field. (Adapted from Ref. [15] with permission from the American Physical Society.)

to be $R_{\text{max}} = 0.5\lambda\!\!\!^{-} = 50$ nm. In Figure 9, we display the amplitudes of the local optical fields in the cross-section of the system for the normal (E_x) and longitudinal (E_z) components of the optical electric field. In Figure 9, far from the tip of the nanoplasmonic waveguide, the optical electric field is mostly transverse, extending in vacuum to distances $\sim \lambda$ where most of the SPP field is propagating. The longitudinal field in the metal is very small, proportional to a factor of $|\varepsilon_d/\varepsilon_m| \ll 1$, as should be from the boundary conditions. Therefore, the guide itself is clearly seen in panel (a) as the acute triangular region of low fields. As SPPs move toward the tip, the SPP fields start to localize at the metal surface, and simultaneously, their wavelength is progressively reducing and amplitude growing. Because the very tip is not included, the singularity point of the fields does not show in these figures. Even with this truncation, the field magnitudes grow significantly at small $|z|$. The transverse (x) component grows by an order of magnitude as the SPPs approach the tip of the guide, while the longitudinal (z) component, which is very small far from the tip, grows relatively much stronger. Close to the tip, both these components are on the same order of magnitude, as is expected for the localized excitations. This growth in magnitude

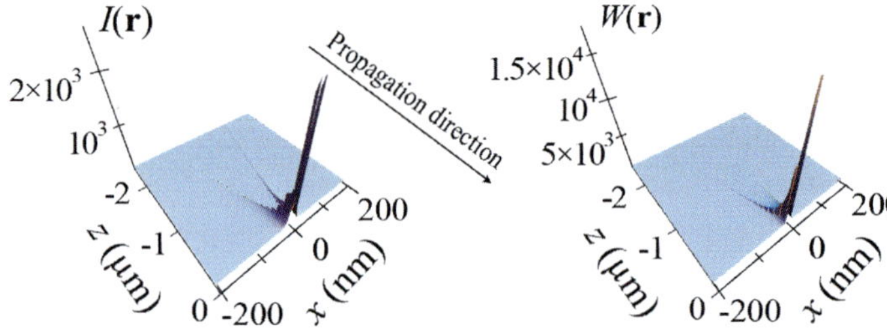

(a) (b)

Figure 10. (a) Intensity $I(\mathbf{r})$ and (b) energy density $W(\mathbf{r})$ of the local optical electric field in the xy plane of the system. The magnitudes are relative to those of the excitation wave. (Adapted from Ref. [15] with permission from the American Physical Society.)

is concurrent with the energy localization in 3D and the significant reduction of the wavelength, which are due to the dramatic slowing down of the SPPs. Note the SPPs in Figure 9 are not standing but running waves; the fields shown represent an instantaneous snapshot of those waves.

In Figure 10, we show the spatial behavior of the relative intensity defined as $I(\mathbf{r}) = \mathbf{E}^2(\mathbf{r})$, and the energy density [36] of the local optical electric field, $W(\mathbf{r}) = \frac{1}{4\pi}\frac{d\omega\varepsilon(\mathbf{r},\omega)}{d\omega}|\mathbf{E}(\mathbf{r})|^2$. The intensity grows by more than three and energy density by four orders of magnitude at the tip. If these fields were used to induce SERS, it would be enhanced by a factor of $G^{\mathrm{SERS}} \sim I(\mathbf{r})^2 \sim 10^7$. This is likely to be below the enhancement $G^{\mathrm{SERS}} \sim 10^{10} - 10^{12}$ believed to be needed for single-molecule SERS, but still sufficient for extremely high sensitivity up to a-few-molecule sensitivity.

Further enhancement of SERS by many orders of magnitude can be achieved by positioning a resonant nanolens [11] at the tip of the nanoplasmonic taper. Note that $|\mathbf{E}|^2$ is concentrated at the outer surface of the metal nanowire, with the exponential decay far away from this surface. In contrast, a metal nanoparticle excited by an external field would produce dipolar local fields where $|\mathbf{E}|^2 \propto r^{-6}$. If a nanoparticle or a cluster thereof (nanolens) is positioned at the tip of a plasmonic adiabatic taper, then the enhancement is increased multiplicatively. Such a configuration — a nanolens at the tip of a taper — was experimentally implemented and used for nanoscopy of single molecule fluorescence with a record-setting combination of high sensitivity and spatial resolution [14].

The importance of the general concept of adiabatic nanofocusing was already well recognized in the original publication in Ref. [15] where a concept and phenomenon of adiabatic nanofocusing in three dimensions were introduced. This phenomenon constitutes a foundation of high-efficiency coupling of the far-field radiation to the near-field zone where the energy concentration occurs at the tip of a smoothly tapered metal nanoplasmonic waveguide. This causes the local field increase by three orders of magnitude in intensity and four orders in energy density. The stopping of SPPs is asymptotic, i.e. they need logarithmically divergent time to reach the tip, which mathematically is the point of an essential singularity. Similar phenomena are likely to exist for a hollow tapered waveguides, in particular subwavelength holes, which was used to produce high harmonics on the nanoscale [72]. The adiabatic nanofocusing finds various applications in nano-optics and nanotechnology where greatly enhanced local optical fields are required, in particular, for probing, spectroscopy, and detection, including SERS, nanomodification and nanofabrication, and also in heat-assisted magnetic recording.

The adiabatic compression is a non-resonant effect whose spectral range is that of SPPs on metal surfaces; it spans over the entire visible and near-infrared range and can be extended to mid-infrared and even THz range [73]. Correspondingly, it is an ultra-fast, femtosecond or even subfemtosecond phenomenon, which can be coherently controlled by pulse shaping [74].

There has been a large body of both theoretical and experimental work done using the principle of adiabatic nanofocusing (or, synonymously, nanoconcentration).

Historically, the first observations of the adiabatic nanofocusing were made using 2D tapers [76–78]. This has provided a proof of principle and has shown that it is possible to transfer optical energy from microscale to nanoscale with a throughput of up to 50%.

Several main configurations for the adiabatic nanofocusing have been developed. The most common of them is the original one using a plasmonic conic taper with a coupler that converts the far-field radiation into SPPs. We will briefly discuss its numerous applications in the discussion at the end of this chapter. Note that an alternative configuration, which is much less common, is a tapered metal–dielectric–metal nanowaveguide [79–81].

Experimentally developed applications of the adiabatic nanofocusing include scattering and nonlinear nanoscopy, which are free from background, with a high spatial resolution [16–19]. The ultrafast nature of the adiabatic concentration has been used to achieve simultaneously nanoscopic spatial resolution and femtosecond temporal resolution [19–21, 82–89]. A unique application of the adiabatic nanofocusing is a new type of nanoscopy with chemical vision based on a highly efficient (with a quantum yield of 10%) hot electron generation [90].

The adiabatic nanofocusing has opened up a new and highly promising route for SERS nanoscopy and SERS ultra-sensing with a record-setting sensitivity [22–24, 30, 49, 91, 92]. It fully uses the advantages of the adiabatic nanofocusing: high efficiency of energy transfer to the nanoscale, absence of background, stability of the adiabatic hot spot, wide spectral range, etc. This probably is a root to the future development of the SERS nanoscopy, sensing and detection.

References

1. Fleischmann, M, Hendra, P. J. and McQuillan, A. J. (1974). Raman spectra of pyridine adsorbed at a silver electrode. *Chem. Phys. Lett.* **26**, pp. 163–166.
2. Jeanmaire, D. L. and Van Duyne, R. P. (1977). Surface Raman spectroelectrochemistry Part I. Heterocyclic, aromatic, and aliphatic amines adsorbed on the anodized silver electrode. *J. Electroanal. Chem.* **84**, pp. 1–20.
3. Albrecht, M. G. and Creighton, J. A. (1977). Anomalously intense Raman spectra of pyridine at a silver electrode. *J. Am. Chem. Soc.* **99**, pp. 5215–5217.
4. Moskovits, M. (1985). Surface-enhanced spectroscopy. *Rev. Mod. Phys.* **57**, pp. 783–826.
5. Maier, S. A., Brongersma, M. L., Kik, P. G., Meltzer, S., Requicha, A. A. G. and Atwater, H. A. (2001). Plasmonics — A route to nanoscale optical devices. *Adv. Mater.* **13**, pp. 1501–1505.
6. Stockman, M. I., Shalaev, V. M., Moskovits, M., Botet, R. and George, T. F. (1992). Enhanced Raman-scattering by fractal clusters — scale-invariant theory. *Phys. Rev. B* **46**, pp. 2821–2830.

7. Kneipp, K., Wang, Y., Kneipp, H., Perelman, L. T., Itzkan, I., Dasari, R. and Feld, M. S. (1997). Single molecule detection using surface-enhanced Raman scattering (SERS). *Phys. Rev. Lett.* **78**, pp. 1667–1670.
8. Nie, S. M. and Emery, S. R. (1997). Probing single molecules and single nanoparticles by surface-enhanced Raman scattering. *Science* **275**, pp. 1102–1106.
9. Dieringer, J. A., Lettan, R. B., Scheidt, K. A. and Van Duyne, R. P. (2007). A frequency domain existence proof of single-molecule surface-enhanced Raman Spectroscopy. *J. Am. Chem. Soc.* **129**, pp. 16249–16256.
10. Morton, S. M., Silverstein, D. W. and Jensen, L. (2011). Theoretical studies of plasmonics using electronic structure methods. *Chem. Rev.* **111**, pp. 3962–3994.
11. Li, K., Stockman, M. I. and Bergman, D. J. (2003). Self-similar chain of metal nanospheres as an efficient nanolens. *Phys. Rev. Lett.* **91**, pp. 227401–227404.
12. Dai, J., Cajko, F., Tsukerman, I. and Stockman, M. I. (2008). Electrodynamic effects in plasmonic nanolenses. *Phys. Rev. B* **77**, pp. 115411–115419.
13. Kneipp, J., Li, X., Sherwood, M., Panne, U., Kneipp, H., Stockman, M. I. and Kneipp, K. (2008). Gold nanolenses generated by laser ablation-efficient enhancing structure for surface enhanced Raman scattering analytics and sensing. *Anal. Chem.* **80**, pp. 4247–4251.
14. Höppener, C., Lapin, Z. J., Bharadwaj, P. and Novotny, L. (2012). Self-similar gold-nanoparticle antennas for a cascaded enhancement of the optical field. *Phys. Rev. Lett.* **109**, pp. 017401–017404.
15. Stockman, M. I. (2004). Nanofocusing of optical energy in tapered plasmonic waveguides. *Phys. Rev. Lett.* **93**, pp. 1374041–1374044.
16. Ropers, C., Neacsu, C. C., Elsaesser, T., Albrecht, M., Raschke, M. B. and Lienau, C. (2007). Grating-coupling of surface plasmons onto metallic tips: A nano-confined light source. *Nano Lett.* **7**, pp. 2784–2788.
17. Ropers, C., Neacsu, C. C., Raschke, M. B., Albrecht, M., Lienau, C. and Elsaesser, T. (2008). Light confinement at ultrasharp metallic tips. *Japan. J. Appl. Phys.* **47**, pp. 6051–6054.
18. Sadiq, D., Shirdel, J., Lee, J. S., Selishcheva, E., Park, N. and Lienau, C. (2011). Adiabatic nanofocusing scattering-type optical nanoscopy of individual gold nanoparticles. *Nano Lett.* **11**, pp. 1609–1613.
19. Schmidt, S., Piglosiewicz, B., Sadiq, D., Shirdel, J., Lee, J. S., Vasa, P., Park, N., Kim, D. S. and Lienau, C. (2012). Adiabatic nanofocusing on ultrasmooth single-crystalline gold tapers creates a 10-nm-sized light source with few-cycle time resolution. *ACS Nano* **6**, pp. 6040–6048.
20. Sigle, W., Talebi, N., Guo, S., Esmann, M., Becker, S. F., Vogelgesang, R., Lienau, C. T. and van Aken, P. A. (2015). Real-space imaging of plasmonic modes of gold tapers by EFTEM and EELS. *Microsc. Microanal.* **21**, pp. 2221–2222.
21. Becker, S. F., Esmann, M., Yoo, K., Groß, P., Vogelgesang, R., Park, N. and Lienau, C. (2016). Gap-plasmon enhanced nanofocusing near-field microscopy. *ACS Photon.* **3**, pp. 223–232.
22. De Angelis, F., Patrini, M., Das, G., Maksymov, I., Galli, M., Businaro, L., Andreani, L. C. and Di Fabrizio, E. (2008). A hybrid plasmonic-photonic

nanodevice for label-free detection of a few molecules. *Nano Lett.* **8**, pp. 2321–2327.

23. De Angelis, F., Das, G., Candeloro, P., Liberale, C., Mecarini, F., Pujia, A., Di Fabrizio, E. (2010). Adiabatic compression of surface plasmon polaritons for label-free few molecules detection by means of Raman scattering. *Microelectron. Eng.* **87**, pp. 1312–1314.
24. De Angelis, F., Das, G., Candeloro, P., Patrini, M., Galli, M., Bek, A., Lazzarino, M., Maksymov, I., Liberale, C., Andreani, L. C. *et al.* (2010). Nanoscale chemical mapping using three-dimensional adiabatic compression of surface plasmon polaritons. *Nature Nanotechnol.* **5**, pp. 67–72.
25. Angelis, F. D., Gentile, F., Das, F. M. G., Moretti, M., Candeloro, P., Coluccio, M. L., Cojoc, G., Accardo, A., Liberale, C., Zaccaria, R. P. *et al.* (2011). Breaking the diffusion limit with super-hydrophobic delivery of molecules to plasmonic nanofocusing SERS structures. *Nature Photon.* **5**, pp. 682–687.
26. Gentile, F., Coluccio, M. L., Toma, A., Alabastri, A., Zaccaria, R. P., Das, G., De Angelis, F., Candeloro, P., Liberale, C., Perozziello, G. *et al.* (2013). Plasmonics and super-hydrophobicity: A new class of nano-bio-devices. In Shahbazyan, T. and Stockman, M. (eds.), *Plasmonics: Theory and Applications.* Vol. 15, pp. 501–524.
27. Perozziello, G., Catalano, R., Francardi, M., Rondanina, E., Pardeo, F., Angelis F. D., Malara, N., Candeloro, P., Morrone, G. and Fabrizio, E. D. (2013). A microfluidic device integrating plasmonic nanodevices for Raman spectroscopy analysis on trapped single living cells. *Microelectron. Eng.* **111**, pp. 185–188.
28. Das, G., Battista, E., Manzo, G., Causa, F., Netti, P. and Di Fabrizio, E. (2015). Large scale plasmonic nanocones array for spectroscopy detection. *ACS Appl. Mater. Inter.* **7**, pp. 23579–23604.
29. Di Fabrizio, E., Schlucker, S., Wenger, J., Regmi, R., Rigneault, H., Calafiore, G., West, M., Cabrini, S., Fleischer, M., van Hulst, N. F. *et al.* (2016). Roadmap on biosensing and photonics with advanced nano-optical methods. *J. Opt.* **18**. 063003.
30. Berweger, S., Atkin, J. M., Olmon, R. L. and Raschke, M. B. (2010). Adiabatic tip-plasmon focusing for nano-Raman spectroscopy. *J. Phys. Chem. Lett.* **1**, pp. 3427–3432.
31. Raschke, M. B., Berweger, S., Atkin, J. M. and Olmon, R. L. (2010). Adiabatic tip-plasmon focusing for nano-Raman spectroscopy, *J. Phys. Chem. Lett.* **1**, pp. 3427–3432.
32. Berweger, S., Atkin, J. M., Olmon, R. L. and Raschke, M. B. (2012). Light on the tip of a needle: Plasmonic nanofocusing for spectroscopy on the nanoscale. *J. Phys. Chem. Lett.* **3**, pp. 945–952.
33. Chen, X., Dong, B. and Balogun, O. (2016). Near-field photothermal heating with a plasmonic nanofocusing probe. *Inter. J. Thermophys.* **37**, pp. 1–9.
34. Bergman, D. J. and Stroud, D. (1992). Properties of macroscopically inhomogeneous media. In Ehrenreich, H. and Turnbull, D. (eds.), Vol. 46. *Solid State Physics.* Academic Press, pp. 148–270.

35. Stockman, M. I. (2011). Nanoplasmonics: Past, present, and glimpse into future. *Opt. Express* **19**, pp. 22029–22106.
36. Landau, L. D. and Lifshitz, E. M. (1984). *Electrodynamics of Continuous Media.* Lifshitz, E. M. and Pitaevskii, L. P. (eds.), Vol. 8, 2nd edn. Oxford and New York, Pergamon.
37. Bergman, D. J. and Stockman, M. I. (2003). Surface plasmon amplification by stimulated emission of radiation: Quantum generation of coherent surface plasmons in nanosystems. *Phys. Rev. Lett.* **90**, pp. 0274021–0274024.
38. Johnson, P. B. and Christy, R. W. (1972). Optical constants of noble metals. *Phys. Rev. B* **6**, pp. 4370–4379.
39. Stockman, M. I. (2006). Electromagnetic theory of SERS. In Kneipp, K., Moskovits, M. and Kneipp, H. (eds.), *Surface Enhanced Raman Scattering.* Springer, pp. 47–66.
40. Weitz, D. A. and Oliveria, M. (1984). Fractal structures formed by kinetic aggregation of aqueous gold colloids. *Phys. Rev. Lett.* **52**, pp. 1433–1436.
41. Kolb, M., Botet, R. and Julienne, J. (1983). Scaling of kinetically growing clusters. *Phys. Rev. Lett.* **51**, pp. 1123–1126.
42. Stockman, M. I., Pandey, L. N., Muratov, L. S. and George, T. F. (1994). Giant fluctuations of local optical fields in fractal clusters. *Phys. Rev. Lett.* **72**, pp. 2486–2489.
43. Stockman, M. I., Bergman, D. J. and Kobayashi, T. (2004). Coherent control of nanoscale localization of ultrafast optical excitation in nanosystems. *Phys. Rev. B* **69**, pp. 054202–054210.
44. Novotny, L. and Stranick, S. J. (2006). Near-field optical microscopy and spectroscopy with pointed probes. *Annu. Rev. Phys. Chem.* **57**, pp. 303–331.
45. Hartschuh, A., Sanchez, E. J., Xie, X. S. and Novotny, L. (2003). High-resolution near-field Raman microscopy of single-walled carbon nanotubes. *Phys. Rev. Lett.* **90**, pp. 095501–095504.
46. Johnson, T. W., Lapin, Z. J., Beams, R., Lindquist, N. C., Rodrigo, S. G., Novotny, L. and Oh, S.-H. (2012). Highly reproducible near-field optical imaging with sub-20-nm resolution based on template-stripped gold pyramids. *ACS Nano* **6**, pp. 9168–9174.
47. Anderson, N., Hartschuh, A. and Novotny, L. (2007). Chirality changes in carbon nanotubes studied with near-field Raman spectroscopy. *Nano Lett.* **7**, pp. 577–582.
48. Anderson, N., Anger, P., Hartschuh, A. and Novotny, L. (2006). Subsurface Raman imaging with nanoscale resolution. *Nano Lett.* **6**, pp. 744–749.
49. Verma, P. (2017). Tip-enhanced raman spectroscopy: Technique and recent advances. *Chem. Rev.* **117**, pp. 6447–6466.
50. Yu, J., Saito, Y., Ichimura, T., Kawata, S. and Verma, P. (2013). Far-field free tapping-mode tip-enhanced Raman microscopy. *Appl. Phys. Lett.* **102**.
51. Ichimura, T., Hayazawa, N., Hashimoto, M., Inouye, Y. and Kawata, S. (2004). Tip-enhanced coherent anti-stokes raman scattering for vibrational nanoimaging. *Phys. Rev. Lett.* **92**, pp. 220801–220804.
52. Novotny, L. and Hecht, B. (2012). *Principles of Nano-Optics.* Cambridge University Press, Cambridge, New York.

53. Wang, Z. J., Pan, S. L., Krauss, T. D., Du, H. and Rothberg, L. J. (2003). The structural basis for giant enhancement enabling single-molecule Raman scattering. *Proc. Natl. Acad. Sci. USA* **100**, pp. 8638–8643.
54. Das, G., Alrasheed, S., Coluccio, M. L., Gentile, F., Nicastri, A., Candeloro, P., Cuda, G., Perozziello, G. and Di Fabrizio, E. (2016). Few molecules SERS detection using nanoLens based plasmonic nanostructure: Application to point mutation detection. *RSC Adv.* **6**, pp. 107916–107923.
55. Bidault, S., Abajo, F. J. G. D. and Polman, A. (2008). Plasmon-based nanolenses assembled on a well-defined DNA template. *J. Am. Chem. Soc.* **130**, pp. 2750–2751.
56. Ding, B., Deng, Z., Yan, H., Cabrini, S., Zuckermann, R. N. and Bokor, J. (2010). Gold nanoparticle self-similar chain structure organized by DNA origami. *J. Am. Chem. Soc.* **132**, pp. 3248–3249.
57. Kravets, V. G., Zoriniants, G., Burrows, C. P., Schedin, F., Casiraghi, C., Klar, P., Geim, A. K., Barnes, W. L. and Grigorenko, A. N. (2010). Cascaded optical field enhancement in composite plasmonic nanostructures. *Phys. Rev. Lett.* **105**, p. 246806.
58. Kravets, V. G., Zoriniants, G., Burrows, C. P., Schedin, F., Geim, A. K., Barnes, W. L. and Grigorenko, A. N. (2010). Composite Au nanostructures for fluorescence studies in visible light. *Nano Lett.* **10**, pp. 874–879.
59. De Angelis, F., Liberale, C., Coluccio, M. L., Cojoc, G. and Di Fabrizio, E. (2011). Emerging fabrication techniques for 3D nano-structuring in plasmonics and single molecule studies. *Nanoscale* **3**. 2011, pp. 2689–2696.
60. Babiaczyk, W. I., Bonella, S., Ciccotti, G., Coluccio, M. L., Gentile, F. and Di Fabrizio, E. (2012). Silver self aggregation in a nanodevice for enhanced Raman spectroscopy: Experiments vs. simplified modeling via molecular dynamics. *Nanoscale.* **4**, pp. 2362–2371.
61. Coluccio, M. L., Gentile, F., Das, G., Perozziello, G., Malara, N., Alrasheed, S., Candeloro, P. and Fabrizio, E. D. (2015). From nucleotides to DNA analysis by a SERS substrate of a self similar chain of silver nanospheres, *J. Opt.* **17**, p. 114021.
62. Lloyd, J. A., Ng, S. H., Liu, A. C. Y., Zhu, Y., Chao, W., Coenen, T., Etheridge, J., Gomez, D. E. and Bach, U. (2017). Plasmonic nanolenses: Electrostatic self-assembly of hierarchical nanoparticle trimers and their response to optical and electron beam stimuli. *ACS Nano* doi: 10.1021/acsnano.6b07336, pp. 1–33.
63. Stockman, M. I., Faleev, S. V. and Bergman, D. J. (2001). Localization versus delocalization of surface plasmons in nanosystems: Can one state have both characteristics? *Phys. Rev. Lett.* **87**, pp. 167401–167404.
64. Kong, J. A. (1990). *Electromagnetic Wave Theory*, 2nd edn. Wiley, New York.
65. Keilmann, F. (1999). Surface-polariton propagation for scanning near-field optical microscopy application. *J. Microscopy* **194**, pp. 567–570.
66. Nerkararyan, K. V. (1997). Superfocusing of a surface polariton in a wedge-like structure. *Phys. Lett. A* **237**, pp. 103–105.
67. Babajanyan, A. J., Margaryan, N. L. and Nerkararyan, K. V. (2000). Superfocusing of surface polaritons in the conical structure. *J. Appl. Phys.* **87**, pp. 3785–3788.

68. Stockman, M. I. (2011). Erratum: Nanofocusing of optical energy in tapered plasmonic waveguides. *Phys. Rev. Lett.* **106**, p. 019901.
69. Cohen-Tannoudji, C., Diu, B. and Laloë, F. (1977). *Quantum Mechanics.* Wiley, New York.
70. Larkin, I. A., Stockman, M. I., Achermann, M. and Klimov, V. I. (2004). Dipolar emitters at nanoscale proximity of metal surfaces: Giant enhancement of relaxation in microscopic theory. *Phys. Rev. B*, **69**, pp. 121401–121404.
71. Liebsch, A. (1985). Screening properties of a metal surface at low frequencies and finite wave vectors. *Phys. Rev. Lett.* **54**, pp. 67–70.
72. Park, I.-Y., Kim, S., Choi, J., Lee, D.-H., Kim, Y.-J., Kling, M. F., Stockman, M. I., Kim, S.-W. (2011). Plasmonic generation of ultrashort extreme-ultraviolet light pulses. *Nature Photon.* **5**, pp. 677–681.
73. Rusina, A., Durach, M., Nelson, K. A. and Stockman, M. I. (2008). Nanoconcentration of terahertz radiation in plasmonic waveguides. *Opt. Exp.* **16**, pp. 18576–18589.
74. Durach, M., Rusina, A., Stockman, M. I. and Nelson, K. (2007). Toward full spatiotemporal control on the nanoscale. *Nano Lett.* **7**, pp. 3145–3149.
75. Gramotnev, D. K., Vogel, M. W. and Stockman, M. I. (2008). Optimized nonadiabatic nanofocusing of plasmons by tapered metal rods. *J. Appl. Phys.* **104**, pp. 034311–034318.
76. Verhagen, E., Kuipers, L. and Polman, A. (2007). Enhanced nonlinear optical effects with a tapered plasmonic waveguide. *Nano Lett.* **7**, pp. 334–337.
77. Verhagen, E., Polman, A. and Kuipers, L. (2008). Nanofocusing in laterally tapered plasmonic waveguides. *Opt. Exp.* **16**, pp. 45–57.
78. Verhagen, E., Spasenovic, M., Polman, A. and Kuipers, L. (2009). Nanowire plasmon excitation by adiabatic mode transformation. *Phys. Rev. Lett.* **102**, pp. 203901–203904.
79. Bao, W., Melli, M., Caselli, N., Riboli, F., Wiersma, D. S., Staffaroni, M., Choo, H., Ogletree, D. F., Aloni, S., Bokor, J. *et al.* (2012). Mapping local charge recombination heterogeneity by multidimensional nanospectroscopic imaging. *Science* **338**, pp. 1317–1321.
80. Choo, H., Kim, M.-K., Staffaroni, M., Seok, T. J., Bokor, J., Cabrini, S., Schuck, P. J., Wu, M. C. and Yablonovitch, E. (2012). Nanofocusing in a metal-insulator-metal gap plasmon waveguide with a three-dimensional linear taper. *Nat. Photon.* **6**, pp. 838–844. doi: 10.1038/nphoton.2012.277.
81. Bao, W., Staffaroni, M., Bokor, J., Salmeron, M. B., Yablonovitch, E., Cabrini, S., Weber-Bargioni, A. and Schuck, P. J. (2013). Plasmonic near-field probes: A comparison of the campanile geometry with other sharp tips. *Opt. Exp.* **21**, pp. 8166–8176.
82. Berweger, S., Atkin, J. M., Xu, X. G., Olmon, R. L. and Raschke, M. B. (2011). Femtosecond nanofocusing with full optical waveform control. *Nano Lett.* **11**, pp. 4309–4313.
83. Kravtsov, V., Atkin, J. M. and Raschke, M. B. (2013). Group delay and dispersion in adiabatic plasmonic nanofocusing. *Opt. Lett.* **38**, pp. 1322–1324.

84. Bechtel, H. A., Muller, E. A., Olmon, R. L., Martin, M. C. and Raschke, M. B. (2014). Ultrabroadband infrared nanospectroscopic imaging. *Proc. Natl. Acad. Sci. USA* **111**, pp. 7191–7196.
85. Kravtsov, V., Ulbricht, R., Atkin, J. and Raschke, M. B. (2016). Plasmonic nanofocused four-wave mixing for femtosecond near-field imaging. *Nature Nanotechnol.* **11**, p. 459.
86. Muller, M., Kravtsov, V., Paarmann, A., Raschke, M. B. and Ernstorfer, R. (2016). Nanofocused plasmon-driven sub-10 fs electron point source. *ACS Photon.* **3**, pp. 611–619.
87. Schmidt, S., Engelke, P., Piglosiewicz, B., Esmann, M., Becker, S. F., Yoo, K., Park, N., Lienau, C. and Gross, P. (2013). Wave front adaptation using a deformable mirror for adiabatic nanofocusing along an ultrasharp gold taper. *Opt. Exp.* **21**, pp. 26564–26577.
88. Piglosiewicz, B., Vogelsang, J., Schmidt, S., Park, D. J., Groß, P. and Lienau, C. (2014). Electron photoemission and acceleration from sharp gold nanotapers in the strong-field, few-cycle regime. *Quantum Matter* **3**, pp. 297–306.
89. Vogelsang, J., Robin, J., Nagy, B. J., Dombi, P., Rosenkranz, D., Schiek, M., Gross, P. and Lienau, C. (2015). Ultrafast electron emission from a sharp metal nanotaper driven by adiabatic nanofocusing of surface plasmons. *Nano Lett.* **15**, pp. 4685–4691.
90. Giugni, A., Torre, B., Toma, A., Francardi, M., Malerba, M., Alabastri, A., Proietti Zaccaria, R., Stockman, M. I. and Di Fabrizio, E. (2013). Hot-electron nanoscopy using adiabatic compression of surface plasmons. *Nat. Nano* **8**, pp. 845–852.
91. Gentile, F., Das, G., Coluccio, M. L., Mecarini, F., Accardo, A., Tirinato, L., Tallerico, R., Cojoc, G., Liberale, C., Candeloro, P. *et al.* (2009). Ultra low concentrated molecular detection using super hydrophobic surface based biophotonic devices. *Microelectron. Eng.* **87**, pp. 798–801.
92. Umakoshi, T., Saito, Y. and Verma, P. (2016). Highly efficient plasmonic tip design for plasmon nanofocusing in near-field optical microscopy. *Nanoscale* **8**, pp. 5634–5640.

Chapter 2

Frontiers in Electromagnetic Mechanism of SERS

Tamitake Itoh

Health Technology Research Center,
National Institute of Advanced Industrial Science and Technology (AIST),
Takamatsu, Kagawa 761-0395, Japan
tamitake-itou@aist.go.jp

1. Introduction

The SERS phenomenon was first reported by Fleischmann *et al.* in 1974 [1] and interpreted as enhanced Raman scattering in 1977 by two other groups [2, 3]. In these studies [1–3], electrochemically roughened silver electrodes were used to enhance the Raman signals of pyridine. The mechanism of the signal enlargement was first explained as "fractal effects", i.e. an increase in the surface area due to the roughness. However, the fractal effects were denied because the estimated Raman enhancement factors cannot be explained by the increase in the surface area, and the electromagnetic and chemical interactions between the roughened metal surfaces and molecules absorbed on the surfaces were considered to be important for the Raman enhancement [4, 5]. In 1997, a silver nanoparticle (NP) dispersion was applied to SERS measurements and an enormous enhancement by factors of $\sim 10^{10}$ and $\sim 10^{14}$ were achieved by two groups, respectively [6, 7]. Considering SM fluorescence detections which require cross-section of 10^{-16} cm^2, the enhancement factors of $\sim 10^{10}$ is enough to detect a single fluorescent dye molecule by SERS because the SERS cross-section (resonance Raman cross-section $\sim 10^{-24}$ cm^2 $\times$ $\sim 10^{10}$) is 100 times larger than the fluorescence cross-section. Furthermore, the enhancement factors of $\sim 10^{14}$ enables single-molecule SERS (SM-SERS) detections even under non-resonant Raman excitation conditions. These phenomena, called SM-SERS [8], have significantly impacted the study of the SERS enhancement mechanism and helped us to quantitatively understand the fundamentals of SERS.

To clarify the fundamentals of SERS, researchers have developed two types of enhancement mechanisms: one is the EM mechanism and another is the chemical enhancement (CE) mechanism [4, 9]. The EM enhancement, which is caused by the coupling of the incident and Raman EM fields with plasmon resonances of metallic nanostructures, is thought to be the main fundamental one in the Raman enhancement. The CE mechanism is explained as resonance Raman effect by the shifting of the Raman scattering in non-resonance to that in resonance through the formation of charge-transfer complexes between adsorbed molecules and metal surfaces [10]. CE enhancement is thought to be far smaller than the EM enhancement, but the CE mechanism attracts attention for its potential to achieve plasmonic chemistry, also called "plasmon-modified chemical reactions", on the surfaces of plasmonic metal [10].

In this chapter, we first focus on the EM mechanism of SERS and extend the mechanism to other linear and nonlinear optical responses, e.g. surface-enhanced fluorescence (SEF), surface-enhanced hyper Raman scattering (SEHRS), and surface-enhanced coherent anti-Stokes Raman scattering (SECARS). Next, we discuss the limitation of the EM mechanism from the points of view of EM field intensities and confinement of the fields and then show the frontiers in the EM mechanism, e.g. ultrafast dynamics, strong coupling, and the field-gradient effect by introducing the breakdown of Fermi's golden rules. We then conclude this chapter with a summary and outlook section.

2. The Electromagnetic Mechanism

In 1977, the explanation of SERS phenomena by the EM mechanism was given. Moskovits first proposed that plasmon resonance of metal nanostructure, e.g. NPs, enhances the Raman signals of molecules on the NPs [4]. Plasmon is the collective motion of conduction electrons, which induce depolarization to cancel the external electric field [11, 12]. When the phase of the depolarization cannot follow the phase of the external field and the phase difference between the depolarization and external field reaches $\pi/2$, they become resonant [11, 12]. Thanks to the rich conduction electrons of specific metals, e.g. gold, silver or copper, and deviation of their resonance energy from interband transition energy of these metals, their plasmon resonance highly increases mode density of states of light near such metal surfaces. The high mode density enables the enlargement of probabilities of interactions between the light and a molecule. Kerker *et al.* applied the enlargement of the interaction probability by plasmon resonance to explain SERS [13]. The enhancement model was both theoretically and experimentally applied to explain the largest enhancement factors $\sim 10^{10}$ located in gaps of an NP dimer, namely, "hotspots", by Metiu, Inoue, Xu, and their collaborators [14–16]. The enhancement factor required for SM-SERS $\sim 10^{14}$ can be explained by the EM mechanism as a product of the enhancement factors of SERS ($<10^{10}$) and resonance Raman effect ($<10^{4}$) at the gaps of an NP dimer.

We discuss the formulation and experimental evaluation of SERS. A Raman process is composed of a two-step interaction of light with a molecule, i.e. an excitation process and an emission process. Thanks to the broad linewidth of plasmon resonance, the resonance can enhance both the Raman excitation and Raman emission probabilities [10]. The Raman excitation (emission) enhancement factor $|M_1|^2$ ($|M_2|^2$) is described as a ratio of EM field intensities with and without an increase in the mode density of states of light by plasmon resonance. Thus, we have [10]

$$|M_1(\lambda_\mathrm{L}, d_\mathrm{av})|^2 |M_2(\lambda, d_\mathrm{av})|^2 = \left|\frac{E_\mathrm{loc}(\lambda_\mathrm{L}, d_\mathrm{av})}{E_\mathrm{I}(\lambda_L)}\right|^2 \times \left|\frac{E_\mathrm{loc}(\lambda, d_\mathrm{av})}{E_\mathrm{I}(\lambda)}\right|^2, \tag{1}$$

where E_I and E_loc indicate the amplitudes of the incident and enhanced local electric field, respectively, λ_L and λ denote the wavelengths of the incident and emission light, respectively, and d_{av} is the "effective" distance between a molecule and metal surface. Here, the term "effective" means that d_{av} includes both separation and orientation of the molecule. Thus, the SERS cross-section $\sigma_{\mathrm{SERS}}(\lambda_{\mathrm{L}}, \lambda)$ is given by [10]

$$\sigma_{\mathrm{SERS}}(\lambda_{\mathrm{L}}, \lambda, d_{\mathrm{av}}) = |M_1(\lambda_{\mathrm{L}}, d_{\mathrm{av}})|^2 |M_2(\lambda, d_{\mathrm{av}})|^2 \sigma_{\mathrm{RS}}(\lambda_{\mathrm{L}}, \lambda), \quad (2)$$

where $\sigma_{\mathrm{RS}}(\lambda_{\mathrm{L}}, \lambda)$ is the Raman scattering cross-section of a molecule in free space. The EM mechanism has been experimentally investigated by many researchers.

Here, we introduce our step-by-step investigation of the EM mechanism using silver NP dimers, which dominantly show dipolar plasmon resonance and a rhodamine 6G (R6G) molecule located at a hotspot of the dimer. The detection of dipolar plasmons by dark-field microspectroscopy is important for reducing complexity in the investigation, e.g. various strong couplings among plasmon modes in NP aggregates [17]. First, the EM coupling of a plasmon and a molecular dipole was observed by measuring a common polarization dependence of the plasmon resonance spectra and SERS spectra from identical silver NP dimers having an R6G molecule at the hotspot [18]. This observation is important for identifying the plasmon involved with the EM enhancement. Second, the Raman enhancement by plasmon resonance was confirmed by showing the relationship between SERS intensities and quality (Q) factors of plasmon resonance because the EM enhancement factor is identical to the Purcell factor $F \sim (\lambda/n)^3 Q/V$, showing temporal confinement effect Q and spatial confinement effect $1/V$ of light, where n is a refractive index of the space of a hotspot and V is an electric field volume of the a hotspot [19]. The value of a Q factor is obtained by dividing the resonance angular frequency ω_{p} with resonance linewidth $\Delta\omega_{\mathrm{p}}$ and means how many times the electric field oscillates at a hotspot (how many times the electric field interacts with the molecule) compared with a free space. Thus, the hotspot with a higher Q factor can more efficiently excite a molecule and can also efficiently de-excite the excited molecule. This confirmation is important because non-radiative plasmons may also work to de-enhance Raman intensity [20]. In other words, the confirmation showed that the "enhancement" is indeed induced by the plasmon resonance. Third, the excitation enhancement factor $|M_1|^2$ in Equation (1) was evaluated using the excitation laser energy dependence of the SERS intensities [21]. Fourth, the emission enhancement factor $|M_2|^2$ in Equation (1) was evaluated as the plasmon-induced spectral modulation of the SERS intensities from the points of dimer-by-dimer variations in plasmon resonance and surrounding medium-by-medium variations in refractive indexes [22–24]. Finally, the experimental values for $\sigma_{\text{SERS}}(\lambda_{\text{L}}, \lambda)$ are quantitatively reproduced by the product of $|M_1|^2|M_2|^2$ and $\sigma_{\text{RS}}(\lambda_{\text{L}}, \lambda)$ as in Equation (2). $|M_1|^2|M_2|^2$s were calculated with a finite-difference time-domain (FDTD) method using silver NP dimers and an R6G molecule located in the hotspots [25]. This reproduction of experimentally obtained SERS spectra by the calculation results of $|M_1|^2|M_2|^2$ and $\sigma_{\text{RS}}(\lambda_{\text{L}}, \lambda)$ demonstrates that the EM mechanism is dominant in SERS phenomena. Figure 1 shows the excitation laser energy dependence of the reproduced SERS spectra in Ref. [25]. The strange SERS spectral properties, e.g. combination and overtone line intensities comparable to fundamental line intensities and anti-Stokes intensities higher than Stokes intensities, are quantitatively reproduced by the EM mechanism. The key for this successful reproduction is that a dipolar plasmon reserves, and temporally and spatially concentrates the excitation light on a hotspot, where the excitation light changes into Raman light, and then radiates the Raman light from the hotspot by inverse process of excitation. Note that this reproduction should be reconsidered if NP aggregates showing the strong coupling between radiative and non-radiative plasmons are used (Fano resonance effect) for SERS measurements [26].

The success of reproduction of SERS spectra as well as SERS cross-section by the EM mechanism indicates that the EM mechanism can also be applied to other surface-enhanced spectra of linear and nonlinear optical responses. Thus, the equation illustrating the EM enhancement of SERS (Equation (2)) is modified here to illustrate the enhancement of SEF, SEHRS, and SECARS.

Fluorescence is an incoherent process composed of an excitation process and an emission process. The actual excited electron loses

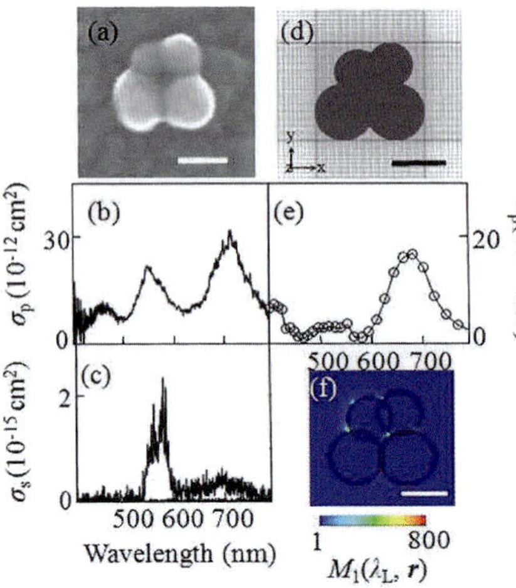

(a)
(d)
(b)
(e)
(c)
(f)
30
0
20
0
2
0
500 600 700
500 600 700
Wavelength (nm)
1
800
$M_1(\lambda_L, \boldsymbol{r})$

(A)

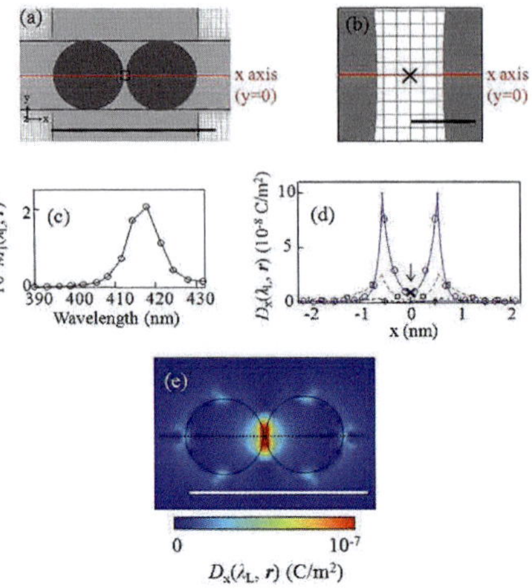

(a)
(b)
x axis
(y=0)
(c)
(d)
390 400 410 420 430
Wavelength (nm)
x (nm)
(e)
0
10^{-7}
$D_x(\lambda_L, \boldsymbol{r})$ (C/m²)

(B)

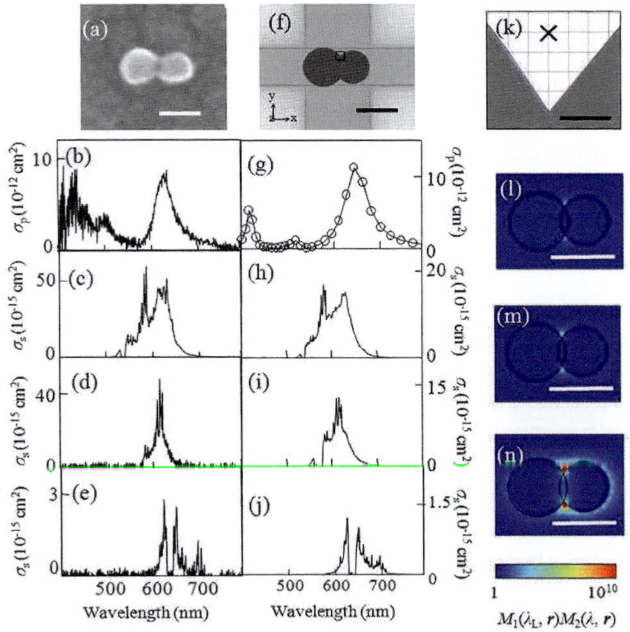

(a)
(f)
(k)
(b)
(g)
(l)
(c)
(h)
(m)
(d)
(i)
(n)
(e)
(j)
500 600 700
Wavelength (nm)
1
10^{10}
$M_1(\lambda_L, \boldsymbol{r})M_2(\lambda, \boldsymbol{r})$

(C)

its coherence and then decays to a ground state. During the decay process, the emission process further branches into two processes: a radiative emission process and non-radiative emission process. In the fluorescence process, the formulations of EM enhancement of the excitation process and radiative emission process are common to those of SERS because both are enhanced by identical radiative plasmon [27]. Thus, the enhancement factor of radiative emission process is $|M_2|^2$ in Equation (1). However, the enhancement factor of non-radiative emission process $|M_\mathrm{d}|^2$, which is assumed to be unmodified by the metal, is largely different from $|M_2|^2$ because both radiative and non-radiative plasmon can contribute to the enhancement factor of non-radiative emission process. Thus, $|M_\mathrm{d}|^2$ is thought to be much larger than $|M_2|^2$ within optical near-fields. $|M_\mathrm{d}|^2$ is well known to be identical to a quenching factor of fluorescence of dye molecules on a metal surface by energy transfer. Thus, SEF cross-section $\sigma_\mathrm{SEF}(\lambda_\mathrm{L}, \lambda)$ is given by [28]

$$\sigma_\mathrm{SEF}(\lambda_\mathrm{L}, \lambda, d_\mathrm{av}) = \frac{|M_1(\lambda_\mathrm{L}, d_\mathrm{av})|^2 |M_2(\lambda, d_\mathrm{av})|^2}{|M_\mathrm{d}(d_\mathrm{av})|^2} \sigma_\mathrm{FL}(\lambda_\mathrm{L}, \lambda), \quad (3)$$

where $\sigma_\mathrm{FL}(\lambda_\mathrm{L},\lambda)$ is fluorescent cross-section of a molecule in free space and $|M_\mathrm{d}|^2$ is enhancement factor of the non-radiative emission process. In the case of SM SERS conditions, $|M_\mathrm{d}|^2$ is much larger than $|M_2|^2$ by a factor of $\sim 10^2$ due to the contribution of both local and non-local effects [28, 29]. The local effect is induced by the responses of bulk electrons of metal and the non-local effect is induced by the responses of surface electrons within Debye radius of approximately several 0.1 nm [30].

Figure 1. (A) Experimental observation of shape of silver NP dimer, its plasmon resonance spectrum, and SERS spectrum detected. Calculations of expected plasmon resonance spectrum according to the shape of dimer are also depicted using the FDTD method. (B) Detailed description for calculating SERS spectrum of single rhodamine 6G (R6G) molecule placed on a nanogap of silver NP dimer. Electromagnetic fields of silver NP dimer and R6G molecule simulated by FDTD method. (C) Reconstruction of SERS spectrum of single R6G molecule from silver NP dimer. Spectrum of plasmon resonance and SERS experimentally observed are well fitted to those theoretically simulated. Taken from Ref. [25].

Hyper Raman scattering, by which one can measure Raman forbidden vibration spectra, is a coherent process composed of a two-photon excitation process and one-photon emission process. The two-photon excitation process is the two-fold interaction between the light and a molecule. Thus, the EM enhancement factor of two-photon excitation is expressed as a product of $|M_1|^2$ and $|M_1|^2$. Thus, the SEHRS cross-section $\sigma_{\mathrm{SEHRS}}(\lambda_{\mathrm{L}}, \lambda)$ is given by [31]

$$\sigma_{\mathrm{SEHRS}}(\lambda_{\mathrm{L}}, \lambda, d_{\mathrm{av}}) = |M_1(\lambda_{\mathrm{L}}, d_{\mathrm{av}})|^4 |M_2(\lambda, d_{\mathrm{av}})|^2 \sigma_{\mathrm{HRS}}(\lambda_{\mathrm{L}}, \lambda), \quad (4)$$

where $\sigma_{\mathrm{SEHRS}}(\lambda_{\mathrm{L}},\lambda)$ is hyper-Raman cross-section of a molecule in a free space. Thanks to $|M_1|^4$, the total enhancement factor of SEHRS reaches the value of 10^{15} ($10^5 \times 10^5 \times 10^5$), enabling SM SEHRS detection even under the excitation power of several 100 kW/cm^2 using a cw laser beam [31]. Equation (4) is also evaluated as the plasmon-induced spectral modulation of the SEHRS intensities regarding dimer-by-dimer variations in plasmon resonance [32].

Coherent anti-Stokes Raman scattering is a coherent process composed of a two-photon excitation process and a two-photon emission process. Thus, the SECARS cross-section $\sigma_{\mathrm{SECARS}}(\lambda_{\mathrm{p}},\lambda)$ is given by [33]

$$\begin{aligned}\sigma_{\mathrm{SECARS}}&(\lambda_p, \lambda_s, \lambda_{as}, d_{\mathrm{av}}) \\ &= |M_1(\lambda_p, d_{\mathrm{av}})|^4 |M_1(\lambda_s, d_{\mathrm{av}})|^2 | \\ &\quad M_2(\lambda_{as}, d_{\mathrm{av}})|^2 \sigma_{\mathrm{CARS}}(\lambda_p, \lambda_s, \lambda_{as}, d_{\mathrm{av}}), \end{aligned} \quad (5)$$

where $\sigma_{\mathrm{CARS}}(\lambda_{\mathrm{L}},\lambda)$ is coherent anti-Stokes Raman cross-section of a molecule in a free space, and λ_p, λ_s and ω_{as} denote the wavelengths of the pump, Stokes and anti-Stokes light, respectively. The formulation of EM enhancement indicates higher order nonlinear responses obtain higher enhancement factors. Indeed, single molecule detections of SECARS have been reported under a peak excitation power of the pump beam approximately several 0.1 GW/cm^2 using a femtosecond laser beam [33, 34].

3. Limitations of the Electromagnetic Mechanism

The EM mechanism describes that plasmon resonance at hotspots realizes strong EM enhancement mainly due to small electric field volume ~1 nm^3 at the hotspots. The values of $|M_1|^2|M_2|^2$s up to 10^{10} at hotspots enable SM SERS detections under a realistic excitation intensity condition approximately several W/cm^2 [25]. However, one may imagine that there are some limitations in EM enhancement, i.e. an upper limitation of the electric field intensity to which molecules can linearly respond and a lower limitation of the electric field volume at the hotspots. Indeed, several experiments and calculations have implicitly or explicitly indicated such limitations [27, 35]. Here, these limitations are explained by changing the parameters describing SERS spectra, i.e. by increasing $|M_1|^2 I_{\text{ex}}$, where I_{ex} is excitation laser intensity [27], and by decreasing electric field volume V, which is confined at hotspots [35]. The limitation by increasing I_{ex} can be explained by the deviation of SERS spectra and their intensities from those predicted by using the EM mechanism considering vibrational pumping effect and Rabi oscillation effect. These limitations have been theoretically studied by Xu *et al.* [27].

First, the limitation by increasing I_{ex} due to the vibrational pumping effect is explained. If the excitation rate becomes comparable to the vibrational damping rate, the SERS excitation process may start from the vibrationally excited state in an electrically ground state because the molecule is mainly found to be in the vibrationally excited state. Such a vibrational pumping effect in SERS process was first observed as a quadratic increase in anti-Stokes SERS intensity against I_{ex}, and even Stokes SERS intensity linearly increases against I_{ex} [36]. Xu *et al.* estimated the value of I_{ex}, realizing such effect using a silver NP dimer and an R6G molecule located at its hotspot [27]. The excitation rate for a molecule located at the dimer junction is roughly estimated to be $|M_1|^2\sigma_{\text{A}}\phi_{\text{ex}}$, where σ_{A} is an absorption cross-section (not σ_{RS}) and ϕ_{ex} is photon flux. Using typical vibrational damping rate $\gamma_{\text{vib}} \sim 10^{12}$ s^{-1}, $|M_1|^2 \sim 10^5$ at hotspots, and $\sigma_{\text{A}} \sim 10^{-16}$ cm^2 for an R6G molecule at the excitation laser energy $\hbar\omega_{\text{L}} = 2.45$ eV, $\phi_{\text{ex}} \sim 10^{23}$ photons/(cm^{2}s) corresponding to $I_{\text{ex}} \sim 5$ kW/cm^2 makes the value of $|M_1|^2\sigma_{\text{A}}\phi_{\text{ex}}$ comparable to 10^{12} s^{-1}, resulting in anomaly of large anti-Stokes SERS intensity compared with Stokes SERS intensity shown as increasing in AS/S in Figure 2 against I_{ex}. For such I_{ex}, Stokes/anti-Stokes ratios of SERS spectral

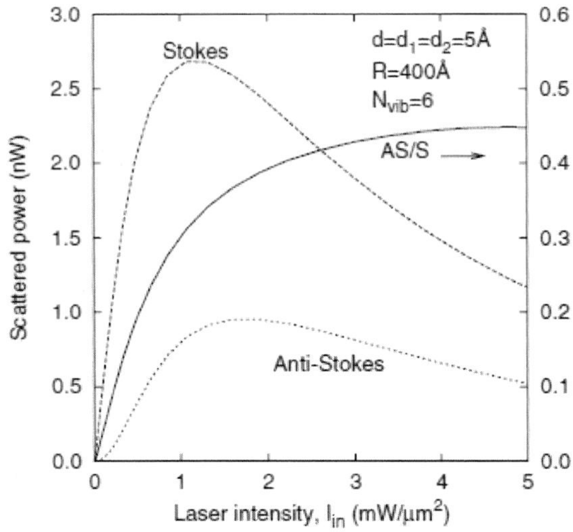

Figure 2. Stokes and anti-Stokes Raman scattering power (for low temperature) as well as their ratio as a function of I_{in} calculated with the parameter values given in Ref. [16]. Taken from Ref. [27].

shapes largely deviated from those predicted by the EM mechanism by Equation (2) [37, 38].

Second, the limitation by increasing I_{ex} due to the Rabi oscillation effect is explained by using the common dimer. When a molecule is excited by a coherent laser beam, the molecule cyclically absorbs photons and re-emits them by stimulated emission [39]. The frequency of this cycle is called Rabi frequency, and the cycle of absorption and re-emission is called Rabi oscillation. Using the two level system, the Rabi frequency under the rigorous resonant condition can be derived as $\Omega_R = M_1 p_0 E_0/\hbar$, where p_0 is a transition dipole moment of an R6G molecule and E_0 is incident electric field amplitude [27, 39]. If Ω_R becomes comparable to the dephasing rate of the electronic transition dipole $\gamma_{\text{ph}} \sim 10^{14}$ s^{-1}, the cycle of absorption and emission cannot be neglected, resulting in a decrease in SERS intensity by the portion of the stimulated emitted light intensity. Xu *et al.* estimated the value of I_{ex} realizing such effect. The $p_0 = el_{\text{dip}}$, where e is the elementary charge of an electron $\sim$1.602 $\times$ 10^{-19} C and l_{dip} is a dipole length $\sim$0.1 nm for an R6G molecule. Using $|M_1|^2 \sim 10^5$

at hotspots, $E_0 = 1.5 \times 10^5$ V/m, which corresponds to I_{ex} of ~500 kW/cm^2, making $\Omega_{\text{R}} \sim 10^{14}$ s^{-1}. The value of Ω_{R} is comparable to $\gamma_{\text{ph}} \sim 10^{14}$ s^{-1}, resulting in a decrease in SERS intensity, even increasing in the region of ~1 mW/μm^2 $< I_{\text{in}}(I_{\text{ex}})$ in Figure 2. For such I_{ex}, SERS intensities largely deviated from those predicted by Equation (2). The estimated values of I_{ex} showing such limitations of EM mechanism are in the range from 5 kW/cm^2 to 500 kW/cm^2. The range is much higher than standard excitation conditions for SM SERS experiments, approximately several W/cm^2 [25]. Several research groups have carried out tests to find out experimental evidence of these phenomena [40].

The limitation by decreasing electric field volume V can be explained by delocalizing unscreened metal surface electrons within the Debye radius <0.5 nm [30]. The upper limit of $|M_1|^2$ and $|M_2|^2$ has been studied by Savage *et al.* [35], who called the limit "plasmon quantum limit". According to this limit, the electric field volume at the hotspot cannot be smaller than the delocalizing volume of the unscreened metal surface electron gas. Savage *et al.* also reported that a quantum-based remodeling of the plasmons is necessary in the case of a nanogap of dimers below 0.3 nm because surface electrons of an NP can coherently tunnel to another NP during forced oscillation by incident light. In fact, Savage *et al.* experimentally and theoretically investigated the tunneling effect in sub-nanometer metal gaps and demonstrated that the coherent tunneling effect becomes important at a scale of 0.3 nm for the gaps corresponding to $V = 1.7 \times 10^{-8}(\lambda/n)^3$ as shown in Figure 3 [35]. The limitation is summarized as (i) the lowest value of the electric field volume and (ii) the appearance of new plasmon oscillations of tunneling electrons. These points cannot be derived from the EM mechanism assuming metal surfaces as dielectric surfaces. In the assumption of the EM mechanism, the tunneling-out of conduction electrons, which loosens the electric field confinement and determines the lowest limit of the electric field volume, does not exist. The interactions among molecular dipoles and such tunneling electrons should be considered for the optical responses of the molecules near the metal surface within ~0.5 nm, i.e. the Debye radius.

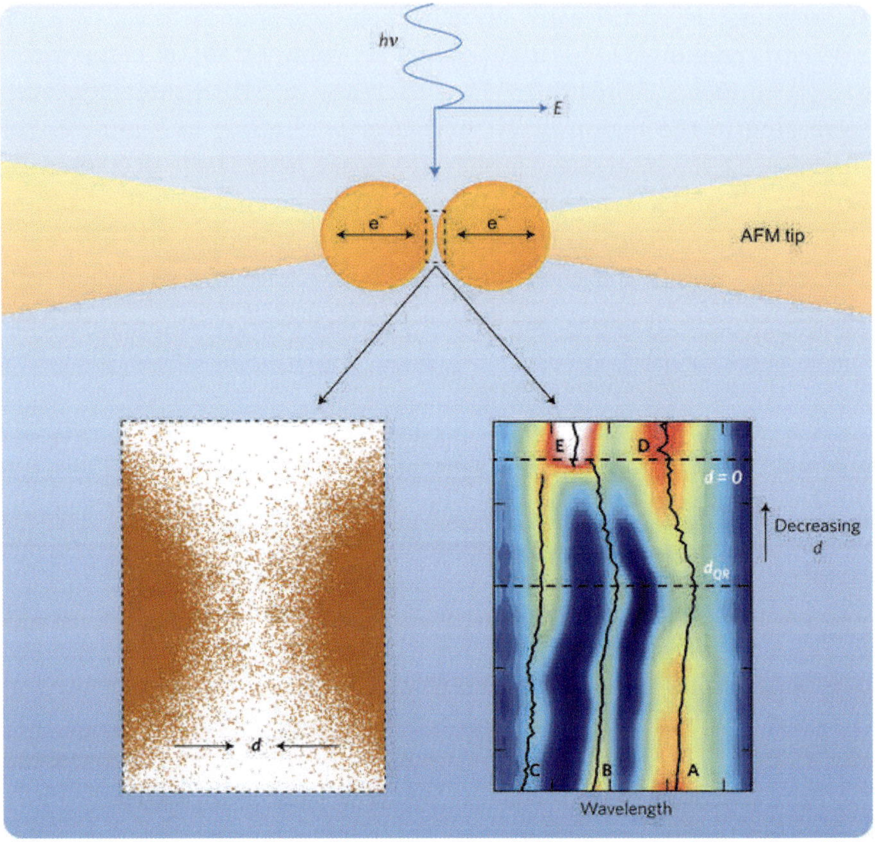

Figure 3. Quantum tunneling regime at sub-nanometer metallic gap. Classical plasmonic modes shown at bottom right; breakdown and strong coupling from quantum tunneling effect arises. Taken from Ref. [35].

4. Beyond the Limitation of the Electromagnetic Mechanism

The limitations of the EM mechanism induced by nonlinear optical responses and unscreened metal surface electrons are explained above. However, there are still problems in the explanation of SERS phenomena exclusively using the EM mechanism as follows: (i) the spectral properties in SEF, which appears as background of SERS, cannot be reproduced by the EM mechanism [41, 42], (ii) the plasmon resonance spectra change significantly after losing SERS activity

[41, 43], and (iii) strange Raman lines frequently sometimes appear in the SERS spectra [41, 44]. These problems cannot be explained by the EM mechanism and may indicate a breakdown of some approximations in the EM mechanism. These problems are realistic challenges currently under discussion. In this section, we introduce these challenges with respect to future perspectives in SERS.

Several approximations are implicitly assumed in the EM mechanism. The validity of the approximations is not yet established [41]. The problems (i)–(iii) above can be explained as a breakdown of approximations in Fermi's golden rule, which shows the transition rate (probability of transition per unit time) from one energy eigenstate, $|i\rangle$, of a quantum system to another energy eigenstate, $|f\rangle$, by the perturbation induced by the interaction between the light and a molecule H' as follows [45]:

$$P \propto \rho\langle i|H'|f\rangle, \tag{6}$$

where P is the transition rate, P is the mode density of the field, and $\langle i|H'|f\rangle$ is the transition element by the perturbation H' between the final and initial states. Equation (6) assumes these approximations in the EM mechanism as follows: (i) the radiative de-excitation rate enhanced by, $|M_2|^2$ becomes slower than the vibrational decay rate, (ii) the coupling rate between plasmon resonance and molecular resonance becomes slower than the dephasing rates of both resonance, and (iii) the steepened local electric field gradient at the hotspots is much larger than the size of the molecular electronic structures. The breakdowns in these approximations (i)–(iii) are explained as follows:

(i) SEF spectra from hotspots generating SM SERS signals are imperfectly reproduced by the EM mechanism. In other words, the product of a fluorescence spectrum and a plasmon resonance spectrum in Equation (3) does not correctly reproduce the SEF spectral shape unlike SERS. The reason for the failure of reproduction is due to the ultra-fast radiative de-excitation by $|M_2|^2$. When the value of ρ, which is enlarged by $|M_2|^2$ at the hotspots of silver NP dimers, allows a de-excitation rate greater than the molecular vibrational decay rate in the electronic excited state, the molecule emits light from the vibrational excited state in the electronic excited state. This light emission means that the molecular electronic dynamics deviate from those in a free space by plasmon resonance. This deviation indicates the breakdown of the well-known molecular electronic dynamics called Kasha's rule, in which molecules are assumed to emit light by electronic transition. It has been investigated as the ultra-fast emission of light within several tens of femtoseconds [29].

The total radiative and non-radiative decay rates of a molecule in a free space are $\Gamma_{r0}(= \int \gamma_{r0}(\lambda)d\lambda)$ and $\Gamma_{nr0s}(= \int \gamma_{nr0}(\lambda)d\lambda)$, respectively, where $\gamma_{r0}(\lambda)d\lambda$ and $\gamma_{nr0}(\lambda)d\lambda$ are the decay rates at λ. EM vacuum fluctuation yields the decay rates from $|i\rangle$ to $|f\rangle$ due to spontaneous emission [27, 42, 46]. Let us estimate the ultra-fast emission for the typical SM SERS system composed of the dimer and an R6G molecule located at the dimer junction. The total radiative and non-radiative decay rates enhanced by plasmon resonance are $\Gamma_{\mathrm{Rad}}(= \int \gamma_{r0}(\lambda)|M_2(\lambda, d_{\mathrm{av}}|^2 d\lambda)$ and $\Gamma_{\mathrm{ET}}(= \int \gamma_{r0}(\lambda)|M_{\mathrm{ET}}(\lambda, d_{\mathrm{av}})|^2 d\lambda)$, respectively. For a molecule, the total internal relaxation rate $\Gamma_{\mathrm{int}} \sim 10^{12}\ \mathrm{s}^{-1}$ is far larger than the total decay rate $\Gamma_{r0}+\omega_{nr0} \sim 10^{8-9}\ \mathrm{s}^{-1}$ [42, 46]. Thus, the fluorescence spectrum of a molecule in a free space is a radiative transition from the bottom of the electronic excited state, which is independent of the excitation laser energy. Under the condition of the conventional SEF, i.e. $\Gamma_{\mathrm{int}} \gg \Gamma_{\mathrm{Rad}} + \Gamma_{\mathrm{ET}} + \Gamma_{nr0}$, the excitation and fluorescence transitions are nearly identical to those for a molecule in a free space. Thus, this type of SEF spectra can be expressed as a product of $\gamma_{nr0}(\lambda)$ and $|M_2(\lambda, d_{\mathrm{av}})|^2$ [42, 46]. However, $|M_2|^2 \sim 10^5$ at hotspots realizes $\Gamma_{\mathrm{int}} < \Gamma_{\mathrm{Rad}}$. Under this condition, SEF cannot be explained in the same manner as conventional SEF because it has a component emitting from the vibrational excited state in the electronic excited state, before relaxing to the bottom of the electronic excited state S_1. The emission indicates that the highest energy of the ultra-fast SEF spectra is blue-shifted from the fluorescence of a molecule in a free space and becomes dependent on the excitation laser energy [42, 46].

Some experimental works on ultra-fast SEF are reviewed here. Le Ru *et al.*'s study may be the most important work regarding the theoretical and experimental aspects of ultra-fast SEF [46]. They developed a theoretical model to evaluate ultra-fast SEF and treated ultra-fast SEF as a failure in the reproduction of the SEF spectra by a product of a fluorescence spectrum and a plasmon resonance spectrum. Ultra-fast SEF has also been investigated from the single nanogap between a scanning tunnel microscope (STM) tip and a gold surface as ultra-fast SEF from vibrational excited states in S_1 of porphyrin molecules located inside nanogap [47]. Cho *et al.* showed ultra-fast SEF using a system: the resonant whispering-gallery plasmonic nanocavities comprising CdS–SiO_2-Ag core–shell nanowires [48]. They tuned the cavity size to match the whispering-gallery resonances and achieved ultra-fast SEF. They also estimated the radiative rate enhancement to be $>10^3$ and predicted ultra-fast lifetimes at the sub-picosecond scale [49]. Itoh *et al.* experimentally averaged out the plasmonic modulation in SEF of dye molecules by measuring large Ag NP aggregates which contain a large number of hotspots [42]. They discovered the excitation laser energy dependence of the ultra-fast SEF spectra regarding the spectral blue-shifts of ~400 meV, the suppression of the anti-Stokes intensity by a factor of several tens, and the super-broadening of the Stokes regions around 1.0 eV as shown in Figure 4. In the framework of the EM effects, the properties are comprehensively explained by the direct emission from the vibrational excited states in S_1 before the electron relaxation to the bottom of S_1 [42].

(ii) When the coupling rate between plasmon and molecular excitonic resonances becomes comparable to the dephasing rates of both resonances, both resonance states coherently exchange excitation energy, resulting in the breakdown of the weak coupling approximation. Note that for weak coupling, the approximations of both resonances are independently treated. The ρ and $\langle i|H'|f\rangle$ interfere with each other, and the two resonances form hybridized resonances. This breakdown is realized by a small electric field volume whose inverse is proportional to the coupling rate, and it can be investigated as a vacuum Rabi splitting by tuning the plasmonic resonance energy [50]. Note that photon number in a hotspot of a dimer is so small that Rabi splitting by light is too small to generate measureable

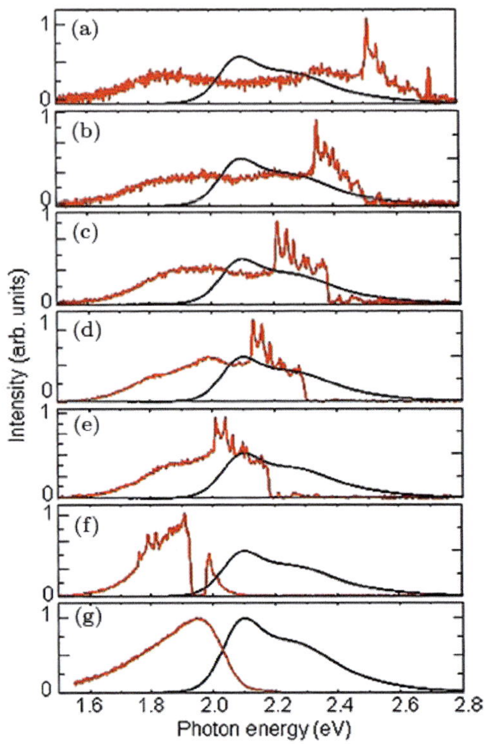

Figure 4. (a)–(f) Excitation laser energy dependence of SEF spectra of crystal violet (CV) dye from large silver NP aggregates excited with (a) 2.71, (b) 2.54, (c) 2.41, (d) 2.33, (e) 2.21, and (f) 1.96 eV (red lines) and absorption spectrum of CV molecules in a free space (black lines). Detailed structures around excitation laser lines are SERS bands. (g) Absorption (black line) and fluorescence (red lines) spectrum of CV molecules in aqueous solution ($\sim 10^{-6}$ M). Taken from Ref. [42].

Ω_R under common SM SERS excitation conditions approximately several W/cm^2 as estimated in Section 3.

The coupling energy $\hbar g$ (g is the coupling rate) is described as an inner product of electric field induced by vacuum fluctuation E_{vac} and the transition dipole moment between the electric states $p_0 = -e\langle i|x|f\rangle$ $(= -el_{\text{dip}})$ as $\hbar g = |p_0 E_{\text{vac}}|$, where $|i\rangle$ and $|f\rangle$ are initial and final states, respectively. E_{vac} is derived by the relationship between the zero-point energy and the electric field energy integrating through the mode volume $\int \varepsilon_0 E_{\text{vac}}^2 dV = 1/2\hbar\omega$

as $E_{\text{vac}} = \sqrt{\frac{\hbar\omega}{2\varepsilon_0 V}}$ [39]. Thus, we obtain $g = \sqrt{\frac{p_0^2\omega}{2\varepsilon_0 \hbar V}}$ when plasmon and molecular excitonic resonance frequencies are equal to be $\omega(=\frac{c}{\lambda}2\pi)$. Let us estimate the value of $\hbar g$ for the typical SM SERS system composed of the dimer and an R6G molecule located at the dimer junction. The value of g can be rewritten using the oscillator strength (OS) of the local electronic resonance of a molecule $f = 2m\omega l_{\text{dip}}^2/(e^2\hbar)$ as [51]

$$g = \sqrt{\frac{1}{4\pi\varepsilon_r\varepsilon_0}\frac{\pi e^2 f}{mV}}, \tag{7}$$

where ε_{r} is the relative permittivity 1.77 and ε_0 is the vacuum permittivity $\sim$8.85 $\times$ 10^{-12} F/m, f for an R6G molecule is 0.69 [27], e is elementary charge of electron $\sim$1.60 $\times$ 10^{-19} C, m is free electron mass $\sim$9.11 $\times$ 10^{-31} kg, and V is effective mode volume of the plasmonic cavity at a dimer crevasse. Let us try to estimate V for the reported value of splitting of plasmon resonance spectra $\sim$200 meV [43]. One can know that V should be $\sim 2.0 \times 10^{-6}(\lambda/n)^3$ at $\lambda = 600$ nm and $n = 1.33$. The value of V is adequately larger than theoretical minimum $V = 1.7 \times 10^{-8}(\lambda/n)^3$ at the crevasse of the dimer as explained above [35], indicating that the reported spectral splitting is reasonably explained by a strong coupling between the plasmon and the excitonic resonances. The effective mode volumes $2.0 \times 10^{-6}(\lambda/n)^3$ are roughly converted into $|M_1|^2$ of 5.0×10^5 using Purcell factor $F = (\lambda/n)^3 Q/V$, considering low Q factors of plasmonic resonance <10. The estimation implies that SM SERS can be a realistic platform to evaluate the strong coupling system [43].

In 2003, Itoh *et al.* implicitly observed the strong coupling between the plasmons and molecular excitons of R6G located in hotspots of a single dimer [52]. They implicitly explained the investigation by the refractive index effect within the resonance region [53]. The explanation is identical to vacuum Rabi splitting as mentioned by Ambjörnsson *et al.* [54]. In 2013, Schlather *et al.* explicitly investigated the strong coupling between the plasmons and molecular excitons of J-aggregates located in hotspots of a single dimer [50]. They reported the anti-crossing of the hybridized resonance in the dispersion curves at the exciton transition energy and Rabi splitting ranging from 230 to 400 meV. Cade *et al.* demonstrated a strong coupling between the plasmon resonances of nanostructured silver films and the molecular excitons of a monolayer of R6G dye adsorbed on the films [55]. They also reported an anticrossing behavior in the dispersion curve of the hybridized resonance, yielding Rabi-splitting energy of 400 meV. Nagasawa *et al.* indicated the relations between the plasmonic resonance and SERS by tuning the plasmonic resonance energy using collective silver NP dimers, which exhibit Rabi-splitting properties [56]. By tuning the plasmon resonance energy, they showed the anticrossing of the hybridized energy a maximum Rabi-splitting energy of 390 meV. Itoh *et al.* used a near-SM SERS system as a probe for local fields to quantify observations of strong coupling at the single-molecule level [43]. During SERS quenching, temporal spectral changes were found in both elastic light scattering and SERS, as shown in Figure 5. A coupled oscillator model [57] composed of a plasmon and a molecular dipole consistently reproduces the changes in the elastic light scattering, indicating that the changes are due to the breaking up of the strong coupling by molecular dissipation. These investigations indicate that SM-SERS systems are useful for probing the dynamics of strong coupling.

(iii) When the steepened local electric field gradient at the hotspots becomes comparable to the size of the molecular electronic structures, the excitation of multi-poles such as a quadrupole, which is negligible in most cases of excitations by light in a free space, is significant. Owing to this unique photo-excitation property, the perturbation term induced by the light M cannot be approximated by a linear function, resulting in changes to the integrated value $\langle i|H'|f\rangle$ that determines the selection rule for the molecular electronic excitation. This failure of the approximation of the linear function indicates the breakdown of the selection rule by the fine local electric field gradient. This breakdown has been investigated by an observation of forbidden transition modes [58].

Theoretical aspects of the breakdown of the selection rule by the local electric field gradient were developed by several groups using Buckingham's work [59]. This breakdown of the selection rules was

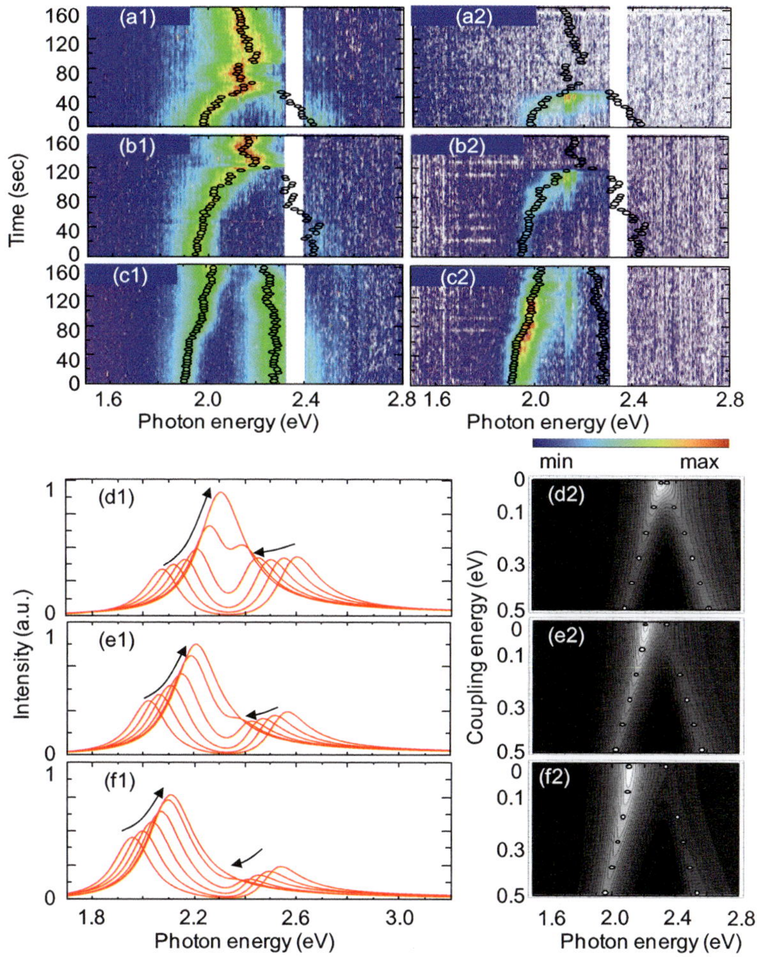

Figure 5. Strong coupling shown in SM-SERS system: (a1)–(c1) Images of temporal changes in elastic light scattering spectra for three silver NP dimers. Lower and higher energy lines in the elastic light scattering spectra are indicated by circles. (a2)–(c2) Images of temporal changes in SERS with SEF spectra for three silver NP dimers. The maximum positions in (a1)–(c1) are indicated by circles. Dimers for (a1)–(c1) correspond to those for (a2)–(c2). The white bars indicate loss by a laser notch filter. (d1)–(f1) Calculated spectra of elastic light scattering using $\hbar g$ from 500 to 0 meV with interval of 100 meV. The insets show widths of lower energy line against coupling energy. The values of original plasmonic resonance energy of (d1)–(f1) are 2.3, 2.2, and 2.1 eV, respectively. The value of original plasmonic resonance line-width of (d1)–(f1) is 220 meV. (d2)–(f2) Contour images of the changes in elastic light scattering. Peak positions of lower and higher energy lines are indicated by circles. Taken from Ref. [43].

reproduced using the transition dipole moment of the excitons without approximating H' in Equation (1) as a linear function. Here, we use explanation by Ayars *et al.* using Buckingham's theory [60]. The transitions in vibration levels due to coupling with an electric field are described by the perturbation Hamiltonian in Equation (6) $H' = -\boldsymbol{\mu}\boldsymbol{E}$, where $\boldsymbol{\mu}$ is the dipole moment, and $\boldsymbol{E}$ is the electric field of light. The dipole moment induced in a molecule placed in $\boldsymbol{E}$ is given as

$$\mu_\alpha = \mu_\alpha^p + \alpha_{\alpha\beta}E_\beta + \frac{1}{3}A_{\alpha\beta\gamma}\frac{\partial E_\beta}{\partial\gamma} + \cdots, \tag{8}$$

where $\{\alpha, \beta, \gamma\}$ correspond to $\{x, y, z\}$, and the tensor convention is assumed for the repeated subscripts [59]. Here, μ_α^p is the permanent dipole moment, and $\alpha_{\alpha\beta}$ and $\mathrm{A}_{\alpha\beta\gamma}$ are the electric dipole–dipole and electric dipole–quadrupole polarizabilities, respectively [58]. Ayars *et al.* reported that the lowest order of the field-gradient effect contributes to the second term in Equation (8) [60]. Equation (8) can be shortened as

$$\mu_\alpha = \mu_\alpha^p + \alpha_{\alpha\beta}E_\beta + \cdots. \tag{9}$$

Vibrational spectroscopy obtains signals from the fluctuating dipole moment related to the coordinate of the nucleus Q. Thus, using partial differentiation, Equation (9) is expanded by Q as

$$\mu_\alpha = (\mu_\alpha^p)_0 + \left(\frac{d\mu_\alpha^p}{dQ}\right)_0 Q + (\alpha_{\alpha\beta}E_\beta)_0 + \left(\frac{d\alpha_{\alpha\beta}E_\beta}{dQ}\right)_0 Q + \cdots. \tag{10}$$

The terms related to Q are collected as

$$\mu_\alpha = \left(\left(\frac{d\mu_\alpha^p}{dQ}\right)_0 + \left(\frac{d\alpha_{\alpha\beta}}{dQ}\right)_0 E_\beta + \alpha_{\alpha\beta}\left(\frac{dE_\beta}{dQ}\right)_0 + \cdots\right)Q. \tag{11}$$

The terms without Q are discarded because we cannot observe them by vibrational spectroscopy. The first and second terms in Equation (11) are infrared (IR) absorption and Raman, respectively. The third term is called the gradient-field Raman (GFR) term and will be important for the location like the hotspot because $(\frac{dE_\beta}{dQ})_0$ in

Equation (11) may not be negligible in SERS hotspots. In a free space, the ratio of the GFR term to the Raman term becomes the order of $2\pi a/\lambda$, where a and λ represent an atomic size of $l_{\text{dip}} \sim 0.1$ nm for an R6G molecule [27] and a wavelength of ∼500 nm, respectively. Thus, this ratio is $\sim 10^{-3}$, and the GFR contribution is insignificant. The FDTD calculations of the EM fields at a SERS hotspot suggest that the intensity of the local electric fields changes by ∼50% within 0.5 nm [25]. Thus, this ratio is $\sim 10^{-1}$ for an R6G molecule, indicating the breakdown of the selection rule in SERS may be still negligible because of the much weaker than normal Raman lines. According to Equation (11), the GFR signal should be large when the polarizability is large, e.g. for ionic bonding, resulting in a similar selection rule as is employed for infrared spectroscopy.

Ayars *et al.* experimentally demonstrated the GFR spectra of a potassium titanyl phosphate sample using near-field scanning optical microscopy (NSOM) as a function of the dependence of the GFR spectra on the distance between the NSOM probe and the surface [60]. Single-walled carbon nanotubes are useful for evaluating GFR signals owing to their large size and strong resonance with light. Indeed, Takase *et al.* demonstrated that the field-gradient effect can induce forbidden electronic transitions (not vibrational transitions) in SERS signals from carbon nanotubes as shown in Figure 6 [44] and observed a forbidden resonance Raman transition using the carbon nanotubes. This work shows that the nanogap not only enhances the EM-field intensity but also increases the optical excitation channels, including the excitation of forbidden modes beyond the selection rules. The demonstration of the importance of the field-gradient effect by Takase *et al.* will pave the way for novel applications of SERS. The recent advances of the field-gradient effect are well summarized by Christine *et al.* [44].

5. Summary and Outlook

Section 1 provided an overview of the EM mechanism in SERS from the beginning of its discovery, while Section 2 summarized the elucidation of the EM mechanism in SERS, and Section 3 discussed the limitations of the EM mechanism in SERS. Note that the analysis of

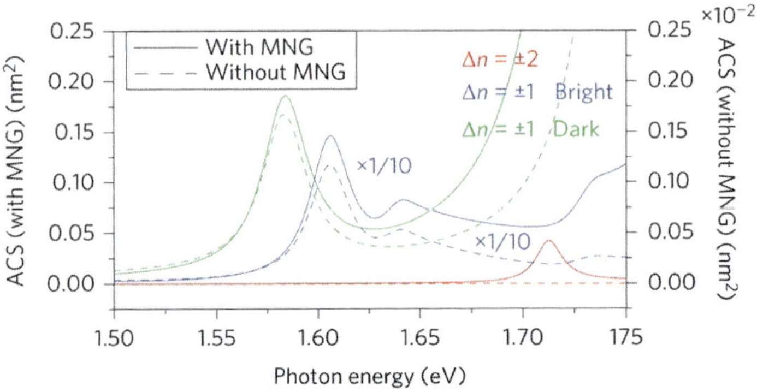

(a)

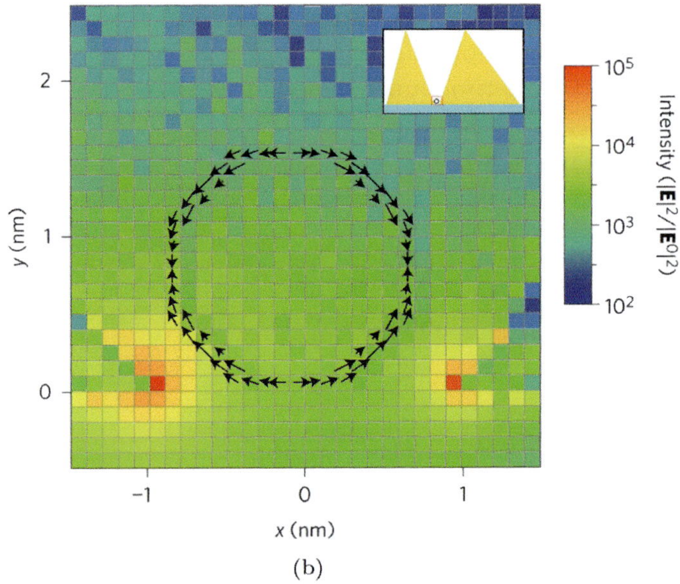

(b)

Figure 6. Selection-rule breakdown as indicated by a numerical calculation. (a) Absorption cross-section with (solid lines, left-hand side scale) and without (dotted lines, right-hand side scale) the metallic nanogap, as given by the numerical calculation. Results indicated by the solid and dotted blue lines are reduced by a factor of 10. (b) Cross-sectional view of the calculated intensity of the EM field, including both the longitudinal and transverse components, under resonant excitation conditions of the E14S (E41S) exciton (see Ref. [44, Figure 1]). The calculated quadrupole polarization of the E14S exciton (see Ref. [44, Figure 1]) is indicated by black arrows. The grids show the cell size in the present numerical calculation. Taken from Ref. [44].

the relations among the SERS spectra, plasmon resonance spectra, and shapes of silver NP dimers is indispensable for the quantitative evaluation of the SERS EM enhancement mechanism [25]. Section 4 analyzed the necessity for re-examining the EM mechanism in SERS with respect to three breakdowns in the approximations of Fermi's golden rule: the breakdown of the molecular electronic dynamics by the ultra-fast plasmonic de-excitation, that of the weak coupling between the plasmon and molecular electron by the strong EM enhancement, and that of the selection rule of SERS by the field-gradient effect. We expect that these breakdowns, if intentionally controlled, will allow the creation of novel plasmonic spectroscopic fields. We believe that the elucidation of SERS mechanism also contributes to making SERS and other plasmonic applications reliable and better suited as practical analytical tools [61–64].

Acknowledgments

The author thanks the significant contribution of Dr. Yuko S Yamamoto to this work. This work was supported by the Japan Society for the Promotion of Science (JSPS) KAKENHI Grant-in-Aid for Scientific Research (C) Number 20510111, and (B) Number 26286066.

References

1. Fleischmann, M., Hendra, P. J. and McQuillan, A. J. (1974). Raman spectra of pyridine adsorbed at a silver electrode. *Chem. Phys. Lett.* **26**, pp. 163–166.
2. Albrecht, M. G. and Creighton, J. A. (1977). Anomalously intense Raman spectra of pyridine at a silver electrode. *J. Am. Chem. Soc.* **99**, pp. 5215–5217.
3. Jeanmaire, D. L. and Van Duyne, R. P. (1977). Surface Raman spectroelectrochemistry: Part I. heterocyclic, aromatic, and aliphatic amines adsorbed on the anodized silver electrode. *J. Electroanal. Chem.* **84**, pp. 1–20.
4. Moskovits, M. (1978). Surface roughness and the enhanced intensity of Raman scattering by molecules adsorbed on metals. *J. Chem. Phys.* **69**, pp. 4159–4161.
5. Moskovits, M. (1985). Surface-enhanced spectroscopy. *Rev. Mod. Phys.* **57**, pp. 783–826.
6. Nie, S. and Emory, S. R. (1997). Probing single molecules and single nanoparticles by surface-enhanced Raman scattering. *Science* **275**, pp. 1102–1106.

7. Kneipp, K., Wang, Y., Kneipp, H., Perelman, L. T., Itzkan, I., Dasari, R. R. and Feld, M. S. (1997). Single molecule detection using surface-enhanced Raman scattering (SERS). *Phys. Rev. Lett.* **78**, pp. 1667–1670.
8. Xu, H., Aizpurua, J., M. Käll and Apell, P. (2000). Electromagnetic contributions to single-molecule sensitivity in surface-enhanced Raman scattering. *Phys. Rev. E* **62**, pp. 4318–4324.
9. Campion, A. and Kambhampati, P. (1998). Surface-enhanced Raman scattering. *Chem. Soc. Rev.* **27**, pp. 241–250.
10. Yamamoto, Y. S., Ozaki, Y. and Itoh, T. (2016). Why and how do the shapes of surface-enhanced Raman scattering spectra change? Recent progress from mechanistic studies. *J. Raman Spectrosco. USA* **47**, pp. 77–88.
11. Kreibig, U. and Vollmer, M. (1995). *Optical Properties of Metal Clusters*, Vol. 25, Springer Series in Materials Science. Springer, Berlin.
12. Bohren, C. F. and Huffman, D. R. (1983). *Absorption and Scattering of Light by Small Particles.* Wiley, New York, USA.
13. Wang, D. S. and Kerker, M. (1981). Enhanced Raman scattering by molecules adsorbed at the surface of colloidal spheroids. *Phys. Rev. B* **24**, p. 1777.
14. Aravind, P. K., Nitzan, A. and Metiu, H. (1981). The interaction between electromagnetic resonances and its role in spectroscopic studies of molecules adsorbed on colloidal particles or metal spheres. *Surf. Sci.* **110**, p. 189.
15. Inoue, M. and Ohtaka, K. (1983). Surface enhanced Raman scattering by metal spheres. I. cluster effect. *J. Phys. Soc. Jpn.* **52**, p. 3853.
16. Xu, H., Bjerneld, E. J., Käll, M. and Börjesson, L. (1999). Spectroscopy of single hemoglobin molecules by surface enhanced Raman scattering. *Phys. Rev. Lett.* **83**, p. 4357.
17. Itoh, T., Asahi, T. and Masuhara, H. (2001). Femtosecond light scattering spectroscopy of single gold nanoparticles. *Appl. Phys. Lett.* **79**, p. 1667.
18. Itoh, T., Hashimoto, K. and Ozaki, Y. (2003). Polarization dependences of surface plasmon bands and surface-enhanced Raman bands of single Ag nanoparticles. *Appl. Phys. Lett.* **2003**, p. 2274.
19. Itoh, T., Biju, V., Ishikawa, M., Kikkawa, Y., Hashimoto, K., Ikehata, A. and Ozaki, Y. (2006). Surface-enhanced resonance Raman scattering and background light emission coupled with plasmon of single Ag nanoaggregates. *J. Chem. Phys.* **124**, p. 134708.
20. Sun, G. and Khurgin, J. B. (2012). Origin of giant difference between fluorescence, resonance, and nonresonance Raman scattering enhancement by surface plasmons. *Phys. Rev. A* **85**, p. 063410.
21. Itoh, T., Hashimoto, K., Ikehata, A. and Ozaki, Y. (2004). Changes in excitation profiles of surface-enhanced resonance Raman scattering induced by changes in surface plasmon resonance of single Ag nano-aggregates. *Chem. Phys. Lett.* **389**, pp. 225–229.
22. Itoh, T., Yoshida, K., Biju, V., Kikkawa, Y., Ishikawa, M. and Ozaki, Y. (2007). Second enhancement in surface-enhanced resonance Raman scattering revealed by an analysis of anti-Stokes and Stokes Raman spectra. *Phys. Rev. B* **76**, p. 085405.

23. Yoshida, K., Itoh, T., Biju, V., Ishikawa, M. and Ozaki, Y. (2009). Experimental evaluation of the twofold electromagnetic enhancement theory of surface-enhanced resonance Raman scattering. *Phys. Rev. B* **79**, p. 085419.
24. Yoshida, K., Itoh, T., Biju, V., Ishikawa, M. and Ozaki, Y. (2009). Spectral shapes of surface-enhanced resonance Raman scattering sensitive to the refractive index of media around single Ag nanoaggregates. *Appl. Phys. Lett.* **95**, p. 263104.
25. Yoshida, K., Itoh, T., Tamaru, H., Biju, V., Ishikawa, M. and Ozaki, Y. (2010). Quantitative evaluation of electromagnetic enhancement in surface-enhanced resonance Raman scattering from plasmonic properties and morphologies of individual Ag nanostructures. *Phys. Rev. B* **81**, p. 115406.
26. Kleinman, S. L., Sharma, B., Blaber, M. G., Henry, A. I., Valley, N., Freeman, R. G., Natan, M. J., Schatz, G. C. and Van Duyne, R. P. (2013). Structure enhancement factor relationships in single gold nanoantennas by surface-enhanced Raman excitation spectroscopy. *J. Am. Chem. Soc.* **135**, p. 301.
27. Xu, H., Wang, X.-H., Persson, M. P., Xu, H. Q., Käll, M. and Johansson, P. (2004). Unified treatment of fluorescence and Raman scattering processes near metal surfaces. *Phys. Rev. Lett.* **93**, p. 243002.
28. Itoh, T., Iga, M., Tamaru, H., Yoshida, K., Biju, V. and Ishikawa, M. (2012). Quantitative evaluation of blinking in surface enhanced resonance Raman scattering and fluorescence by electromagnetic mechanism. *J. Chem. Phys.* **136**, p. 024703.
29. Galloway, C. M., Etchegoin, P. G. and Le Ru, E. C. (2009). Ultrafast nonradiative decay rates on metallic surfaces by comparing surface-enhanced Raman and fluorescence signals of single molecules. *Phys. Rev. Lett.* **103**, p. 063003.
30. Larkin, I. A., Stockman, M. I., Achermann, M. and Klimov, V. I. (2004). Dipolar emitters at nanoscale proximity of metal surfaces: Giant enhancement of relaxation in microscopic theory. *Phys. Rev. B* **69**, p. 121403.
31. Itoh, T., Yoshikawa, H., Yoshida, K., Biju, V. and Ishikawa, M. (2009). Evaluation of electromagnetic enhancement of surface enhanced hyper Raman scattering using plasmonic properties of binary active sites in single Ag nanoaggregates. *J. Chem. Phys.* **130**, p. 214706.
32. Itoh, T., Yoshikawa, H., Yoshida, K., Biju, V. and Ishikawa, M. (2010). Spectral variations in background light emission of surface-enhanced resonance hyper Raman scattering coupled with plasma resonance of individual silver nanoaggregates. *J. Chem. Phys.* **133**, p. 124704.
33. Zhang, Y., Zhen, Y. R., Neumann, O., Day, J. K., Nordlander, P. and Halas, N. J. (2014). Coherent anti-Stokes Raman scattering with single-molecule sensitivity using a plasmonic Fano resonance. *Nat. Comm.* **5**, p. 4424.
34. Yampolsky, S., Fishman, D. A., Dey, S., Hulkko, E., Banik, M., Potma, E. O. and Apkarian, V. A. (2014). Seeing a single molecule vibrate through time-resolved coherent anti-Stokes Raman scattering. *Nat. Photon.* **8**, pp. 650–656.
35. Savage, K. J., Hawkeye, M. M., Esteban, R., Borisov, A. G., Aizpurua, J. and Baumberg, J. J. (2012). Revealing the quantum regime in tunneling

plasmonics. *Nature* **491**, pp. 574–577; van Hulst, N. F. (2012). Plasmon quantum limit exposed. *Nat. Nanotech.* **7**, pp. 775–777.

36. Kneipp, K., Wang, Y., Kneipp, H., Itzkan, I., Dasari, R. R. and Feld, M. S. (1996). Population pumping of excited vibrational states by spontaneous surface-enhanced Raman scattering. *Phys. Rev. Lett.* **76**, pp. 2444–2447.
37. Kneipp, K., Kneipp, H., Corio, P., Brown, S. D. M., Shafer, K., Motz, J., Perelman, L. T., Hanlon, E. B., Marucci, A., Dresselhaus, G. and Dresselhaus, M. S. (2000). Surface-enhanced and normal stokes and anti-Stokes Raman spectroscopy of single-walled carbon nanotubes. *Phys. Rev. Lett.* **84**, pp. 3470–3473.
38. Haslett, T. L., Tay, L. and Moskovits, M. (2000). Can surface-enhanced Raman scattering serve as a channel for strong optical pumping? *J. Chem. Phys.* **113**, pp. 1641–1646.
39. Fox, M. (2006). *Quantum Optics: An Introduction.* Oxford Master Series in Physics, Oxford.
40. Roelli, P., Galland, C., Piro, N. and Kippenberg, T. J. (2015). Molecular cavity optomechanics as a theory of plasmon-enhanced Raman scattering. *Nat. Nanotech.* **11**, pp. 164–169.
41. Yamamoto, Y. S., Ozaki, Y. and Itoh, T. (2014). Recent progress and frontiers in the electromagnetic mechanism of surface-enhanced Raman scattering. *J. Photochem. Photobio. C* **21**, pp. 81–104.
42. Itoh, T., Yamamoto, Y. S., Tamaru, H., Biju, V., Murase, N. and Ozaki, Y. (2013). Excitation laser energy dependence of surface-enhanced fluorescence showing plasmon-induced ultrafast electronic dynamics in dye molecules. *Phys. Rev. B* **87**, p. 235408.
43. Itoh, T., Yamamoto, Y. S., Tamaru, H., Biju, V., Wakida, S. and Ozaki, Y. (2014). Single-molecular surface-enhanced resonance Raman scattering as a quantitative probe of local electromagnetic field: The case of strong coupling between plasmonic and excitonic resonance. *Phys. Rev. B* **89**, p. 195436.
44. Takase, M., Ajiki, H., Mizumoto, Y., Komeda, K., Nara, M., Nabika, H., Yasuda, S., Ishihara, H. and Murakoshi, K. (2013). Selection-rule breakdown in plasmon-induced electronic excitation of an isolated single-walled carbon nanotube. *Nat. Photon.* **7**, pp. 550–554.
45. Dirac, P. A. M. (1927). The quantum theory of the emission and absorption of radiation. *Proc. Roy. Soc.* **A114**, p. 243.
46. Le Ru, E. C., Etchegoin, P. G., Grand, J., Félidj, N., Aubard, J. and Lévi, G. (2007). Mechanisms of spectral profile modification in surface-enhanced fluorescence. *J. Phys. Chem. C* **111**, p. 16076.
47. Dong, Z. C., Zhang, X. L., Gao, H. Y., Luo, Y., Zhang, C., Chen, L. G., Zhang, R., Tao, X., Zhang, Y., Yang, J. L. and Hou, J. G. (2010). Generation of molecular hot electroluminescence by resonant nanocavity plasmons. *Nat. Photon.* **4**, pp. 50–54.
48. Cho, C. H., Aspetti, C. O., Turk, M. E., Kikkawa, J. M., Nam, S. W. and Agarwal, R. (2011). Tailoring hot-exciton emission and lifetimes in semiconducting nanowires via whispering-gallery nanocavity plasmons. *Nat. Mater.* **10**, pp. 669–675.

49. Cho, C. H., Aspetti, C. O., Park, J. and Agarwal, R. (2013). Silicon coupled with plasmon nanocavities generates bright visible hot luminescence. *Nat. Photon.* **7**, pp. 285–289.
50. Schlather, A. E., Large, N., Urban, A. S., Nordlander, P. and Halas, N. J. (2013). Near-field mediated plexcitonic coupling and giant Rabi splitting in individual metallic dimers. *Nano Lett.* **13**, pp. 3281–3286.
51. Andreani, L. C., Panzarini, G. and Gerard, J. M. (1999). Strong-coupling regime for quantum boxes in pillar microcavities: Theory. *Phys. Rev. B* **60**, pp. 13276–13279.
52. Itoh, T., Hashimoto, K., Ikehata, A. and Ozaki, Y. (2003). Direct demonstration for changes in surface plasmon resonance induced by surface-enhanced Raman scattering quenching of dye molecules adsorbed on single Ag nanoparticles. *App. Phys. Lett.* **83**, pp. 5557–5559.
53. Itoh, T., Kikkawa, Y., Biju, V., Ishikawa, M., Ikehata, A. and Ozaki, Y. (2006). Variations in steady-state and time-resolved background luminescence from surface-enhanced resonance Raman scattering-active single Ag nanoaggregates. *J. Phys. Chem. B* **110**, pp. 21536–21544.
54. Ambjörnsson, T., Mukhopadhyay, G., Apell, S. P. and Käll, M. (2006). Resonant coupling between localized plasmons and anisotropic molecular coatings in ellipsoidal metal nanoparticles. *Phys. Rev. B* **73**, p. 085412.
55. Cade, N. I., Ritman-Meer, T. and Richards, D. (2009). Strong coupling of localized plasmons and molecular excitons in nanostructured silver films. *Phys. Rev. B* **79**, No. 241404.
56. Nagasawa, F., Takase, M. and Murakoshi, K. (2014). Raman enhancement via polariton states produced by strong coupling between a localized surface plasmon and dye excitons at metal nanogaps. *J. Phys. Chem. Lett.* **5**, pp. 14–19.
57. Wu, X., Gray, S. K. and Pelton, M. (2010). Quantum-dot-induced transparency in a nanoscale plasmonic resonator. *Opt. Express* **18**, pp. 23633–23645.
58. Sass, J. K., Neff, H., Moskovits, M. and Holloway, S. (1981). Electric field gradient effects on the spectroscopy of adsorbed molecules. *J. Phys. Chem.* **85**, pp. 621–623.
59. Buckingham, A. D. (1967). Permanent and induced molecular moments and long-range intermolecular forces. *Adv. Chem. Phys.* **12**, pp. 107–142.
60. Ayars, E. J., Hallen, H. D. and Jahncke, C. L. (2000). Electric field gradient effects in Raman spectroscopy. *Phys. Rev. Lett.* **85**, pp. 4180–4183.
61. Yamamoto, Y. S., Ishikawa, M., Ozaki, Y. and Itoh, T. (2014). Fundamental studies on enhancement and blinking mechanism of surface-enhanced Raman scattering (SERS) and basic applications of SERS biological sensing. *Front. Phys.* **9**, pp. 31–46.
62. Yamamoto, Y. S., Hasegawa, K., Hasegawa, Y., Takahashi, N., Kitahama, Y., Fukuoka, S., Murase, N., Baba, Y., Ozaki, Y. and Itoh, T. (2013). Direct conversion of silver complexes to nanoscale hexagonal columns on copper alloy for plasmonic applications. *Phys. Chem. Chem. Phys.* **15**, pp. 14611–14615.

63. Yamamoto, Y. S., Hirano, K., Ishido, T., Yasui, T., Murase, N., Baba, Y. and Itoh, T. (2013). Plasmonic staining of DNA molecules with photo-induced Ag nanoparticles monitored using dark-field microscopy. *Phys. Chem. Chem. Phys.* **15**, pp. 10316–10320.
64. Yamamoto, Y. S., Fujime, Y., Takahashi, N., Nakanishi, S. and Itoh, T. (2016). Formation mechanism of plasmonic silver nanohexagonal particles made by galvanic displacement reaction. *RSC Adv.* **6**, pp. 31454–31461.

Chapter 3

Plasmon-Supported Two-Photon Excited Vibrational Sensing and Imaging

Janina Kneipp* and Katrin Kneipp

Department of Chemistry,
Humboldt Universität zu Berlin,
10099 Berlin, Germany
**janina.kneipp@chemie.hu-berlin.de*

1. Introduction

In 1928, it was predicted theoretically that a molecule might absorb two photons simultaneously, thus allowing electronic transitions that could not be reached by the energy of a single photon [1]. Two-photon excitation can also occur for elastic and inelastic scattering of photons on molecular vibrations resulting in hyper Rayleigh and hyper Raman scattering (HRS), respectively. HRS is the two-photon excited analogue of Raman scattering (RS). During HRS, two photons are scattered simultaneously, and scattering signals appear shifted relative to the second harmonic of the excitation light. Simultaneous absorption or scattering of two photons is very unlikely, and a high concentration of photons in time and space is a prerequisite for observing those phenomena.

As a two-photon excited process, HRS follows different selection rules than RS [2]. In systems of high symmetry, RS and HRS signals are mutually excluded. Dipole-active modes are always active in HRS. In particular, HRS can also access "silent" vibrational modes that are forbidden in both RS and also infrared absorption. As two-photon excited absorption/fluorescence does, two-photon excited scattering signals also depend on the excitation intensity to the power of two. Due to this nonlinear dependence, two-photon excited signals come from much smaller volumes compared to linear excitation. The confined excitation volume allows an efficient optical sectioning of samples and suggests two-photon excited spectroscopic effects for imaging.

In addition to processes based on simultaneous spontaneous or incoherent absorption or scattering of two photons, nonlinear optical effects can also be obtained due to coherent interaction of photons. For example, second harmonic generation (SHG) represents the coherent counterpart to hyper Rayleigh scattering and results in a coherent signal at doubled frequency of the excitation laser. In stimulated or coherent Raman probing, two optical fields coherently drive a vibrational mode [3]. This results in loss or gain in the interacting coherent fields. In a different experimental modality, a third coherent field can probe the coherent molecular vibration [4]. The availability of picosecond excitation enabled Raman-based vibrational pump-probe experiments [5]. Coherent Raman methods again became the focus of interest during the last decade [6–8].

While coherent nonlinear processes result in relatively strong and geometrically directed optical signals, spontaneous nonlinear optical effects suffer from small cross-sections. This applies particularly for two-photon excited RS. HRS appears at scattering cross-sections on the order of 10^{-65} $cm^4 s$ $photon^{-1}$, 35 orders of magnitude smaller than cross-sections of "normal" one-photon-excited RS and 15 orders of magnitude below typical two-photon absorption cross-sections. On the other hand, HRS can provide unique structural information and benefits from several experimental advantages over one-photon excitation. Therefore, despite their inherently extremely small cross-sections, it would be of great interest to exploit two-photon excited processes also for vibrational spectroscopy and imaging. In this way, high local fields, which appear due to field concentration and confinement in the vicinity of plasmonic nanostructures, can open up interesting opportunities for creating favorable conditions for nonlinear Raman probing and for harnessing HRS for analytical applications.

In Section 2, we discuss nonlinear incoherent and coherent Raman processes under the condition of plasmon support and show that enhanced local fields of plasmonic nanostructures can provide essential help for developing nonlinear vibrational spectroscopy. In Section 3, we focus on the application of surface-enhanced HRS (SEHRS) for vibrational probing and imaging with main emphasis on biologically relevant targets. Section 4 gives a brief summary.

2. Surface-Enhanced Nonlinear Raman Spectroscopy

Figure 1 illustrates different inelastic scattering processes in which photons $h\nu_L$ interact with vibrational quantum states of matter $h\nu_M$.

In the following, we briefly discuss Raman processes shown in Figure 1, for the case that they take place in the enhanced local fields of plasmonic nanostructures [9]. We describe this effect by frequency dependent field enhancement factors $A(\nu)$ with

$$|A(\nu)|^2 = \frac{|E(\nu)|^2}{|E^{(0)}(\nu)|^2}, \tag{1}$$

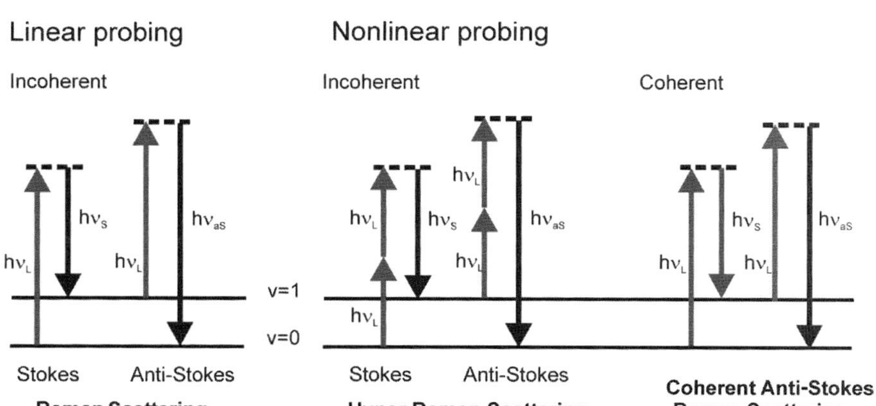

Figure 1. Raman processes for probing vibrational transitions displayed in an energy level diagram, $v = 0$ and $v = 1$ describe vibrational ground and first excited vibrational states.

where $E(\nu)$ are the local optical fields and $E^{(0)}(\nu)$ are the same fields in the absence of the metal nanostructure. In general, plasmon-supported field enhancement explains and describes SERS and related surface-enhanced effects displayed in Figure 1 very well. But, since the early days of SERS, there are experimental observations which cannot be understood in an electromagnetic field enhancement model, and which suggest additional "chemical" or "electronic" enhancement. Such enhancement effects are related to specific interactions between the molecule and the metal surface, where transfer/sharing of charges between molecule and substrate is considered as the most likely underlying basic process [10–12]. In the following, we describe chemical contributions in the framework of a molecule–metal complex which exhibits a larger intrinsic Raman cross-section $\sigma_{\text{ads}}^{\text{RS}}$ compared to that of the molecule without coupling to the metal $\sigma_{\text{free}}^{\text{RS}}$.

2.1. *Surface-enhanced Raman and hyper Raman scattering (SERS and SEHRS)*

Alternatively to an inelastic scattering process of one or two photons on vibrational quantum states, RS and HRS can be also described by radiation from an induced dipole

$$P_{\text{ind}} = \alpha E + \beta EE + \cdots \tag{2}$$

with α and β as polarizability and hyper polarizability, respectively. The first linear term describes Rayleigh and RS, the second term describes *hyper* Rayleigh and *hyper* RS. The cross-sections σ^{RS} and σ^{HRS} are proportional to the square of the change of α and β, respectively, with the change in vibrational coordinate Q.

The number of Stokes photons per second $n_{\text{S}}^{\text{SERS}}$ in SERS can be calculated as [13]

$$n_S^{\text{SERS}} = N_0 \sigma^{\text{SERS}} n_L, \tag{3a}$$

with an effective SERS cross-section

$$\sigma^{\text{SERS}} = \sigma_{\text{ads}}^{\text{RS}} |A(\nu_L)|^2 |A(\nu_S)|^2. \tag{3b}$$

The SERS enhancement factor G_{SERS} for Stokes scattering is determined by the ratio of the effective SERS cross-section σ^{SERS} to the normal Raman cross-section $\sigma_{\text{free}}^{\text{RS}}$

$$G_{\text{SERS}} = \frac{\sigma_{\text{ads}}^{\text{RS}}}{\sigma_{\text{free}}^{\text{RS}}} |A(\nu_L)|^2 |A(\nu_S)|^2. \tag{4}$$

Assuming that the field enhancement is roughly equal at excitation and scattering frequencies, the electromagnetic enhancement scales with the field enhancement factors $A(\nu_L)$ to the power of four.

HRS can be enhanced in an analogous fashion to normal RS. First observation of surface-enhanced HRS was reported in 1982 [14, 15]. In analogy to Equations (3) and (4), the number of surface-enhanced hyper Raman Stokes photons n^{SEHRS} can be written as

$$n^{\text{SEHRS}} = N_0\, \sigma^{\text{SEHRS}} n_L^2, \tag{5a}$$

with an effective surface-enhanced hyper Raman cross-section

$$\sigma^{\text{SEHRS}} = \sigma_{\text{ads}}^{\text{HRS}} |A(\nu_L)|^4 |A(\nu_{\text{hS}})|^2 \tag{5b}$$

and an enhancement factor for SEHRS as

$$G_{\text{SEHRS}} = \frac{\sigma_{\text{ads}}^{\text{HRS}}}{\sigma_{\text{free}}^{\text{HRS}}} |A(\nu_L)|^4 |A(\nu_{\text{HS}})|^2. \tag{6}$$

$\sigma_{\text{ads}}^{\text{HRS}}/\sigma_{\text{free}}^{\text{HRS}}$ describes the chemical enhancement effect. Theoretical approaches using electronic structure methods allow a more quantitative description of chemical contributions of SEHRS [16, 17]. Studies on pyridines show that the chemical enhancement factors for SEHRS can be larger than for SERS with 10^2–10^4 for SEHRS versus 10^1–10^2 for SERS [16, 18]. The electromagnetic enhancement for SEHRS scales with $A(\nu)$ to the power of six. Optimum structures should provide high field enhancement in two relatively widely separated ranges of the electromagnetic spectrum. The design and the optimization of plasmonic nanostructures for SEHRS have been discussed, including suggestions to improve SEHRS signals versus second harmonic background [19]. The strong plasmonic enhancement operative in SEHRS enables to measure two-photon excited vibrational spectra at very low concentrations of the target molecule down to the limit of single-molecule detection [13, 20], to detect HRS at low-intensity cw excitation [21] and to collect hyper Raman spectra also at the anti-Stokes side [22, 23]. Overall, plasmonic support can overcome the low-intensity problems inherent in HRS and suggest SEHRS as powerful two-photon spectroscopic tool for vibrational sensing and imaging [23–25], particularly also for structurally sensitive probing of biologically relevant molecules [26, 27] and of complex biological structures such as cultured cells and plant material [23, 28]. We come back to SEHRS for vibrational sensing and imaging in Section 3 of this chapter.

2.2. *Surface-enhanced pumped anti-Stokes Raman scattering and observation of "hot" Raman transitions*

Anti-Stokes RS is proportional to the number of molecules in the first excited vibrational state N_1. This number N_1, relative to the number of molecules in the vibrational ground state N_0 is determined by the Boltzmann factor. This results in much weaker anti-Stokes Raman signals than Stokes signals. Several effects can give rise to a relative increase of anti-Stokes signals compared to Stokes above the Boltzmann ratio, particularly also in SERS. In addition to obvious dependence of anti-Stokes to Stokes SERS signal (aS/S) ratios on temperature [29], the plasmon resonance spectrum can give rise to different field enhancement factors for Stokes and anti-Stokes scattering. This different electromagnetic enhancement as well as molecular and/or charge-transfer resonances which particularly support anti-Stokes or Stokes scattering [30–33] disturb temperature-dependent aS/S ratios. Moreover, in case of molecular resonant and pre-resonant RS, ultrafast internal conversion after electronic excitation and/or fluorescence act as main source for the population of excited vibrational states and can give rise to increased anti-Stokes signals [34]. Deviations from expected anti-Stokes signals in SERS were also discussed as result of a single-beam pumped coherent anti-Stokes Raman effect [35, 36].

Here, we will discuss surface-enhanced pumped anti-Stokes Raman scattering (SEPARS). In this process, a surface-enhanced non-resonant spontaneous Raman Stokes process with an effective cross-section σ^{SERS} populates the first excited vibrational levels in addition to thermal population [37–40]. Under stationary conditions and in a weakly saturating intensity regime ($\exp(-h\nu_M/kT) \leq \sigma_{\mathrm{S}}^{\mathrm{SERS}}\tau_1\, n_{\mathrm{L}} \ll 1$), the anti-Stokes signal n_{aS}^{SERS} can be estimated according to [37]

$$n_{\mathrm{aS}}^{\mathrm{SERS}} = N_0\, e^{-\frac{h\nu_M}{kT}}\, \sigma^{\mathrm{SERS}}\, n_L + N_0(\sigma^{\mathrm{SERS}})^2\, \tau_1\, n_L^2. \qquad (7a)$$

The first term describes anti-Stokes scattering related to thermal population, and the second term anti-Stokes signals related to the "pumped" population. One photon populates the excited vibrational state, a second photon generates the anti-Stokes scattering. τ_1 is the lifetime of the excited vibrational state. SEPARS shows a quadratic dependence on the excitation intensity while the Stokes signal shows a linear dependence. This provides evidence that the observed pumping is the result of spontaneous RS. SEPARS can be described by an effective two-photon cross-section [37],

$$\sigma^{\mathrm{SEPARS}} = (\sigma^{\mathrm{SERS}})^2\, \tau_1. \qquad (7b)$$

The obtained Raman pumping using *cw* excitation at a relatively low power level suggests plasmon-enhanced coherent Raman processes as a simple and efficient way to pump vibrational levels using pulsed excitation lasers [37]. This vibrational pumping up to saturation has been observed in recent experiments on plasmon supported CARS carried out with ultrashort pulses [41].

Based on experimentally observed anti-Stokes to Stokes signal ratios, and assuming vibrational lifetimes on the order of 10 ps, effective non-resonant SERS cross-sections σ^{SERS} must be on the order of 10^{-16} cm^2 [37]. Effective two-photon cross sections σ^{SEPARS} can be inferred to be about 10^{-43} cm^{4}s. This is about 7 orders of magnitude larger than typical cross-sections for two-photon excited fluorescence and also exceeds typical SEHRS cross-sections. The large effective cross-section can be explained by the nature of the two-photon process which exploits the first excited vibrational level as a real intermediate state.

As in formulas (3b) and (5b), we can split chemical and electromagnetic enhancement and write the effective surface-enhanced cross-section for pumped anti-Stokes scattering as

$$\sigma^{\mathrm{SEPARS}} = (\sigma_{\mathrm{ads}}^{\mathrm{RS}})^2 \tau_1 |A(\nu_L)|^4 |A(\nu_S)|^2 |A(\nu_{aS})|^2. \tag{7c}$$

Two-photon cross-sections according to Equation (7c), by replacing $A(\nu_{\mathrm{aS}})$ by $A(\nu_S)$, also describe the observation of so-called "hot" molecular transitions from first $v = 1$ to second $v = 2$ excited vibrational levels [37, 42]. These nonlinear SERS Stokes signals appear at slightly smaller Stokes shifts compared to $v = 0$ to $v = 1$ transitions usually observed in SERS. By this way, this plasmon-supported pumped spontaneous Stokes scattering permits to gather information about the anharmonicity of the electronic ground state potentials of the molecule, information which is usually not accessible from RS.

2.3. *Surface-enhanced stimulated or coherent Raman spectroscopy*

There are different options for coherent nonlinear Raman probing. Coherent anti-Stokes Raman scattering (CARS) might be currently the most prominent technique [43]. During CARS, an excitation laser (ν_L) and a Stokes laser (ν_S) create a coherent molecular vibration. The higher frequency field then interacts with the coherently excited molecular vibration and generates a strong coherent anti-Stokes signal. The nonlinear polarization, which drives the generation of CARS can be written as

$$P^{\mathrm{CARS}} \propto \mathrm{X}^{(3)}\, E_L(\nu_L)\, E_L(\nu_L)\, E_S^*(-\nu_S). \tag{8}$$

The three optical fields interact through the third-order nonlinear susceptibility $\mathrm{X}^{(3)}$, also called Raman susceptibility. $\mathrm{X}^{(3)}$ has resonances at the vibrational frequencies, and therefore CARS can probe vibrational modes of a molecule. In general, CARS requires phase matching between the interacting lasers which is always fulfilled when the CARS is generated in a tight focus [43]. The number of generated CARS photons n_{CARS} depends on the photon flux density of the laser field n_L to the power of two and of the Stokes field n_S linearly.

$$n_{\text{CARS}} \propto |\text{X}^{(3)}|^2\, n_L^2\, n_S. \tag{9}$$

Since the signal depends upon $|\text{X}^{(3)}|^2$, CARS shows dispersive line shapes due to the interference between the resonant vibrational contribution and the non-resonant electronic contribution to $\text{X}^{(3)}$.

CARS gives rise to strong directed signals and therefore, it provides very suitable conditions for imaging without requiring plasmonic support [7, 44]. However, coherent Raman methods can also essentially benefit from plasmonic field enhancement. Due to the four interacting coherent optical fields, surface-enhanced coherent Raman spectroscopy (SECARS) signals scale with local fields to the power of eight. Enhancement factors for the coherent nonlinear Raman processes can be written as

$$G_{\text{CARS}} = \frac{|\text{X}^{(3)}_{\text{ads}}|^2}{|\text{X}^{(3)}_{\text{free}}|^2}|A(\nu_L)|^4|A(\nu_S)|^2|A(\nu_{\text{aS}})|^2. \tag{10}$$

Here, we consider that also $\text{X}^{(3)}$ of a molecule adsorbed on a metal nanostructure can be influenced by “chemical” interactions between molecule and metal. Note that Im $\text{X}^{(3)}$ is directly related to the spontaneous Raman cross-section [45].

SECARS experiments have been reported for probing vibrations of different molecules and also single-wall carbon nanotubes [46–51] and can achieve single molecule sensitivity [52, 53]. SECARS in combination with tip-enhanced Raman spectroscopy allows vibrational imaging at the nanometer scale [54].

In addition to CARS, other nonlinear coherent Raman methods performed in plasmon enhanced local fields are also topics of growing interest [55]. Recently, surface-enhanced femtosecond stimulated Raman spectroscopy (SESRS) has been reported [56, 57]. SECARS and SESRS used in pump–probe experiments combine high spatial and temporal resolution with high sensitivity and can provide new insights into molecular dynamics and chemical processes on the surface of plasmonic structures [57]. Optimized plasmonic support enables pump–probe coherent Raman experiment at the single molecule level [58] and on individual nanostructures [41]. Ultrafast

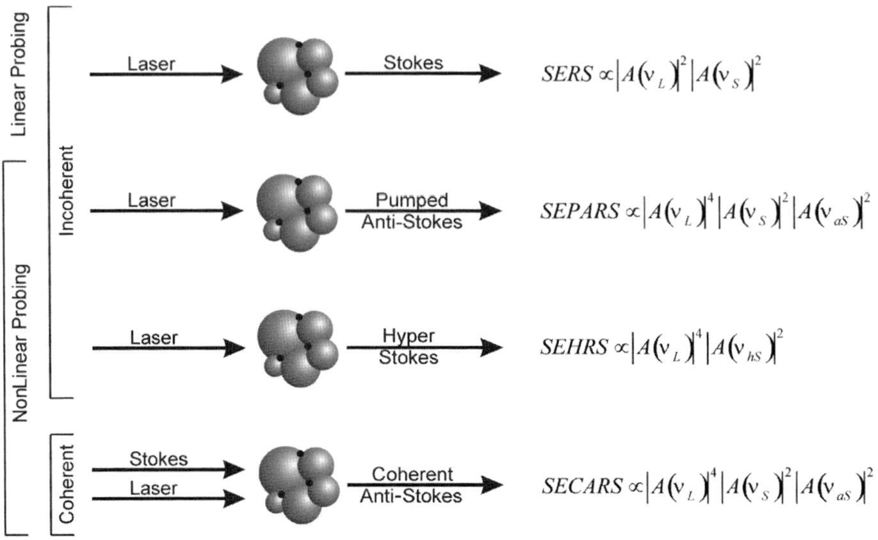

Figure 2. Plasmon-supported coherent and incoherent Raman processes for vibrational probing (adapted from Ref. [9] with permission from Springer Heidelberg).

surface-enhanced stimulated RS in combination with tip-enhanced Raman spectroscopy (TERS) is briefly discussed in the chapter by Rick van Duyne and co-workers.

Figure 2 summarizes different plasmon-supported Raman processes. SERS and SEHRS depend on field enhancement factors to the power of four and six, respectively, while SEPARS and SECARS benefit from field enhancement to the power of eight. While in SEPARS population of the excited vibrational state occurs due to spontaneous SERS Stokes scattering followed by surface-enhanced spontaneous anti-Stokes scattering [37], ultrafast SECARS has been discussed as saturated surface-enhanced SRS step followed by surface-enhanced spontaneous anti-Stokes RS [41]. This spontaneous anti-Stokes process can explain the missing dispersive line shapes in some SECARS experiments [41] and also the linear dependence of the SECARS signal on the number of molecules [52], even though a coherent anti-Stokes signal should show a quadratic dependence on this number.

From a practical spectroscopic point of view, SEPARS and SECARS provide vibrational information at the high energy side of the excitation laser, which is free from fluorescence background. In general, SEPARS is experimentally simple to perform using only one excitation laser, in contrast to SECARS that requires a "biharmonic" pump. However, note that experimental observation of SEPARS requires strong surface enhancement [37] while the strong and directed CARS signal is very valuable for microscopy and imaging also without plasmonic support [43].

3. Vibrational Sensing and Imaging using SEHRS

In order to use nonlinear vibrational spectroscopy for the characterization of molecular structure and composition, depending on the particular problem, several challenges need to be addressed. They include (i) realization of high molecular structural sensitivity, (ii) high signal levels to achieve high detection sensitivity, e.g. for fast imaging or high throughput screening, (iii) confined probed volumes, specifically for imaging at high spatial resolution, and (iv) Raman processes that are suitable for ultrafast vibrational pump–probe experiments. While coherent nonlinear Raman methods have a great potential regarding the last point (iv), SEHRS can excellently address all of the three first items. In contrast to coherent Raman methods, which are restricted to selected Raman lines that match the frequency difference of two excitation lasers, SEHRS provides information about the entire vibrational spectrum.

Due to different selection rules for one- and two-photon excitation, HRS can add new spectral information beyond RS and in this way improve sensing and imaging. The enhanced local fields that are responsible for SEHRS result in strong scattering signals that enable high detection sensitivity. Additionally, the field confinement automatically determines small probed volumes. Both, small probed volumes and surface-enhanced high signals, suggest SEHRS as a method for vibrational imaging. As will be discussed, surface-enhanced hyper Raman hyperspectral mapping can improve contrast in vibrational imaging, especially when it is combined with other spectral information, e.g. from SERS images. Finally, two-photon excitation allows the use of longer excitation wavelengths. In particular, biological samples profit from near-infrared (NIR) excitation

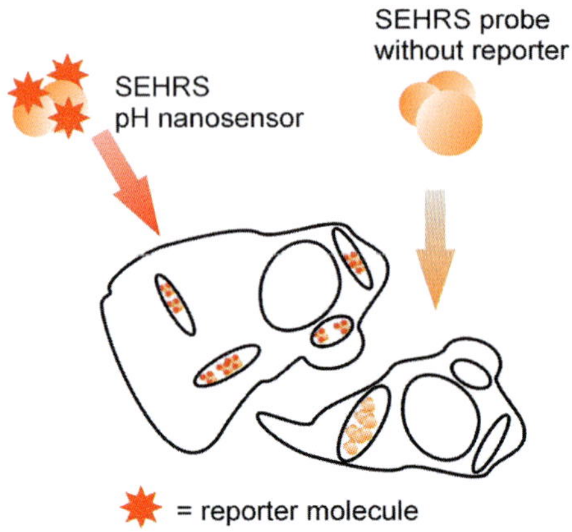

Figure 3. Schematic of a SEHRS nanosensor carrying a reporter molecule and a SEHRS probe. Both can be brought to specific regions in a complex sample, such as a cell and provide information about the molecules in their close proximity or other chemical parameters, such as pH.

due to reduced phototoxicity and better propagation of NIR photons through cells and tissues. Ideally, one-photon and two-photon NIR-excited processes (NIR-SERS and SEHRS) combined together will give different insights into molecular symmetry and in a comprehensive way could be used to investigate different molecules and molecule–nanostructure interactions.

In general, SEHRS, similar to SERS, can be used for both sensing by using the spectrum of a reporter molecules, as well as for probing the structure of molecules interacting with the surface of the plasmonic nanostructures (see Figure 3 for a schematic). In the following, we illustrate the spectroscopic potential of SEHRS using the spectra of important biomolecules, in particular nucleobases and amino acids. As an example of SEHRS nanosensing, pH-sensitive SEHRS spectra of a reporter molecule are discussed. For a model system, using the spectral fingerprints of two very similar molecules, an approach to SEHRS hyperspectral imaging is shown. The examples will show that two-photon excited SERS can give additional vibrational information and a better characterization of molecules and materials, in combination with the advantages of NIR excitation and the high lateral resolution of plasmonics-based spectroscopy.

3.1. *SEHRS spectra of biomolecules*

When plasmonic nanoparticles are introduced into lysosomes of macrophage cells, SEHRS signals can be obtained [23]. The SEHRS spectra look different from the one-photon excited SERS spectra from the same cells. They show bands that can be assigned to biomolecules typical of the phago-lysosomal compartment of the cells, namely, nucleic acids or nucleotides, amino acids, and proteins.

In order to understand the complex signatures of a whole cell, several different types of biomolecules were studied recently. All of them show electronic transitions in the UV. High repetition rate mode locked picosecond pulses at 1064 nm are used for two-photon excitation. For nucleic acids and amino acid molecules, this means non-resonant excitation and an experiment that relies entirely on surface enhancement. For the non-resonant SEHRS spectra of the DNA base adenine, effective two-photon cross-sections were shown to be comparable or even better than those of two-photon fluorescence of typical chromophores [23]. The SEHRS spectra of the nucleobases, obtained in solutions of silver nanoparticles, differ greatly from the SERS spectra, due to the different selection rules of the one- and two-photon excited Raman process. Band assignments in the spectra can be based on SERS data and on infrared spectra, and are discussed in detail in Ref. [26]. All five nucleobases show relatively low intensities of the symmetric ring breathing mode [26]. This band is very strong in SERS but compared to other bands only weak or medium in the SEHRS spectra. As an example, Figure 4 displays the SEHRS spectrum of cytosine obtained with citrate-stabilized silver nanoparticles, where the ring breathing mode at 799 cm^{-1} is strongly diminished compared to the SERS spectrum shown at the bottom.

Spectral contributions from the citrate molecules stabilizing the plasmonic silver nanostructures are evident in the SERS spectra (see e.g. the bands at 900 cm^{-1}, 828 cm^{-1}, and 952 cm^{-1} in the lower spectrum of Figure 4), and almost absent from the corresponding SEHRS spectra.

In Figure 5, the SEHRS and SERS spectra of the four aromatic amino acids are shown.

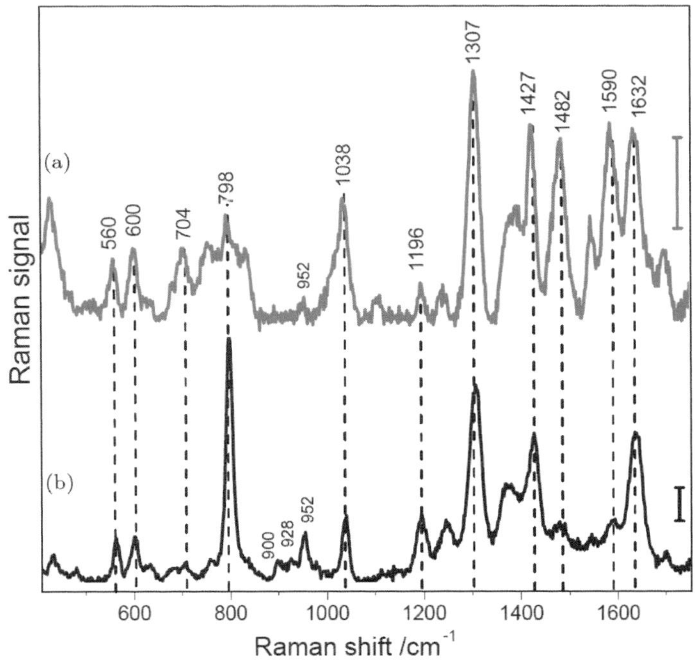

Figure 4. (a) SEHRS and (b) SERS spectrum of cytosine obtained with citrate reduced silver nanoparticles. Excitation: 1064 nm (a) and 532 nm (b), photon flux density: 4.7×10^{28} photons $cm^{-2}\,s^{-1}$ (a) and 1.4×10^{27} photons $cm^{-2}\,s^{-1}$ (b), acquisition time: 40 s (a) and 1 s (b), scale bars: 10 cps (a) and 500 cps (b), and cytosine concentration: 5×10^{-5} M. (Reprinted from Ref. [26] with permission from American Chemical Society.)

There, in a similar fashion to the spectra of the nucleic acids, the ring breathing modes are also much less pronounced than in the respective SERS spectra.

The two-photon excited spectra of the nucleobases and the amino acids are very sensitive with respect to molecule–nanoparticle interactions. When spectra of uracil and thymine, which interact with silver nanostructures in their deprotonated forms, are acquired at alkaline pH, conclusions could be drawn on the respective tautomeric form interacting with the silver surface [26]. In a similar fashion, the SEHRS spectra of the amino acids histidine and tyrosine varied greatly when the concentration of the molecules was slightly changed, and small changes in interaction and orientation of these

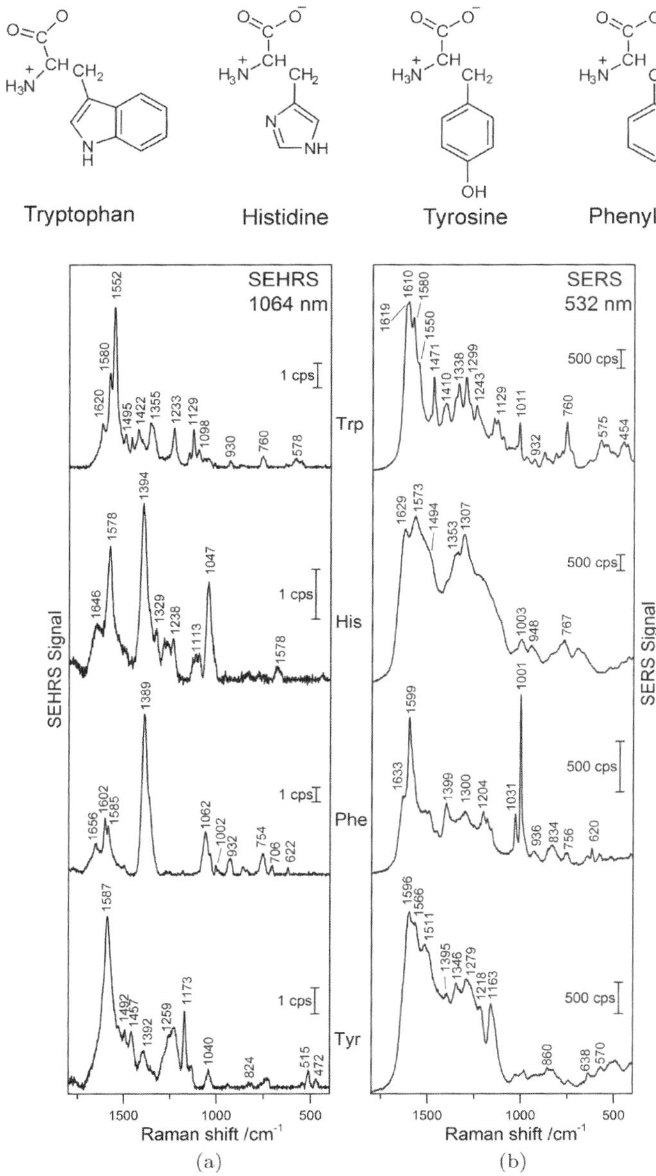

Figure 5. SEHRS spectra (a) and SERS spectra (b) of the amino acids tryptophan, histidine, phenylalanine, and tyrosine (from top to down) obtained with silver nanoaggregates. Excitation: 1064 nm (a) and 532 nm (b), photon flux density: 3×10^{28} photons cm^{-2} s^{-1} (a) and 6×10^{23} photons cm^{-2} s^{-1} (b), acquisition time: 60 s (a) and 5 s (b); concentration: 2×10^{-3} M; pH 10. (Reprinted from Ref. [59] with permission from American Chemical Society.)

two molecules were found [59]. From a general perspective, the high sensitivity of the SEHRS spectra with respect to surface environmental changes can help to improve our understanding of molecule–nanostructure interactions and of consequent phenomena, including SERS itself. It strongly suggests SEHRS as probe of amino acid–nanoparticle interactions. Specifically, when using both SEHRS and SERS, different interaction sites and/or different interacting molecular species can be probed. As will be shown in the next section, it also improves sensing by SEHRS reporter molecules.

3.2. *Nanoscale pH meters*

An interesting field of plasmon enhanced RS with a particular focus on biological applications deals with multifunctional, mobile SERS nanosensors with gold nanoparticles as basic building blocks [60] (see also the chapter by Mostafa El-Sayed and Sajanlal Panikkanvalappil).

SERS sensors measure Raman spectra of molecules residing in their enhanced local field and by this way, for example after delivering them into cells, they can probe cellular structures and cellular processes on the molecular level. (see also the chapter by El Sayed). pH is an important microenvironmental parameter, e.g. in biological samples such as cells and tissues, or in microfluidic systems, and the SERS spectrum of several molecules was shown to depend on pH. SERS spectra of 4-mercaptobenzoic acid (pMBA) show such pH dependence [61]. pH sensing using pMBA SERS spectra relies on changes in relative intensities of Raman lines that are caused by both protonation and deprotonation of the molecule [24, 62–65]. Spectra of pMBA on gold nanospheres indicated variation of pH in the endosomal system of cultured cells between 6.8 and 5.4 in endosomes of different ages [64].

SEHRS spectra of pMBA are very similar to SERS spectra, but the relative signal strengths of the bands in the spectrum are changed. As was reported for the SERS spectra, when pH increases, the relative band intensities of vibrations associated with the carboxyl group change relative to those of the aromatic ring [63, 64]. In the SEHRS spectra, upon deprotonation, the intensity of the

COO- band at 1365 cm^{-1} increases, and the C=O stretching vibration at 1685 cm^{-1} decreases. Recently, a better understanding of the two-photon excited SEHRS spectrum pf pMBA could be obtained. SEHRS spectra of pMBA were acquired between pH 2 and 12 under non-resonant conditions, also together with the visible and near-infrared (1064 nm) excited SERS spectra [24]. We find in the SEHRS spectra smaller Raman shifts by 5–15 cm^{-1} for several bands, when comparing them to one-photon excited SERS spectra. Investigating four different types of silver nanoparticles, it was noticed that these shifts are not specific for one nanoparticle type [24]. It is very likely that these bands indicate probing of different vibrations in SERS and in SEHRS due to the different selection rules [66], and that the shifts are in fact indicative of different adsorption species. For other molecules, it was shown as well that SEHRS is very sensitive to surface potential [22] and to local surface environmental changes [66], more than SERS. For high pH values, new band components appear, such as asymmetric broadening and new shoulders, supporting the conclusion that different adsorption species are probed in SERS and SEHRS. Probing orientation of the reporter molecule pMBA provides an additional indicator of changes in pH and makes SEHRS pH nanosensors even more sensitive than SERS probing alone. As was shown ideally, a combination of SEHRS with NIR-excited SERS could be used.

3.3. *Multivariate analysis of SEHRS spectra and hyperspectral imaging*

In order to achieve separation of SEHRS reporter molecules or an identification of spectra from bioorganic compounds in mixtures, multivariate methods that consider many bands in the spectra are clearly of advantage. Automated identification and classification will be crucial for localizing SEHRS nanoprobes in complex microscopic environments and for generating image contrast based on the full spectral information that they can provide.

Principal components analysis (PCA) is a multivariate technique that has found application in fields such as pattern recognition [67] and image compression [68], and has widely been applied for finding patterns in data of high dimension. In PCA, a data matrix is projected into a variance-weighted coordinate system. This orthogonal linear transformation (or eigenvalue problem) transforms a number of possibly correlated variables of the spectral dataset into a smaller number of uncorrelated variables such that the greatest variance by any projection of the data comes to lie on the first coordinate [69]. In the case of SERS probes, the full exploitation of their multiplexing capabilities was shown for identification of reporter spectra and different spectral contributions from the biomolecules in cells using PCA and hierarchical clustering [70]. Furthermore, for SERS data, automated pattern recognition tools, in particular artificial neural networks (ANN), can also be applied. In pollen tissue, analysis of the SERS data with an artificial neural network showed that intrinsic biochemical information of the pollen cells can be utilized for the discrimination of different plant species [71]. As suggested in recent studies on using support vector machines for SERS-based cancer [72] and virus [73] detection, the application of supervised learning based on pattern recognition to SERS data could in general be very promising.

As was shown for SERS, and is also the case in SEHRS, variation in a set of data of cells or other samples can be very different compared to variation in datasets of typical normal Raman or infrared data of similar samples. While in the latter, most of the variance in the spectra is based on biochemistry, and usually covered using a limited number of principal components, e.g. 5 or 10, variation in SERS and SEHRS data can have many different origins, molecular ones, as well as variation due to, e.g. specific interaction of molecules with the plasmonic substrate, or heterogeneity of the substrates. For example, since different nanoparticles or nanoaggregates are residing in the focal volume at the time each spectrum is acquired, the variance within the spectra from one sample can be very high. In the work that was done on extracts of plant (pollen) tissue, heterogeneity in SERS datasets was studied, and PCA was shown to provide a very good means to assess this heterogeneity [71, 74]. To investigate the potential of SEHRS in multivariate mapping of large areas, a well-characterized type of silver nanoparticle surface, [75] was

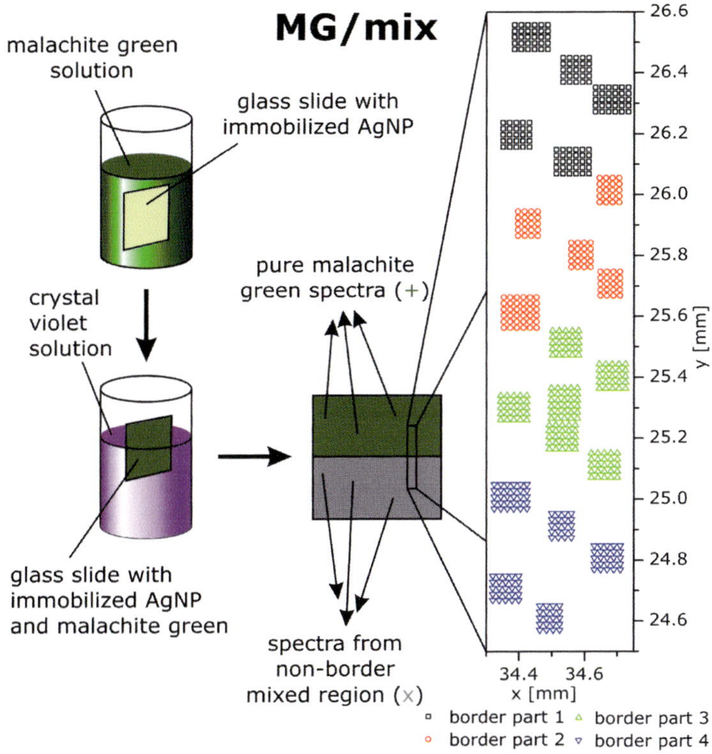

Figure 6. Schematic of the preparation of macroscopic samples that contain a border region of microscopic heterogeneity in the distribution of malachite green (MG) and crystal violet (CV) on immobilized silver nanoparticles (Copyright 2016, Published by the PCCP Owner Societies).

used for first PCA-based imaging experiments [25]. SEHRS spectra were excited in resonance with two very similar reporter molecules, crystal violet (CV) and malachite green (MG). Samples were prepared with distinguishable areas for SEHRS hyperspectral mapping by immersing slides with immobilized silver nanoparticles first in MG and subsequently in CV (schematics shown in Figure 6).

The region marked with the rectangle in the center is a border region between pure MG (green) and MG+CV mix (grey), and has a microscopic heterogeneity that can be investigated by SEHRS mapping. Within this heterogeneous region, microscopic maps were collected.

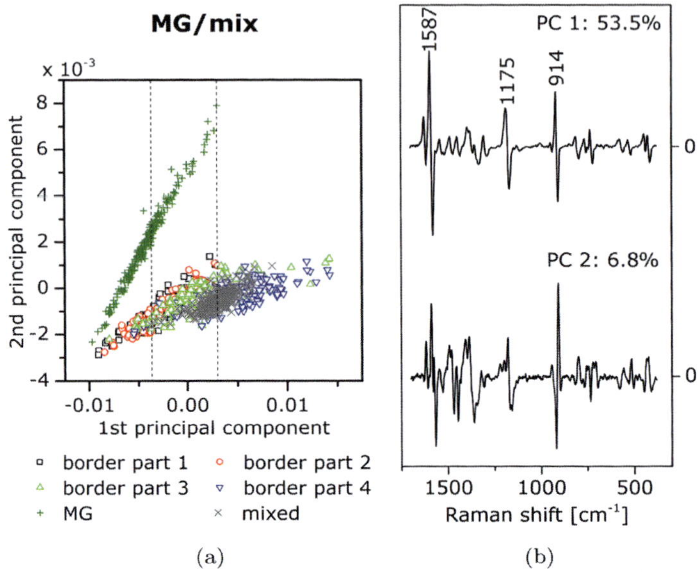

Figure 7. (a) Scores and (b) loadings of the first two principal components (PC) resulting from the principal component analysis (PCA) of the vector-normalized first derivatives of SEHRS spectra over the range 380–1700 cm^{-1} of the MG/mix samples. Symbols and colors correspond to the subregions in Figure 6. Spectra from the two non-border regions on each sample (MG and mixed, respectively) were also included in the PCA and served as references. The dashed lines in a demarcate the ranges of score values of the first PC, which were used for the differentiation between different types of spectra in Figure 6. Variance in percent that is explained by each PC is indicated in the loadings plot in b. Reproduced with permission from Ref. [25] (Copyright 2016, Published by the PCCP Owner Societies).

The SEHRS spectra were analyzed with PCA. The score plot in Figure 7(a) indicates the formation of groups of spectra that are similar with respect to a specific principal component. The loadings (Figure 7(b)) give information about those bands in the spectra that are responsible for their separation with respect to the corresponding principal component. The scores of the PCA of the SEHRS spectra of an MG/mix sample yield clear separation of the MG spectra (green "+") from the mixed spectra (grey "x") mostly due to the variation represented by the first principal component (PC) with a small contribution of the second PC. The loadings show a large variance in the spectra around 914 cm^{-1}, where the SEHRS spectra of CV and MG display bands at different frequencies.

The first PC also indicates differences around 1587 cm^{-1} and 1175 cm^{-1}, where all of the SEHRS spectra show intense phenyl bands (Figure 7(b)).

PCA of the SEHRS spectra enables separation of the different regions of the samples. In the scores plots (Figure 7(a)), the direction of changes in the spectra from the MG/mix border regions along the first PC was used to reconstruct hyperspectral maps (Figure 8(a)). Similar analysis was also performed with the SERS spectra (Figure 8(b)).

The color scale of the maps was set according to the score values of the spectra, arbitrarily separating the axis of the first PC into three ranges. Threshold score values for the separation are presented by the maximum and minimum score values of the spectra of regions with only MG and mixed regions very far from the border region: The left range in Figure 7(a) covers those spectra that can be unequivocally assigned to MG, the right range covers those from the mixed (non-border) region. In the middle, SEHRS spectra from the border, mixed, and single-analyte regions overlap. In the upper part of the border region, several SEHRS spectra are comparable to pure MG spectra (green pixels), while the rest of the SEHRS spectra and most of the SERS spectra in this part have score values within the middle range. In the lower part of the map, more spectra are assigned to the mixed region (Figure 8, grey pixels). The differences within the border region are very small. The separation of the score values based on the values for non-border spectra was arbitrary in this case and needs to be defined by each respective experiment. Also, the principal component(s) chosen for imaging are specific for each particular imaging problem. For each future application of such a hyperspectral imaging approach, the parameters for the construction of the color scale have to be determined based on corresponding reference spectra. The fact that classification of pixels in the SEHRS maps can differ from the SERS maps (Figures 8(a) and 8(b)) holds great promise to exploit the extensive chemical information of complementary SEHRS and SERS imaging.

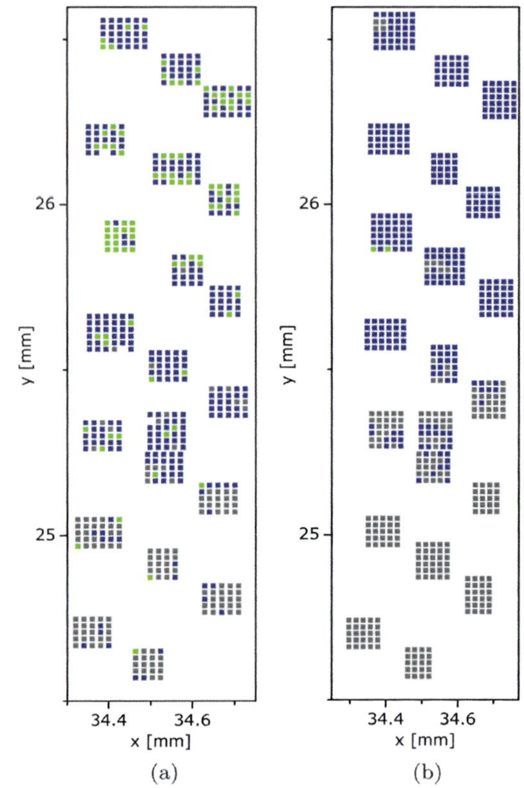

Figure 8. False-color (a) SEHRS and (b) SERS map of the border region of the MG/mix sample shown schematically in Figure 6. The maps are based on the scores of the respective first principle component of the PCA of the SEHRS spectra (Figure 7(a)) and of the SERS spectra (PCA not shown here). The x and y axes of the maps display the absolute coordinates of the x, y-stage that was used for the SEHRS and SERS mapping experiments in order to demonstrate the overlap of the microscopic regions that were probed by SEHRS and SERS. Color code: Assignment according to 1st PC score. Green: MG, blue: between mix and MG, grey: mix. x and y scales show absolute positions in millimeter of the x,y-stage for comparison of the two maps. Reproduced with permission from Ref. [25] (Copyright 2016, Published by the PCCP Owner Societies).

4. Summary

Incoherent or spontaneous two-photon excited HRS as well as stimulated coherent Raman processes enable nonlinear Raman probing. Both spontaneous as well as stimulated RS benefit from enhanced local fields. SEHRS can provide unique vibrational information and combines the advantages of two-photon spectroscopy with the structural information of vibrational spectroscopy and the high-sensitivity and nanometer-scale local confinement of plasmon-supported spectroscopy. In addition to their great potential in sensing and imaging, two-photon Raman processes can be powerful tools for studying the plasmonic near field and the interaction between molecules and plasmonic nanostructures. This applies particularly also to surface-enhanced coherent Raman probing, which can be used in ultrafast time resolved studies.

References

1. Goeppert-Mayer, M. (1929). Über die Wahrscheinlichkeit des Zusammenwirkens zweier Lichtquanten in einem Elementarakt. *Die Naturwissenschaften* **17**, p. 932.
2. Denisov, V. N., Mavrin, B. N. and Podobedov, V. B. (1987). Hyper-Raman scattering by vibrational excitations in crystals, glasses and liquids. *Phys. Rep.-Rev. Sec. Phys. Lett.* **151**, pp. 1–92.
3. Garmire, E., Pandarese, F. and Townes, C. H. (1963). Coherently driven molecular vibrations and light modulation. *Phys. Rev. Lett.* **11160**, *Phys. Rev. Lett.* **11**, p. 160.
4. Bloembergen, N. (1967). The stimulated Raman effect. *Am. J. Phys.* **35**, p. 989.
5. Laubereau, A. and Kaiser, W. (1978). Vibrational dynamics of liquids and solids investigated by picosecond light pulses. *Rev. Mod. Phys.* **50**, p. 607.
6. Freudiger, C. W., Min, W., Saar, B. G., Lu, S., Holtom, G. R., He, C. W., Tsai, J. C., Kang, J. X. and Xie, X. S. (2008). Label-free biomedical imaging with high sensitivity by stimulated Raman scattering microscopy. *Science* **322**, pp. 1857–1861.
7. Evans, C. L. and Xie, X. S. (2008). Coherent anti-Stokes Raman scattering microscopy: Chemical imaging for biology and medicine. *Annu. Rev. Anal. Chem.* **1**, pp. 883–909.
8. Umapathy, S., Lakshmanna, A. and Mallick, B. (2009). Ultrafast Raman loss spectroscopy. *J. Raman Spectrosc.* **40**, pp. 235–237.
9. Kneipp, K., Kneipp, H. and Kneipp, J. (2013). Plasmonics for enhanced vibrational signatures. In Shahbazyan, T. V. and Stockman, M. I. (eds.), *Plasmonics: Theory and Applications.* Springer, pp. 103–124.
10. Otto, A. (1984). Surface- enhanced Raman scattering: 'Classical' and 'chemical' origins. In Cardona, M. and Guntherodt, G. (eds.), *Light Scattering in Solids IV. Electronic Scattering, Spin Effects, SERS and Morphic Effects.* Springer-Verlag, pp. 289–418.
11. Campion, A. and Kambhampati, P. (1998). Surface-enhanced Raman scattering. *Chem. Soc. Rev.* **27**, pp. 241–250.

12. Jensen, L., Aikens, C. M. and Schatz, G. C. (2008). Electronic structure methods for studying surface-enhanced Raman scattering. *Chem. Soc. Rev.* **37**, pp. 1061–1073.
13. Kneipp, K., Kneipp, H., Itzkan, I., Dasari, R. R., Feld, M. S. and Dresselhaus, M. S. (2002). Nonlinear Raman probe of single molecules attached to colloidal silver and gold clusters. In *Optical Properties of Nanostructured Random Media*. Springer-Verlag Berlin, pp. 227–247.
14. Murphy, D. V., Von Raben, K. U., Chang, R. K. and Dorain, P. B. (1982). Surface-enhanced hyper-Raman scattering from SO/sub 3//sup 2-/ adsorbed on Ag powder. *Chem. Phys. Lett.* **85**, pp. 43–47.
15. Baranov, A. V. and Bobovich, Y. S. (1982). Super-enhanced hyper-Raman scattering from dyes adsorbed on colloidal silver particles. *JETP Lett.* **36**, pp. 339–343.
16. Valley, N., Jensen, L., Autschbach, J. and Schatz, G. C. (2010). Theoretical studies of surface enhanced hyper-Raman spectroscopy: The chemical enhancement mechanism. *J. Chem. Phys.* **133**, p. 8.
17. Turley, H. K., Hu, Z. W., Silverstein, D. W., Cooper, D. A., Jensen, L. and Camden, J. P. (2016). Probing two-photon molecular properties with surface-enhanced hyper-raman scattering: A combined experimental and theoretical study of crystal violet. *J. Phys. Chem. C* **120**, pp. 20936–20942.
18. Golab, J. T., Sprague, J. R., Carron, K. T., Schatz, G. C. and van Duyne, R. P. (1988). A surface enhanced hyper-Raman scattering study of pyridine adsorbed onto silver: experiment and theory. *J. Chem. Phys.* **88**, pp. 7942–7951.
19. Butet, J. and Martin, O. J. F. (2015). Surface-enhanced hyper-Raman scattering: A new road to the observation of low energy molecular vibrations. *J. Phys. Chem. C* **119**, pp. 15547–15556.
20. Milojevich, C. B., Mandrell, B. K., Turley, H. K., Iberi, V., Best, M. D. and Camden, J. P. (2013). Surface-enhanced hyper-Raman scattering from single molecules. *J. Phys. Chem. Lett.* **4**, pp. 3420–3423.
21. Itoh, T., Ozaki, Y., Yoshikawa, H., Ihama, T. and Masuhara, H. (2006). Hyper-Rayleigh scattering and hyper-Raman scattering of dye-adsorbed silver nanoparticles induced by a focused continuous-wave near-infrared laser. *Appl. Phys. Lett.* **88**, p. 084102.
22. Nie, S., Lipscomb, L. A. and Yu, N. T. (1991). Surface-enhanced hyper-Raman spectroscopy. *Appl. Spectros. Rev.* **26**, pp. 203–276.
23. Kneipp, J., Kneipp, H. and Kneipp, K. (2006). Two-photon vibrational spectroscopy for biosciences based on surface-enhanced hyper-Raman scattering. *Proc. Nat. Acad. Sci.* **103**, pp. 17149–17153.
24. Guhlke, M., Heiner, Z. and Kneipp, J. (2015). Combined near-infrared excited SEHRS and SERS spectra of pH sensors using silver nanostructures. *Phys. Chem. Chem. Phys.* **17**, pp. 26093–26100.
25. Guhlke, M., Heiner, Z. and Kneipp, J. (2016). Surface-enhanced hyper-Raman and Raman hyperspectral mapping. *Phys. Chem. Chem. Phys.* **18**, pp. 14228–14233.

26. Madzharova, F., Heiner, Z., Guhlke, M. and Kneipp, J. (2016). Surface-Enhanced Hyper-Raman Spectra of Adenine, Guanine, Cytosine, Thymine, and Uracil. *J. Phys. Chem. C* **120**, pp. 15415–15423.
27. Guhlke, M., Heiner, Z. and Kneipp, J. (2016). Surface-enhanced Raman and surface-enhanced hyper-Raman scattering of thiol-functionalized carotene. *J. Phys. Chem. C* **120**, pp. 20702–20709.
28. Palanco, M. E., Mogensen, K. B., Guhlke, M., Heiner, Z., Kneipp, J. and Kneipp, K. (2016). Templated green synthesis of plasmonic silver nanoparticles in onion epidermal cells suitable for surface-enhanced Raman and hyper-Raman scattering. *Beilstein J. Nanotechnol.* **7**, pp. 834–840.
29. Pozzi, E. A., Zrimsek, A. B., Lethiec, C. M., Schatz, G. C., Hersam, M. C. and Van Duyne, R. P. (2015). Evaluating single-molecule Stokes and anti-Stokes SERS for nanoscale thermometry. *J. Phys. Chem. C* **119**, pp. 21116–21124.
30. Brolo, A. G., Sanderson, A. C. and Smith, A. P. (2004). Ratio of the surface-enhanced anti-Stokes scattering to the surface-enhanced Stokes-Raman scattering for molecules adsorbed on a silver electrode. *Phys. Rev. B* **69**, p. 045424.
31. Itoh, T., Yoshida, K., Biju, V., Kikkawa, Y., Ishikawa, M. and Ozaki, Y. (2007). Second enhancement in surface-enhanced resonance Raman scattering revealed by an analysis of anti-Stokes and Stokes Raman spectra. *Phys. Rev. B* **76**, p. 085405.
32. dos Santos, D. P., Andrade, G. F. S., Brolo, A. G. and Temperini, M. L. A. (2011). Fluctuations of the Stokes and anti-Stokes surface-enhanced resonance Raman scattering intensities in an electrochemical environment. *Chem. Commun.* **47**, pp. 7158–7160.
33. dos Santos, D. P., Temperini, M. L. A. and Brolo, A. G. (2016). Single-molecule surface-enhanced (Resonance) Raman scattering (SE(R)RS) as a probe for metal colloid aggregation state. *J. Phys. Chem. C* **120**, pp. 20877–20885.
34. Kozich, V. and Werncke, W. (2010). The vibrational pumping mechanism in surface-enhanced Raman scattering: A subpicosecond time-resolved study. *J. Phys. Chem. C* **114**, pp. 10484–10488.
35. Lefrant, S., Buisson, J. P., Mevellec, J. Y., Baltog, L. and Baibarac, M. (2008). Single-beam pumped coherent anti-Stokes Raman scattering on carbon nanotubes. *Phys. Status Solidi B, Basic Solid State Phys.* **245**, pp. 2221–2224.
36. Lefrant, S., Buisson, J. P., Mevellec, J. Y., Baibarac, M. and Baltog, I. (2011). Nonlinear and resonance effects in carbon nanotube structures. *Opt. Mater.* **33**, pp. 1410–1414.
37. Kneipp, K., Yang, W., Kneipp, H., Itzkan, I., Dasari, R. R. and Feld, M. S. (1996). Population pumping of excited vibrational states by spontaneous surface-enhanced Raman scattering. *Phys. Rev. Lett.* **76**, pp. 2444–2447.
38. Kneipp, K., Kneipp, H., Kartha, V. B., Manoharan, R., Deinum, G., Itzkan, I., Dasari, R. R. and Feld, M. S. (1998). Detection and identification of a single DNA base molecule using surface- enhanced Raman scattering (SERS). *Phys. Rev. E* **57**, pp. R6281–6284.

39. Kneipp, K., Kneipp, H., Corio, P., Brown, S. D. M., Shafer, K., Motz, J., Perelman, L. T., Hanlon, E. B., Marucci, A. and Dresselhaus, G. *et al.* (2000). Surface-enhanced and normal Stokes and anti-Stokes Raman spectroscopy of single-walled carbon nanotubes. *Phys. Rev. Lett.* **84**, pp. 3470–3473.
40. Kneipp, K., Kneipp, H., Itzkan, I., Dasari, R. R. and Feld, M. S. (1999). Ultrasensitive chemical analysis by Raman spectroscopy. *Chem. Rev.* **99**, pp. 2957–2975.
41. Crampton, K. T., Zeytunyan, A., Fast, A. S., Ladani, F. T., Alfonso-Garcia, A, Banik, M., Yampolsky, S., Fishman, D. A., Potma, E. O. and Apkarian, V. A. (2016). Ultrafast coherent Raman scattering at plasmonic nanojunctions. *J. Phys. Chem. C* **120**, pp. 20943–20953.
42. Kneipp, K., Kneipp, H., Manoharan, R., Itzkan, I., Dasari, R. R. and Feld, M. S. (1998). Near-infrared surface-enhanced Raman scattering can detect single molecules and observe "hot" vibrational transitions. *J. Raman Spectrosc.* **29**, pp. 743–747.
43. Cheng, J. X., Volkmer, A. and Xie, X. S. (2002). Theoretical and experimental characterization of coherent anti-Stokes Raman scattering microscopy. *J. Opt. Soc. Am. B, Opt. Phys.* **19**, pp. 1363–1375.
44. Lu, F. K., Basu, S., Igras, V., Hoang, M. P., Ji, M. B., Fu, D., Holtom, G. R., Neel, V., Freudiger, C. W. and Fisher, D. E. *et al.* (2015). Label-free DNA imaging *in vivo* with stimulated Raman scattering microscopy (Vol. 112, p. 11624, 2015). *Proc. Nat. Acad. Sci. USA* **112**, pp. E5902–E5902.
45. Bloembergen, N. (1982). Nonlinear optics and spectroscopy. *Rev. Mod. Phys.* **54**, pp. 685–695.
46. Liang, E., Weippert, A., Funk, J., Materny, A. and Kiefer W. (1994). Experimental observation of surface-enhanced coherent anti-Stokes Raman scattering. *Chem. Phys. Lett.* **227**, pp. 115–120.
47. Baltog, I., Baibarac, M. and Lefrant, S. (2005). Coherent anti-Stokes Raman scattering on single-walled carbon nanotubes and copper phthalocyanine thin films excited through surface plasmons. *J. Opt. Pure Appl. Opti.* **7**, pp. 632–639.
48. Baltog, I., Baibarac, M. and Lefrant, S. (2008). "Single-beam pumped" coherent anti-Stokes Raman scattering on carbon nanotubes thin films excited through surface plasmons. *Phys. E Low-Dimen. Syst. Nanostruct.* **40**, pp. 2380–2385.
49. Addison, C. J., Konorov, S. O., Brolo, A. G., Blades, M. W. and Turner, R. F. B. (2009). Tuning gold nanoparticle self-assembly for optimum coherent anti-Stokes Raman scattering and second harmonic generation response. *J. Phys. Chem. C* **113**, pp. 3586–3592.
50. Steuwe, C., Kaminski, C. F., Baumberg, J. J. and Mahajan, S. (2011). Surface enhanced coherent anti-Stokes Raman scattering on nanostructured gold surfaces. *Nano Lett.* **11**, pp. 5339–5343.
51. Schlucker, S., Salehi, M., Bergner, G., Schutz, M., Strobel, P., Marx, A., Petersen, I., Dietzek, B. and Popp, J. (2011). Immuno-surface-enhanced coherent anti-Stokes Raman scattering microscopy: Immunohistochemistry

with target-specific metallic nanoprobes and nonlinear Raman microscopy. *Anal. Chem.* **83**, pp. 7081–7085.
52. Koo, T. W., Chan, S. and Berlin, A. A. (2005). Single-molecule detection of biomolecules by surface-enhanced coherent anti-Stokes Raman scattering. *Opt. Lett.* **30**, pp. 1024–1026.
53. Zhang, Y., Zhen, Y. R., Neumann, O., Day, J. K., Nordlander, P. and Halas, N. J. (2014). Coherent anti-Stokes Raman scattering with single-molecule sensitivity using a plasmonic Fano resonance. *Nature Commun.* **5**, p. 7.
54. Ichimura, T., Hayazawa, N., Hashimoto, M., Inouye, Y. and Kawata, S. (2004). Tip-enhanced coherent anti-Stokes Raman scattering for vibrational nanoimaging. *Phys. Rev. Lett.* **92**, p. 220801.
55. Aeschlimann, M., Brixner, T., Fischer, A., Kramer, C., Melchior, P., Pfeiffer, W., Schneider, C., Struber, C., Tuchscherer, P. and Voronine, D. V. (2011). Coherent two-dimensional nanoscopy. *Science* **333**, pp. 1723–1726.
56. Frontiera, R. R. and Mathies, R. A. (2011). Femtosecond stimulated Raman spectroscopy. *Laser Photon. Rev.* **5**, pp. 102–113.
57. Gruenke, N. L., Cardinal, M. F., McAnally, M. O., Frontiera, R. R., Schatza, G. C. and Van Duyne, R. P. (2016). Ultrafast and nonlinear surface-enhanced Raman spectroscopy. *Chem. Soci. Rev.* **45**, pp. 2263–2290.
58. Yampolsky, S., Fishman, D. A., Dey, S., Hulkko, E., Banik, M., Potma, E. O. and Apkarian, V. A. (2014). Seeing a single molecule vibrate through time-resolved coherent anti-Stokes Raman scattering. *Nature Photon.* **8**, pp. 650–656.
59. Madzharova, F., Heiner, Z. and Kneipp, J. (2017). Surface enhanced hyper-Raman scattering of the amino acids tryptophan, histidine, phenylalanine, and tyrosine. *J. Phys. Chem. C*, doi: 10.1021/acs.jpcc.6b10905.
60. Kneipp, J., Kneipp, H., Wittig, B. and Kneipp, K. (2010). Novel optical nanosensors for probing and imaging live cells, *Nanomedi. Nanotechnol. Biol. Med.* **6**, pp. 214–226.
61. Michota, A. and Bukowska, J. (2003). Surface-enhanced Raman scattering (SERS) of 4-mercaptobenzoic acid on silver and gold substrates. *J. Raman Spectrosc.* **34**, pp. 21–25.
62. Bishnoi, S. W., Rozell, C. J., Levin, C. S., Gheith, M. K., Johnson, B. R., Johnson, D. H. and Halas, N. J. (2006). All-optical nanoscale pH meter. *Nano Lett.* **6**, pp. 1687–1692.
63. Kneipp, J., Kneipp, H., Wittig, B. and Kneipp, K. (2007). One- and two-photon excited optical pH probing for cells using surface-enhanced Raman and hyper-Raman nanosensors. *Nano Lett.* **7**, pp. 2819–2823.
64. Kneipp, J., Kneipp, H., Wittig, B. and Kneipp, K. (2010). Following the dynamics of pH in endosomes of live cells with SERS nanosensors. *J. Phys. Chem. C* **114**, pp. 7421–7426.
65. Palanco, M. E., Mogensen, K. B. and Kneipp, K. (2016). Raman spectroscopic probing of plant material using SERS. *J. Raman Spectrosc.* **47**, pp. 156–161.

66. Hulteen, J. C., Young, M. A. and Van Duyne, R. P. (2006). Surface-enhanced hyper-Raman scattering (SEHRS) on Ag film over nanosphere (FON) electrodes: Surface symmetry of centrosymmetric adsorbates. *Langmuir* **22**, pp. 10354–10364.
67. Bereton, R. G. (2006). Pattern recognition. In *Chemometrics: Data Analysis for the Laboratory and Chemical Plant*. John Wiley & Sons Ltd, pp. 183–269.
68. Clausen, C. and Wechsler, H. (2000). Color image compression using PCA and backpropagation learning. *Pattern Recogn.* **33**, pp. 1555–1560.
69. Pearson, K. (1901). On lines and planes of closest fit to a system of points in space. *London, Edinburgh, Philos. Mag. J. Sci.* **6**, pp. 559–572.
70. Matschulat, A., Drescher, D. and Kneipp, J. (2010). Surface-enhanced Raman scattering hybrid nanoprobe multiplexing and imaging in biological systems. *ACS Nano* **4**, pp. 3259–3269.
71. Seifert, S., Merk, V. and Kneipp, J. (2016). Identification of aqueous pollen extracts using surface enhanced Raman scattering (SERS) and pattern recognition methods. *J. Biophoton.* **9**, pp. 181–189.
72. Li, S.-X., Zeng, Q.-Y., Li, L.-F., Zhang, Y.-J., Wan, M.-M., Liu, Z.-M., Xiong, H.-L., Guo, Z.-Y. and Liu, S.-H. (2013). Study of support vector machine and serum surface-enhanced Raman spectroscopy for noninvasive esophageal cancer detection. *J. Biomedi. Opt.* **18**.
73. Negri, P., Choi, J. Y., Jones, C., Tompkins, S. M., Tripp, R. A. and Dluhy, R. A. (2014). Identification of virulence determinants in influenza viruses. *Anal. Chem.* **86**, pp. 6911–6917.
74. Joseph, V., Gensler, M., Seifert, S., Gernert, U., Rabe, J. P. and Kneipp, J. (2012). Nanoscopic properties and application of mix-and-match plasmonic surfaces for microscopic SERS. *J. Phys. Chem. C* **116**, pp. 6859–6865.
75. Joseph, V., Engelbrekt, C., Zhang, J., Gernert, U., Ulstrup, J. and Kneipp, J. (2012). Characterizing the kinetics of nanoparticle-catalyzed reactions by surface-enhanced Raman scattering. *Angew. Chem. Int. Ed.* **51**, pp. 7592–7596.

Chapter 4

Plasmonically Enhanced Elastic and Inelastic Light Scattering for Real-Time Study of Molecular Cell Functions

Sajanlal R. Panikkanvalappil and Mostafa A. El-Sayed*

Laser Dynamics Laboratory
School of Chemistry and Biochemistry
Georgia Institute of Technology
Atlanta, Georgia 30332-0400, USA
**melsayed@gatech.edu*

1. Introduction

Interfacing the unique physiochemical properties of plasmonic nanoparticles with biotechnology and molecular cell biology has paved the way for unprecedented growth of biological research in all aspects. This has resulted in the development of powerful analytical tools for cellular imaging, labeling, sensing, as well as various innovative techniques in theranostics. In biology, most aspects of cell functions are tightly regulated by inter and intracellular signaling processes as well as the dynamic and complex molecular interactions of a myriad of chemical signaling species including proteins, lipids, nucleic acids, primary and secondary metabolites, etc. The success of these biomolecular events largely depends on timely and precise manipulation of bonding interaction between these molecules. Consequently, visualizing complex biomolecular events from their molecular perspective in real time received greater attention as it can enhance our understanding on the cell responses, thereby developing more efficient diagnostic and therapeutic techniques. Over the past few decades, the unique physiochemical properties of plasmonic NPs have been widely explored for addressing many challenging issues in molecular cell biology [1–8]. The appropriate size match of NPs with various biological molecules makes them suitable candidates for many biological applications. The innovative technologies based on these extremely small particles offer a new dimension to molecular cell biology research in the prospect of visualizing complex biomolecular events at the nanoscale [9–18]. Among the plasmonic NPs, gold nanoparticles (AuNPs) have been extensively used for developing various bioimaging, diagnostics, and therapeutic techniques owing to their unique characteristics such as localized surface plasmon resonance (LSPR), enhanced light scattering, ease in synthesizing NPs of different sizes and shapes, apparent low toxicity and facile surface chemistry allowing conjugation with various ligands for biofunctionalization [4, 19–22].

In this chapter, we address the recent developments and applications of nanotechnology-based techniques, with main emphasis on studies conducted in our laboratory, which combines elastic and inelastic light scattering of AuNPs to detect and characterize molecular cell functions. We refer the reader to a number of excellent reviews that discuss the immediate applications of plasmonic NPs in the area of molecular cell biology in greater detail [1–6, 23–29].

2. Light Scattering by Plasmonic Nanoparticles

The basis of nanotechnology lies in the fact that properties of materials change dramatically when their size is reduced to the nanometer scale. At the nanoscale, physiochemical properties of nanomaterials are mainly dependent on their spatial confinement of electrons. The unique properties of plasmonic NPs to absorb and scatter light of specific wavelengths of the electromagnetic spectrum have been widely explored for developing innovative technologies in nanomedicine such as resonance light scattering imaging [30], photoacoustic imaging [31, 32], multiphoton imaging [33], surface-enhanced Raman spectroscopy [34], plasmonic photothermal therapy [35], and so forth.

Light scattering has been differentiated generally into elastic or inelastic scattering. During elastic scattering (e.g. Rayleigh scattering) of light, the scattered light waves retain the same wavelength as the incident light. However, inelastic light scattering occurs when the wavelength of the scattered light shifts relative to the incident wavelength. The frequency shift can occur either by losing energy (Stokes shift) or by gaining energy (anti-Stokes shift) of the scattering substance. An example of inelastic light scattering is Raman scattering [36]. The unique properties of light such as high speed, directionality, and non-invasiveness make them attractive for studying highly sensitive biological systems.

The interaction of light with plasmonic nanoparticles results in both absorption and scattering of light at the plasmon resonance frequency, which contributes to the broad extinction spectrum of the nanoparticle [37]. During the absorption processes, the nanoparticle will utilize the energy of incident light and convert it into heat. However, when scattering occurs, the oscillating electrons reradiate electromagnetic field at the frequency of their oscillation. Even though studies on the interaction of gold colloids with light dated back to 19th century (Faraday), Gustav Mie proposed and developed a theory in 1908 for studying the scattering and absorption phenomena of light by spherical particles [38–42]. Later, in 1912, Gans extended Mie theory to predict the distinct surface plasmon bands for spheroidal metal nanoparticles of both oblate and prolate shapes [43, 44]. Interaction of light with plasmonic nanoparticles and subsequent excitation and relaxation of electrons result in a number of important linear and nonlinear optical effects [45]. Due to this unique behavior of free electrons, metals can display localized surface plasmon resonance (LSPR) when their sizes are reduced to the nanoscale [45–50]. Briefly, the LSPR is a phenomenon exhibited by the plasmonic nanoparticles, where the conduction band electrons near a metal–dielectric interface of nanoparticle get excited when they are illuminated by light and undergo a collective and coherent oscillation of the free electrons relative to the lattice of positive nuclei at the frequency of the incident light. This creates a localized surface plasmon as the electrons resonate together, which will

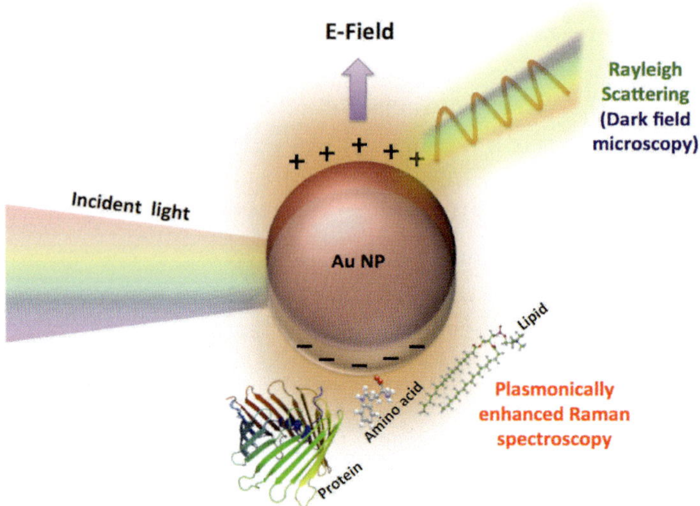

Figure 1. Schematic representation showing the collective oscillation of the free conduction band electrons in plasmonic AuNPs upon interaction with light.

result in the generation of a dipole around the nanoparticle surface that can switch along the direction of the electric field of the light [51, 52]. This enhanced field is several orders of magnitude above the incident field, which brings novel properties to the plasmonic nanoparticles and ultimately for the development of many innovative spectroscopic and microscopic techniques (Figure 1). Especially, noble metal nanoparticles made of gold and silver can strongly interact with visible light. As a result, they show intense absorptions and scatterings with absorption cross-sections orders of magnitude larger than those of typical organic dyes [53].

The effectiveness of plasmonic nanoparticle as contrast agents for bioimaging mainly depends on their optical properties, especially scattering. In order to develop highly sensitive light scattering-based spectroscopic and microscopic techniques, plasmonic nanoparticles with high scattering cross-section are more desirable. The molar extinction coefficient of plasmonic nanoparticles increases linearly with increasing volume of the particles. It is known that increasing the size of spherical nanoparticles from 20 to 80 nm can largely enhance the magnitude of extinction as well as the relative contribution of scattering to the extinction [53]. These extinction coefficients are typically 4–5 orders of magnitude higher than those for the very strong absorbing organic dye molecules, which makes them suitable for biological applications [51, 53]. One of the most important characteristics, which make plasmonic nanoparticles suitable for biological applications, is their ability to enhance the electromagnetic field of the incident radiation in the proximity of their surface. The enhanced electromagnetic field generated near the surface of plasmonic nanoparticles can interact with cells and biomolecules within it, which can be utilized for improving the sensitivity of many sensing and imaging techniques.

3. Biomolecular Targeting Using Plasmonic Nanoparticles for Spectroscopic and Imaging Applications

One of the most intriguing applications of plasmonic nanomaterials in biology is its unique capability to understand the biological system at the nanoscale level. Though the size of the mammalian cells are in the range of ~10–20 μm, a variety of subcellular compartments usually are in the submicron and nanoregime, which control most of the cellular functions. Nanotechnology provides new possibilities in designing and developing functional and biocompatible materials having appropriate sizes that can work within the size typical of biological molecules or completely bypass biological hurdles. The extremely small size of nanoparticles helps them to precisely target very specific cell types, tissues, and even specific subcellular organelle such as cell nucleus, cell membrane, cytosol, endosomes, mitochondria, and so on [54].

In plasmonics, the intensity of scattered light is extremely sensitive to the size and aggregation state of the nanoparticles [55, 56]. In order to improve the efficiency of plasmonic nanoparticles-based light scattering spectroscopic and imaging techniques, precise targeting of specific organs, tissues, and cells with nanoparticles is inevitable as it can increase the overall concentration of nanoparticles at the target site. This can significantly enhance the plasmonic field coupling and scattering intensity [56]. The facile surface chemistry of AuNPs makes them suitable for functionalizing them with several biotargeting moieties such as proteins, peptides, small molecules, etc. Among them, receptor-mediated cellular targeting has been widely explored for increased localization of nanoparticles on the cell surface, thereby using it in various therapeutic and diagnostic applications [35, 54, 57].

In many active disease-related cells, certain receptors are overexpressed when compared to the healthy ones. In such cases, actively targeting the overexpressed receptors on the unhealthy cells using properly conjugated nanoparticles can help develop sensitive diagnostic techniques. Few examples of such receptors are folate receptor (FR), transferrin receptor (TfR), epidermal growth factor receptor (EGFR), asialoglycoprotein, G-protein coupled receptor (GPCR), low-density lipoprotein receptor (LDLR), and chemokine [58–60]. One of the main advantages of active targeting is that it can minimize off-site cellular toxicity. In order to strengthen the extent of interaction of nanoparticles with cells and subsequent internalization, more specific targeting of peptide functionalized nanoparticles is necessary as it can precisely target specific biomolecules associated with the organelles. Figure 2 shows an illustration of various steps involved in the nuclear targeting of plasmonic nanoparticles. This includes the binding of ligands with specific cell surface receptors and forms a receptor–ligand complex, which facilitates their directional movement toward clathrin-coated pits in the cell membrane [61]. While, targeting ligand 1 on the nanoparticle surface enhances their selective binding with membrane receptors and subsequent endocytosis of nanoparticles, nuclear targeting ligand 2 accelerates their accumulation at the nuclear region via selective interaction with nuclear import receptors [62]. Augmented localization of the nanoparticles will increase their scattering cross-section due to the formation of hot spots via enhanced plasmonic field coupling, which ultimately improves their imaging utility [56].

4. Cellular Imaging Using Plasmonically Enhanced Elastic (Rayleigh) Light Scattering

Plasmonic nanoparticles-based optical imaging techniques, designed to visualize molecular cell functions, have received significant

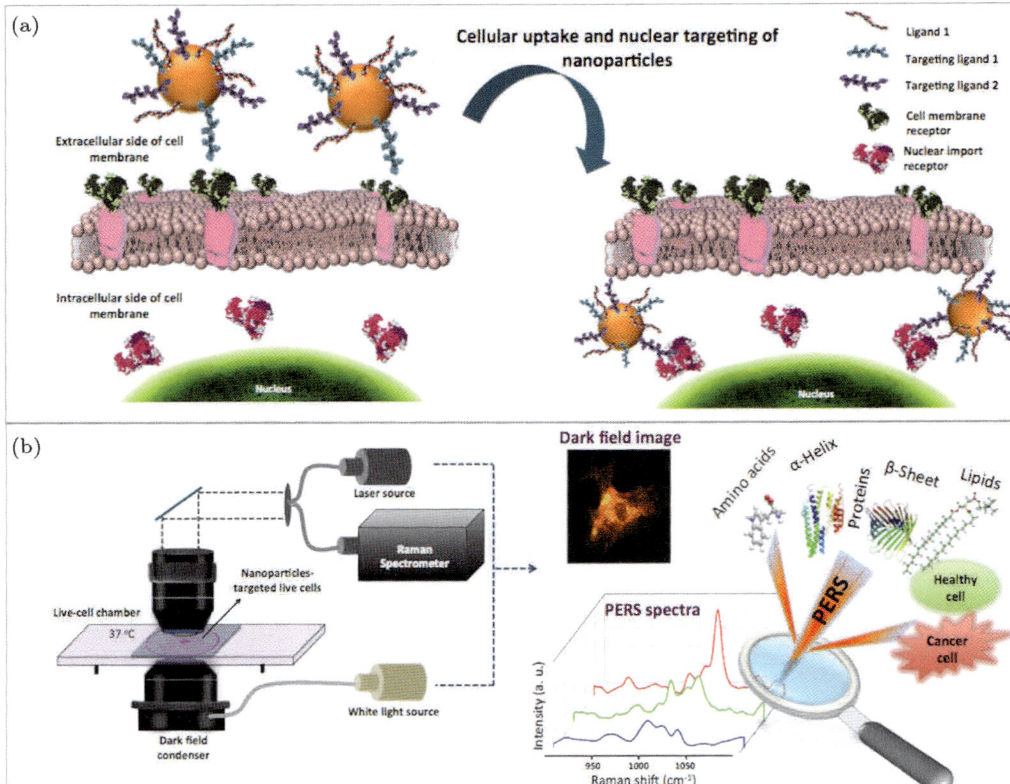

Figure 2. (a) Illustration showing the various steps involved in cellular uptake and nuclear targeting of plasmonic nanoparticle. (b) Schematic showing instrumental setup used to monitor the molecular cell function in real time using dark-field microscopy and PERS. Typical dark-field image of a nanoparticles-targeted nMLF cell and PERS spectra are also given.

attention of biomolecular researchers in the last two decades. While extremely small size of nanoparticles helps them to easily enter into the cells and interact with sub-cellular structures, their unique optical properties make them good candidates for visualizing the cellular and sub-cellular level biomolecular events in real time. The recent advancements in nanotechnology have paved the way for development of many innovative AuNPs-based bioimaging techniques such as confocal scanning reflectance microscopy [63], multiphoton plasmon resonance microscopy [64], optical coherence microscopy [65, 66], photoacoustic imaging [32], etc. Among these techniques, imaging elastic scattering of light (Rayleigh scattering) using dark-field microscopy and their immediate applications in studying cell functions will be discussed in the following section. The dark-field microscopic technique is an extremely simple and inexpensive technique which uses the plasmonically enhanced elastic light scattering property of plasmonic nanoparticles (i.e. Rayleigh scattering) resulting from the excitation of their LSPR [30, 67, 68]. This has been widely used for many biological applications such as cellular imaging [30, 65, 69–71], imaging of microbial cells and their functions [72–74], intracellular trafficking of nanoparticles [3, 8, 10, 15, 75–79], and so on.

AuNPs can scatter light of different wavelengths in the electromagnetic spectrum when illuminated with white light at appropriate angles. The wavelength distribution of the light in this case is determined by the shape and size of the nanoparticles. This color scattering property offers the potential for labeling studies with a white-light source.

The light scattering by individual metal nanoparticle was first observed by Richard Zsigmondy and Henry Siedentopf back in 1900s by using an ultra-microscope configured for dark-field illumination [81]. Zsigmondy's technique enabled the extensive study of optical properties of colloids in detail and has enormously influenced the development of modern nanotechnology [82]. In 2005, our group explored the unique capability of this technique in conjunction with plasmonic nanoparticles made of gold for developing a sensitive and reliable diagnostics technique in oncology [30]. Using this technique,

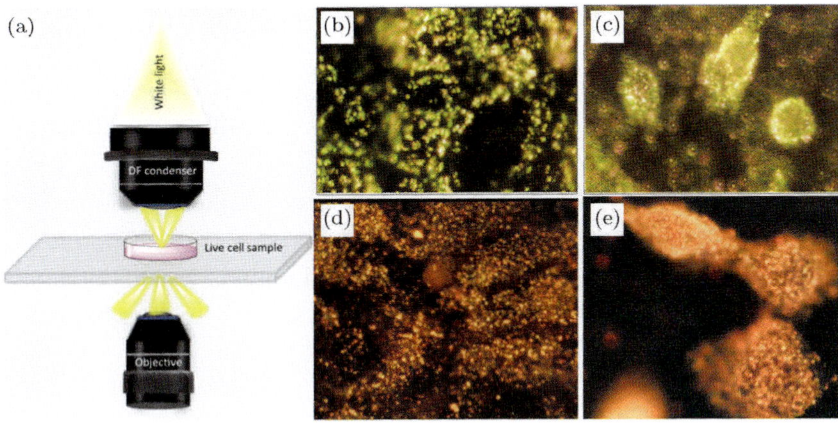

Figure 3. (a) Schematic diagram of a traditional dark-field microscope setup for acquiring Rayleigh scattering from nanoparticle-internalized live cells. (b) and (d) are the Rayleigh scattering images of healthy HaCaT cells (b) and cancerous HSC cells (c) incubated with anti-EGFR conjugated AuNSs acquired using dark-field microscopy [30]. (d) and (e) show the scattering images collected using conjugated AuNRs from HaCaT and HSC cells, respectively [80]. The wavelength of the strongly scattered light is dependent on the size and shape of the AuNPs.

the cancer cells, which are incubated with anti-epidermal growth factor receptor (anti-EGFR) conjugated spherical AuNPs, were imaged by using a very simple and inexpensive conventional microscope coupled with a dark-field condenser (Figures 3(b) and 3(c)). Here, only the light which is scattered by the sample under lateral illumination will reach the microscope objective because the white light will be guided through a dark-field condenser that blocks the centered light rays from reaching the sample [83]. Therefore, the scattering object shines and appears bright against the dark background and the images will be collected using a color-CCD camera. Even though a broad white-light source will excite the AuNPs, only the light frequencies corresponding to SPR will be strongly scattered by the nanoparticles and will appear as bright spots. Their color will be dependent on their size and shape, which provide a relatively large contrast to the complex scattering from the inherent cellular structure. A significant difference in the light scattering pattern was observed when the cells were incubated with anti-EGFR antibodies conjugated AuNPs. While the noncancerous human keratinocytes

(HaCaT) cells were poorly labeled by the conjugated nanoparticles, they are bound homogeneously and specifically to the surface of the human oral squamous carcinoma (HSC) cancer cells and appeared brighter in the dark-field images (Figures 3(b) and 3(c)). This difference has been used to distinguish between cancerous and non-cancerous cells. The high binding affinity of conjugated nanoparticles to the cancer cells has been attributed to the specific binding of over-expressed EGFR on the cancer cells with the anti-EGFR antibodies on the conjugated AuNP surfaces [30]. A similar technique, which uses anti-EGFR conjugated gold nanorods (AuNRs), has also been used as novel contrast agents for imaging cancer cells [35]. In this case, the strongly scattered red light from AuNRs in dark field was observed and was used to visualize and diagnose cancer cells from healthy cells (Figures 3(d) and 3(e)). Using this technique, the scattering spectral information can also be obtained by placing a spectrometer into the dark-field microscope such that the spectrometer's entrance slit position is adjusted to the image plane of the microscope. Scattered light comprised different wavelengths collected in the objective and dispersed by the grating within the spectrometer to a cooled detector. In later studies, our group and other research groups have followed this technique extensively by using a large number of nanoparticles of various sizes and shapes such as nanoshells [84], nanorods [35], nanostars [85], etc. for the same purpose.

Plasmonically enhanced elastic scattering properties of AuNPs can also be used for probing cell behavior and subcellular localization dynamics of nanoparticles in real time. Our group has shown that AuNRs conjugated with a nuclear localization sequence (NLS) peptide can greatly enhance their uptake by both malignant and non-malignant cells [86]. Furthermore, we have shown that the presence of the cell penetrating arginine–glycine–aspartic acid peptide (RGD) sequence on the AuNP surface can facilitate the internalization of the nanoparticles into the cancer cell cytoplasm via receptor-mediated endocytosis (Figure 4(a)), whereas the presence of a nuclear localization peptide sequence, NLS, on the same RGD-conjugated nanoparticle allows the localization of nanoparticles around the nuclear vicinity [62].

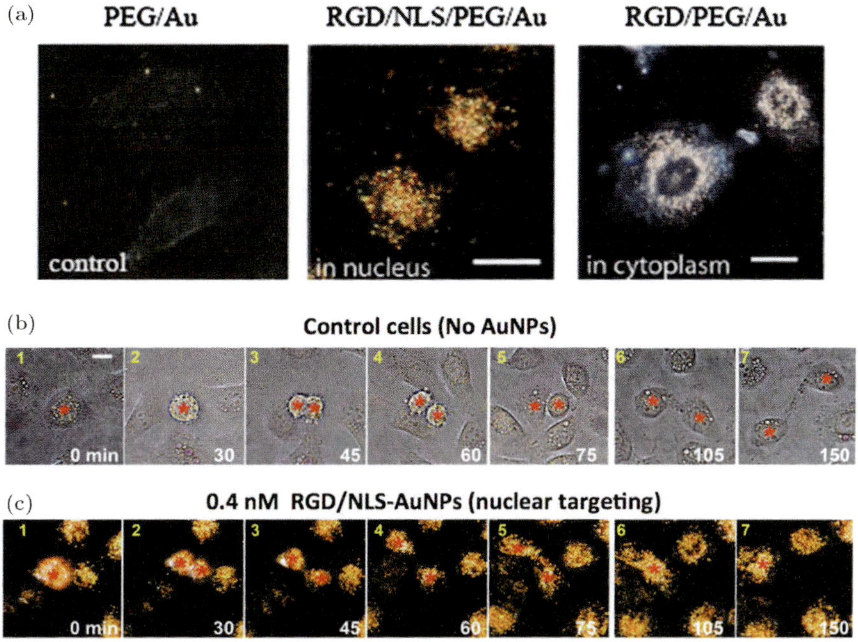

Figure 4. (a) Dark-field images of HSC cancer cells after overnight incubation in cell culture medium at 37°C with 0.1 nM PEG conjugated (left), RGD/NLS/PEG conjugated (middle) and RGD/PEG conjugated (right) AuNPs [70]. Even though the cellular uptake of PEG conjugated AuNPs is very low (b), RGD/NLS/PEG conjugated nanoparticles largely accumulated in cell nuclear region and the RGD/PEG conjugated ones accumulated in cytoplasm. Scale bar: 10 μm. (c) Real-time dark-field images of cell division of HSC cells showing an apparent cytokinesis arrest (b4) followed by binucleate cell formation (b6, b7) in the presence of 0.4 nM RGD/NLS-AuNSs [62]. This phenomenon was not observed in untreated cancer cells (b1-7) or cancer cells under other conditions. Red stars in the cell indicate the nuclei. Scale bar: 10 μm.

As the RGD/NLS conjugation on the nanoparticle surface promotes the uptake and nuclear targeting of nanoparticles by malignant cells, the cellular events can be easily visualized by using dark-field imaging by taking advantage of the increased subcellular localization of AuNPs (formation of denser nanoparticles cluster) and resulting enhancement in the Rayleigh scattering of cellular internalized AuNPs [56]. Using this technique, we made a novel observation that nuclear targeting of AuNPs in cancer cells (at specific nanoparticle concentration) can induce cytokinesis failure, leading to the arrest of complete cell division and subsequent cell death through apoptotic pathways (Figure 4(b)) [10, 62]. This has been visualized in real time by using dark-field imaging technique. Plasmonically enhanced Rayleigh scattering imaging technique has further been utilized for imaging the morphological changes in cells during cell death. In a recent study, we employed AuNPs as light-scattering probes to compare the relative efficacy of three chemotherapeutic drugs on HSC cells [56]. The morphological changes associated with cell death such as cell membrane blebbing and shrinkage and an increase in white-light scattering could clearly be observed in the plasmonically enhanced dark-field images. Moreover, from the Rayleigh scattering spectra, we have obtained the time profile of cell death as well as cell death enhancement factors from the anticancer drugs.

5. Understanding Cell Functions Using Plasmonically Enhanced Inelastic Light Scattering: PERS

Precise understanding of biomolecular events associated with biological processes at the nanoscale level significantly influenced the development of nanomedicine. In biomedical research, real-time evaluation of biochemical aspects of various biomolecular events associated with cellular function can enhance our knowledge about the cell responses, thereby enabling the development of more efficient diagnostic and therapeutic techniques against many diseases. It has been shown that the enhanced electromagnetic field produced by the plasmon(s) decays from a distance of ~10 nm away from the nanoparticle surface [48]. Any molecule within this distance of the nanoparticle surface will experience these large fields. The intensities of elastic and inelastic scattering can be enhanced to several orders of magnitude by this enhanced plasmonic field, which has been used for developing innovative spectroscopic and microscopic techniques. Surface-enhanced Raman scattering (SERS) [28, 29, 87, 88] is an example of one such technique, where the inherently weak Raman scattering signals of probe molecules will be enhanced by many folds of magnitude when they are bound or in very close vicinity to the plasmonic field of the nanoparticles. Even though an enhanced Raman scattering from pyridine adsorbed onto a roughened silver electrode surface in solution was first observed by Fleischmann *et al.* (though not recognized as SERS) in 1974 [89], Van Duyne *et al.* identified it independently in 1977 [87] and found that the large intensities observed are due to the enhancement of the scattered intensity of Raman signal by the roughened plasmonic metal surface. In 1979, Van Duyne named this phenomenon as "surface-enhanced Raman spectroscopy" [90, 91]. As the signal enhancement of Raman scattering is largely due to the enhanced electromagnetic field associated with plasmonic nanoparticles, this phenomenon can also be termed as "plasmonically enhanced Raman scattering (PERS)" [10, 11, 14, 18, 92–95].

As the biomolecular events are highly complex and dynamic in nature, simultaneous visualization of such chemical and conformational modifications is extremely challenging and requires high-throughput methods for their precise analysis. The high signal amplification in PERS and non-invasive nature of this technique has generated excitement in the visualization of molecular fingerprint snapshots as well as structural and functional modifications associated with various biomolecules with high spatial and temporal resolution [10, 18, 34, 97–102]. In the targeted plasmonically enhanced Raman spectroscopic studies of the cellular functions, as the enhanced Raman signal originates from the nano-environment around the nanoparticle surface, it can provide biomolecular information about the region of interest within the live cells surrounding the nanoparticles.

Another important application of PERS in molecular cell biology is its capability to probe intracellular environment through non-specific and organelle-specific targeted AuNPs [10, 18, 34, 71, 80, 98, 103]. Kneipp *et al.* have demonstrated the feasibility of PERS in live cell imaging and measured native constituents within a single viable cell using colloidal gold particles [34, 96]. In these studies, PERS signatures indicative of chemical changes in the surrounding environment of the AuNPs during their uptake were probed from single living cells in real time. Furthermore, they used a non-specific intracellular targeting approach, which uses 4-mercaptobenzoic acid (pMBA)-attached AuNPs to probe the pH of the endosomal compartments in a living NIH/3T3 cell and imaged it by using surface-enhanced

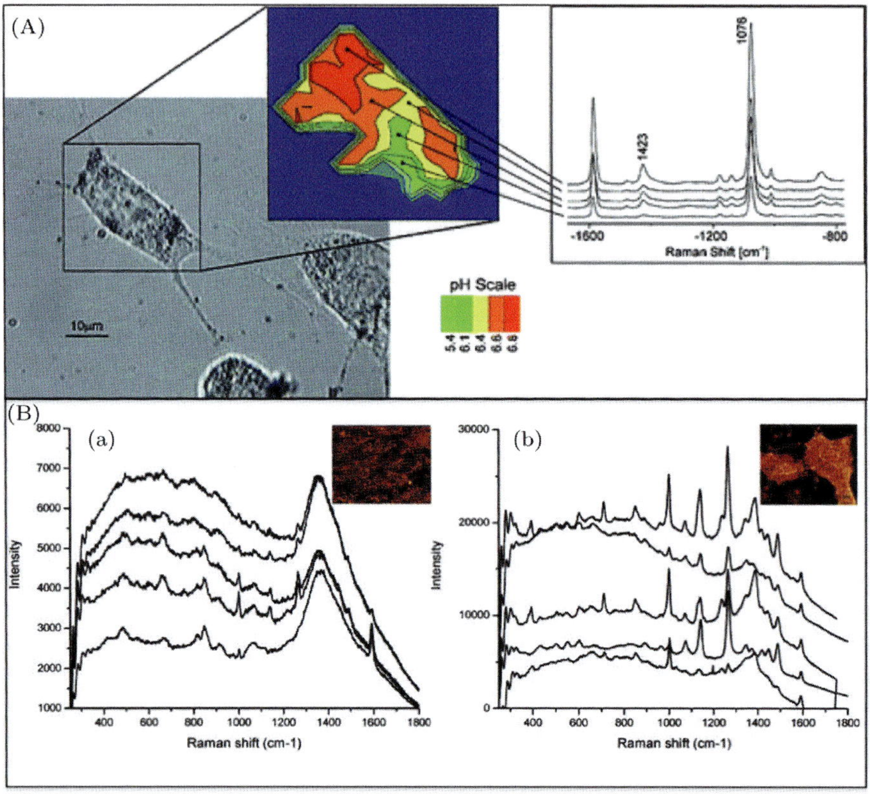

Figure 5. (A) PERS image and spectra collected from viable NIH/3T3 cells. The endosomal compartments with different pH have been probed with PERS by using pMBA-modified Au nanosensor. Lysosomal accumulations can be observed as black spots in the light microscope image [96]. (B) PERS spectra and corresponding dark field images collected from healthy HaCaT (a) and cancerous HSC (b) cells [80]. In each case, spectra are collected from five cells. In comparison to low PERS signal intensity from healthy cells, a high degree of cancer-specific spectral features is seen in cancer cell spectra as it contains the high density of PERS-enhancing particles within it.

hyper-Raman scattering (Figure 5(A)) [103]. Here, the characteristic Raman vibrations of pH-sensitive pMBA were analyzed in order to determine and to image the intracellular pH values ranging from pH 6.8 to 5.4. In our early study on *in vitro* diagnostics of cancer, we showed that anti-EGFR conjugated AuNRs on the HSC cancer cells can give a greatly enhanced Raman spectra in comparison to the healthy HaCaT cells (Figure 5(B)) [80]. This has been attributed to the high surface plasmon field of the assembled anti-EGFR conjugated AuNRs on the cancer cells surface due to the relative overexpression of EGFR and the formation of homogeneous electromagnetic fields. These observed SERS fingerprints, unique to the cancer cells, could be used as diagnostic signatures for the cancer cells.

Another direction we are pursuing in our lab is the combination of elastic scattering with inelastic scattering, specifically targeted plasmonically enhanced Rayleigh/Raman scattering, to visualize the cell functions in real time. Here, we simultaneously utilized the imaging capability of Rayleigh scattering microscopy and the unique advantages of PERS to investigate the biomolecular information associated with different cellular processes such as cell division, mitosis, drug efficacy, and cell death mechanism [10, 11, 13–15, 18, 83, 99].

Recently, using this technique (using RGD/NLS-AuNSs), we investigated the molecular mechanism of cell cycle regulation in cancer cells as it goes through its entire cell cycle phases (until the cell divides). This was conducted by recording both, Rayleigh/Raman images and spectra from a viable single cell in real time (Figure 6(A)) [18]. In this technique, dark-field Rayleigh scattering images are used to locate the intracellular position of the plasmonic AuNSs. This will allow one to obtain PERS spectra corresponding to specific intracellular location within a single cell with high signal-to-noise ratios. Furthermore, the plasmon-enhanced properties of gold nanocubes (AuNCs) together with PERS have been utilized to unravel the complex biomolecular processes involved in various stages of mitosis and AuNPs-induced cytokinesis failure [10]. From this study, it was found that relatively high rate of conversion of mitotic proteins from their α-helix structure to β-sheet conformation is likely in the cancer cells compared to healthy cells during meta, ana, and telophases (Figure 6(B)) of mitosis. Unique biochemical modifications to the lipid and amino acid moieties, associated with the observed protein conformational modifications, were also identified. The role of abnormal conformational modifications of mitotic proteins on the development of anomalous mitotic activities were further confirmed by

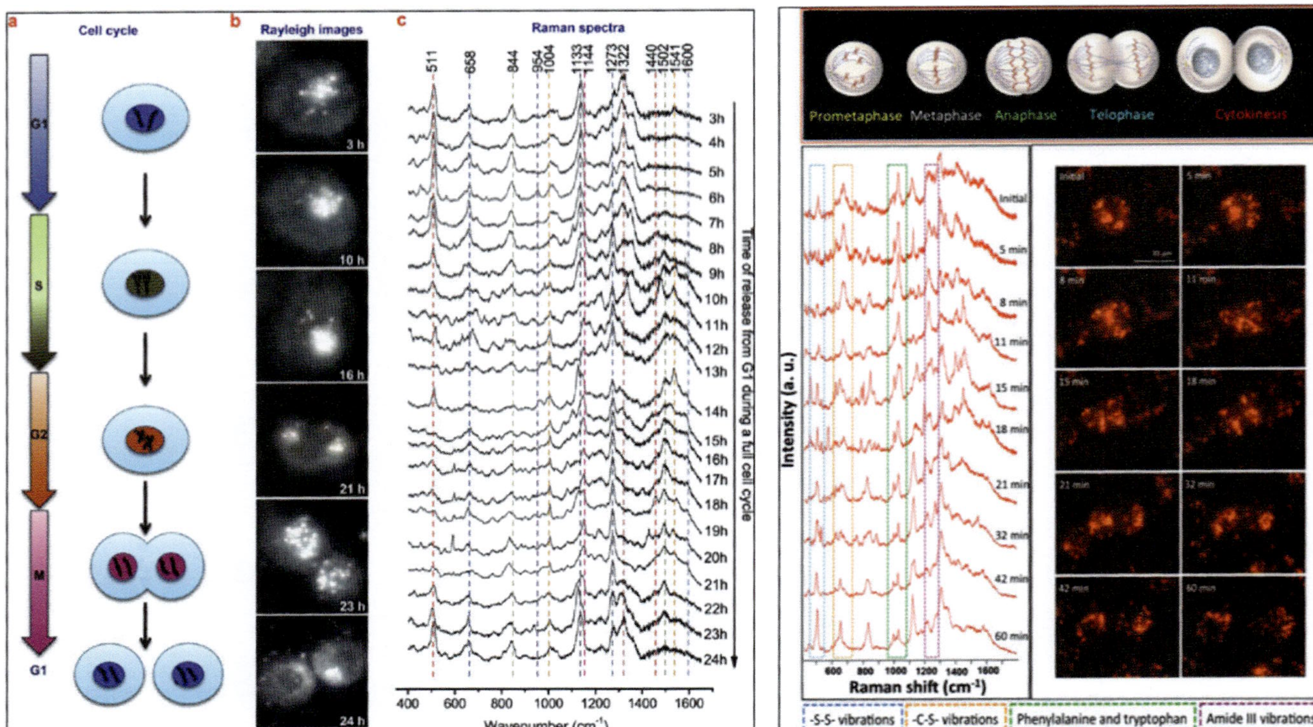

(A) (B)

Figure 6. (A) Real-time dark-field images and spectra of the targeted nucleus during the complete cycle of a single HSC cell: (a) the cell cycle; (b) the plasmonically enhanced Rayleigh images; (c) the PERS [18]. (B) Schematic representation of various stages of mitosis. Time-dependent PERS spectra collected from the HSC cells at different stages of mitosis and corresponding DF images collected at respective times are also given [10].

looking at AuNPs-induced cytokinesis failure in cancer cells. These findings provided the vast possibilities of using PERS in real-time tracking of subtle biomolecular modifications indicative of cellular dysfunctions, which could provide new insights about the role of protein conformation dynamics during cell function in the development of many diseases.

Fingerprinting biochemical changes associated with cellular responses to external stimuli such as drugs, radiations, etc. can provide vital information on the dynamics of biological processes and their defense mechanisms. This is essential in selecting and optimizing effective therapeutic strategies. PERS is a vital tool toward this end as it enables non-invasive and continuous monitoring of biomolecular events from their molecular perspective in real time. Recently, our group reported an easy PERS-based technique to study the relative efficacy of commonly used anti-cancer drugs by monitoring the time profile of spectral signals after drug administration until cell death. In this study, the relative efficacy and dynamics of single cell death of two popular chemotherapy drugs, cisplatin and 5-fluorouracil, on HSC cells were evaluated (Figure 7(A)) [104].

PERS spectra revealed that three Raman vibration bands at around 500, 1000, and 1585 cm^{-1} are associated with the cellular death dynamics. Though 500 cm^{-1} bands did not correlate well with drug efficacy, the bands observed at 1000 and 1585 cm^{-1} appeared as highly resolved bands after the drug treatments and the intensity variations of these bands signify the cell death dynamics. Additionally, plasmonically enhanced Rayleigh/Raman spectroscopic technique has been used to directly monitor the morphological and molecular events during the apoptotic cell death in real time [11]. During the apoptotic cell death of HSC cells, they exhibited characteristic morphology changes such as cell shrinkage, and blebbing, which has been imaged using dark-field microscopy. The shrunken apoptotic cells exhibited strong white Rayleigh scattered light. The PERS spectra of HSC cells during apoptosis showed many new Raman bands including intense bands associated with the cell death at 1000 and 1585 cm^{-1}, which was absent in the control cells (Figure 7(B)).

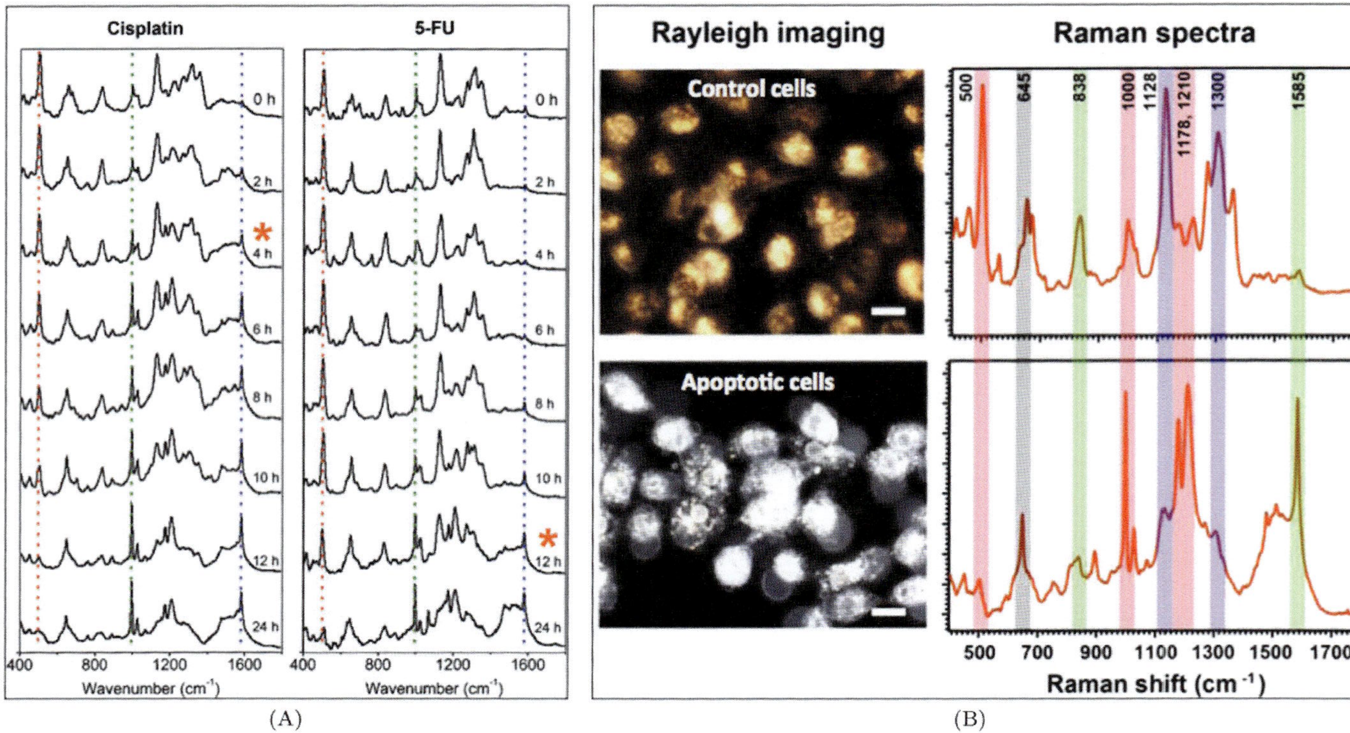

Figure 7. (A) Real-time SERS spectra of HSC cells treated for 24 h with cisplatin and 5-FU [104]. The identified "SERS death" bands are marked at 1000 and 1585 cm^{-1}. The stars indicate the cell death, when the spectra remain constant. (B) Plasmonically enhanced Rayleigh scattering images and corresponding Raman spectra of untreated viable control cells and 100 μM H_2O_2 treated apoptotic cells [11].

More recently, the cellular responses to UV-C irradiation, both intracellular nanoparticle motion and activation of a cellular defense mechanism in cancer cells, were successfully investigated by our group in real time by using targeted AuNC reporters, PERS, and dark-field microscopy [15]. In this study, we were able to visualize the UV-C light-induced photolytic cleavage of disulfide bonds present in cytoplasmic proteins and subsequent deactivation of protein activity and induction of apoptotic cell death. These results provided new insights in understanding the mechanism of thiol metabolism associated with cellular defense against externally applied cellular stress. Furthermore, PERS has been used to understand the mechanism associated with plasmonic photothermal cell death of cancer cells using targeted AuNSs [99]. The results suggest that heat generated from the aggregated AuNSs absorbing near-infrared laser light of sufficient powers caused modifications in the protein and lipid structures within the cell and ultimately led to cell death. At high laser intensity (9.4 mW), the protein and lipid began to denature upon laser exposure, which was obvious in the PERS spectra. In a very recent study [14], in conjunction with fluorescence and proteomic analysis, we explored the unique capability of PERS in identifying subtle modification in biomolecules during hyperoxic conditions in neonatal mouse lung fibroblasts (nMLFs). In this study, using real-time PERS, we made the novel observation that hyperoxia induces intracellular acidification in nMLF, which significantly affects the structure and conformation of proteins and subsequently induces modifications in the Raman vibrations corresponding to aromatic amino acids such as phenylalanine and tryptophan, as well as cysteine moieties. These results point toward the fact that PERS has the potential to provide new insights on the role of intracellular acidification and associated biomolecular modification under hypertoxic conditions during the pathogenesis of many diseases.

Acknowledgment

We wish to acknowledge the NSF-DMR grant (1206637) for the financial support of this work.

References

1. Jain, P. K., Huang, X., El-Sayed, I. H. and El-Sayed, M. A. (2008). Noble metals on the nanoscale: Optical and photothermal properties and some applications in imaging, sensing, biology, and medicine. *Acc. Chem. Res.* **41**, pp. 1578–1586.
2. Eustis, S. and El-Sayed, M. A. (2006). Why gold nanoparticles are more precious than pretty gold: Noble metal surface plasmon resonance and its enhancement of the radiative and nonradiative properties of nanocrystals of different shapes. *Chem. Soc. Rev.* **35**, pp. 209–217.
3. Alkilany, A. M., Lohse, S. E. and Murphy, C. J. (2013). The gold standard: Gold nanoparticle libraries to understand the nano-bio interface. *Acc. Chem. Res.* **46**, pp. 650–661.
4. Saha, K., Agasti, S. S., Kim, C., Li, X. and Rotello, V. M. (2012). Gold nanoparticles in chemical and biological sensing. *Chem. Rev.* **112**, pp. 2739–2779.
5. Yang, X., Yang, M., Pang, B., Vara, M. and Xia, Y. (2015). Gold nanomaterials at work in biomedicine. *Chem. Rev.* **115**, pp. 10410–10488.
6. Peer, D., Karp, J. M., Hong, S., Farokhzad, O. C., Margalit, R. and Langer, R. (2007). Nanocarriers as an emerging platform for cancer therapy. *Nat. Nanotechnol.* **2**, pp. 751–760.
7. Zheng, Y. B., Kiraly, B., Weiss, P. S. and Huang, T. J. (2012). Molecular plasmonics for biology and nanomedicine. *Nanomedicine* (London) **7**, pp. 751–770.
8. Xiao, L. and Yeung, E. S. (2014). Optical imaging of individual plasmonic nanoparticles in biological samples. *Annu. Rev. Anal. Chem. (Palo Alto Calif)* **7**, pp. 89–111.
9. Jun, Y. W., Sheikholeslami, S., Hostetter, D. R., Tajon, C., Craik, C. S. and Alivisatos, A. P. (2009). Continuous imaging of plasmon rulers in live cells reveals early-stage caspase-3 activation at the single-molecule level. *Proc. Natl. Acad. Sci. USA* **106**, pp. 17735–17740.
10. Panikkanvalappil, S. R., Hira, S. M., Mahmoud, M. A. and El-Sayed, M. A. (2014). Unraveling the biomolecular snapshots of mitosis in healthy and cancer cells using plasmonically-enhanced Raman spectroscopy. *J. Am. Chem. Soc.* **136**, pp. 15961–15968.
11. Kang, B., Austin, L. A. and El-Sayed, M. A. (2014). Observing realtime molecular event dynamics of apoptosis in living cancer cells using nuclear-targeted plasmonically enhanced Raman nanoprobes. *ACS Nano.* **8**, pp. 4883–4892.
12. Howes, P. D., Chandrawati, R. and Stevens, M. M. (2014). Bionanotechnology: Colloidal nanoparticles as advanced biological sensors. *Science* **346**, p. 1247390.
13. Kang, B., Afifi, M. M., Austin, L. A. and El-Sayed, M. A. (2013). Exploiting the nanoparticle plasmon effect: Observing drug delivery dynamics in single cells via Raman/fluorescence imaging spectroscopy. *ACS Nano.* **7**, pp. 7420–7427.

14. Panikkanvalappil, S. R., James, M., Hira, S. M., Mobley, J., Jilling, T., Ambalavanan, N. and El-Sayed, M. A. (2016). Hyperoxia induces intracellular acidification in neonatal mouse lung fibroblasts: Real-time investigation using plasmonically enhanced Raman spectroscopy. *J. Am. Chem. Soc.* **138**, pp. 3779–3788.
15. Panikkanvalappil, S. R., Hira, S. M. and El-Sayed, M. A. (2016). Elucidation of ultraviolet radiation-induced cell responses and intracellular biomolecular dynamics in mammalian cells using surface-enhanced Raman spectroscopy. *Chem. Sci.* **7**, pp. 1133–1141.
16. Kumar, S., Harrison, N., Richards-Kortum, R. and Sokolov, K. (2007). Plasmonic nanosensors for imaging intracellular biomarkers in live cells. *Nano Lett.* **7**, pp. 1338–1343.
17. Choi, Y., Kang, T. and Lee, L. P. (2009). Plasmon resonance energy transfer (pret)-based molecular imaging of cytochrome C in living cells. *Nano Lett.* **9**, pp. 85–90.
18. Kang, B., Austin, L. A. and El-Sayed, M. A. (2012). Real-time molecular imaging throughout the entire cell cycle by targeted plasmonic-enhanced Rayleigh/Raman spectroscopy. *Nano Lett.* **12**, pp. 5369–5375.
19. Sperling, R. A., Rivera Gil, P., Zhang, F., Zanella, M. and Parak, W. J. (2008). Biological applications of gold nanoparticles. *Chem. Soc. Rev.* **37**, pp. 1896–1908.
20. Giljohann, D. A., Seferos, D. S., Daniel, W. L., Massich, M. D., Patel, P. C. and Mirkin, C. A. (2010). Gold nanoparticles for biology and medicine. *Angew. Chem. Int. Ed. Engl.* **49**, pp. 3280–3294.
21. Dreaden, E. C., Austin, L. A., Mackey, M. A. and El-Sayed, M. A. (2012). Size matters: Gold nanoparticles in targeted cancer drug delivery. *Ther. Deliv.* **3**, pp. 457–478.
22. Panikkanvalappil, S. R. and El-Sayed, M. A. (2014). Gold-nanoparticle-decorated hybrid mesoflowers: An efficient surface-enhanced Raman scattering substrate for ultra-trace detection of prostate specific antigen. *J. Phys. Chem. B* **118**, pp. 14085–14091.
23. Murphy, C. J., Gole, A. M., Hunyadi, S. E., Stone, J. W., Sisco, P. N., Alkilany, A., Kinard, B. E. and Hankins, P. (2008). Chemical sensing and imaging with metallic nanorods. *Chem. Commun. (Camb.)*, **5**, pp. 544–557.
24. Dykman, L. and Khlebtsov, N. (2012). Gold nanoparticles in biomedical applications: Recent advances and perspectives. *Chem. Soc. Rev.* **41**, pp. 2256–2282.
25. Rosi, N. L. and Mirkin, C. A. (2005). Nanostructures in biodiagnostics. *Chem. Rev.* **105**, pp. 1547–1562.
26. Qian, X. M. and Nie, S. M. (2008). Single-molecule and single-nanoparticle sers: From fundamental mechanisms to biomedical applications. *Chem. Soc. Rev.* **37**, pp. 912–920.
27. Kneipp, K., Kneipp, H. and Kneipp, J. (2006). Surface-enhanced Raman scattering in local optical fields of silver and gold nanoaggregatess — from single-molecule Raman spectroscopy to ultrasensitive probing in live cells. *Accounts Chem. Res.* **39**, pp. 443–450.

28. Stiles, P. L., Dieringer, J. A., Shah, N. C. and Van Duyne, R. P. (2008). Surface-enhanced Raman spectroscopy. *Annu. Rev. Anal. Chem.* (*Palo Alto Calif*) **1**, pp. 601–626.
29. Schatz, G. C., Young, M. A. and Van Duyne, R. P. (2006). Electromagnetic mechanism of sers. *Top Appl. Phys.* **103**, pp. 19–45.
30. El-Sayed, I. H., Huang, X. and El-Sayed, M. A. (2005). Surface plasmon resonance scattering and absorption of anti-Egfr antibody conjugated gold nanoparticles in cancer diagnostics: Applications in oral cancer. *Nano Lett.* **5**, pp. 829–834.
31. Agarwal, A., Huang, S. W., O'Donnell, M., Day, K. C., Day, M., Kotov, N. and Ashkenazi, S. (2007). Targeted gold nanorod contrast agent for prostate cancer detection by photoacoustic imaging. *J. Appl. Phys.* **102**(6), p. 064701.
32. Mallidi, S., Larson, T., Tam, J., Joshi, P. P., Karpiouk, A., Sokolov, K. and Emelianov, S. (2009). Multiwavelength photoacoustic imaging and plasmon resonance coupling of gold nanoparticles for selective detection of cancer. *Nano Lett.* **9**, pp. 2825–2831.
33. Durr, N. J., Larson, T., Smith, D. K., Korgel, B. A., Sokolov, K. and Ben-Yakar, A. (2007). Two-photon luminescence imaging of cancer cells using molecularly targeted gold nanorods, *Nano Lett.* **7**, pp. 941–945.
34. Kneipp, K., Haka, A. S., Kneipp, H., Badizadegan, K., Yoshizawa, N., Boone, C., Shafer-Peltier, K. E., Motz, J. T., Dasari, R. R. and Feld, M. S. (2002). Surface-enhanced Raman spectroscopy in single living cells using gold nanoparticles. *Appl. Spectrosc.* **56**, pp. 150–154.
35. Huang, X., El-Sayed, I. H., Qian, W. and El-Sayed, M. A. (2006). Cancer cell imaging and photothermal therapy in the near-infrared region by using gold nanorods. *J. Am. Chem. Soc.* **128**, pp. 2115–2120.
36. Raman, C. V. and Krishnan, K. S. (1928). A new type of secondary radiation. *Nature* **121**, pp. 501–502.
37. Bohren, C. F. and Huffman, D. R. (2007). Extinction, in absorption and scattering of light by small particles, Wiley-VCH Verlag GmbH, pp. 286–324.
38. Mie, G. (1908). Beiträge Zur Optik Trüber Medien, Speziell Kolloidaler Metallösungen. *Annalen der Physik* **330**, pp. 377–445.
39. Berne, B. J. (1976). *Dynamic Light Scattering: With Applications to Chemistry, Biology, and Physics*, Berne, B. J. and Pecora, R. (eds.). Wiley, New York.
40. Bohren, C. F. and Huffman, D. R. (2007). Rayleigh-Gans Theory. *Absorption and Scattering of Light by Small Particles.* Wiley-VCH Verlag GmbH, pp. 158–165.
41. Pradeep, T. (2012). *Textbook of Nanoscience and Nanotechnology*, p. 1 online resource. McGraw-Hill Professional.
42. Cao, A. (2003). Light scattering, recent applications. *Anal. Lett.* **36**, pp. 3185–3225.
43. Gans, R. (1912). Über Die Form Ultramikroskopischer Goldteilchen. *Annalen der Physik* **342**, pp. 881–900.

44. Sajanlal, P. R. and Pradeep, T. (2000). Gold nanoparticles, *Kirk-Othmer Encyclopedia of Chemical Technology*. John Wiley & Sons, Inc., pp. 1–22.
45. El-Sayed, M. A. (2001). Some interesting properties of metals confined in time and nanometer space of different shapes. *Acc. Chem. Res.* **34**, pp. 257–264.
46. Link, S. and El-Sayed, M. A. (1999). Spectral properties and relaxation dynamics of surface plasmon electronic oscillations in gold and silver nanodots and nanorods. *J. Phys. Chem. B* **103**, pp. 8410–8426.
47. Haynes, C. L. and Van Duyne, R. P. (2001). Nanosphere lithography: A versatile nanofabrication tool for studies of size-dependent nanoparticle optics. *J. Phys. Chem. B* **105**, pp. 5599–5611.
48. Haes, A. J., Zou, S. L., Schatz, G. C. and Van Duyne, R. P. (2004). A nanoscale optical biosensor: The long range distance dependence of the localized surface plasmon resonance of noble metal nanoparticles. *J. Phys. Chem. B* **108**, pp. 109–116.
49. Chang, W. S., Willingham, B., Slaughter, L. S., Dominguez-Medina, S., Swanglap, P. and Link, S. (2012). Radiative and nonradiative properties of single plasmonic nanoparticles and their assemblies. *Acc. Chem. Res.* **45**, pp. 1936–1945.
50. Hooshmand, N., Panikkanvalappil, S. R. and El-Sayed, M. A. (2016). Effects of the substrate refractive index, the exciting light propagation direction, and the relative cube orientation on the plasmonic coupling behavior of two silver nanocubes at different separations. *J. Phys. Chem. C* **120**, pp. 20896–20904.
51. Link, S. and El-Sayed, M. A. (2000). Shape and size dependence of radiative, non-radiative and photothermal properties of gold nanocrystals. *Int. Rev. Phys. Chem.* **19**, pp. 409–453.
52. Eustis, S. and El-Sayed, M. A. (2006). Why gold nanoparticles are more precious than pretty gold: Noble metal surface plasmon resonance and its enhancement of the radiative and nonradiative properties of nanocrystals of different shapes. *Chem. Soc. Rev.* **35**, pp. 209–217.
53. Jain, P. K., Lee, K. S., El-Sayed, I. H. and El-Sayed, M. A. (2006). Calculated absorption and scattering properties of gold nanoparticles of different size, shape, and composition: Applications in biological imaging and biomedicine. *J. Phys. Chem. B* **110**, pp. 7238–7248.
54. Field, L. D., Delehanty, J. B., Chen, Y. and Medintz, I. L. (2015). Peptides for specifically targeting nanoparticles to cellular organelles: Quo vadis? *Acc. Chem. Res.* **48**, pp. 1380–1390.
55. Jain, P. K., Qian, W. and El-Sayed, M. A. (2006). Ultrafast electron relaxation dynamics in coupled metal nanoparticles in aggregates. *J. Phys. Chem. B* **110**, pp. 136–142.
56. Aioub, M., Kang, B., Mackey, M. A. and El-Sayed, M. A. (2014). Biological targeting of plasmonic nanoparticles improves cellular imaging via the enhanced scattering in the aggregates formed. *J. Phys. Chem. Lett.* **5**, pp. 2555–2561.

57. Zhang, S., Gao, H. and Bao, G. (2015). Physical principles of nanoparticle cellular endocytosis. *ACS Nano* **9**, pp. 8655–8671.
58. Xu, S., Olenyuk, B. Z., Okamoto, C. T. and Hamm-Alvarez, S. F. (2013). Targeting receptor-mediated endocytotic pathways with nanoparticles: Rationale and advances. *Adv. Drug Deliv. Rev.* **65**, pp. 121–138.
59. Yameen, B., Choi, W. I., Vilos, C., Swami, A., Shi, J. and Farokhzad, O. C. (2014). Insight into nanoparticle cellular uptake and intracellular targeting. *J. Control Release* **190**, pp. 485–499.
60. Bareford, L. M. and Swaan, P. W. (2007). Endocytic mechanisms for targeted drug delivery. *Adv. Drug Deliv. Rev.* **59**, pp. 748–758.
61. Sorkin, A. and Puthenveedu, M. A. (2013). Clathrin-mediated endocytosis. In Yarden, Y. and Tarcic, G. (eds)., *Vesicle Trafficking in Cancer*. Springer New York, pp. 1–31.
62. Kang, B., Mackey, M. A. and El-Sayed, M. A. (2010). Nuclear targeting of gold nanoparticles in cancer cells induces DNA damage, causing cytokinesis arrest and apoptosis. *J. Am. Chem. Soc.* **132**, pp. 1517–1519.
63. Sokolov, K., Follen, M., Aaron, J., Pavlova, I., Malpica, A., Lotan, R. and Richards-Kortum, R. (2003). Real-time vital optical imaging of precancer using anti-epidermal growth factor receptor antibodies conjugated to gold nanoparticles. *Cancer Res.* **63**, pp. 1999–2004.
64. Yelin, D., Oron, D., Thiberge, S., Moses, E. and Silberberg, Y. (2003). Multiphoton plasmon-resonance microscopy. *Opt. Exp.* **11**, pp. 1385–1391.
65. Gobin, A. M., Lee, M. H., Halas, N. J., James, W. D., Drezek, R. A. and West, J. L. (2007). Near-infrared resonant nanoshells for combined optical imaging and photothermal cancer therapy. *Nano Lett.* **7**, pp. 1929–1934.
66. Cang, H., Sun, T., Li, Z. Y., Chen, J. Y., Wiley, B. J., Xia, Y. N. and Li, X. D. (2005). Gold nanocages as contrast agents for spectroscopic optical coherence tomography. *Opt. Lett.* **30**, pp. 3048–3050.
67. Sonnichsen, C., Geier, S., Hecker, N. E., von Plessen, G., Feldmann, J., Ditlbacher, H., Lamprecht, B., Krenn, J. R., Aussenegg, F. R., Chan, V. Z. H., Spatz, J. P. and Moller, M. (2000). Spectroscopy of single metallic nanoparticles using total internal reflection microscopy. *Appl. Phys. Lett.* **77**, pp. 2949–2951.
68. Schultz, S., Smith, D. R., Mock, J. J. and Schultz, D. A. (2000). Single-target molecule detection with nonbleaching multicolor optical immunolabels. *Proc. Natl. Acad. Sci. USA* **97**, pp. 996–1001.
69. Murphy, C. J., Gole, A. M., Stone, J. W., Sisco, P. N., Alkilany, A. M., Goldsmith, E. C. and Baxter, S. C. (2008). Gold nanoparticles in biology: Beyond toxicity to cellular imaging. *Acc. Chem. Res.* **41**, pp. 1721–1730.
70. Qian, W., Huang, X., Kang, B. and El-Sayed, M. A. (2010). Dark-field light scattering imaging of living cancer cell component from birth through division using bioconjugated gold nanoprobes. *J. Biomed. Opt.* **15**, p. 046025.
71. Ba, H., Rodriguez-Fernandez, J., Stefani, F. D. and Feldmann, J. (2010). Immobilization of gold nanoparticles on living cell membranes upon controlled lipid binding. *Nano Lett.* **10**, pp. 3006–3012.
72. Ueno, H., Nishikawa, S., Iino, R., Tabata, K. V., Sakakihara, S., Yanagida, T. and Noji, H. (2010). Simple dark-field microscopy with nanometer spatial

precision and microsecond temporal resolution. *Biophys. J.* **98**, pp. 2014–2023.
73. Kyriacou, S. V., Brownlow, W. J. and Xu, X. H. (2004). Using nanoparticle optics assay for direct observation of the function of antimicrobial agents in single live bacterial cells. *Biochemistry* **43**, pp. 140–147.
74. Nallathamby, P. D., Lee, K. J., Desai, T. and Xu, X. H. (2010). Study of the multidrug membrane transporter of single living pseudomonas aeruginosa cells using size-dependent plasmonic nanoparticle optical probes. *Biochemistry* **49**, pp. 5942–5953.
75. Gibbs-Flournoy, E. A., Bromberg, P. A., Hofer, T. P., Samet, J. M. and Zucker, R. M. (2011). Darkfield-confocal microscopy detection of nanoscale particle internalization by human lung cells. *Part Fibre Toxicol*, **8**, p. 2.
76. Zhang, L., Li, Y., Li, D. W., Jing, C., Chen, X., Lv, M., Huang, Q., Long, Y. T. and Willner, I. (2011). Single gold nanoparticles as real-time optical probes for the detection of nadh-dependent intracellular metabolic enzymatic pathways. *Angew. Chem. Int. Ed. Engl.* **50**, pp. 6789–6792.
77. Wan, X. Y., Zheng, L. L., Gao, P. F., Yang, X. X., Li, C. M., Li, Y. F. and Huang, C. Z. (2014). Real-time light scattering tracking of gold nanoparticles-bioconjugated respiratory syncytial virus infecting Hep-2 cells. *Sci. Rep.* **4**, p. 4529.
78. Yang, J. A., Phan, H. T., Vaidya, S. and Murphy, C. J. (2013). Nanovacuums: Nanoparticle uptake and differential cellular migration on a carpet of nanoparticles. *Nano Lett.* **13**, pp. 2295–2302.
79. Wang, S. H., Lee, C. W., Chiou, A. and Wei, P. K. (2010). Size-dependent endocytosis of gold nanoparticles studied by three-dimensional mapping of plasmonic scattering images. *J. Nanobiotechnol.* **8**, p. 33.
80. Huang, X., El-Sayed, I. H., Qian, W. and El-Sayed, M. A. (2007). Cancer cells assemble and align gold nanorods conjugated to antibodies to produce highly enhanced, sharp, and polarized surface Raman spectra: A potential cancer diagnostic marker. *Nano Lett.* **7**, pp. 1591–1597.
81. Siedentopf, H. and Zsigmondy, R. (1902). Visualisation and determination of size of ultra microscopic particles, with special use of goldrubin glasses. *Annalen Der Physik* **10**, pp. 1–39.
82. Mappes, T., Jahr, N., Csaki, A., Vogler, N., Popp, J. and Fritzsche, W. (2012). The invention of immersion ultramicroscopy in 1912-the birth of nanotechnology? *Angew. Chem. Int. Ed. Engl.* **51**, pp. 11208–11212.
83. Austin, L. A., Kang, B. and El-Sayed, M. A. (2015). Probing molecular cell event dynamics at the single-cell level with targeted plasmonic gold nanoparticles: A review. *Nano Today* **10**, pp. 542–558.
84. Loo, C., Hirsch, L., Lee, M. H., Chang, E., West, J., Halas, N. and Drezek, R. (2005). Gold nanoshell bioconjugates for molecular imaging in living cells. *Opt. Lett.* **30**, pp. 1012–1014.
85. Aaron, J., de la Rosa, E., Travis, K., Harrison, N., Burt, J., Jose-Yacaman, M. and Sokolov, K. (2008). Polarization microscopy with stellated gold nanoparticles for robust, in-situ monitoring of biomolecules. *Opt. Express* **16**, pp. 2153–2167.

86. Oyelere, A. K., Chen, P. C., Huang, X., El-Sayed, I. H. and El-Sayed, M. A. (2007). Peptide-conjugated gold nanorods for nuclear targeting. *Bioconjug Chem.* **18**, pp. 1490–1497.
87. Jeanmaire, D. L. and Vanduyne, R. P. (1977). Surface Raman spectroelectrochemistry 1. Heterocyclic, aromatic, and aliphatic-amines adsorbed on anodized silver electrode. *J. Electroanal. Chem.* **84**, pp. 1–20.
88. Moskovits, M. (1985). Surface-enhanced spectroscopy. *Rev. Mod. Phys.* **57**, pp. 783–826.
89. Fleischmann, M., Hendra, P. J. and Mcquilla, A. J. (1974). Raman-spectra of pyridine adsorbed at a silver electrode. *Chem. Phys. Lett.* **26**, pp. 163–166.
90. Van Duyne, R. P. (1979). Laser excitation of Raman scattering from adsorbed molecules on electrode surfaces A2. In Moore, C. Bradley (eds.), *Chemical and Biochemical Applications of Lasers.* Academic Press, pp. 101–185.
91. Haynes, C. L., Yonzon, C. R., Zhang, X. Y. and Van Duyne, R. P. (2005). Surface-enhanced raman sensors: Early history and the development of sensors for quantitative biowarfare agent and glucose detection. *J. Raman Spectrosc.* **36**, pp. 471–484.
92. Ding, S. Y., Yi, J., Li, J. F., Ren, B., Wu, D. Y., Panneerselvam, R. and Tian, Z. Q. (2016). Nanostructure-based plasmon-enhanced Raman spectroscopy for surface analysis of materials. *Nature Rev. Mater.* **1**, p. 16021.
93. Aroca, R. F. (2013). Plasmon enhanced spectroscopy, *Phys. Chem. Chem. Phys.* **15**, pp. 5355–5363.
94. Roelli, P., Galland, C., Piro, N. and Kippenberg, T. J. (2016). Molecular cavity optomechanics as a theory of plasmon-enhanced Raman scattering. *Nat. Nanotechnol.* **11**, pp. 164–169.
95. Jiang, S., Zhang, Y., Zhang, R., Hu, C., Liao, M., Luo, Y., Yang, J., Dong, Z. and Hou, J. G. (2015). Distinguishing adjacent molecules on a surface using plasmon-enhanced Raman scattering. *Nat. Nanotechnol.* **10**, pp. 865–869.
96. Kneipp, J., Kneipp, H., McLaughlin, M., Brown, D. and Kneipp, K. (2006). *In vivo* molecular probing of cellular compartments with gold nanoparticles and nanoaggregates. *Nano Lett.* **6**, pp. 2225–2231.
97. Zavaleta, C. L., Smith, B. R., Walton, I., Doering, W., Davis, G., Shojaei, B., Natan, M. J. and Gambhir, S. S. (2009). Multiplexed imaging of surface enhanced Raman scattering nanotags in living mice using noninvasive Raman spectroscopy. *Proc. Natl. Acad. Sci. USA* **106**, pp. 13511–13516.
98. Ando, J., Fujita, K., Smith, N. I. and Kawata, S. (2011). Dynamic SERS imaging of cellular transport pathways with endocytosed gold nanoparticles. *Nano Lett.* **11**, pp. 5344–5348.
99. Aioub, M. and El-Sayed, M. A. (2016). A real-time surface enhanced Raman spectroscopy study of plasmonic photothermal cell death using targeted gold nanoparticles. *J. Am. Chem. Soc.* **138**, pp. 1258–1264.
100. Palonpon, A. F., Ando, J., Yamakoshi, H., Dodo, K., Sodeoka, M., Kawata, S. and Fujita, K. (2013). Raman and SERS microscopy for molecular imaging of live cells. *Nat. Protoc.* **8**, pp. 677–692.

101. Panikkanvalappil, S. R., Mahmoud, M. A., Mackey, M. A. and El-Sayed, M. A. (2013). Surface-enhanced Raman spectroscopy for real-time monitoring of reactive oxygen species-induced DNA damage and its prevention by platinum nanoparticles. *ACS Nano* **7**, pp. 7524–7533.
102. Panikkanvalappil, S. R., Mackey, M. A. and El-Sayed, M. A. (2013). Probing the unique dehydration-induced structural modifications in cancer cell DNA using surface enhanced Raman spectroscopy. *J. Am. Chem. Soc.* **135**, pp. 4815–4821.
103. Kneipp, J., Kneipp, H., Wittig, B. and Kneipp, K. (2007). One- and two-photon excited optical Ph probing for cells using surface-enhanced Raman and hyper-Raman nanosensors. *Nano Lett.* **7**, pp. 2819–2823.
104. Austin, L. A., Kang, B. and El-Sayed, M. A. (2013). A new nanotechnology technique for determining drug efficacy using targeted plasmonically enhanced single cell imaging spectroscopy. *J. Am. Chem. Soc.* **135**, pp. 4688–4691.

Chapter 5

Deep-Ultraviolet Surface- and Tip-Enhanced Raman Spectroscopy

Atsushi Taguchi

Department of Applied Physics,
Osaka University, Suita,
Osaka 565-0871, Japan
taguchi@ap.eng.osaka-u.ac.jp

1. Introduction

Single molecule detection via surface-enhanced Raman scattering (SERS) was originally demonstrated using dye molecules having absorption in the visible wavelength region as samples to take advantage of the resonance Raman effect [1, 2]. Not only dye molecules but also a variety of molecules have intrinsic absorption in the deep-ultraviolet (DUV) spectral region. Here in this chapter, DUV and UV regions are defined as wavelengths in the range of 200–300 nm and 300–400 nm, respectively. Excited by DUV light, such molecules exhibit the resonance Raman scattering effect, which can increase the Raman scattering cross-section to as high as $\sim 10^8$ [3]. The ubiquity of molecular absorptions in the DUV region is the principal advantage of using DUV light as an excitation source of SERS, which can realize ultrasensitive molecular detection for a variety of molecules.

In this chapter, the progress in DUV-SERS and tip-enhanced Raman spectroscopy (TERS) is described. First, the history of this relatively new topic in SERS research field is briefly reviewed. Then, the reason why DUV is interesting is described from the viewpoint of Raman spectroscopy. The material aspect of DUV-SERS metals and photodegradation of samples induced by DUV irradiation will also be discussed. Finally, some issues to be overcome and future perspectives are discussed.

2. DUV-SERS and TERS

The first experiment on SERS in the DUV was reported in 2007 [4]. Using a wavelength of 244 nm for Raman excitation, the group led by Popp measured the Raman scattering of a crystal violet aqueous solution placed on an aluminum thin film. The crystal violet, also known as dye molecule, has an absorption band of around 250 nm, originating from benzene rings in the molecular structure. The wavelength of 244 nm used for Raman excitation nearly matches with the DUV absorption of crystal violet, allowing the resonance Raman spectrum to be measured. In the report, they could detect two-orders-of-magnitude lower concentration of crystal violet solution on aluminum compared to one placed on a quartz substrate. They also observed a change in the resonance Raman spectral shape when the molecules were placed on aluminum, which they attributed to a manifestation of the surface enhancement effect of aluminum. Before this report, the shortest wavelength reported for SERS experiments was 325 nm, in which transition metals such as rhodium and ruthenium were used as metal surfaces [5,6].

Proteins and DNAs have characteristic absorption bands between 200 and 300 nm in the DUV, depending on the molecules. Taguchi *et al.* reported the detection of adenine molecules via DUV-SERS using an aluminum thin film [7]. By using an aluminum tip, the authors also demonstrated tip enhancement in the DUV with an excitation wavelength of 266 nm. Figure 1 shows the SEM image of aluminum tip, prepared by thermal deposition of aluminum onto a silicon cantilever probe under high vacuum condition. From the SEM image, the size of the aluminum grains at the tip apex was around 10–20 nm. DUV light was focused on the sample surface with an objective lens made of quartz. The Raman scattered light was collected with the same objective in a backscattering geometry and delivered to a

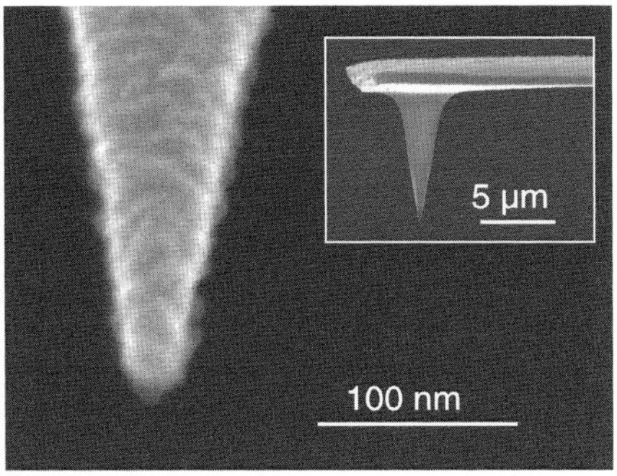

Figure 1. SEM image of aluminum-coated probe, showing a nanoparticle (NP) attached at the probe apex. The deposition thickness was 25 nm. The inset shows the corresponding low-magnification image. (Reprinted with permission from Ref. [7]. Copyright 2009 John Wiley & Sons.)

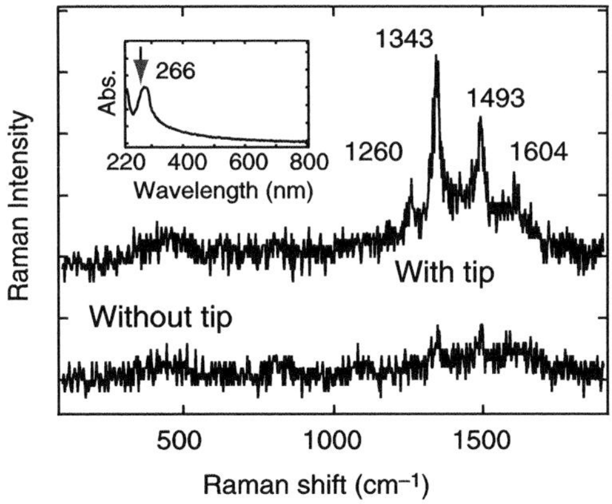

Figure 2. DUV-TERS spectra of adenine nanocrystals excited with 266 nm laser wavelength measured with and without tip. The inset shows the absorption spectrum of adenine nanocrystals. (Reprinted with permission from Ref. [7]. Copyright 2009 John Wiley & Sons.)

spectrometer [7]. The apex of the aluminum-coated tip was set in contact with the sample surface using an atomic force microscope apparatus. Figure 2 shows the measured Raman spectra of adenine nanocrystals with and without the tip. When the tip contacted the sample, Raman peaks were strongly enhanced. The inset of Figure 2 shows the separately measured absorption spectrum of the adenine nanocrystals. The Raman excitation wavelength (266 nm) used in the experiment was closely matched to the maximum absorption peak of adenine nanocrystals. The observed spectra of the adenine nanocrystals were consistent with the reported resonance Raman spectrum of the deoxyribonucleotides of adenine (dAMP) for the excitation wavelength of 266 nm [8–10]. This experiment firstly demonstrated the plasmonic enhancement of DUV resonance Raman scattering from biologically relevant molecules. The quartz peaks observed around 500 cm^{-1} remained unchanged with and without the tip, indicating the localized character of the tip-enhancement effect. The estimated enhancement factor was 1.3×10^3 in this experiment.

After those successful demonstrations of DUV-SERS and TERS, aluminum suddenly became a hot material in the research fields of SERS and related surface plasmon photonics [11]. It is interesting to note that the surface plasmon was first discovered on aluminum surface through electron energy loss experiments [12]. However, people recognized that, in the visible wavelength, much more stronger plasmon resonance is obtained with gold and silver. Therefore, not much attention has been paid to aluminum. After a long time, the demonstrations of DUV-SERS/TERS redirected researchers' interest to aluminum.

Clear resonance modes of localized plasmons are observable using isolated NPs rather than roughly corrugated films. Ekinci *et al.* fabricated an aluminum NPs array and used it to demonstrate DUV-SERS [13]. The uniformly sized NPs exhibited well-defined plasmon resonances, which helped clarify the contribution of the DUV plasmon to the observed SERS effect [13]. They deposited a 1 nm-thick adenine film onto a 140 nm-wide aluminum NP array and onto a fused silica. The Raman scattering intensity of adenine on aluminum nanoparticles was significantly increased compared with that on fused silica. The excitation laser wavelength was 257 nm, which matches with the observed quadrupolar plasmon resonance in a 140 nm-diameter aluminum nanoparticle. They reported stronger contribution by quadrupolar resonance than dipole for the observed SERS enhancement. They used extreme-UV interference lithography (λ = 13.5 nm) to fabricate aluminum nanoparticle array [14].

Such a short wavelength is required to achieve high spatial resolution in lithography to precisely control the diameter of nanoparticles. Taguchi *et al.* demonstrated a convenient technique to control the diameter of aluminum nanoparticles for controlling localized plasmon resonance in the DUV [15]. Other types of aluminum nanostructures, including nanovoids [16] and bow-tie antenna (a pair of triangles whose tips are facing each other across a nanogap) [17], have also been fabricated and used to demonstrate DUV-SERS.

Indium was found to be another promising material for plasmonic enhancement in the DUV [18, 19]. Figure 3(a) shows the SEM image of indium deposited on fused silica by thermal evaporation. The fabricated indium particles were polycrystalline, having an average size of 95 nm with a separation of several nanometers. By adjusting the deposition mass thickness, the size of the indium nanoparticles can be changed to tailor the resonance peak wavelength [19]. The Raman scattering spectrum of adenine (thickness 1 nm) coated on indium nanoparticles excited by 266 nm-wavelength laser light is shown in Figure 3(b). The Raman intensity of the adenine on the indium nanoparticles was strongly enhanced compared with that on the fused silica. The Raman enhancement factor was estimated to be ~100.

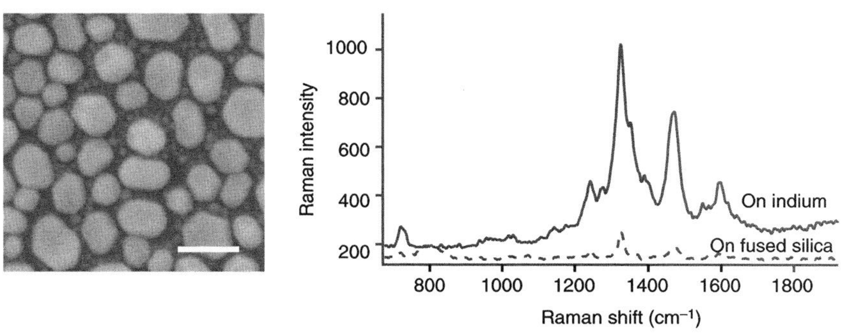

(a) (b)

Figure 3. (a) SEM image of indium-coated quartz substrate prepared by vacuum thermal vapor deposition with mass thickness of 25 nm. The scale bar is 200 nm. (b) Raman spectra of 1-nm thick adenine films deposited on the indium-coated substrate and on a bare fused silica substrate. (Reprinted with permission from Ref. [19]. Copyright 2014 American Chemical Society.)

3. Why DUV?

In this section, we examine the use of DUV light as an excitation source of Raman scattering from the aspect of Raman spectroscopy, specifically, for bioanalysis and imaging. Figure 4(a) shows Raman spectra of HeLa cells measured with excitation wavelengths of 532 and 244 nm. The spectral shape and the observed peaks obtained with DUV excitation (244 nm) are completely different from that obtained with visible excitations (532 nm). With visible excitation,

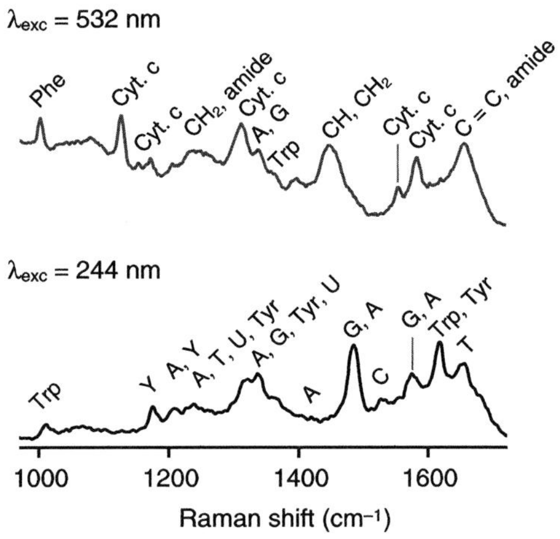

(a)

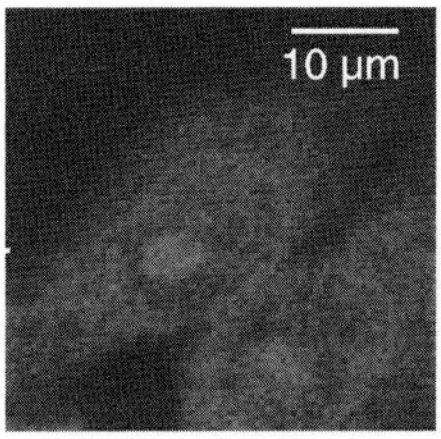

(b)

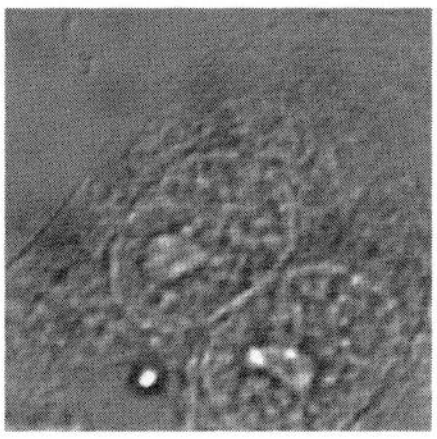

(c)

Figure 4. (a) Raman spectra of HeLa cells excited with 532 nm in visible wavelength (upper panel) and 244 nm in DUV wavelength (lower panel). (b) DUV Raman image of HeLa cells reconstructed with Raman bands of guanine and adenine at 1490 cm^{-1} measured in liquid with an excitation wavelength of 257 nm. (c) Bright field image corresponding to (b) observed with visible wavelength.

the peaks coming from lipids, proteins, nucleic-acid bases, and metalloproteins are all simultaneously observed in the Raman spectrum. In contrast, the Raman spectrum excited with 244 nm contains only the bands of nucleic-acid bases and amino acids. This is because these molecules have absorption at the excitation wavelength of 244 nm and the resonance Raman enhancement results. By selecting Raman excitation wavelength, it becomes possible to selectively sense a specific molecule of interest out of a mixture of heterogeneous substances, as is the case of biological cells.

The selective detection of ribonucleic acid (RNA) in biological cells due to the resonance Raman enhancement is demonstrated in Figure 4(b) by a DUV Raman image. The image, representing the distribution of guanine and adenine, was directly reconstructed using the peak intensity of 1490 cm^{-1} excited with 257 nm laser wavelength. With 257 nm laser excitation, the Raman peaks from guanines and adenines are highlighted, while the Raman peaks of amino acids are relatively suppressed [20]. It is clearly seen that RNAs are localized in the nucleolus and the cytoplasm. The bright field image of the corresponding cells is shown in Figure 4(c) for comparison. Shapes and positions of the nucleolus as well as the whole cell morphologies are consistent with each other. Analysis of the DUV Raman spectra suggests that guanine and adenine have different concentration distributions in a cell [20]. This kind of information is hardly obtained with visible Raman spectroscopy, where mathematical analysis, such as principle component analysis, will be required to extract DNA components from the complex Raman spectrum as seen in Figure 4(a).

4. Plasmonic Metals for DUV-SERS and TERS

In this section, we will look more deeply into the plasmonic material for DUV-SERS. Since the discovery of SERS in 1974 [21], noble metals such as silver and gold have been the predominant choice of metals for obtaining SERS enhancement. This is because, among the practically available metals, these metals support strong plasmon resonances in the visible and near-infrared spectral regions with minimal absorption loss. However, at DUV wavelengths, silver and gold

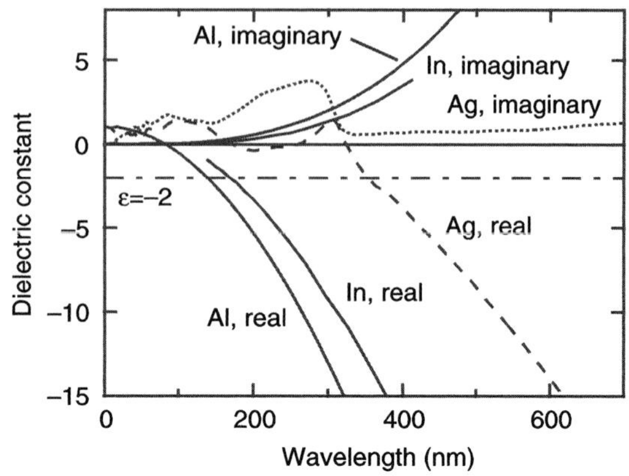

Figure 5. Dielectric functions of aluminum [22], indium [23], and silver [22]. Dashed dotted line represents $\varepsilon = -2$. (Reprinted from Ref. [25] with permission of Springer.)

lose their metallic character and behave as dielectrics, which means that no plasmonic enhancement can be obtained in the DUV region. In place of gold and silver, aluminum is the choice of metal to obtain plasmonic enhancement in DUV [4, 7].

The dielectric function of aluminum is shown in Figure 5 [22]. It has been shown that localized surface plasmon resonance of metal nanoparticles occurs at the wavelength at which the real part of the dielectric function of the metal becomes less than −2 [24]. In aluminum, this region lies at wavelengths longer than 140 nm. At the same time, the strength of plasmon resonance is governed by the value of the imaginary part of the dielectric function describing absorption loss: a smaller value provides stronger electromagnetic field enhancement. The imaginary part of the dielectric function in aluminum becomes smaller at shorter wavelengths. Thus, it is anticipated that aluminum nanoparticles will provide strong plasmon resonance at wavelengths between 140 nm and 300 nm. The localized surface plasmon resonance in the DUV was first experimentally observed by Ekinci *et al.* at a wavelength of 270 nm with 40 nm aluminum nanoparticles fabricated by EUV lithography [14]. For reference, the dielectric function of silver is also shown in Figure 5 [22]. Unlike aluminum, the plasmonic region of silver lies at wavelengths

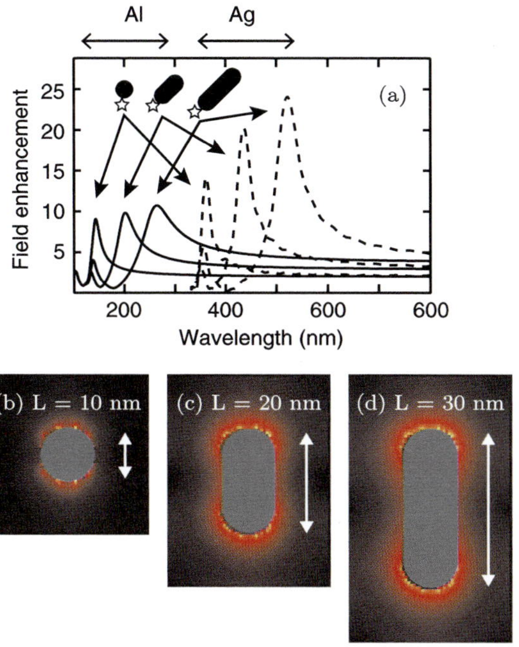

Figure 6. (a) Calculated localized plasmon resonance at the ends (indicated by stars) of a nanosphere (10 nm diameter) and nanorods (20 and 30 nm long by 10 nm wide) made of aluminum and silver. The solid and dashed lines represent the spectra for aluminum and silver, respectively. (b)–(d) Calculated field distributions near the aluminum nanoparticles in the plasmon resonance of the 10-nm diameter nanosphere, 20-nm long nanorod, and the 30-nm long nanorod, respectively. (Reprinted from Ref. [25] with permission of Springer.)

longer than 350 nm. In the DUV, silver does not meet the criteria (real part of dielectric function ≤ -2) to support localized surface plasmon resonance.

In order to examine surface plasmons in aluminum nanoparticles, nearfield enhancement given by aluminum nanorods is calculated and shown in Figure 6(a). The lengths of the nanorods are 10, 20 and 30 nm, and the width is 10 nm. The nanorods are excited by a linearly polarized plane wave with the polarization parallel to the rod long axis. The spectra are evaluated at the ends of the nanorods (indicated by the stars in the figure inset). For comparison, field enhancement obtained by silver nanorods having the same dimensions is also shown. The calculation results clearly demonstrate that nanorods made of aluminum have plasmon resonance in the DUV, whereas nanorods made of silver have plasmon resonance in the visible region. The difference in resonance wavelength arises from the difference in dielectric functions between aluminum and silver, as shown in Figure 5. It is also seen in the calculation results that the resonance wavelength shifts toward longer wavelength as the nanorod length increases. This means that the resonance wavelength of localized surface plasmons is controllable by selecting the size of the aluminum nanoparticles. Figures 6(b)–6(d) show the calculated field distributions near the nanoparticles at the peak wavelength of the plasmon resonance. Electromagnetic fields are localized at the ends of the nanorods. The calculations were performed using a finite-difference time-domain (FDTD) method [26], using the dielectric function shown in Figure 5.

In addition to aluminum, the dielectric function of indium is also shown in Figure 5 [23]. Indium behaves as a plasmonic metal at wavelengths down to 170 nm. The imaginary part of the dielectric function is reasonably small, similar to that of aluminum. DUV-SERS using indium has also been demonstrated [19].

4.1. *Aluminum*

To obtain plasmon resonance in the DUV region, the size of aluminum nanoparticles needs to be considerably small. For example, a 40-nm aluminum nanodisc was shown to have the plasmon resonance at 270 nm [14]. Techniques to fabricate such small nanoparticles include EUV lithography [14] and electron beam lithography [27]. Nanosphere lithography has been developed as a facile technique to fabricate metal nanoparticle arrays having desired size without the need of an expensive apparatus [28]. In this technique, spherical beads of polystyrene (PS) were self-assembled to form a hexagonal close-packed (hcp) monolayer on a substrate, which is then used as a mask for metal deposition to produce metal nanoparticle arrays on a substrate. The width of the fabricated nanoparticles is determined by the diameter of the PS beads. Using nanosphere lithography, Van Duyne and co-workers fabricated aluminum nanotriangle arrays with varying widths from 280 to 590 nm and reported that the plasmon resonances of those particles appear from 390 to 806 nm depending on the width [29]. To further decrease the size of nanoparticles, Taguchi *et al.* applied microwave heating to thermally expand the aligned beads before the metal deposition [15]. By heating up the PS beads, the diameter of the PS beads is increased while the center of the bead remained fixed, resulting in a gradual shrinking of the mask apertures surrounded by three adjacent beads. Figure 7(a) shows the SEM images of the expanded PS beads after microwave heating with different heating times (0, 90, 100, and 110 s). The sizes of the gaps between PS beads decrease with the increase in microwave heating time. The resultant aluminum nanoparticles produced with the heated masks are also shown in the lower panels in Figure 7(a). The PS beads that were heated for a longer time produced structures having smaller lateral sizes. The widths of the fabricated nanoparticles were 80, 71, 59 and 50 nm for the mask heated for 0, 90, 100, and 110 s, respectively. These results demonstrate that aluminum nanoparticles as small as several tens of nanometers can be fabricated by nanosphere lithography combined with the mask heating technique.

The measured UV–visible extinction spectra of those fabricated nanoparticles are shown in Figure 7(b) by solid lines. The extinction spectra showed peaks at 342, 314, 295 and 270 nm for the nanoparticles with the widths of 80, 71, 59 and 50 nm, respectively. For the particles obtained with longer heating times, i.e. smaller particles, the plasmon resonance wavelength moved toward deeper UV wavelengths. The overlaid circular dots are the calculated extinction spectra of aluminum nanoparticles with the same widths and heights. In the calculations, a native oxide layer of 2 nm thickness was assumed. The agreement between the measurements and calculations is excellent. The shortest wavelength achieved was 270 nm, while further reduction of the resonance wavelength would be possible by making heating durations longer.

What shortest wave lengths of the localized plasmon resonance can we achieve by reducing the size of aluminum nanoparticles? It is anticipated that the thickness of the native oxide layer surrounding

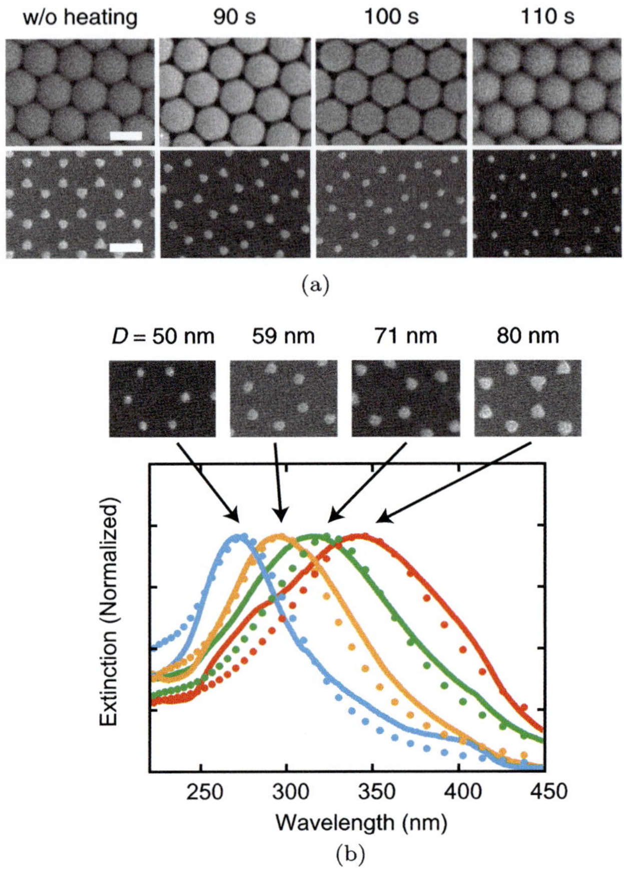

Figure 7. (a) SEM images of the aligned PS beads that were heated in a microwave oven (upper panel) and the resultant aluminum nanoparticle arrays fabricated using the heated PS masks (lower panel), demonstrating the shrinkage of the gaps between aligned PS beads upon microwave heating. The heating times were varied from 0, 90, 100 and 110 s. The deposition thickness was 30 nm. The average width of the fabricated nanoparticles were 80, 71, 59 and 50 nm, respectively, for the mask heated for 0, 90, 100 and 110 s. The scale bar is 300 nm. (b) Extinction spectra of the aluminum nanoparticles fabricated through PS masks shown in (a) (solid lines). The widths of the nanoparticles are shown in the figure. The spectra are normalized. The overlaid circular dots are the calculated extinction spectra of aluminum nanoparticles with the same widths and heights. In the calculations, a native oxide layer of 2 nm thickness was assumed. (Reproduced from Ref. [15], with the permission of AIP Publishing.)

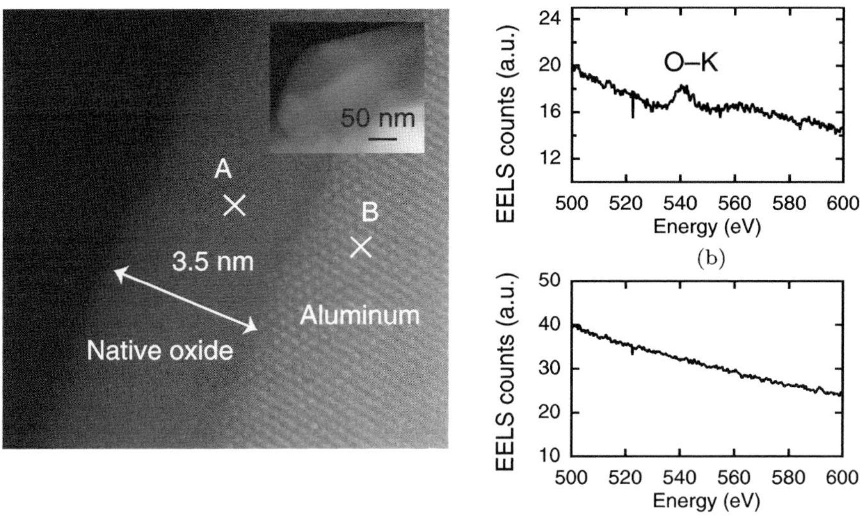

(a) (c)

Figure 8. (a) High-resolution scanning transmission electron microscopy (STEM) image of the surface of an aluminum nanoparticle deposited on a quartz using thermal evaporation. The inset shows the low-magnification STEM image of the observed particle. (b) Electron energy loss spectroscopy (EELS) spectra measured at point A in (a). (c) EELS spectra measured at point B in (a).

the aluminum nanoparticles determines the smallest size of nanoparticle, i.e. the plasmon resonance wavelength. The localized plasmon resonance energy as high as 6.8 eV was reported for oxide-free 10-nm aluminum nanoparticles placed under ultrahigh vacuum condition (~180 nm) [30]. However, in the atmospheric condition, formation of a native oxide is inevitable.

Figure 8(a) shows a high-resolution scanning transmission electron microscopy (STEM) image of the surface of an aluminum nanoparticle deposited on quartz via thermal evaporation. In the inner bulk side of the particle, atomic lattice is clearly seen, which is the evidence that the inner part of the particle is metallic. The surface of the particle is uniformly covered with a layer having no lattice fringe. Figures 8(b) and 8(c) show the electron energy loss spectroscopy (EELS) spectra measured at points A and B, respectively, shown in Figure 8(a). The peak of oxygen–K (O–K) appears in the EELS spectrum at point A, while the O–K peak is absent in the spectrum obtained at point B. This result indicates that the surrounding layer is an oxide of aluminum. The thickness of the oxide layer was 3.5 nm as measured from the high-resolution STEM image. This value is also consistent with the reported values around 2–4 nm [29, 15, 30]. Since near-field intensity quickly decays when the distance from the metal surface increases, the thickness of the oxide layer can affect the SERS enhancement achievable.

Unlike gold and silver, aluminum has a large negative redox potential, which makes it difficult to chemically reduce atomic aluminum from ionic form. Halas and colleagues have recently developed a technique to chemically synthesize high purity aluminum nanocrystals in solution [31]. Currently, the reported DUV-SERS studies worked on a substrate prepared by physical vapor deposition such as vacuum evaporation. High-quality aluminum nanocrystal in solution will provide another modality in applications of DUV-SERS, for example, SERS in solution and SERS in biological cell [32, 33].

4.2. *Indium*

As shown in Figure 5, indium is another promising material for plasmonic enhancement in the DUV. Judging from the value of the dielectric function, indium has smaller absorption loss compared with aluminum between 200 and 300 nm. This makes indium an attractive metal for DUV-SERS. DUV-SERS using indium nanoparticles has been demonstrated [19]. It has been also found that indium deposited on a fused silica substrate tends to form discrete nanoparticles separated by several nanometers [34]. The ease of obtaining discrete nanoparticle morphology is another attraction of indium, since isolated individual particle works as an antenna to couple with propagating incident light for the excitation of localized mode of plasmon. It is also anticipated that the nanometric gap between the particles works as a hot spot and can produce a strong SERS enhancement, also known as gap-mode SERS.

A drawback of indium compared to aluminum is the optical absorption loss by the naturally formed indium oxide on the surface. While aluminum oxide is a transparent material having no absorption between 200 nm and 5 μm in wavelength [22], indium oxide shows absorption below 300 nm [35]. Even a thin layer of indium oxide (e.g. 2 nm thickness) can reduce the plasmonic near-field enhancement of indium nanoparticles. The native oxide gradually grows in thickness under atmospheric condition [36], which means indium nanoparticles should be used quickly after they are taken out of the vacuum evaporation chamber.

5. Photodegradation of Sample Induced by DUV Irradiation

While DUV excitation of the resonant molecules enhances the Raman scattering cross-section, organic molecules suffer from photodegradation induced by DUV irradiation, which becomes a problematic issue in the detection of trace amount of molecules with DUV-SERS. Figure 9 shows the Raman spectra of desiccated HeLa cells with accumulating exposure durations excited with 244-nm laser wavelength [37]. In the spectrum obtained for the exposure duration of 0.25 s, characteristic spectral peaks of nucleotide bases (guanine and adenine) and aromatic amino acids (tryptophane and tyrosine) are observed. As the exposure duration increases, the characteristic

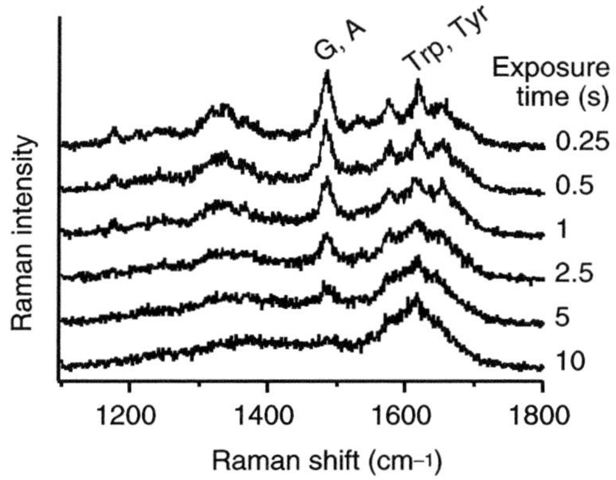

Figure 9. Raman spectra of desiccated cells obtained at different durations of sample exposure (0.25, 0.5, 1, 2.5, 5, 7.5, 10 and 12.5 s) at 16 μW/m^2 irradiation with an excitation wavelength of 244 nm. Each spectrum is shifted vertically for clarity. (Reproduced from Ref. [37] with the permission of OSA Publishing.)

bands decay, while a broad peak becomes apparent at around 1600 cm^{-1} originating from polymeric hydrogenated amorphous carbon [37]. Rate analysis revealed that the peak of aromatic amino acids decays faster than that of nucleotide bases. The underlying mechanism of photodegradation process involves several simultaneously progressing reaction pathways. One of the major reaction pathways is the ionization and dimerization of photoexcited molecules after absorbing DUV light. Removal of the excited energy from the DUV-absorptive molecules is possible using trivalent ions in the lanthanide group as an energy quencher [38]. Generation of reactive oxygen species (ROS) under DUV exposure is another pathway for photodegradation, since ROS is a strong oxidant to degrade molecules. To suppress the generation of ROS, argon purge to the sample chamber is a potential solution [38]. Water is transparent at wavelengths longer than 200 nm and has no significant effect on the Raman spectroscopy in the DUV region.

6. Summary and Future Perspectives

By using aluminum and indium, it is possible to obtain SERS enhancement in the DUV spectral region. Several types of aluminum nanostructures have been proposed, including rough films, nanoparticles, bow-tie antennas, and sharp tip. Since the development of DUV-SERS is still in its infancy, only a few different molecules — mostly adenine — have been tested as samples for DUV-SERS. Organic materials and inorganic materials such as wide gap materials are also interesting samples to be examined with DUV-SERS. Photodegradation of the sample induced by DUV light is a major stumbling block in finding practical applications of DUV-SERS. Achieving higher SERS enhancement will be beneficial for reducing DUV photodamage because this will make it possible to reduce the Raman excitation laser power required for detecting the Raman signal. Issues that remain to be addressed in order to achieve strong plasmon resonance in aluminum include the effect of purity of the deposited aluminum [11] and the design and fabrication of efficient nanostructures for higher and reproducible SERS and TERS [39].

Acknowledgment

The author would like to thank Dr. N. Taguchi at the National Institute of Advanced Industrial Science and Technology (AIST) for the STEM analysis.

References

1. Nie, S. and Emery, S. (1997). Probing single molecules and single nanoparticles by surface-enhanced Raman scattering. *Science* **275**, pp. 1102–1106.
2. Kneipp, K., Wang, Y., Kneipp, H., Perelman, L., Itzkan, I., Dasari, R. and Feld, M. (1997). Single molecule detection using surface-enhanced Raman scattering (SERS). *Phys. Rev. Lett.* **78**, pp. 1667–1670.
3. Asher, S. A. (2002). *Ultraviolet Raman spectrometry.* In Chalmers, J. M. and Griffiths, P. R. (eds.), *Handbook of Vibrational Spectroscopy.* John Wiley & Sons Ltd., Chichester.
4. Dörfer, T., Schmitt, M. and Popp, J. (2007). Deep-UV surface-enhanced Raman scattering. *J. Raman Spectrosc.* **38**, pp. 1379–1382.
5. Ren, B., Lin, X., Yang, Z., Liu, G., Aroca, R., Mao, B. and Tian, Z. (2003). Surface-enhanced Raman scattering in the ultraviolet spectral region: UV- SERS on rhodium and ruthenium electrodes. *J. Am. Chem. Soc.* **125**, pp. 9598–9599.
6. Lin, X. F., Ren, B., Yang, Z. L., Liu, G. K. and Tian, Z. Q. (2005). Surface-enhanced Raman spectroscopy with ultraviolet excitation. *J. Raman Spectrosc.* **36**, pp. 606–612.
7. Taguchi, A., Hayazawa, N., Furusawa, K., Ishitobi, H. and Kawata, S. (2009). Deep-UV tip-enhanced Raman scattering. *J. Raman Spectrosc.* **40**, pp. 1324–1330.
8. Fodor, S., Rava, R. P., Hays, T. R. and Spiro, T. G. (1985). Ultraviolet resonance Raman-spectroscopy of the nucleotides with 266-, 240-, 218- and 200-nm pulsed laser excitation. *J. Am. Chem. Soc.* **107**, pp. 1520–1529.
9. Blazej, D. C. and Peticolas, W. L. (1977). Ultraviolet resonant Raman spectroscopy of nucleic acid components. *Natl. Acad. Sci. USA* **74**, pp. 2639–2643.
10. Ziegler, L. D., Hudson, B., Strommen, D. P. and Peticolas, W. L. (1984). Resonance Raman spectra of mononucleotides obtained with 266 and 213 nm ultraviolet radiation. *Biopolymers* **23**, pp. 2067–2081.
11. Knight, M. W., King, N. S., Liu, L., Everitt, H. O., Nordlander, P. and Halas, N. J. (2014). Aluminum for plasmonics, *ACS Nano* **8**, pp. 834–840.
12. Powell, C. and Swan, J. (1959). Origin of the characteristic electron energy losses in aluminum. *Phys. Rev.* **115**, pp. 869–875.
13. Jha, S. K., Ahmed, Z., Agio, M., Ekinci, Y. and Löffler, J. F. (2012). Deep-UV surface-enhanced resonance Raman scattering of adenine on aluminum nanoparticle arrays. *J. Am. Chem. Soc.* **134**, pp. 1966–1969.
14. Ekinci, Y., Solak, H. H. and Löffler, J. F. (2008). Plasmon resonances of aluminum nanoparticles and nanorods. *J. Appl. Phys.* **104**, p. 083107.

15. Taguchi, A., Saito, Y., Watanabe, K., Yijian, S. and Kawata, S. (2012). Tailoring plasmon resonances in the deep-ultraviolet by size-tunable fabrication of aluminum nanostructures. *Appl. Phys. Lett.* **101**, p. 081110.
16. Sigle, D. O., Perkins, E., Baumberg, J. J. and Mahajan, S. (2013). Reproducible deep-UV SERRS on aluminum nanovoids. *J. Phys. Chem. Lett.* **4**, pp. 1449–1452.
17. Li, L., Fang Lim, S., Puretzky, A. A., Riehn, R. and Hallen, H. D. (2012). Near-field enhanced ultraviolet resonance Raman spectroscopy using aluminum bow-tie nano-antenna. *Appl. Phys. Lett.* **101**, p. 113116.
18. Mcmahon, J. M., Schatz, G. C. and Gray, S. K. (2013). Plasmonics in the ultraviolet with the poor metals Al, Ga, In, Sn, Tl, Pb and Bi. *Phys. Chem. Chem. Phys.* **15**, pp. 5415–5423.
19. Kumamoto, Y., Taguchi, A., Honda, M., Watanabe, K., Saito, Y. and Kawata, S. (2014). Indium for deep-ultraviolet surface-enhanced resonance Raman scattering. *ACS Photon.* **1**, pp. 598–603.
20. Kumamoto, Y., Taguchi, A., Smith, N. I. and Kawata, S. (2012). Deep ultraviolet resonant Raman imaging of a cell. *J. Biomed. Opt.* **17**, p. 076001.
21. Fleischmann, M., Hendra, P. J. and Aj, M. (1974). Raman spectra of pyridine adsorbed at a silver electrode. *Chem. Phys. Lett.* **26**, pp. 163–166.
22. Palik, E. D. (1997). *Handbook of Optical Constants of Solids.* Academic Press, San Diego.
23. Lemonnier, J. C., Jezequel, G. and Thomas, J. (1975). Optical properties in the far UV and electronic structure of indium films. *J. Phys. C, Solid State Phys.* **8**, pp. 2812–2818.
24. Bohren, C. F. and Huffman, D. R. (1983). *Absorption and Scattering of Light by Small Particles.* Wiley, New York.
25. Ozaki, Y. and Kawata, S. (2015). *Far- and Deep-Ultraviolet Spectroscopy.* Springer, Tokyo.
26. Taflove, A. (1995). *Computational Electrodynamics: The Finite Difference Time Domain Method.* Artech House, London.
27. Knight, M. W., Liu, L., Wang, Y., Brown, L., Mukherjee, S., King, N., Everitt, H., Nordlander, P. and Halas, N. J. (2012). Aluminum plasmonic nanoantennas. *Nano Lett.* **12**, pp. 6000–6004.
28. Jensen, T. R., Malinsky, M. D., Haynes, C. L. and Van Duyne, R. (2000). Nanosphere lithography: Tunable localized surface plasmon resonance spectra of silver nanoparticles. *J. Phys. Chem. B* **104**, pp. 10549–10556.
29. Chan, G. H., Zhao, J., Schatz, G. C. and Van Duyne, R. (2008). Localized surface plasmon resonance spectroscopy of triangular aluminum nanoparticles. *J. Phys. Chem. C* **112**, pp. 13958–13963.
30. Bisio, F., Proietti Zaccaria, R., Moroni, R., Maidecchi, G., Alabastri, A., Gonella, G., Giglia, A., Andolfi, L., Nannarone, S., Mattera, L. and Canepa, M. (2014). Pushing the high-energy limit of plasmonics. *ACS Nano* **8**, pp. 9239–9247.
31. McClain, M. J., Schlather, A. E., Ringe, E., King, N. S., Liu, L., Manjavacas, A., Knight, M. W., Kumar, I., Whitmire, K. H., Everitt, H. O.,

Nordlander, P. and Halas, N. J. (2015). Aluminum nanocrystals. *Nano Lett.* **15**, pp. 2751–2755.

32. Fujita, K., Ishitobi, S., Hamada, K., Smith, N. I., Taguchi, A., Inouye, Y. and Kawata, S. (2009). Time-resolved observation of surface-enhanced Raman scattering from gold nanoparticles during transport through a living cell. *J. Biomed. Opt.* **14**, p. 024038.
33. Ando, J., Fujita, K., Smith, N. I. and Kawata, S. (2011). Dynamic SERS imaging of cellular transport pathways with endocytosed gold nanoparticles. *Nano Lett.* **11**, pp. 5344–5348.
34. Dragan, A. I. and Geddes, C. D. (2010). Indium nanodeposits: A substrate for metal-enhanced fluorescence in the ultraviolet spectral region. *J. Appl. Phys.* **108**, p. 094701.
35. Hamberg, I. and Granqvist, C. G. (1983). Dielectric function of undoped In_2O_3. *Thin Solid Films* **105**, pp. L83–L86.
36. Schoeller, H. and Cho, J. (2009). Oxidation and reduction behavior of pure indium. *J. Mater. Res.* **24**, pp. 386–393.
37. Kumamoto, Y., Taguchi, A., Smith, N. I. and Kawata, S. (2011). Deep UV resonant Raman spectroscopy for photodamage characterization in cells, *Biomed. Opt. Express* **2**, pp. 927–936.
38. Kumamoto, Y., Fujita, K., Smith, N. I. and Kawata, S. (2016). Deep-UV biological imaging by lanthanide ion molecular protection, Biomed. *Opt. Exp.* **7**, pp. 158–170.
39. Taguchi, A., Yu, J., Verma, P. and Kawata, S. (2015). Optical antennas with multiple plasmonic nanoparticles for tip-enhanced Raman microscopy. *Nanoscale* **7**, pp. 17424–17433.

Chapter 6

Lithographically Prepared SERS-Active Substrates with Well-Defined Gaps Below 1 nm

Kenneth Crozier[*,†,§] and Wenqi Zhu[‡]

[*]*Department of Electrical and Electronic Engineering, University of Melbourne, Victoria 3010, Australia*

[†]*School of Physics, University of Melbourne, Victoria 3010, Australia*

[‡]*School of Engineering and Applied Sciences, Harvard University, 29 Oxford Street, Cambridge, Massachusetts 02138, USA*

[§]*kcrozier@unimelb.edu.au*

1. Introduction

Metallic structures with nanogap features have proven to be highly effective for SERS, enabling an extreme level of sensitivity, down to that of single molecules [1–3]. This sensitivity comes from localized surface plasmon resonances supported by nanogaps, in which the enhancement of the electric field intensity can be as high as $\sim 10^4$. In most previous works, the general practice to model such plasmonic resonances involved the use of classical electromagnetic theory. In this chapter, a regime is considered where quantum mechanical effects become important. It has only recently become widely appreciated among the nanooptics community that phenomena such as electron tunneling [4–9] and non-locality [10–13] can appear as feature sizes approach atomic length scales. Here, we review our work [14] that shows unambiguously that electron tunneling at optical frequencies emerges for metallic nanostructures containing nanogaps with widths down to a few Angstroms. Of particular significance for SERS is that this work experimentally demonstrates, for the first time to the best of our knowledge, that the maximum plasmonic enhancement that can be achieved is limited by the emergence of electron tunneling.

This chapter is organized as follows. We begin by describing a method for the lithographic fabrication of gold dimers with Angstrom-scale gaps. We next describe a method for measuring the (far-field) scattering spectra of the gold dimers with gaps ranging from 10 nm down to a few Angstroms. We discuss the results, which demonstrate the impact of electron tunneling. We next describe our method for studying the impact of electron tunneling upon plasmonic near-field enhancement via SERS. This demonstration that electron tunneling limits the maximum achievable plasmonic enhancement is likely to be important for many other applications in nanooptics in addition to SERS.

2. Fabrication of Dimers with Angstrom-Scale Gaps

Electron beam lithography (EBL) allows plasmonic nanostructures to be produced with flexible designs and with controllable gap widths. Dimer structures are particularly useful for studying quantum mechanical effects in plasmonic nanogaps. They consist of pairs of metallic nanoparticles separated by small gaps. The fabrication of dimers by electron beam lithography faces difficulties for gap sizes below 10 nm [15], although nanogaps of this size have been demonstrated using techniques that include advanced lithographic tools [16], special substrates [17] and the use of sacrificial layers [18]. Nonetheless, gaps below 1 nm are extremely difficult to attain using the conventional method of EBL, evaporation and the lift-off process. This is due to the resolution of the electron beam resist and other factors [19]. We demonstrate here a two-step EBL method (Figure 1) that can produce metallic dimers separated by Angstrom-scale gaps. Our dimers comprise pairs of gold disks with diameters of 90 nm and thickness 20 nm, on a silicon nitride (SiN, 30 nm thick) window, with thin titanium (1 nm thickness) as the adhesion layer.

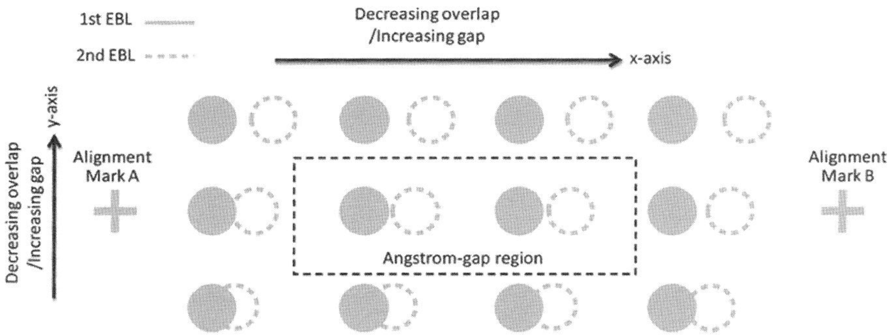

Figure 1. Schematic diagram illustrating two-step EBL method for the fabrication of plasmonic dimers with Angstrom-scale gaps. Patterns produced by first EBL step are shown as solid objects. Patterns produced in second EBL step (aligned to the first) are shown as dashed objects.

The EBL tool employed is an Elionix F-125. For the first EBL exposure, we start by locating the SiN window using the scanning electron microscopy (SEM) mode of the tool. We then pattern a nanodisk array with an overall extent of $300\,\mu m \times 300\,\mu m$, which is slightly larger than the SiN window. The nanodisks are in a square array (periodicity of $4\,\mu m$). The exposed pattern also includes two alignment markers. The sample is developed, and titanium and gold are evaporated. Lift-off is performed using acetone. The second EBL exposure is aligned to the first and contains another nanodisk array. This is not a square array however. Rather, the locations of the nanodisks are such that perfect alignment would result in an array of dimers with gaps ranging from $-40\,nm$ (overlapped) to $+40\,nm$. The gap width is designed to increase by 0.05 nm for each successive dimer, from the lower left corner to the upper right corner of the pattern. This results in dimers with Angstrom-scale gaps being achieved in the center of the pattern. We use alignment marks A and B to minimize translational and rotational errors. Typical alignment and rotational errors are below 10 nm and 1 mrad, respectively, for state-of-the-art EBL systems [20]. Such errors make it inevitable that dimers with Angstrom scale gaps are not necessarily produced at the very center of the pattern. Nonetheless, such gaps will be produced at other places in the pattern. In the next section, we discuss their characterization.

3. Electron Microscope Characterization of Fabricated Dimers

As discussed, the dimers are formed on 30-nm thick SiN membranes. This is done to enable their characterization by transmission electron microscopy (TEM). Top-view TEM images of four typical dimers are shown in Figures 2(a)–2(d). These dimers are termed Dimers I–IV, respectively. The widths of the gaps of these dimers are determined from the magnified views of Figures 2(e)–2(h). The high resolution of these images is confirmed by the fact that the atomic lattice of gold can be observed. It can be seen that there are small particles around the dimer, which is consistent with the previous reports made for similar structures [17]. Here, we select dimers that have no such small particles in the smallest region of the gap.

To determine the widths of the gaps in a consistent manner, we employ the following process. One possible source of error is that of parallax error from sample tilting [21]. To avoid this, 10 TEM images are taken for each sample, with the TEM sample holder being rotated within ±30° about an axis that is in the plane of the substrate and

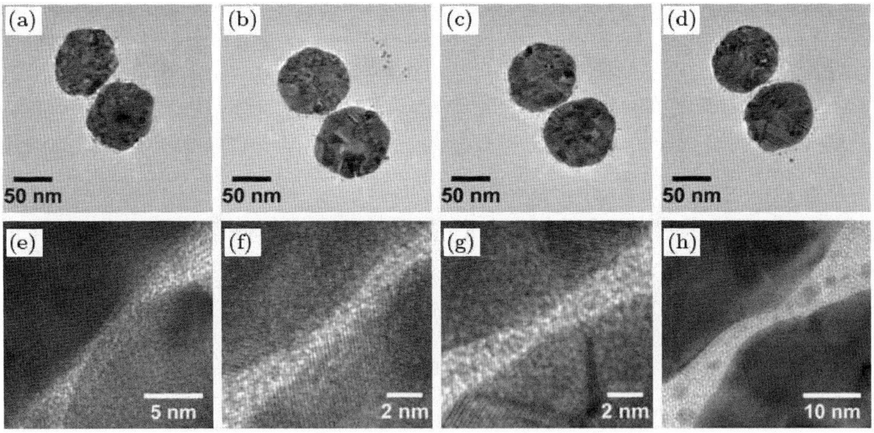

Figure 2. TEM images of dimers with atomic length-scale gap width. (a)–(d) Top-view TEM images of four representative dimers, Dimers I–IV, with gradually increasing gap width. (e)–(h) High-resolution magnified-view TEM images of gap regions of dimers of panels (a)–(d). In panel (e), nanoparticles touch, while panels (f)–(h) have gap widths of 2.0 Å, 6.7 Å, and 5.8 nm, respectively.

perpendicular to the dimer long axis. We find the width of the gap for each image. The gap width is defined as the distance between the closest two points on the two disks. The gap width of a particular dimer is then taken to be the largest gap width measured from this set of TEM images. From this method, the nanodisks of Dimer I (Figure 2(e)) are found to be touching, while Dimers I–IV (Figures 2(f)–2(h)) are found to have gap widths of 2.0 Å, 6.7 Å, and 5.8 nm, respectively.

4. Modeling the Plasmonic Response of Nanoparticle Dimers

As discussed in the Section 1, the overarching theme of this chapter concerns the emergence of quantum mechanical effects in plasmonic dimers with sub-nanometer gaps. Understanding this phenomenon is greatly facilitated by the ability to simulate the plasmonic response of dimers using a model that does not incorporate such effects and using another model that does. The former is the classical electrodynamics model (CEM), while for the latter, we employ the quantum corrected model (QCM, [5]). In this section, we describe these modeling approaches.

We begin by discussing the method used for the CEM simulations. The geometry of the simulated structure is chosen to match that of the fabricated dimer, i.e. two gold nanodisks with diameters of 90 nm and thicknesses of 20 nm. The permittivity of gold is taken as being described by a Drude model, with $\varepsilon_\infty = 1, \omega_p = 1.94 \times 10^{15}$ Hz, and $\gamma = 2.14 \times 10^{13}$ Hz. The TEM membrane on which the dimer sits is modeled as a 30 nm thick layer of SiN with a refractive index of $n = 2$. As discussed earlier, titanium is used to facilitate adhesion of the dimer to the membrane. TEM imaging indicates that the titanium layer is slightly (~2 nm) wider than the gold structure. We therefore model the titanium adhesion layer as two nanodisks with diameters of 94 nm and thicknesses of 1 nm that sit beneath the gold nanodisks. Our simulations are performed with the finite difference time domain (FDTD) method implemented in a commercial software package (Lumerical, Figure 3(a)). The permittivity of titanium is from a multi-coefficient fitting model provided by the software

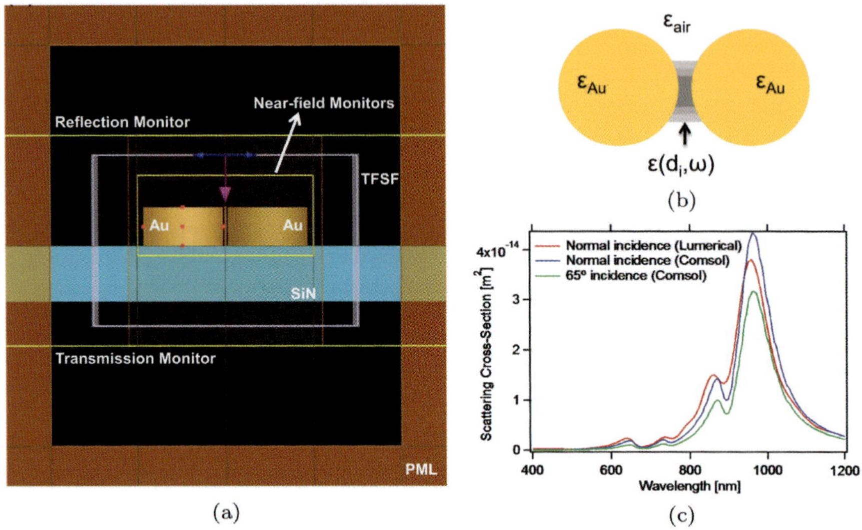

Figure 3. Simulation configurations of plasmonic dimer. (a) Side view of simulation domain in Lumerical. (b) Scheme of Drude-like tunneling region in QCM. (c) Comparison between normal incidence and oblique incidence.

package. To minimize the computational workload, we make use of the symmetry in the plane of the substrate. To simulate the case of an isolated dimer, which is appropriate due to the large spacing in the fabricated array, the boundaries of the computational domain contain perfectly matched layers (PMLs) to absorb outgoing waves. A total-field scattered-field (TFSF) source is used to realize a linearly polarized plane wave incident on the dimer. In our experiments, dark-field scattering spectra are measured as a means for understanding the plasmonic response of the dimers. In the simulations, therefore, we place two power monitors outside the TFSF region to measure the scattering spectrum of the dimer. The meshing is as follows. The dimer structure is enclosed by a box that has a uniformly spaced mesh of 0.5 nm. The mesh outside the box is non-uniform. For dimers with gaps narrower than 1 nm, a more finely spaced mesh is generated within the gap region, such that the gap width corresponds to 4 unit cells. To determine the SERS enhancement factor (EF) predicted by these simulations, it is necessary to know the electric fields generated on the surfaces of the dimer. We therefore use a 3D monitor with the same dimensions as the uniform mesh box to record all field data. The surface-averaged SERS EF is then found using

$$\mathrm{EF} = \iint_A |E|^4 ds \Big/ \iint_A ds,$$

where the surface A comprises all surfaces that are at a distance of 1 nm from both the top surfaces and the side walls of the dimer.

The method we use for incorporating quantum mechanical effects is the QCM approach introduced by Esteban *et al.* [5]. This approach is summarized below and enables the effect of electron tunneling to be included in classical electromagnetic calculations. Electron tunneling represents a charge-transfer channel across the gap, with charges of opposite sign neutralizing each other. To implement the QCM approach, the vacuum permittivity of the gap region is replaced by a Drude model that captures the tunneling process. As indicated in Figure 3(b), different parts of the gap have different gap distances l, meaning that l is a function of (x,y). For a certain gap distance l, the relative permittivity of the gap region is given by

$$\varepsilon(l,\omega) = 1 - \frac{\omega_g^2}{\omega(\omega + i\gamma_g(l))}, \tag{1}$$

where ω_g is the effective plasma frequency and taken to be the plasma frequency of gold, based on the consideration that the model should resemble the Drude model of gap for gap separation $l = 0$, γ_g is the effective damping frequency and is a function of gap distance l. The imaginary part of the permittivity is related to the conductivity in the usual manner. The conductivity is thus defined as

$$\sigma_0(l) = \frac{\omega_g^2}{4\pi\gamma_g(l)}. \tag{2}$$

On the other hand, through full quantum mechanical calculations of the probability of tunneling between two metallic parallel plates, the tunneling conductivity σ_0 at a junction under bias $U = lE$ can be related to the current density J as follows:

$$\sigma_0(l) = l dJ/dU = l\frac{2}{(2\pi)^2}\int_0^{\Omega_F} T(\Omega, l)d\Omega, \tag{3}$$

where Ω_F is the electron Fermi energy, $T(\Omega,l)$ is the energy-dependent electron tunneling probability at a lateral position within the gap for which the local separation is l. Combining Equations (2) and (3) therefore allows the separation-dependent damping frequency to be obtained. When this is performed, it is furthermore found out that γ_g generally increases exponentially with gap separation, i.e.

$$\gamma_g(l) = \gamma_g(l = 0)\exp(ql), \tag{4}$$

where $\gamma_g(l = 0)$ is taken as the damping frequency of gold, based on the consideration that the effective Drude model should resemble that of gold for gap separation $l = 0$, and q is a material-dependent characteristic length from the exponential fit of $\gamma_g(l)$ and takes a value of $q = 2.24\,\text{Å}^{-1}$ for gold.

With the permittivity of the gap region chosen in this way, the QCM approach allows the quantum mechanical effect of electron tunneling to be captured.

It is also important to consider a difference that exists between the illumination conditions used in the simulations and experiments. As described above, our simulation method assumes normal-incidence illumination, while in our dark-field scattering measurements, the white light source is incident at an angle of 65°. This is because the nature of the FDTD method means that oblique incidence is computationally expensive. This leads to the question of the effect of this difference in illumination conditions. To understand this, we use the finite element method (FEM) to simulate the dark-field scattering spectrum of a dimer for the case of oblique incidence (65°). The simulations are performed using the commercial software package COMSOL Multiphysics (RF module). We also simulate the dark-field scattering spectrum of the dimer under normal incidence using FEM. In Figure 3(c), the scattering spectra simulated via FEM (normal and oblique incidence) are compared to the spectrum simulated by FDTD (normal incidence). These simulations are performed using the CEM approach. It can be seen that, for normal incidence, the FEM and FDTD approaches are in general agreement. There are slight differences which we ascribe to differences in collection angles and to differences between the dielectric functions employed. In the FEM simulations, the dielectric function of gold from Ref. [22] is used, while in the FDTD method, the software package (Lumerical) fits the data of Ref. [22] with a multiple coefficient model. It can be seen the spectrum under oblique incidence has a lineshape similar to that under normal incidence, but only differing in intensity. This leads us to conclude that while the FDTD simulation configuration (normal incidence) is not the same as that used in the dark-field scattering experiments (65° incidence), the difference that results is relatively minor.

5. Impacts of Electron Tunneling on Far-Field Properties of Dimer

A home-built dark-field microscope (Figure 4) is used to measure the dark-field scattering spectra of the dimers. White light from a halogen lamp is polarized, then loosely focused on the dimer array by a microscope objective (Nikon, ULWD, magnification 20×, numerical aperture NA 0.25). The polarizer is rotated so that the illumination is linearly polarized along the dimer axis. The illumination is incident at an angle of 65° from the normal to the substrate. Scattered light is collected by an objective lens (Nikon, ELWD, 50×, NA = 0.55) and a tube lens (focal length $f = 200\,\text{mm}$), producing an image of the plasmonic dimer array at the entrance slit of the spectrometer (Acton SP2300). We then replace the entrance slit assembly with a precision pinhole (diameter 200 μm). This allows us to select the light scattered by an individual plasmonic dimer under study. The dark-field scattering spectrum S generated by the spectrometer (wavelength range: 500–1050 nm) is then recorded by the liquid nitrogen cooled charge coupled device (CCD) camera (Princeton Instruments Spec-10). An integration time of 30 s is used. To remove the effect of second-order diffraction from the grating in the spectrometer upon the recorded spectrum, a long pass filter (passes $\lambda \geq 532\,\text{nm}$) is placed in front of the pinhole. We obtain a reference spectrum R by measuring the spectrum with the plasmonic dimer sample replaced with a white light reflectance standard (WS-1-SL, Ocean Optics). We furthermore obtain a background spectrum B by measuring the spectrum from a region on the SiN window that does not contain any dimers. With these spectra measured, the spectrum from the dimer under study is calculated as $(S - B)/R$.

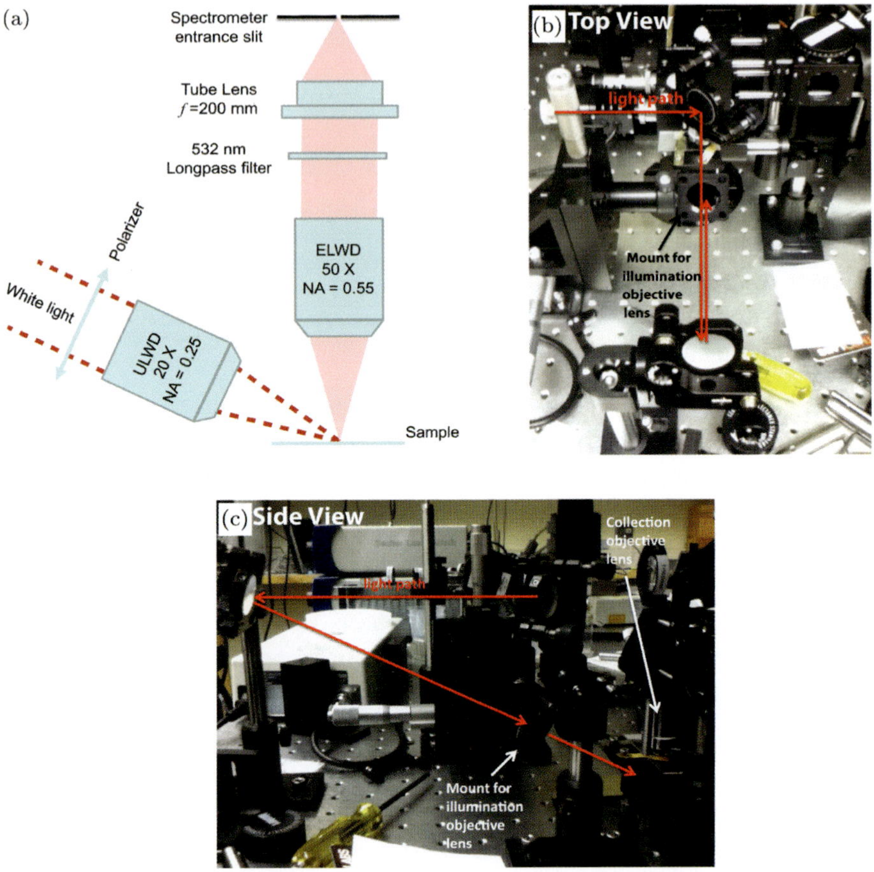

Figure 4. Experimental setup for dark-field scattering spectroscopy: (a) schematic diagram, (b) top view photograph of setup, and (c) side view photograph of setup.

Using the microscope described above, we measure scattering spectra from dimers with gap widths ranging from 9.1 nm down to 2.0 Å. The results are presented in Figure 5. It should be noted that all optical measurements reported in this chapter are performed prior to TEM characterization. This prevents the possibility of carbon contamination affecting the results. CEM and QCM simulations of dark-field scattering spectra are presented in Figures 6(a) and 6(b), respectively.

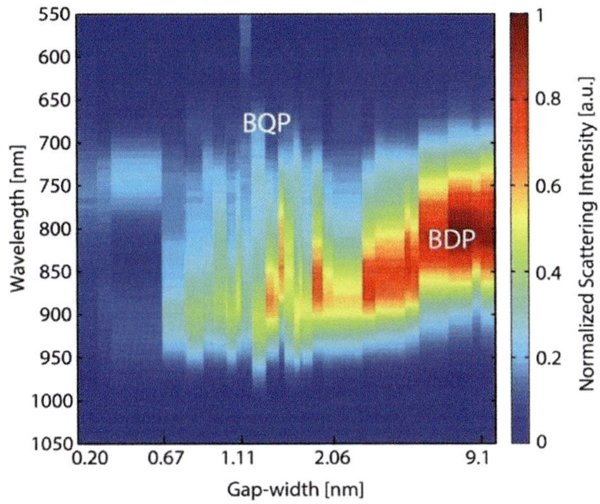

Figure 5. Measured dark-field scattering spectra from dimers with various gap widths. Horizontal axis represents measured gap widths from 2 Å to 9.1 nm in log scale. The scattering intensities are normalized by their largest value.

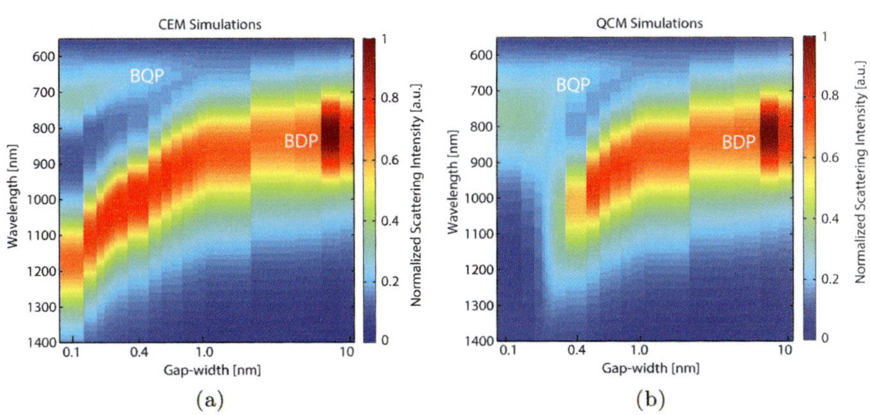

Figure 6. Simulated impacts of electron tunneling on dark-field scattering from dimers with angstrom-scale gaps. (a) Scattering intensities simulated by CEM for gap widths from 1 Å to 10 nm. The scattering intensities are normalized by their largest value. (b) Scattering intensities simulated by QCM for same range of gap widths as panel (a). Again, the scattering intensities are normalized by their largest value.

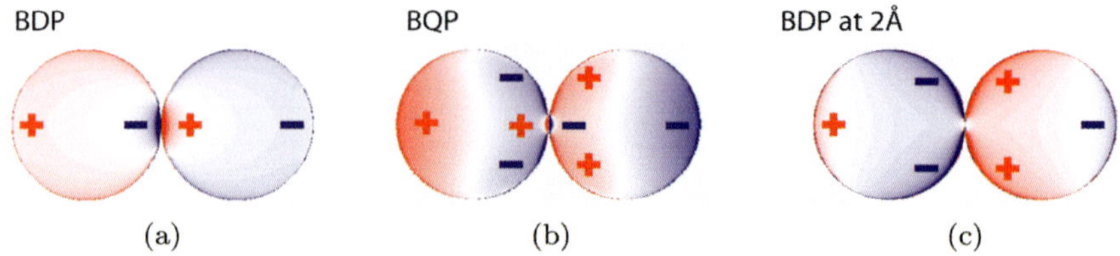

Figure 7. Simulated charge density distributions for three plasmonic modes: (a) Bonding dipole plasmon (BDP) simulated with CEM. (b) Bonding quadrupole plasmon (BQP) simulated with CEM. These are calculated from dimer with gap width of 9 Å. (c) BDP simulated with QCM for dimer with gap width of 2 Å.

We now discuss the results (Figures 5 and 6). According to classical plasmon hybridization theory [23], there are two resonance modes, namely the bonding dipole plasmon (BDP) mode and the bonding quadrupole (BQP) mode. The simulated charge density distributions of these modes are shown in Figure 7. From Figure 5, it can be seen that as the gap width decreases from 9.1 nm to 6.7 Å, the experimentally measured BDP resonance red-shifts from 807 to 892 nm, and the scattering intensity drops to 46% of its peak value. It can be furthermore seen that the measured scattering intensities of the BQP are much weaker than the BDP. Let us now consider the predictions of simulations. The CEM simulations predict that, as the gap width decreases from 10 nm to 7 Å, the BDP scattering intensity drops to 83% of its peak value, while the BDP peak resonance shifts from 800 to 929 nm (Figure 6(a)). It can be seen from Figure 6(b) that the QCM-predicted scattering spectra are very similar to those of the CEM approach over this range of gap widths, due to the fact that the Drude model of the gap region almost resembles vacuum over this range.

We next consider further reductions in gap width. It can be seen that the experimental results (Figure 5) show interesting behaviors for the BDP mode as the gap narrows from 6.7 (Dimer III) to 2.0 Å (Dimer II). The scattering resonance peak no longer red-shifts, and the strength of the scattering becomes comparable to, or even smaller than, that of the BQP around 750 nm. This behavior is substantially different from what is predicted by the CEM approach (Figure 6(a)). However, for gaps smaller than 4 Å, the QCM approach (Figure 6(b)) predicts that the BDP resonance no longer red-shifts and that it drops in intensity, making the BQP mode the dominant resonance in the spectra. From Figure 5, it can be seen that these features are indeed experimentally observed for dimers with gap widths below 6.7 Å. In this regime, electron tunneling represents a charge-transfer channel across the gap, with charges of opposite sign neutralizing each other, resulting in the mode shown in Figure 7(c).

6. Impact of Electron Tunneling on Near-Field Properties of Dimer

We next study the effect of electron tunneling on the plasmonic near-field enhancement provided by dimers with atomic length-scale gaps. This is performed using SERS measurements. We choose this method based on the consideration that SERS EF scales roughly as the fourth power of the near-field enhancement. Self-assembled monolayers (SAMs) of thiophenol are formed on the plasmonic dimers. In Figures 8(a) and 8(b), TEM images of gold dimers with and without the thiophenol SAM are shown, respectively. These dimers are 40 nm thick, and have no adhesion layer. The light gray region (Figures 8(a) and 8(c)) that appears 1 nm wider than the gold dimer thus represents the thiophenol SAM. The gap width of the dimer of Figure 8(c) is determined to be ~5.7 Å. Despite the fact that this gap is very small, it can be seen that the thiophenol molecules fill the gap region (Figure 8(c)). Our estimations of SERS EF thus assume monolayer coverage of thiophenol on all exposed surfaces of the gold dimers.

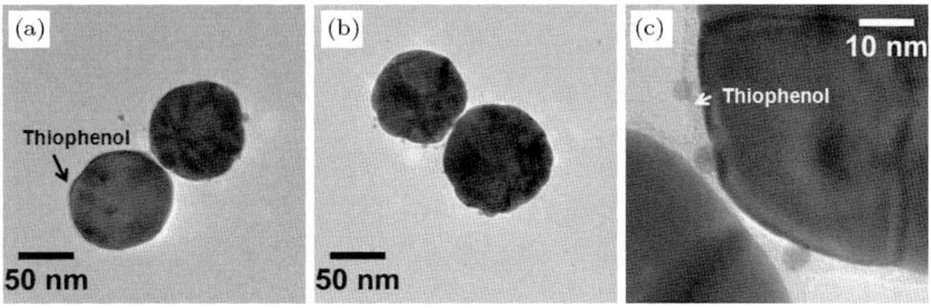

Figure 8. TEM characterization of SAM of thiophenol: (a) gold dimer with thiophenol SAM, (b) gold dimer without thiophenol SAM, and (c) thiophenol SAM within gap region.

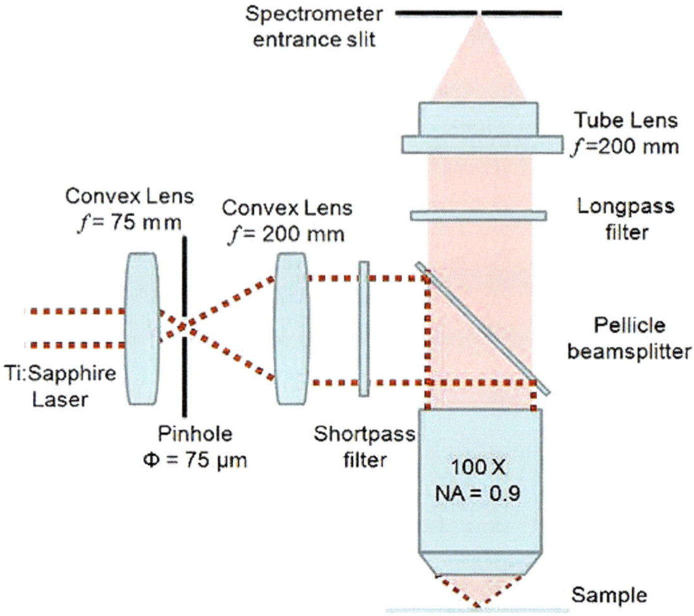

Figure 9. Experimental setup for wavelength-scanning Raman spectroscopy.

SERS measurements are performed using the homebuilt confocal Raman microscope (epiconfiguration) shown in Figure 9. The different dimers have different gap widths and thus different plasmon resonance wavelengths. We therefore measure SERS spectra at 19 different wavelengths (every 10 nm from 705 to 885 nm) for each dimer to ensure that the maximum SERS enhancement is measured. Continuous wave (CW) illumination from a Ti:sapphire laser (Mira 900, Coherent) that is tunable from 700 to 900 nm is used as the excitation. We keep the laser power to below 20 μW to avoid damage to the sample. This permits multiple SERS measurements to be made and also avoids the nonlinear field enhancement that has been predicted to happen at higher intensities [24, 25]. The laser polarization is chosen to be along the dimer axis, i.e. it is consistent with the dark-field measurements from before. The beam from the Ti:sapphire laser is spatially filtered using a precision pinhole (diameter 75 μm). The light transmitted by the pinhole is collimated, passes through a short-pass filter, is reflected by a beam splitter, then is focused by an objective lens (100×, NA = 0.9) into a nearly diffraction limited spot on the sample. The beam splitter is a pellicle beam splitter (CM1-BP108, ThorLabs) with an 8:92 splitting ratio i.e. it directs ~8% of the laser beam power onto the objective. The long axis of the dimer being studied is placed along the polarization axis of the laser beam. The sample is on a piezo-stage, allowing the dimer to be positioned at the center of the beam to maximize the intensity of the Raman scattering. The Stokes Raman scattering is collected by the objective lens, transmitted through the beam splitter with an efficiency of ~92%, passes through long-pass filter, and is focused by the tube lens on the spectrometer entrance slit. The long pass filter also removes the laser line. The Raman spectra are then measured by the spectrometer, which is fitted with a CCD camera (liquid nitrogen cooled). We use integration times of 60 s.

We perform wavelength-scanning Raman measurements for 45 individual dimers. We present results from two representative dimers (Dimers II and III of Figure 2) as Figure 10. Based on the far-field optical measurements (Figure 5) and the high resolution TEM images, we expect that electron tunneling will occur for Dimer II, but not for Dimer III. We use our system (Figure 4) to record dark-field scattering spectra of the dimers with and without the thiophenol SAMs. For both dimers, the SAM formation results in increased scattering intensities but with the resonance lineshapes being maintained (Figures 10(a)–10(b)). Interestingly, it can be seen that the suppression of the BDP of Dimer II, in comparison to Dimer III, is maintained even with the formation of the thiophenol SAM. We thus conclude that electron tunneling through the SAM is still significant for such narrow gaps. The wavelength-scanning Raman measurements are shown in Figures 10(c) and 10(d). Each column of measurements shows the Raman intensity normalized by the laser power. We discuss the results in more detail in the next paragraph.

We measure the SERS EF of the 1074 cm^{-1} Raman line at each laser wavelength for the two dimers. This permits us to quantify the SERS performance. We plot the results in Figures 10(a) and 10(b). Contrary to previous reports [26], our results demonstrate that the wavelength-dependent SERS EF follows the lineshape of

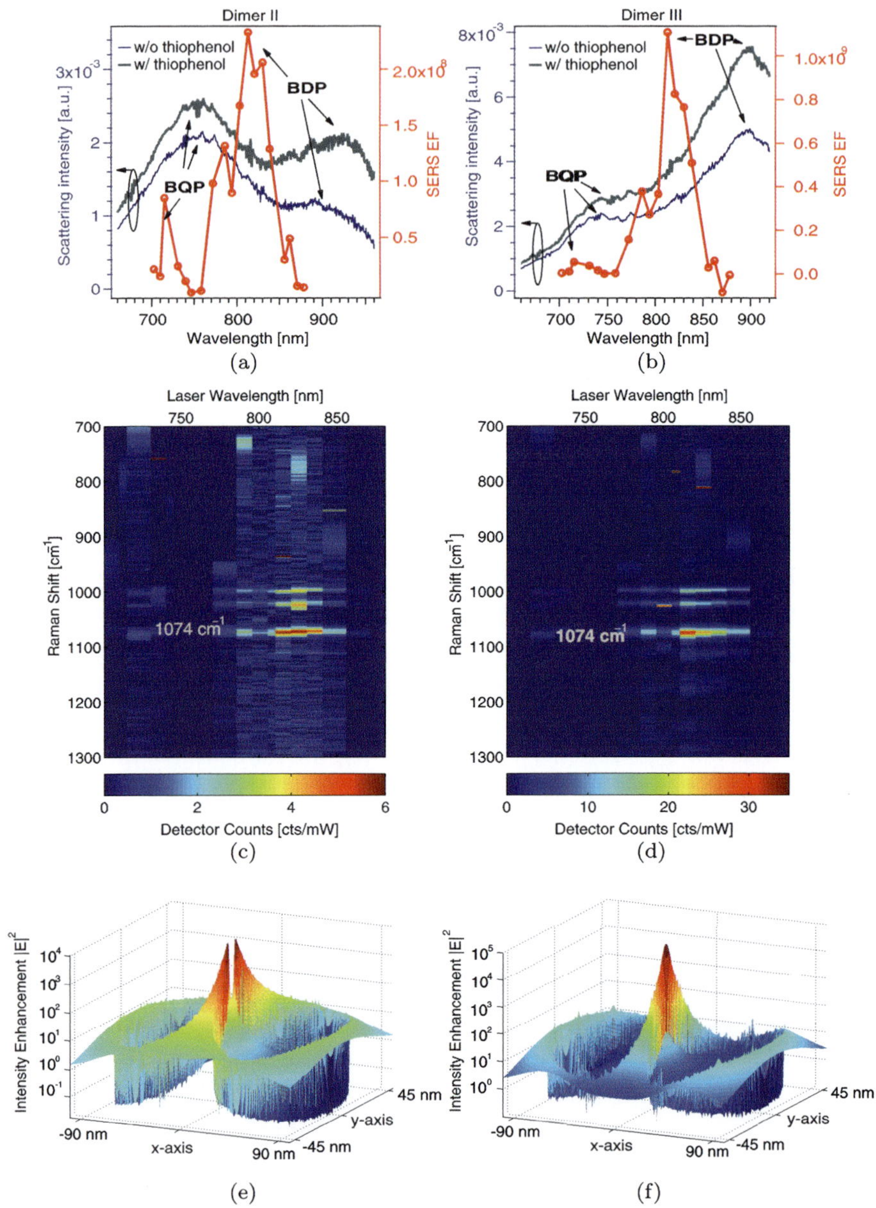
Dimer II
w/o thiophenol
w/ thiophenol
BDP
BQP
Scattering intensity [a.u.]
SERS EF
Wavelength [nm]
(a)
Dimer III
w/o thiophenol
w/ thiophenol
BDP
BQP
Scattering intensity [a.u.]
SERS EF
Wavelength [nm]
(b)
Laser Wavelength [nm]
Raman Shift [cm-1]
1074 cm-1
Detector Counts [cts/mW]
(c)
Laser Wavelength [nm]
Raman Shift [cm-1]
1074 cm-1
Detector Counts [cts/mW]
(d)
Intensity Enhancement |E|2
-90 nm
x-axis
90 nm
-45 nm
y-axis
45 nm
(e)
Intensity Enhancement |E|2
-90 nm
x-axis
90 nm
-45 nm
y-axis
45 nm
(f)

the dark-field scattering spectra of Figures 10(a) and 10(b). It should be noted that our experiments differ from those of Ref. [26] in that ours are obtained with the incident light polarized along the axis of each dimer (both for the dark-field scattering and SERS measurements). The maximum SERS EF for Dimer II (with the 2.0 Å gap) is found to be 2.4×10^8, achieved at a laser wavelength of 815 nm. For Dimer III, the maximum SERS EF is 1.2×10^9, achieved at a laser wavelength of 820 nm. Our results indicate that as the gap widths enter the single-digit Angstrom range, the SERS EF and hence the field enhancement do not increase as predicted by the CEM approach, but instead decreases.

We use the QCM approach to obtain insight into these observations by simulating the electric-field intensity distributions for dimers with gap widths of 2 and 7 Å. The results are shown in Figures 10(e) and 10(f), respectively. For each of these, the intensity distribution is plotted for the illumination wavelength for which the BDP has the largest surface-averaged enhancement. We first consider the dimer with the larger gap width (7 Å). From Figure 10(e), it can be seen that only one hotspot occurs in the narrowest region of the gap, and that the maximum intensity enhancement is 7.0×10^4. This is consistent with the near-field distribution of plasmonic dimers in the classical model [27]. The situation is rather different for the dimer with the narrower gap (2 Å). The QCM simulations (Figure 10(f)) predict a significant reduction in the near-field, with the maximum intensity enhancement being 1.1×10^4. Furthermore, two hotspots occur, and the field at the gap's center is suppressed. For dimers

Figure 10. Dark-field scattering and wavelength-scanning SERS measurements on Dimers II and III, whose gap widths are 2.0 and 6.7 Å, respectively. (a) and (b) Dark-field scattering spectra measured before and after thiophenol SAM formation, along with measured wavelength-dependent SERS EF, for Dimers II and III, respectively. (c) and (d) Intensity maps of Raman scattering from Dimers II and III, respectively. 1074 cm^{-1} thiophenol Raman line is used to calculate SERS EFs. (e) Near-field intensity distribution simulated by QCM for ideal nanodisk dimer structure with gap width of 2.0 Å at its peak resonance wavelength. (f) Near-field intensity distribution simulated by QCM for ideal nanodisk dimer structure with gap width of 7 Å.

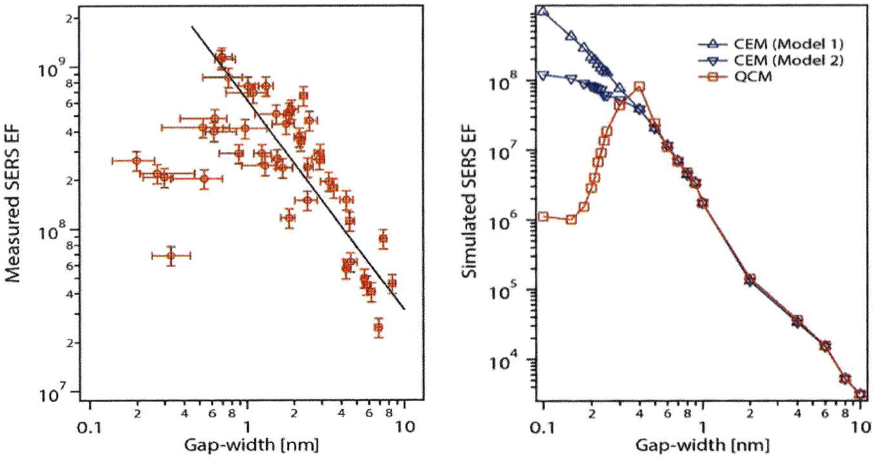

(a) (b)

Figure 11. (a) Maximum SERS EFs for 45 dimers measured by wavelength-scanned SERS, shown as red circles. The horizontal error bars represent the errors in gap width determination. The vertical error bars represents the errors in measuring SERS EFs, which is calculated from the uncertainties in reference Raman measurements for neat thiophenol. The gap widths range from (2.0±0.6) Å to (9.1±0.4) nm. The maximum SERS EF measured is $(1.2\pm0.2)\times10^9$ for Dimer III with gap width of (6.7±1.1) Å. (b) Simulated SERS EFs using both QCM and CEM for gap widths ranging from 1 Å to 10 nm.

with such small gap widths, electron tunneling becomes appreciable, providing an effective "charge-transfer" channel that reduces build-up of charges of opposite signs on the two sides of the gap. This phenomenon sets an upper limit to the achievable field enhancement and thus a limit to SERS EF.

As discussed, we perform SERS measurements on 45 different dimers. Their gap widths range from 9.1 nm down to 2.0 Å. The results are shown as Figure 11(a). For comparison, we use both the CEM and QCM approaches to predict the SERS EFs (Figure 11(b)). It can be seen that there are two regimes in Figure 11(a) that display opposite trends for the SERS EF as a function of gap width. For gap widths in the range from 10 nm to 6.7 Å, it can be seen that the SERS EFs generally increase as the gap width decreases. This is also consistent with the predictions made by the both the QCM and CEM simulations (Figure 11(b)). Let us now consider the range of gap width from 6.7 to 2.0 Å. It can be seen that the measured SERS EF actually decreases, rather than increases, as the gap width drops. Let us now consider the predictions of the CEM and QCM approaches. We use two models for the CEM simulations. In the first ("Model 1"), monolayer coverage of the thiophenol molecules on the gold surfaces is assumed. In the second ("Model 2"), we assume that the thiophenol molecules cannot access the narrowest regions of the gap. It can be seen (Figure 11(b)) that both Models 1 and 2 CEM simulations predict a monotonic increase of the SERS EF as the gap narrows, which is very different to what is revealed by experiments. The QCM simulations, on the other hand, predict a turning point at around 4 Å in the SERS EF vs. gap width characteristics, similar to what is seen experimentally. Our results show that, in addition to reducing the scattering intensity of the BDP mode, the emergence of electron tunneling reduces the SERS EF.

7. Conclusions and Perspectives

In this chapter, we have described a top-down nanofabrication approach that uses lithography to produce plasmonic dimers with gap widths down to the Angstrom range. Using dark-field scattering to measure plasmonic resonances in the far field and using SERS to measure plasmonic enhancement in the near field, we demonstrate the effect of the emergence of electron tunneling at optical frequencies. Our work demonstrates that the maximum achievable plasmonic enhancement is limited by electron tunneling. A long-held assumption in SERS has been that smaller gaps in metallic dimers are always beneficial for field enhancement. Our work runs counter to this.

In conclusion, we note that our work is part of growing research field concerned with quantum mechanical effects in plasmonic structures with gaps below 1 nm [28]. Interest in this topic is motivated by the importance of field enhancement for many plasmonic applications and thus by the curiosity of many researchers regarding its fundamental limits. In addition, possibilities for the control of electron transport afforded by sub-nanometer gaps present a means for tuning the plasmon resonances of a structure [29]. There are also exciting prospects for nanoscale light sources via the coupling between tunneling electrons and photons [30, 31]. In our study, we use electron beam lithography to produce sub-nanometer gaps. Some, other groups have developed innovative alternatives to achieve the same goal. Savage *et al.* demonstrated gap widths in the tunneling regime by pushing gold-coated atomic force microscope (AFM) cantilever tips together, with physical contact indicated by the dc tunneling conductance G exceeding the quantum conductance $G_0 = 2e^2/h$ (where h is Planck's constant and e is the elementary charge). Scholl *et al.* [7, 32] used electron beams to manipulate plasmonic nanoparticles with sizes up to ~25 nm via Coulomb forces. Sub-nanometer gaps have also been achieved via bottom-up approaches based on inorganic 2D van der Waals spacer materials [33] and using small surfactant molecules [34–37]. In our study, we monitor plasmon resonances in the far field using dark-field scattering spectroscopy. Other groups have also employed this method [6, 11], while others have used electron energy loss spectroscopy (EELS) [7, 32]. In our study, we quantify near-field enhancement using SERS. Some, others have employed photoluminescence [38], four-wave mixing [39] and third harmonic generation [40] toward this goal. Despite recent advances in this field, there is much more to be done. There are many fundamental questions to be explored and many potential applications awaiting realization. We thus anticipate that quantum mechanical effects in plasmonic structures with gaps below 1 nm will be a fertile research topic in the years to come.

Acknowledgments

K.B.C. acknowledges funding from the Australian Research Council's Discovery Projects (project number DP150103736) and Future Fellowship (project number FT140100577) funding schemes, and also funding from the Victorian Endowment for Science, Knowledge and Innovation (VESKI).

References

1. Kneipp, K., Wang, Y., Kneipp, H., Perelman, L. T., Itzkan, I., Dasari, R. R. and Feld, M. S. (1997). Single molecule detection using surface-enhanced Raman scattering (SERS). *Phys. Rev. Lett.* **78**, pp. 1667–1670.

2. Nie, S. and Emory, S. R. (1997). Probing single molecules and single nanoparticles by surface-enhanced Raman scattering. *Science* **275**, pp. 1102–1106.
3. Xu, H., Bjerneld, E., Kall, M. and Borjesson, L. (1999). Spectroscopy of single hemoglobin molecules by surface enhanced Raman scattering. *Phys. Rev. Lett.* **83**, pp. 4357–4360.
4. Zuloaga, J., Prodan, E. and Nordlander, P. (2009). Quantum description of the plasmon resonances of a nanoparticle dimer. *Nano Lett.* **9**, pp. 887–891.
5. Esteban, R., Borisov, A. G., Nordlander, P. and Aizpurua, J. (2012). Bridging quantum and classical plasmonics with a quantum-corrected model. *Nat. Commun.* **3**, p. 825.
6. Savage, K. J., Hawkeye, M. M., Esteban, R., Borisov, A. G., Aizpurua, J. and Baumberg, J. J. (2012). Revealing the quantum regime in tunnelling plasmonics. *Nature* **491**, pp. 574–577.
7. Scholl, J. A., García-Etxarri, A., Koh, A. L. and Dionne, J. A. (2013). Observation of quantum tunneling between two plasmonic nanoparticles. *Nano Lett.* **13**, pp. 564–569.
8. Kadkhodazadeh, S., Wagner, J. B., Kneipp, H. and Kneipp, K. (2013). Coexistence of classical and quantum plasmonics in large plasmonic structures with subnanometer gaps. *Appl. Phys. Lett.* **103**, p. 083103.
9. Tame, M. S., McEnery, K. R., Özdemir, Ş. K., Lee, J., Maier, S. A. and Kim, M. S. (2013). Quantum plasmonics. *Nat. Phys.* **9**, pp. 329–340.
10. McMahon, J. M., Gray, S. K. and Schatz, G. C. (2010). Optical properties of nanowire dimers with a spatially nonlocal dielectric function. *Nano Lett.* **10**, pp. 3473–3481.
11. Ciracì, C., Hill, R. T., Mock, J. J., Urzhumov, Y., Fernández-Domínguez, A. I., Maier, S. A., Pendry, J. B., Chilkoti, A. and Smith, D. R. (2012). Probing the ultimate limits of plasmonic enhancement. *Science* **337**, pp. 1072–1074.
12. Toscano, G., Raza, S., Jauho, A., Mortensen, N. A. and Wubs, M. (2012). Modified field enhancement and extinction by plasmonic nanowire dimers due to nonlocal response. *Opt. Express* **20**, pp. 4176–4188.
13. Luo, Y., Fernandez-Dominguez, A. I., Wiener, A., Maier, S. A. and Pendry, J. B. (2013). Surface plasmons and nonlocality: A simple model. *Phys. Rev. Lett.* **111**, p. 093901.
14. Zhu, W. and Crozier, K. B. (2014). Quantum mechanical limit to plasmonic enhancement as observed by surface-enhanced Raman scattering. *Nat. Commun.* **5**, p. 5228.
15. Félidj, N., Aubard, J., Lévi, G., Krenn, J. R., Salerno, M., Schider, G., Lamprecht, B., Leitner, A. and Aussenegg, F. (2002). Controlling the optical response of regular arrays of gold particles for surface-enhanced Raman scattering. *Phys. Rev. B* **65**, p. 075419.
16. Clark, A. W. and Cooper, J. M. (2010). Optical properties of multiple-split nanophotonic ring antennae. *Adv. Mater.* **22**, pp. 4025–4029.
17. Fischbein, M. D. and Drndic, M. (2006). Nanogaps by direct lithography for high resolution imaging and electronic characterization of nanostructures. *Appl. Phys. Lett.* **88**, p. 063116.

18. Zhu, W., Banaee, M. G., Wang, D., Chu, Y. and Crozier, K. B. (2011). Lithographically fabricated optical antennas with gaps well below 10 nm. *Small* **7**, pp. 1761–1766.
19. Manfrinato, V. R., Zhang, L., Su, D., Duan, H., Hobbs, R. G., Stach, E. and Berggren, K. K. (2013). Resolution limits of electron-beam lithography toward the atomic scale. *Nano Lett.* **13**, pp. 1555–1558.
20. Liu, N., Guo, H., Fu, L., Kaiser, S., Schweizer, H. and Giessen, H. (2008). Three-dimensional photonic metamaterials at optical frequencies. *Nat. Mater.* **7**, pp. 31–37.
21. Kern, J., Großmann, S., Tarakina, N. V., Ha, T., Emmerling, M., Kamp, M., Huang, J., Biagioni, P., Prangsma, J. C. and Hecht, B. (2012). Atomic-scale confinement of resonant optical fields. *Nano Lett.* **12**, pp. 5504–5509.
22. Johnson, P. B. and Christy, R. W. (1972). Optical constants of the noble metals. *Phys. Rev. B* **6**, pp. 4370–4379.
23. Nordlander, P., Oubre, C., Prodan, E., Li, K. and Stockman, M. I. (2004). Plasmon hybridization in nanoparticle dimers. *Nano Lett.* **4**, pp. 899–903.
24. Wu, L., Duan, H., Bai, P., Bosman, M., Yang, J. K. W. and Li, E. (2013). Fowler-Nordheim tunneling induced charge transfer plasmons between nearly touching nanoparticles. *ACS Nano* **7**, pp. 707–716.
25. Marinica, D. C., Kazansky, A. K., Nordlander, P., Aizpurua, J. and Borisov, A. G. (2012). Quantum plasmonics: Nonlinear effects in the field enhancement of a plasmonic nanoparticle dimer. *Nano Lett.* **12**, pp. 1333–1339.
26. Kleinman, S. L., Sharma, B., Blaber, M. G., Henry, A.-I., Valley, N., Freeman, R. G., Natan, M. J., Schatz, G. C. and Van Duyne, R. P. (2013). Structure enhancement factor relationships in single gold nanoantennas by surface-enhanced Raman excitation spectroscopy. *J. Am. Chem. Soc.* **135**, pp. 301–308.
27. Mühlschlegel, P., Eisler, H.-J., Martin, O. J. F., Hecht, B. and Pohl, D. W. (2005). Resonant optical antennas. *Science* **308**, pp. 1607–1609.
28. Zhu, W., Esteban, R., Borisov, A. G., Baumberg, J. J., Nordlander, P., Lezec, H. J., Aizpurua, J. and Crozier, K. B. (2016). Quantum mechanical effects in plasmonic structures with subnanometre gaps. *Nat. Commun.* **7**, p. 11495.
29. Benz, F., Tserkezis, C., Herrmann, L. O., de Nijs, B., Sanders, A., Sigle, D. O., Pukenas, L., Evans, S. D., Aizpurua, J. and Baumberg J. J. (2015). Nanooptics of molecular-shunted plasmonic nanojunctions. *Nano Lett.* **15**, pp. 669–674.
30. Lambe, J. and McCarthy, S. (1976). Light emission from inelastic electron tunneling. *Phys. Rev. Lett.* **37**, pp. 923–925.
31. Kern, J., Kullock, R., Prangsma, J., Emmerling, M., Kamp, M. and Hecht, B. (2015). Electrically driven optical antennas. *Nat. Photon.* **9**, pp. 582–586.
32. Scholl, J. A., Garcia-Etxarri, A., Aguirregabiria, G., Esteban, R., Narayan, T. C. Leen, A. L., Aizpurua, J. and Dionne, J. A. (2016). Evolution of plasmonic metamolecule modes in the quantum tunneling regime. *ACS Nano* **10**, pp. 1346–1354.

33. Mertens, J., Eiden, A. L., Sigle, D. O., Huang, F., Lombardo, A., Sun, Z., Sundaram, R. S., Colli, A., Tserkezis, C., Aizpurua, J., Milana, S., Ferrari, A. C. and Baumberg, J. J. (2013). Controlling subnanometer gaps in plasmonic dimers using graphene. *Nano Lett.* **13**, pp. 5033–5038.
34. Taylor, R. W., Lee, T.-C., Scherman, O. A., Esteban, R., Aizpurua, J., Huang, F. M., Baumberg, J. J. and Mahajan, S. (2011). Precise subnanometer plasmonic junctions for SERS within gold nanoparticle assemblies using Cucurbit[n]uril 'Glue'. *ACS Nano* **5**(5), pp. 3878–3887.
35. Bochterle, J., Neubrech, F., Nagao, T. and Pucci, A. (2012). Angstrom-scale distance dependence of antenna-enhanced vibrational signals. *ACS Nano* **6**, pp. 10917–10923.
36. Yu, X., Lei, D. Y., Amin, F., Hartmann, R., Acuna, G. P., Guerrero-Martínez, A., Maier, S. A., Tinnefeld, P., Carregal-Romero, S. and Parak, W. J. (2013). Distance control in-between plasmonic nanoparticles via biological and polymeric spacers. *Nano Today* **8**, pp. 480–493.
37. Tan, S. F., Wu, L., Yang, J. K. W., Bai, P. Bosman, M. and Nijhuis, C. A. (2014). Quantum plasmon resonances controlled by molecular tunnel junctions. *Science* **343**, pp. 1496–1499.
38. Kravtsov, V., Berweger, S., Atkin, J. M. and Raschke, M. B. (2014). Control of plasmon emission and dynamics at the transition from classical to quantum coupling. *Nano Lett.* **14**, pp. 5270–5275.
39. Danckwerts, M. and Novotny, L. (2007). Optical frequency mixing at coupled gold nanoparticles, *Phys. Rev. Lett.* **98**, p. 026104.
40. Hajisalem, G., Nezami, M. S. and Gordon, R. (2014). Probing the quantum tunneling limit of plasmonic enhancement by third harmonic generation. *Nano Lett.* **14**, pp. 6651–6654.

Chapter 7

Hierarchical Porous Plasmonic Nanostructures as New SERS Substrates with Ultra-High Reproducibility and Sensitivity

Dang Yuan Lei

Department of Applied Physics,
The Hong Kong Polytechnic University,
Hong Kong, China
Shenzhen Research Institute,
The Hong Kong Polytechnic University,
Shenzhen, China
dylei@polyu.edu.hk

1. Introduction

Surface-enhanced Raman scattering (SERS) has recently emerged as a versatile vibrational spectroscopic technique for ultrasensitive sensing and imaging with fascinating capability at the single-molecule level [1–3]. The technique can find a wide range of applications potentially in biochemical sensing [4–6], biomedical imaging [7–9], environmental monitoring [10–12] and food safety [13]. In principle, successful implementation of ultrasensitive and reliable SERS-based molecular detection relies on sophisticated design and preparation of metallic nanostructures that simultaneously exhibit strong plasmonic field enhancement, numerous molecular bonding sites, and ultra-high structural stability and signal reproducibility. Realization of these critical features in single plasmonic nanostructures can possibly be achieved by rationally devising and preparing hierarchically arranged 3D multiscale architectures with each structure order possessing a specific functionality [14, 15].

In general, it is believed that periodic arrangement of individual elements, in the form of periodic nanopore arrays [16–20] or ordered nanoparticle arrays [21–23], can produce excellent uniformity and high reproducibility both in structural property and SERS signal. However, sophisticated and costly electron-beam lithography (EBL) or interference lithography often have to be used to make such periodic nanostructures, with the smallest possible feature size on the order of about 10 nm for EBL-made structures and very limited Raman active surface areas [24]. To overcome these limitations and the 2D planar nature inherent to lithography techniques, a great deal of efforts have been made to fabricate 3D SERS substrates by rationally assembling metal nanoparticles via non-lithographic methods [25, 26] or growing porous nanostructures with the use of various prefabricated templates [14, 21]. These 3D SERS substrates in principle can: (i) well preserve long-range ordered structural arrangement or quasi-periodicity, which guarantees high uniformity and reproducibility, (ii) provide larger effective surface areas for particle coverage and molecular adsorption, in comparison to traditional 2D planar substrates, by extending the structures to the third dimension, and (iii) have much larger number of stronger plasmonic "hotspots" with feature sizes down to a few nanometers (such as nanogaps and mesopores) over entire volumes.

This chapter aims at providing a comprehensive review on recent exciting developments in the design and fabrication of hierarchically arranged porous nanostructures (and their complementary structures) based SERS substrates that exhibit excellent detection sensitivity and reproducibility as well as other new functionalities. It will focus on the following three types of recently developed plasmonic nanostructures: (i) individual hierarchical porous nanoparticles, (ii) hierarchically ordered porous arrays, and (iii) hierarchical composite bimetallic core–satellite nanostructures. At the end of this chapter, a brief summary of the results discussed herein and an outlook of future development in 3D hierarchical SERS substrates will be given from the perspective of structural stability, SERS sensitivity, enhancement reproducibility, and detection limit.

2. Hierarchical Porous Nanoparticles for Enhanced SERS

In the past decade, enormous efforts have been made to develop a variety of porous metal nanostructures used as stable and reproducible SERS substrates, including dealloyed nanoporous metal films [27–33], metal nanoparticles decorated photonic crystal fibers [34] and porous alumina membranes [35–40]. These nanostructures share a same feature, that is, single-scale or non-hierarchical architectures in which either nano and mesopores or nanoparticles are uniformly distributed through entire macroscopic bodies, giving rise to large surface areas and abundant inherent plasmonic hotspots. Since the detailed preparation approaches, characterization methods and their unique properties (mechanical, optical, and electrocatalytical) of dealloyed nano and mesoporous metal films, as well as their versatile capabilities in various applications (including SERS) have been comprehensively documented in the literature [27–40], in this section, we will focus our attention on hierarchical dealloyed nanoparticles dressed with numerous mesopores, which have recently been demonstrated to have extraordinary SERS response due to the presence of double-scale plasmonic hotspots and molecular bonding sites.

A variety of dealloying methods, including selective chemical and electrochemical dealloying, have been developed to fabricate various nano and mesoporous metals by leaching a less-stable metal from an alloy. For example, porous gold films with average pore size varying from 5 to 700 nm were fabricated through selective dissolution of Ag from $Ag_{70}Au_{30}$ alloys in HNO_3 [27, 28]; porous copper films were prepared by dealloying either $CuAl_2$ in NaOH [41], and Cu–Mn alloy in sodium chloride electrolyte [42], or $Cu_{30}Mn_{70}$ in hydrogen chloride [29], and porous silver films were obtained by dealloying $Ag_{30}Al_{70}$ in hydrogen chloride [30]. Very recently, the dealloying methods have been modified and applied to noble metal nanoparticles

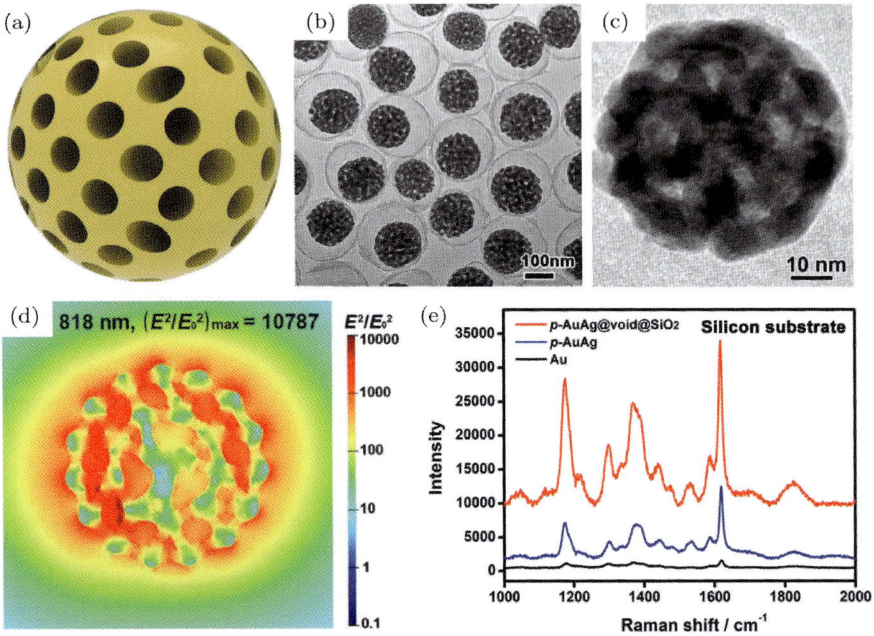

Figure 1. (a) Schematic diagram of a porous metal nanosphere. (b) Low-magnification TEM image of silica-shell-encased porous Au–Ag bimetallic nanospheres. (c) High-magnification TEM image of porous Au–Ag nanoparticles. (d) Simulated electric near-field distribution in a porous Au nanosphere of 60 nm diameter, irradiated by an incident plane wave at wavelength 818 nm, corresponding to the localized surface plasmon resonance of the nanosphere. (e) SERS spectra of crystal violet (10^{-6} M) recorded from silicon substrates deposited with porous Au–Ag alloy@void@SiO_2 nanoparticles (red), porous Ag–Ag nanoparticles (blue), and solid Au nanoparticles (black). The images (b)–(e) are adapted from Ref. [47].

to produce mesopores-incorporated nanoparticles [43–48], which can support both cascaded electromagnetic field enhancement and efficient molecular bonding sites.

Figure 1(a) shows the schematic diagram of a porous nanoparticle that has a total dimension of a few hundreds of nanometers, dressed with abundant pores of a few nanometers. A similar porous architecture has recently been realized in experiment by producing porous Au–Ag alloy nanoparticles encased in ultra-thin hollow silica shells (see Figures 1(b) and 1(c)), with an approach combining successive annealing, etching and dealloying processes [47]. Briefly, Au@Ag@SiO_2 core–shell–shell nanoparticles were prepared by stoichiometric seeded growth of Ag on Au nanoparticles, followed by silica layer coating and high-temperature annealing at 950°C, resulting in fully alloyed Au–Ag nanoparticles. Due to annealing-induced silicate condensation and collapse of their intrinsic micropores, the silica shell was dramatically densified. Consequently, layer-by-layer stripping of the silicate species was achieved by etching the Au–Ag alloy@SiO_2 nanoparticles in an alkaline solution, leading to an ultrathin layer of silica (< 5 nm, see Figure 1(b)). Finally, the structures were dealloyed with concentrated nitric acid to partially dissolve Ag from the alloy nanoparticles, resulting in the porous Au-rich Au–Ag alloy nanoparticles (see Figure 1(c)).

Figure 1(d) shows the numerically simulated electric near-field distribution in a porous Au nanosphere of 60 nm diameter that is used as a model system to unravel the electromagnetic enhancement in the system. As inherited from its hierarchical structure, the porous Au sphere exhibits a double-scale field enhancement, namely, nanoscale field localization at the surface of the nanosphere and mesoscale field concentration inside the pores. This cascaded field enhancement mechanism is distinctive from that in a solid metal nanoparticle in which the electric near-field intensity is amplified only at the surface of the nanoparticle, which limits the accessible areas of plasmonic hotspots and molecular bonding sites. With this unique field enhancement mechanism, the mesoporous Au–Ag alloyed nanoparticles demonstrate significantly improved SERS response in comparison to conventional solid Au nanoparticles of the same size and concentration (see Figure 1(e)). More remarkably, encasing the porous nanoparticles in ultra-thin porous silica shells (to maintain their colloidal stabilization and avoid the use of conventional organic capping ligands) further increases the SERS intensity by more than 2.5 fold, illustrating the critical roles of the mesopores and the clean surface in affording optimal SERS activities [47].

Combining the advantages of both top-down lithographic patterning and bottom-up atomic dealloying, Shih *et al.* have fabricated hierarchical nanoporous gold disks on silicon and silica substrates, a

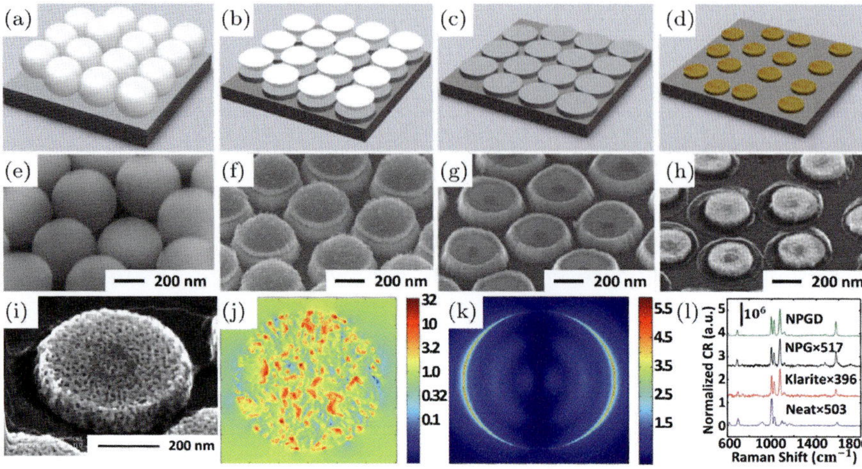

Figure 2. (a)–(d) Sketch of the preparation procedures for porous gold nanodisks and (e)–(h) corresponding SEM images (oblique view) captured after each fabrication step. (i) A high-magnification SEM image of a single nanoporous gold disk of diameter 500 nm. The scale bars in (e)–(i) are 200 nm. (j) and (k) Simulated electric near-field distributions for (j) a porous gold disk and (k) a solid disk both with 500 nm diameter, under plane wave illumination at 785 nm. (l) SERS spectra measured for benzenethiol monolayers self-assembled on a nanoporous gold disk array (green), a porous gold film (black), and a commercial Klarite substrate (red). The normal Raman spectrum of neat benzenethiol solution is shown for comparison (blue). The images (a)–(k) are adapted from Ref. [45]. The image (l) is adapted from Ref. [44].

unique class of metal nanostructures with tunable localized surface plasmon resonance, high-density plasmonic hotspots, and large surface areas [44–46, 48], enabling improved SERS response, enhanced plasmonic refractive index sensing and near-infrared absorption spectroscopy. Figures 2(a)–2(h) show the procedures for preparing such porous gold disks and the SEM images taken after each process. Briefly, a 120-nm thick film of Ag–Au alloy was sputtered either on a silicon or silica substrate, followed with a self-assembled, close-packed monolayer of polystyrene beads (460–1100 nm diameter) atop the film (see Figures 2(a) and 2(e)), which was then exposed to oxygen plasma to shrink the beads and make neighboring beads separated (see Figures 2(b) and 2(f)). The sample was subsequently sputter-etched to transfer the bead pattern into the alloy film, and the beads were then removed by selective dissolution in chloroform (see Figures 2(c) and 2(g)). By dealloying the disks in concentrated nitric acid and rinsing in deionized water, an array of porous gold nanodisks on silicon wafer was obtained (see Figures 2(d) and 2(h)). Suspended colloidal nanostructures can also be produced by sonicating high-density porous nanodisks on silicon wafer [45]. Figure 2(i) shows the high-resolution SEM image of an individual nanoporous disk, which is decorated with a large number of pores with diameter around 10 nm.

Full-wave electromagnetic simulations show that, in comparison to a conventional solid disk, the unique porous structure of the prepared disk can generate uniformly distributed plasmonic hotspots of much larger electric-field enhancement (~6-fold) over the entire disk (see Figures 2(j) and 2(k)). Such extraordinary field enhancement in the nanoporous gold disks is expected to produce improved SERS performance. Figure 2(l) compares the SERS performance of several typical plasmonic substrates and shows that the SERS intensity of benzenethiol monolayers self-assembled on an array of nanoporous gold disks can be boosted up by more than two orders of magnitude compared with a nanoporous gold film, a commercial Klarite substrate as well as neat benzenethiol solution. Note that the SERS enhancement of the nanoporous gold disks can be further increased by adjusting the localized surface plasmon resonance to match the Raman excitation wavelength and increasing the surface density of mesopores.

3. Hierarchical Porous Plasmonic Metamaterial for Reproducible SERS

As mentioned in the Introduction, achieving reproducible SERS response is as important as maximizing SERS sensitivity because it is of critical importance for reliable quantification of a biochemical analyte with a SERS-based biosensor. Assembling hierarchical porous nanostructures in (quasi)periodic arrangement can in principle simultaneously preserve their extraordinary SERS sensitivity and also achieve ultra-high SERS reproducibility. This elegant design concept has recently been realized for the first time by Zhang and his co-workers by fabricating novel 3D plasmonic metamaterial-based SERS substrates, which are hierarchically ordered porous gold membranes comprising densely-packed nanohole channels and uniformly distributed mesopores over the entire membrane [15]. The periodic nanoholes in the plasmonic metamaterial are expected to efficiently harvest the incident light through excitation of their localized surface plasmons, concentrating the captured energy at the edges of the nanoholes and therefore generating strong electromagnetic field enhancement over the whole surface; the incorporation of abundant mesopores in the nanoholes can produce a unique picture of plasmon modes and generate a cascaded electromagnetic field enhancement, thereby resulting in strong Raman-active sites over the entire structure. Additionally, the mesopores also function as effective binding sites for the adsorption of analyte molecules at the hotspots. More importantly, the periodic arrangement of both nanopores and mesopores endows the plasmonic metamaterial with ultra-high SERS reproducibility from spot to spot in the same substrate and uniform signal response from sample to sample.

Figure 3(a) sketches the fabrication recipe for the hierarchically ordered porous plasmonic metamaterial proposed by Zhang and his co-workers [15]. It employs a delicate dual-templating method utilizing a reverse porous PMMA, cast from an anodic aluminum oxide array, as a sacrifice, hard template and non-ionic surfactant $C_{16}EO_8$ as lyotropic liquid-crystal precursor to form hexagonal mesophases in the nanopores of the PMMA template (steps 1–3 in Figure 3(a)). Subsequent electrochemical deposition of gold and final removal of the PMMA template produce a truly hierarchical plasmonic metamaterial consisting of periodic nanoholes decorated with enormous mesopores through the entire structure (steps 4 and 5), with these surface and bulk morphological features unambiguously revealed in the low- and high-magnification SEM and TEM images at different views (Figures 3(b)–3(g)). The top and cross-sectional views of the SEM images in Figures 3(b)–3(e) demonstrate that the plasmonic metamaterial preserves the structural dimensions and morphologies of the alumina template, exhibiting an extremely uniform and highly ordered hexagonal array of nanoholes. The edge-to-edge distance of

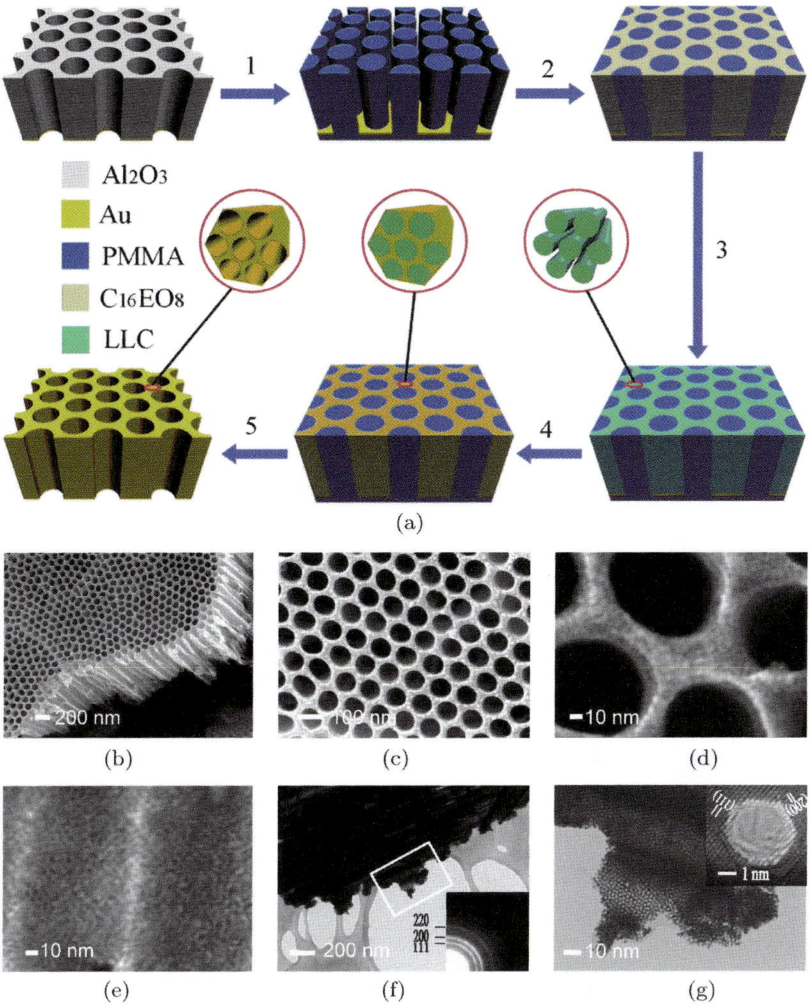

Figure 3. (a) Schematic sketch of the fabrication procedures for hierarchically ordered plasmonic metamaterials: (1) a negative PMMA template was prepared from a porous anodic alumina, followed with (2) injection of lyotropic liquid-crystal precursor and (3) the generation of lyotropic liquid-crystal mesophases in the nanopores of the PMMA template, (4) electrochemical deposition of gold, and (5) dissolution of the PMMA template. Oblique (b) and plain (c) view SEM images of an as-prepared porous gold membrane and its highresolution oblique (d) and cross-sectional (e) view images. (f) TEM image of the membrane and its electronic diffraction pattern (inset). (g) TEM image of the region indicated in (f). The inset shows the highresolution TEM image of an individual mesopore. All the images are adapted from Ref. [15].

neighboring nanoholes and the nanohole diameter are estimated from Figure 3(d) to be 25 and 80 nm, respectively. The cross-sectional views of the TEM images shown in Figures 3(f) and 3(g) clearly uncover the mesoporous feature of the metamaterial, demonstrating well-organized mesopores of diameter ranging from 2 to 3 nm. The bright concentric rings observed in the electron diffraction pattern of the enclosed region in Figure 3(f) can be indexed as the face-centered cubic lattice structure of gold (inset of Figure 3(f)). The high-magnification TEM image of an individual mesopore (inset of Figure 3(g)) shows two lattice spacings of 0.23 and 0.20 nm, which correspond to the {110} and {111} atomic planes of gold, respectively.

To demonstrate the improved SERS performance of the prepared porous plasmonic metamaterial substrate, Figure 4 compares the SERS spectra of benzenethiol (used as a non-resonant Raman analyte) chemisorbed on four different substrates, including a commercial Klarite substrate (1.6 μm width, 0.8 μm depth, and 0.2 μm wall thickness), a mesoporous gold film, a nanohole gold membrane, and the hierarchical porous gold metamaterial [15]. The geometric parameters of the mesoporous film and the nanohole membrane inherit those respective values of the hierarchical porous metamaterial. As can be seen from the SERS spectra, the intensities of the marked characteristic Raman peaks of benzenethiol increase from the Klarite substrate, the mesoporous film and the nanohole membrane to the plasmonic porous metamaterial, thereby substantially corroborating the critical importance of the hierarchical structure that enables a synergetic cascaded field enhancement mechanism.

The above-mentioned dual-enhancement mechanism is further confirmed by comparing the plasmonic electric-field distribution in the three fabricated substrates as shown in Figure 4(b) [15]. In the simulations, the dielectric response of gold was modeled by the measured permittivities [49] that were used for the two single-scale structures (mesoporous film and nanohole membrane). Considering the deep sub-wavelength character and uniform distribution of the mesopores, the dielectric response of the porous gold metamaterial was modeled in the framework of effective medium theory developed by

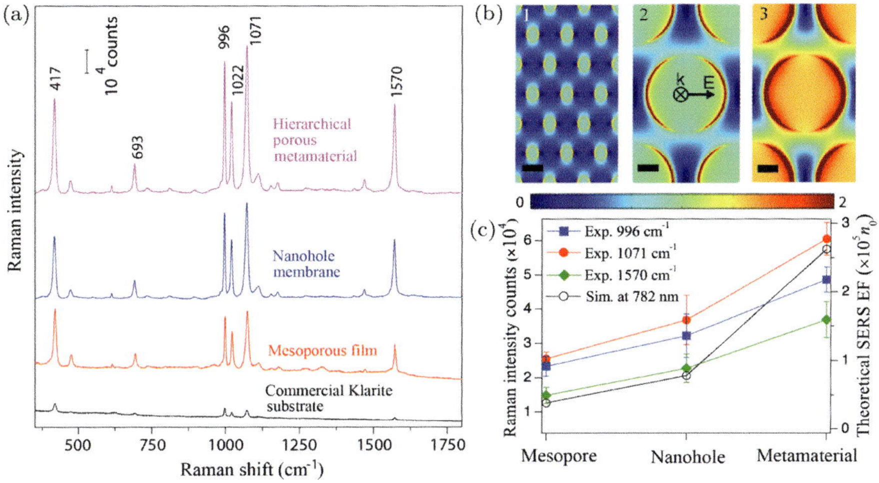

Figure 4. (a) Measured SERS spectra of benzenethiol chemisorbed on a Klarite substrate (black), a mesoporous gold film (red), a gold nanohole membrane (blue), and a hierarchical porous gold metamaterial (pink). The Raman spectra were captured over a $1\,\mu m^2$ area of the samples illuminated at wavelength 782 nm. (b) Simulated normalized electric field (E-field) intensity distribution profiles for the mesoporous film (b1), the nanohole membrane (b2), and the hierarchical metamaterial (b3). The incident direction and polarization (k and E) are sketched in the inset of (b2). The black scale bars are 6, 20, and 20 nm from (b1) to (b3), respectively. (c) Right axis: Raman peak intensities at 996 (squares), 1071 (circles), and $1570\,cm^{-1}$ (diamonds); left axis: simulated average electromagnetic enhancement factors (EF, open circles) at 782 nm for the three structures in (b). The error bars represent the standard deviation of five SERS measurements. n_0 in the left axis is a virtual line density of molecules chemisorbed along the circumferences of the mesopores and nanoholes, and is introduced to normalize the theoretical SERS EF. All the images are adapted from Ref. [15] with color modifications by the authors.

Bruggeman [50]. As can be seen from the results, the electric field strength in the mesoporous film is much weaker than that in the nanohole structure though the former has a larger density of plasmonic hotspots than the latter. This comparison can be understood by the fact that the subwavelength nanoholes can support efficient excitation of localized surface plasmons while the mesopores of a few nanometers are too small to sustain intense plasmon resonances. More remarkably, synergetic combination of ordered mesopores and periodic nanoholes produces a fascinating dual-scale plasmon mode and further increases the electromagnetic enhancement both at the edges of the nanoholes and at the areas between adjacent nanoholes. Here, the plasmonic near-fields undergo a two-step amplification process, where the incident light is effectively collected by the subwavelength nanoholes and the induced evanescent plasmonic fields are further concentrated by the mesopores. Note that the second-step amplification was naturally included in the simulations by the effective permittivity of the porous gold metamaterial. With the calculated electric field maps, the SERS enhancement factors (EF) for the three structures can be calculated with normalization to the incident field intensity and the amount of adsorbed Raman molecules, and the results in Figure 4(c) show good agreement with the trend of the measured Raman intensities.

The SERS reproducibility of the porous plasmonic metamaterial was quantified by determining the spot-to-spot variation and substrate-to-substrate variation in the Raman intensities. As shown in Figure 5(a), the average signal intensity for such typical metamaterial substrate is ca. 62,200 with a spot-to-spot variation coefficient of 21%, while the Klarite substrate shows an average intensity of ca. 4350 with a single variation coefficient of 45%, indicating a significantly higher SERS sensitivity and much better signal reproducibility in the metamaterial SERS substrate. In addition, Figure 5(b) shows that the six hierarchical porous metamaterials with the same nominal geometric parameters and loaded with the same concentration benzenethiol show similar SERS signal intensities, demonstrating remarkable substrate-to-substrate SERS reproducibility. Finally, the detection limit of the porous gold metamaterial is determined by exposing six identical substrates, respectively, to benzenethiol with concentrations varying from 10^{-13} to 10^{-8} M. Figure 5(c) shows that the Raman signal intensities at the four characteristic peaks of benzenethiol diminishes with decreasing loading concentration, following the first-order adsorption kinetics. More interestingly, the log–log plot of SERS intensity and loading concentration gives rise to a slope of 0.322 ± 0.002, suggesting that the hierarchical porous plasmonic metamaterial substrate can provide sufficient active binding sites for the efficient adsorption of the benzenethiol molecules

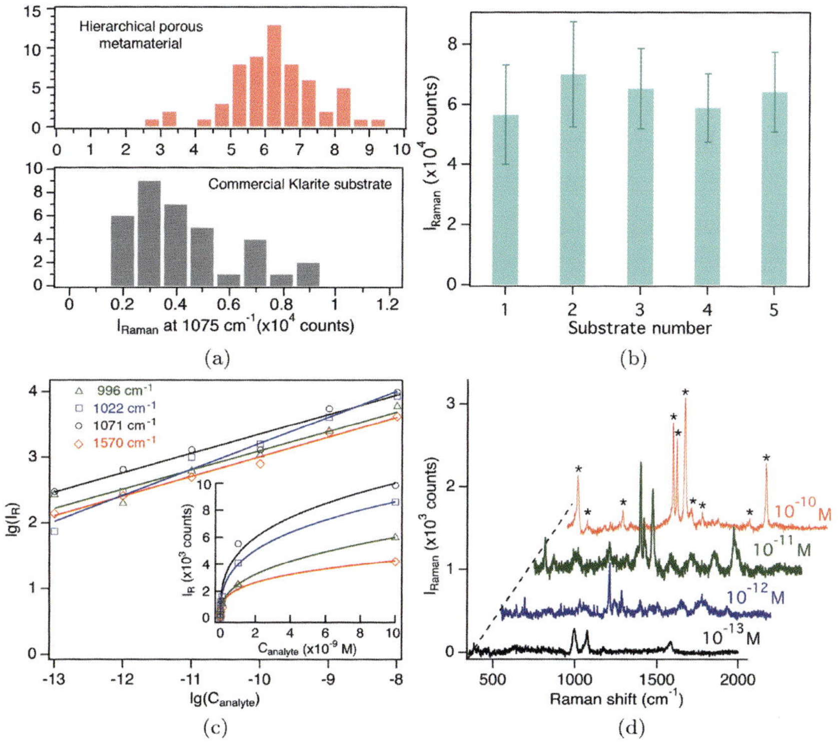

Figure 5. (a) Spot-to-spot Raman intensity variation at the Raman peak $1071\,cm^{-1}$ of benzenethiol adsorbed respectively on the porous gold metamaterial (top) and the Klarite substrate (bottom). (b) Raman signal intensities at the peak $1071\,cm^{-1}$ of benzenethiol measured for six identical substrates. Here, each intensity value (error bar) is the average (standard deviation) of 12 independently measurement results at different spots. (c) Measured Raman signal intensities at the peaks 996 (triangles), 1022 (squares), 1071 (circles), and $1570\,cm^{-1}$ (diamonds) for benzenethiol adsorbed on the gold porous metamaterials as a function of benzenethiol concentration. The inset plots the same set of results for solution concentration varying from 10^{-13} to 10^{-8} M. The symbols are the experimental data points and the solid lines are the fitted curves. (d) Measured SERS spectra of the gold porous metamaterials loaded with benzenethiol of concentration varying from 10^{-13} to 10^{-10} M. All the images are adapted from Ref. [15] with color modifications by the authors.

from the loading solution. Figure 5(d) further shows that the characteristic Raman peaks of benzenethiol at 996, 1071, and $1570\,cm^{-1}$ can still be observed even at a concentration of 10^{-13} M, with the signal-to-noise ratios of 3.3, 3.2 and 1.7 for the three peaks, hence leading to a detection limit of 10^{-13} M for the non-resonant benzenethiol Raman analyte.

4. Hierarchical Metal Nanostructures with Plasmonic Focusing Effect for Improved SERS

Similar to the hierarchically ordered porous plasmonic metamaterial discussed above, a self-similar plasmonic chain consisting of several metal nanospheres with progressively decreasing size and separation can develop a plasmonic nanofocus (electromagnetically hottest spot) in the gap between the smallest nanospheres, just like what a plasmonic nanolens would do. Benefiting from the multiplicative, cascaded effect of its geometry, it was theoretically predicted under the quasi-static approximation that the local field strength at the hottest spot can be enhanced by orders of magnitude in the near-ultraviolet region [51]. Briefly, in such a self-similar silver nanosphere chain, the plasmonic near fields produced by illuminating large spheres can subsequently excite smaller ones through near-field coupling, forming cascaded plasmonic excitations and field amplification. Self-similarity was assumed in their theory by scaling both nanosphere diameter and gap separation with a constant factor much smaller than unity. For instance, in a plasmonic chain comprising three silver nanospheres of radii 45, 15 and 5 nm and interparticle gaps 9 and 3 nm, the local field enhancement in the proximity of the third nanosphere is estimated to be $\sim 10^3$ as shown in Figure 6(a), which may lead to an SERS enhancement factor of 10^{12} when a Raman analyte molecule is positioned at the hottest spot. Although this theoretical field enhancement factor was argued to be overestimated due to the ignorance of spatial dispersion and Laudau damping [52, 53], the geometrically self-similar nanosphere chains have nevertheless provided an excellent model system for achieving extremely high field enhancement at nanoscale spots with their unique cascaded plasmonic excitation and field amplification mechanism.

Following the theoretical prediction, Kravets and co-workers made the first experimental realization (to the best of our knowledge) of the cascaded optical field enhancement in a self-similar nanodisk tower, which consists of three gold nanodisks of different diameters

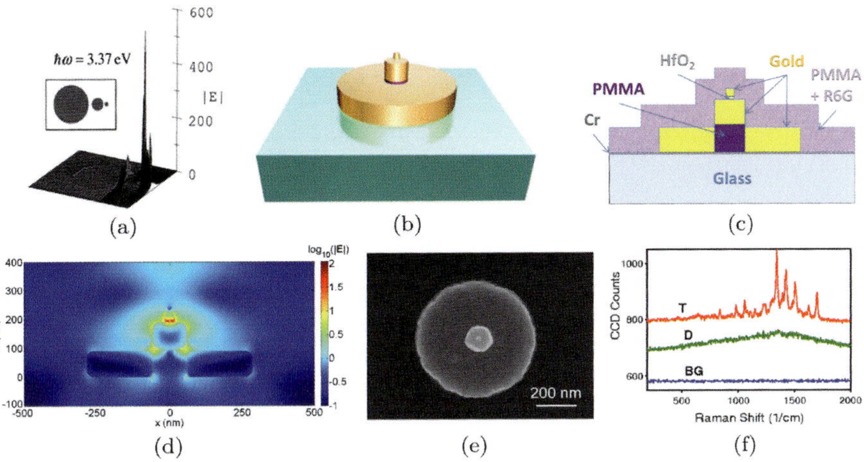

Figure 6. (a) Cascaded optical field enhancement in a plasmonic chain of three silver nanospheres of radii 45, 15 and 5 nm and interparticle gaps 9 nm and 3 nm (inset). The local field strength distribution (normalized to the excitation field) in the equatorial plane of the chain is plotted. The image (a) is adapted from Ref. [51]. (b) Illustration of a self-similar composite nanostructure, a tower-type three-level system. (c) Cross-sectional view of the composite structure coated with a PMMA/dye layer. (d) Simulated field strength (normalized to the peak field magnitude of the incident beam) in the composite structure under illumination by a Gaussian beam at 633 nm. (e) Scanning electron microscope image of the nanostructure comprising three coaxial gold disks fabricated with multistep electron beam lithography. (f) Confocal Raman spectra measured for a triple structure (T, three disks as shown in (b)), a double structure (D, without the top, smallest disk), and a bare substrate (BG). The images (b)–(f) are adapted from Ref. [54] with slight modifications.

co-axially stacked one on top of another and separated by dielectric spacers [54]. Figures 6(b) and 6(c) show the schematic illustration and the cross-sectional view of their structure, where the largest disk of diameter D has at its center a cylindrical hole of diameter d and thickness H filled with a dielectric column produced by exposed PMMA, stacked by a medium-sized gold disk of diameter d and thickness H on the top of the column and a small disk of diameter δ and thickness h placed topmost. Numerical simulations using finite-element modeling show that the local field enhancement in the three-tier tower structure under illumination by a Gaussian beam at 633 nm strongly depends on the thickness of the dielectric spacers separating the disks. Figure 6(d) plots the electric near-field distribution calculated for the experimental geometry ($D = 590\,\text{nm}$, $d = 110\,\text{nm}$, $H = 90\,\text{nm}$, $\delta = 30\,\text{nm}$ and $h = 30\,\text{nm}$), demonstrating a maximum field enhancement up to two orders of magnitude at the gap between the smallest gold disk and the medium-sized one. Figure 6(e) renders the SEM image of the fabricated self-similar composite structure with the above dimensions, which was prepared at the resolution limit and the maximum number of cascades achievable with the electron-beam lithography used in the experiment.

To carry out SERS measurements, the whole tower structure was coated with a layer of dye molecules, rhodamine 6G (R6G), uniformly dispersed in a polymer host (3% PMMA in anisole), as illustrated in Figure 6(c). The R6G concentration in the solution was 10^{-6} M, leading to an average distance of $\sim$120 nm between neighboring molecules in the PMMA film. As a result, the probability of having one R6G molecule within $\sim$30 nm diameter of the top nanodisk was about 20%, indicating that the SERS measurements were at the single-molecule level. Figure 6(f) presents the SERS spectra of R6G measured from the three-tier tower structure, a double structure, and a bare substrate. It is clearly seen that the tower structure shows strong intensity enhancement in the Raman signals while the double structure exhibits no clear characteristic Raman peaks, but a broad fluorescence background. The bare substrate shows no detectable Raman or fluorescence signal of R6G. Note that by optimizing the Raman excitation conditions such as increasing laser power and signal integration time, it was able to collect a weak SERS spectrum for the double structure [54]. Comparing the SERS signals from the triple and double nanostructures under the same excitation conditions, the ratio of the field enhancement observed for the triple and double structures (the cascaded enhancement factor) was estimated to be $\sim 3.6 \pm 0.2$ for $\lambda = 633\,\text{nm}$. This estimated field-enhancement factor is in good agreement with the theoretical factor of 3.4 ± 1.0 determined from the finite-element modeling. Comparing the fluorescence signals from the single (only the bottom disk) and double structures, the cascaded field enhancement for the double structure was estimated to be 30 ± 7 [54]. Therefore, the electric near-field enhancement factor for the triple structure was about 100 compared to the single structure.

5. Hierarchical Bimetallic Composite Nanostructures for Real-time, *in situ* SERS Monitoring of Chemical Reactions

Recently, various hierarchical bimetallic nanoparticles, complementary to the above-discussed porous nanospheres and nanodisks, have been developed and attracted a great deal of research interest in a wide range of applications. Bimetallic nanostructures made of two different metals in the core–satellite form often show better optical/chemical activity and selectivity compared to monometallic counterparts because of their synergetic effects of the two metals and can thus achieve multiple functionalities. Bimetallic core–satellite composite nanoparticles can be fabricated by successive reduction of two types of metallic precursors. A variety of synthesis methods, including seeded growth of metallic precursors and galvanic replacement reaction, have been developed to make different types of bimetallic core–satellite nanostructures. The former method is similar to that used for preparing monometallic nanoparticles, but the metallic precursors are reduced directly on the surface of the preformed particles in the presence of stabilizing agents to prevent aggregation of composite nanostructures [55, 56]. The latter method provides a simple and versatile route for producing bimetallic core–shell–satellite superstructures with controllable geometry and composition [57] and hollow nanostructures with tunable porosity and chemical composition [58, 59].

In an early study, Henglein prepared two types of Au–Pt bimetallic nanoparticles, Au core–Pt shell and Pt core–Au shell, with the successive reduction method [60]. This approach was also employed to prepare PVP-stabilized Pd–Pt bimetallic colloids [61]. The key principle of the method used in Ref. [61] is that an H bond can easily adsorb on a Pd surface to form a metal–H bond, where the H atom works as an effective reducing agent that reduces the Pt ions initially added to a preformed Pd colloidal suspension. Recently, the bimetallic Pd–Pt nanodendrites comprising densely packed Pt branches on the surface of faceted Pd nanocrystals have been fabricated by a modified seeded growth method that is capable of precisely controlling the morphology and composition of the Pd–Pt nanodendrites [62]. In these methods, stabilizing polymers, such as sodium citrate, PVP, CTAB, and PVA, play an important role: they can coordinate with metal ions such that the metal ions can be reduced slowly under mild conditions; additionally, the formation of the polymer–metal ion complex enables flexible control over the morphology of bimetallic nanoparticles by adjusting the reduction order of the two metals through their redox potentials [63].

Since bimetallic nanoparticles consist of both transition metals and noble metals, they thus possess both catalytic activity and plasmonic property, enabling the use of SERS as an optical probe for *in situ* monitoring, in a real-time manner, the process of catalytic reactions. When selecting a bimetallic core–shell nanostructure that exploits plasmonic near fields for SERS enhancement and Pd or Pt nanoparticles for catalytic purpose, the following design considerations should be addressed. (i) To maximize the SERS enhancement, the localized surface plasmon resonance of the hybrid nanostructure has to match with the excitation laser wavelength and cover the Stocks frequency range. (ii) To achieve a high catalytic efficiency, the size of the Pd/Pt nanoparticles coated on noble metal nanocrystals has to be smaller than 5 nm. (iii) To achieve an optimized balance between SERS enhancement and catalytic efficiency, the surface coverage of Pd/Pt nanoparticles on noble metal cores has to be properly controlled because large-area coating of catalytic metals on noble metals can dramatically damp the plasmonic properties of the hybrid structures.

Bao *et al.* presented a facile approach for synthesizing Au@Pt core–satellite nanostructures [56] in which the coverage density of sub-5 nm Pt nanoparticles on the surface of Au nanorods can be easily controlled (Figure 7(a)). By sophisticatedly optimizing the Pt coverage density, they demonstrated that such hybrid nanostructures could achieve unique dual functionalities: the tiny Pt nanoparticles exhibit robust catalytic activity benefiting from the large surface area to volume ratio while the Au nanorods provide dramatically enhanced local electromagnetic fields at the uncoated surface areas upon excitation of their longitudinal localized surface plasmon resonance. Note that the plasmonic response of the hybrid nanostructures gets damped gradually with increasing the surface coverage of Pt as can be seen from the measured absorption spectra and simulated electric-field distribution (Figures 7(b) and 7(c)). Taking advantage of the dual functionalities, the optimized hybrid nanostructures were used to track in real time the Pt-catalyzed 4-NTP to 4-ATP reaction

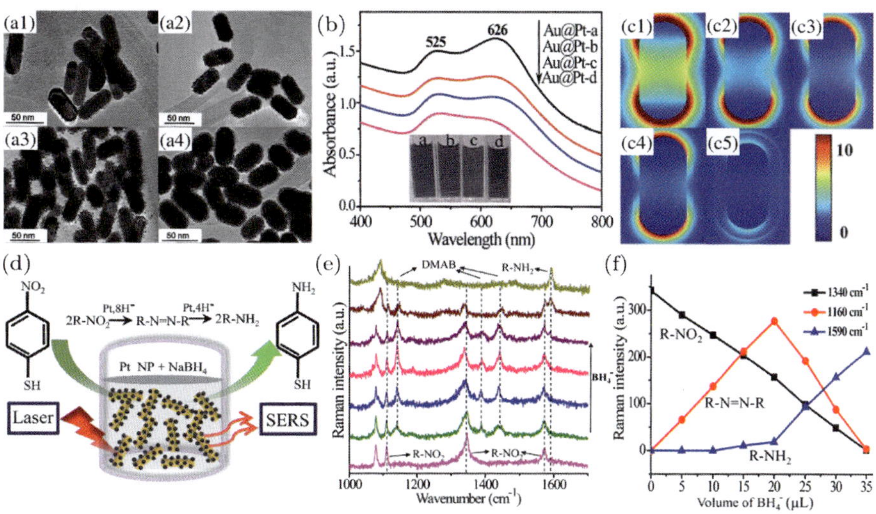

Figure 7. (a1)–(a4) TEM images of Au@Pt core–satellite nanostructures with the surface coverage of Pt nanoparticles on Au nanorods increased from (a1) to (a4). (b) Absorption spectra of the nanostructures shown in (a1)–(a4). (c1)–(c5) Cross-section views of the simulated electric-field intensity enhancement distribution in a bare Au nanorod (c1) and the Au@Pt nanostructures (c2)–(c5) shown in (a1)–(a4). In the simulations, the coverage of Pt nanoparticles was modeled with an effective thin shell of thickness 3 nm in (c2)–(c4) and 6 nm in (c5) as determined from their TEM images shown in (a1)–(a4). (d) Schematic representation of *in situ* SERS monitoring the reduction of 4-NTP to 4-ATP by the Au@Pt hybrid nanostructures in the presence of $NaBH_4$. The reaction is catalyzed by the Pt nanoparticles, while the SERS signals are boosted up by the plasmonic local electric fields on the surface of the Au nanorods. (e) SERS spectra recorded in catalyzing 4-NTP to 4-ATP by the Au@Pt nanohybrids as a function of the added amount of $NaBH_4$. (f) Raman intensities extracted from (e) at the characteristic peaks 1340, 1160 and 1590 cm^{-1}, corresponding to the groups of R–NO_2, R–N=N–R, and R–NH_2, respectively. The images (a)–(f) are adapted from Ref. [56].

by *in situ* measurement of the SERS spectra of the reactant, intermediate and final products with increasing amount of the co-catalyst $NaBH_4$ (Figure 7(d)). As can be seen from Figure 7(e), the Raman spectrum recorded at the beginning (magenta) demonstrates that the reactant is pure 4-NTP since the three Raman signature peaks at 1113, 1345 and 1572 cm^{-1}, corresponding to the group R-NO_2, are clearly observed. Upon gradually adding $NaBH_4$ to the mixed solution of 4-NTP and Au@Pt, the Raman intensities of the three peaks of R-NO_2 start to decrease and three new Raman peaks at 1140, 1385 and 1434 cm^{-1} emerge, corresponding to the vibrational modes of 4,4′-DMAB, an intermediate product of this reduction reaction. Interestingly, the intensities of the three 4,4′-DMAB Raman peaks dramatically increase with further addition of $NaBH_4$ (from green to pink) and then sharply decrease to nearly zero (from purple to dark yellow), further ascertaining its character as an intermediate product. Meanwhile, the three Raman peaks of 4-NTP also diminish and a new peak at 1590 cm^{-1} ascribed to the amino vibrational mode appears and finally dominates the spectrum. The observed variations in the Raman spectra are summarized in Figure 7(f) that plots the Raman intensities of the three characteristic peaks associated respectively with the reactant — 4-NTP, the intermediate product — 4,4′-DMAB, and the final product — 4-ATP as a function of the added amount of $NaBH_4$. The results not only reflect the reaction process but also provide useful quantitative information about the relative concentration of the three substances. The SERS-based spectroscopic technique demonstrated here can be applied to monitor other heterogeneous catalytic reactions and is suitable for studying, both qualitatively and quantitatively, the reaction dynamics of many reduction and oxidation reactions in fuel cells and catalytic water splitting.

Other variations of the above-mentioned bimetallic nanostructures, such as Au@Pd core–shell nanostructures [55], Pt nanoparticles-coated Au nanospheres [64], Au@Pt@Au core–shell–satellite nanoraspberries [57], have also been synthesized for quantitative plasmonic sensing and *in situ* SERS monitoring of Pt-catalyzed reactions. Another interesting type of bifunctional composites, Fe_3O_4@Au/Ag core–satellite nanostructures, can combine SERS activity and magnetic response in a single entity [65]. By incorporating the magnetic functionality with SERS detection, the noble metal nanoparticles-decorated magnetic microspheres can achieve dramatically enhanced SERS sensitivity by dispersing them in an analyte solution to capture the target molecules efficiently and subsequently assembling the structures in a close-packed monolayer with a magnet. Such a magnetically self-assembled monolayer substrate has two types of plasmonic hotspots, namely, the nanogaps between neighboring Au/Ag nanoparticles located at the same Fe_3O_4 microsphere and that at neighboring microspheres [66], which altogether enhance the overall SERS performance of the hybrid structures.

6. Conclusion and Outlook

This chapter summarized recent fascinating developments in the design, synthesis and application of hierarchically arranged porous nanostructures and their complementary structures for ultra-sensitive, reproducible and reliable SERS detection. It has reviewed three types of plasmonic composite nanostructures, including hierarchical porous nanoparticles, hierarchically ordered porous plasmonic metamaterial and hierarchical core–satellite composite nanostructures. With their intrinsic multiscale characters, the hierarchical porous nanoparticles and core–satellite composite nanostructures exhibit cascaded electromagnetic field enhancement that can dramatically enhance the SERS sensitivity; with the (quasi)periodic arrangement of abundant nanoholes and mesopores, the hierarchical porous plasmonic metamaterial exhibits not only an ultra-low detection limit but also, more importantly, ultra-high structural and signal reproducibility.

With regard to the future development of hierarchical porous or core–satellite composite nanostructures, it would be extremely desirable to simultaneously integrate the most important aspects of an SERS substrate, such as structural stability, SERS sensitivity, signal enhancement reproducibility, and detection limit in single plasmonic entities. With regard to experimental fabrication, it is feasible to realize such multiscale, multifunctional plasmonic nanostructures by combining cost-effective lithographic methods (for example, nanosphere lithography and interference lithography) and sophisticated wet-chemistry approaches like atomic dealloying, seeded growth and galvanic replacement. Besides the above considerations, the development of next-generation SERS substrates has to take into account the needs of real-world applications and potential customers, which often require SERS substrates with additional functionalities (for example, mechanical robustness) much beyond the conventionally concerned. This is critical to transition of the SERS detection technique from the lab to the field [67], i.e. from the gentle laboratory environment to real-world harsh environment [68], e.g. on-site and long-time direct sampling, online detection, and in-field detection of warfare agents.

Acknowledgments

DYL acknowledges the financial support from the National Natural Science Foundation of China (Grant No. 11474240) and the State Key Laboratory of Physical Chemistry of Solid Surfaces (Grant No. 201402), and thanks Dr. Jinglai Duan and Mr. Zhiyong Bao for their great help in the preparation of this chapter.

References

1. Kneipp, K., Wang, Y., Kneipp, H., Perelman, L. T., Itzkan, I., Dasari, R. R. and Feld, M. S. (1997). Single molecule detection using surface-enhanced Raman scattering (SERS). *Phys. Rev. Lett.* **78**, pp. 1667–1670.
2. Nie, S. and Emory, S. R. (1997). Probing single molecules and single nanoparticles by surface-enhanced Raman scattering. *Science* **275**, pp. 1102–1106.
3. Xu, H. X., Bjerneld, E. J., Käll, M. and Börjesson, L. (1999). Spectroscopy of single hemoglobin molecules by surface enhanced Raman scattering. *Phys. Rev. Lett.* **83**, pp. 4357–4360.
4. Kneipp, K., Kneipp, H., Itzkan, I., Dasari, R. R. and Feld, M. S. (1999). Ultrasensitive chemical analysis by Raman spectroscopy. *Chem. Rev.* **99**, pp. 2957–2976.
5. Abramczyk, H. and Brozek-Pluska, B. (2013). Raman imaging in biochemical and biomedical applications. Diagnosis and treatment of breast cancer. *Chem. Rev.* **113**, pp. 5766–5781.
6. Schlücker, S. (2014). Surface-enhanced Raman spectroscopy: concepts and chemical applications. *Angew. Chem. Int. Ed.* **53**, pp. 4756–4795.
7. Kneipp, K., Kneipp, H. and Kneipp, J. (2006). Surface-enhanced Raman scattering in local optical fields of silver and gold nanoaggregates-from

single-molecule Raman spectroscopy to ultrasensitive probing in live cells. *Acc. Chem. Res.* **39**(7), pp. 443–508.

8. Kneipp, J., Kneipp, H. and Kneipp, K. (2008). SERS — a single-molecule and nanoscale tool for bioanalytics. *Chem. Soc. Rev.* **37**, pp. 1052–1060.
9. Wang, Y., Yan, B. and Chen, L. (2013). SERS tags: Novel optical nanoprobes for bioanalysis. *Chem. Rev.* **113**, pp. 1391–1428.
10. Halvorson, R. A. and Vikesland, P. J. (2010). Surface-enhanced Raman spectroscopy (SERS) for environmental analyses. *Environ. Sci. Technol.* **44**(20), pp. 7749–7755.
11. Wei, H., Hossein Abtahi, S. M. and Vikesland, P. J. (2015). Plasmonic colorimetric and SERS sensors for environmental analysis. *Environ. Sci., Nano* **2**, pp. 120–135.
12. Yang, S., Dai, X., Stogin, B. B. and Wong, T.-S. (2016). Ultrasensitive surface-enhanced Raman scattering detection in common fluids. *Proc. Natl. Acad. Sci. USA* **113**, pp. 268–273.
13. Craig, A. P., Franca, A. S. and Irudayaraj, J. (2013). Surface-enhanced Raman spectroscopy applied to food safety. *Annu. Rev. Food. Sci. Technol.* **4**, pp. 369–380.
14. Lee, S. Y., Kim, S.-H., Kim, M. P., Jeon, H. C., Kang, H., Kim, H. J., Kim, B. J. and Yang, S.-M. (2013). Freestanding and arrayed nanoporous microcylinders for highly active 3D SERS substrate. *Chem. Mater.* **25**, pp. 2421–2426.
15. Zhang, X., Zheng, Y., Liu, X., Lu, W., Dai, J., Lei, D. Y. and MacFarlane, D. R. (2015). Hierarchical porous plasmonic metamaterials for reproducible ultrasensitive surface-enhanced Raman spectroscopy. *Adv. Mater.* **27**, pp. 1090–1096.
16. Brolo, A. G., Arctander, E., Gordon, R., Leathem, B. and Kavanagh, K. L. (2004). Nanohole-enhanced Raman scattering. *Nano Lett.* **4**, pp. 2015–2018.
17. Baumberg, J. J., Kelf, T. A., Sugawara, Y., Cintra, S., Abdelsalam, M. E., Bartlett, P. N. and Russell, A. E. (2005). Angle-resolved surface-enhanced Raman scattering on metallic nanostructured plasmonic crystals. *Nano Lett.* **5**(11), pp. 2262–2267.
18. Mahajan, S., Abdelsalam, M., Suguwara, Y., Cintra, S., Russell, A., Baumberg, J. and Bartlett, P. (2007). Tuning plasmons on nano-structured substrates for NIR-SERS. *Phys. Chem. Chem. Phys.* **9**, pp. 104–109.
19. Anema, J. R., Brolo, A. G., Marthandam, P. and Gordon, R. (2008). Enhanced Raman scattering from nanoholes in a copper film. *J. Phys. Chem. C* **112**, pp. 17051–17055.
20. Chan, C. Y., Xu, J. B., Waye, M. Y. and Ong, H. C. (2010). Angle resolved surface enhanced Raman scattering (SERS) on two-dimensional metallic arrays with different hole sizes. *Appl. Phys. Lett.* **96**, p. 033104.
21. Zhu, W., Wu, Y., Wang, C., Zhang, M. and Dong, G. (2013). Fabrication of large-area 3-D ordered silver-coated colloidal crystals and macroporous silver films using polystyrene templates. *Nano-Micro Lett.* **5**, pp. 182–190.
22. Adato, R., Yanik, A. A., Amsden, J. J., Kaplan, D. L., Omenetto, F. G., Hong, M. K., Erramilli, S. and Altug, H. (2009). Ultra-sensitive vibrational

spectroscopy of protein monolayers with plasmonic nanoantenna arrays. *Proc. Natl. Acad. Sic.* **106**, pp. 19227–19232.
23. Theiss, J., Pavaskar, P., Echternach, P. M., Muller, R. E. and Cronin S. B. (2010). Plasmonic nanoparticle arrays with nanometer separation for high-performance SERS substrates. *Nano Lett.* **10**(8), pp. 2749–2754.
24. Abu Hatab, N. A., Oran, J. M. and Sepaniak, M. J. (2008). Surface-enhanced Raman spectroscopy substrates created via electron beam lithography and nanotransfer printing. *ACS Nano* **2**, pp. 377–385.
25. Fang, Y., Seong, N.-H. and Dlott, D. D. (2008). Measurement of the distribution of site enhancements in surface-enhanced Raman scattering. *Science* **321**, pp. 388–392.
26. Lu, L., Randjelovic, I., Capek, R., Gaponik, N., Yang, J., Zhang, H. and Eychmüller, A. (2005). Controlled fabrication of gold-coated 3D ordered colloidal crystal films and their application in surface-enhanced Raman spectroscopy. *Chem. Mater.* **17**, pp. 5731–5736.
27. Kucheyev, S. O., Hayes, J. R., Biener, J., Huser, T., Talley, C. E. and Hamza, A. V. (2006). Surface-enhanced Raman scattering on nanoporous Au. *Appl. Phys. Lett.* **89**, p. 053102.
28. Qian, L. H., Yan, X. Q., Fujita, T., Inoue, A. and Chen, M. W. (2007). Surface enhanced Raman scattering of nanoporous gold: smaller pore sizes stronger enhancements. *Appl. Phys. Lett.* **90**, p. 153120.
29. Chen, Z., Qin, Y., Weng, D., Xiao, Q., Peng, Y., Wang, X., Li, H., Wei, F. and Lu, Y. (2009). Design and synthesis of hierarchical nanowire composites for electrochemical energy storage. *Adv. Funct. Mater.* **19**, pp. 3420–3426.
30. Qiu, H., Zhang, Z., Huang, X. and Qu, Y. (2011). Dealloying Ag-Al alloy to prepare nanoporous silver as a substrate for surface-enhanced Raman scattering: effects of structural evolution and surface modification. *Chem. Phys. Chem.* **12**, pp. 2118–2123.
31. Liu, H., Zhang, L., Lang, X., Yamaguchi, Y., Iwasaki, H., Inouye, Y., Xue, Q. and Chen, M. (2011). Single molecule detection from a large-scale SERS-active $Au_{79}Ag_{21}$ substrate. *Sci. Rep.* **1**, p. 112.
32. Fujita, T., Guan, P., McKenna, K., Lang, X., Hirata, A., Zhang, L., Tokunaga, T., Arai, S., Yamamoto, Y., Tanaka, N., Ishikawa, Y., Asao, N., Yamamoto, Y., Erlebacher, J. and Chen, M. (2012). Atomic origins of the high catalytic activity of nanoporous gold. *Nat. Mater.* **11**, pp. 775–780.
33. Chen, C. and Qi Z.-M. (2016). Nanoporous gold films prepared by sputtering-dealloying combination for total internal reflection SERS measurement. *Opt. Mater. Exp.* **6**, pp. 1561–1569.
34. Amezcua-Correa, A., Yang, J., Finlayson, C. E., Peacock, A. C., Hayes, J. R., Sazio, P. J. A., Baumberg, J. J. and Howdle, S. M. (2007). Surface-enhanced Raman scattering using microstructured optical fiber substrates. *Adv. Funct. Mater.* **17**, pp. 2024–2030.
35. Ko, H., Singamaneni, S. and Tsukruk, V. V. (2008). Nanostructured surfaces and assemblies as SERS media. *Small* **4**, pp. 1576–1599.

36. Ko, H., Chang, S. and Tsukruk, V. V. (2009). Porous substrates for label-free molecular level detection of nonresonant organic molecules. *ACS Nano.* **3**, pp. 181–188.
37. Ji N., Ruan, W., Wang, C., Lu, Z. and Zhao, B. (2009). Fabrication of silver decorated anodic aluminum oxide substrate and its optical properties on surface-enhanced Raman scattering and thin film interference. *Langmuir* **25**, pp. 11869–11873.
38. Kodiyath, R., Wang, J., Combs, Z. A., Chang, S., Gupta, M. K., Anderson, K. D., Brown, R. J. and Tsukruk, V. V. (2011). SERS effects in silver-decorated cylindrical nanopores. *Small* **7**, pp. 3452–3457.
39. Shan, D., Huang, L., Li, X., Zhang, W., Wang, J., Cheng, L., Feng, X., Liu, Y., Zhu, J. and Zhang, Y. (2014). Surface plasmon resonance and interference co-enhanced SERS substrate of AAO/Al-based Ag nanostructure arrays. *J. Phys. Chem. C* **118**, pp. 23930–23936.
40. Sui, C., Wang, K., Wang, S., Ren, J., Bai, X. and Bai, J. (2016). SERS activity with tenfold detection limit optimization on a type of nanoporous AAO-based complex multilayer substrate. *Nanoscale* **8**, pp. 5920–5927.
41. Smith, A. J., Tran, T. and Wainwright, M. S. (1999). Kinetics and mechanism of the preparation of Raney® copper. *J. Appl. Electrochem.* **29**, pp. 1085–1094.
42. Keir, D. S. and Pryor, M. J. (1980). The dealloying of copper-manganese alloys. *J. Electrochem. Soc.* **127**, pp. 2138–2144.
43. Wi, J. S., Tominaka, S., Uosaki, K. and Nagao, T. (2012). Porous gold nanodisks with multiple internal hot spots. *Phys. Chem. Chem. Phys.* **14**, pp. 9131–9136.
44. Qi. J., Motwani, P., Gheewala, M., Brennan, C., Wolfe, J. C. and Shih, W. C. (2013). Surface-enhanced Raman spectroscopy with monolithic nanoporous gold disk substrates. *Nanoscale* **5**, pp. 4105–4109.
45. Zhao, F., Zeng, J., Parvez Arnob, M. M., Sun, P., Qi, J., Motwani, P., Gheewala, M., Li, C. H., Paterson, A., Strych, U., Raja, B., Willson, R. C., Wolfe, J. C., Lee, T. R. and Shih, W. C. (2014). Monolithic NPG nanoparticle with large surface area, tunable plasmonics, and high-density internal hot-spots. *Nanoscale* **6**, pp. 8199–8207.
46. Zeng, J. B., Zhao, F. S., Li, M., Li, C. H., Lee, T. R. and Shih, W. C. (2015). Morphological control and plasmonic tuning of nanoporous gold disks by surface modifications. *J. Mater. Chem. C* **3**, pp. 247–252.
47. Liu, K., Bai, Y., Zhang, L., Yang, Z., Fan, Q., Zheng, H., Yin, Y. and Gao, C. (2016). Porous Au-Ag nanospheres with high-density and highly accessible hotspots for SERS analysis. *Nano Lett.* **16**, pp. 3675–3681.
48. Shih, W. C., Santos, G. M., Zhao, F., Zenasni, O. and Arnob, M. M. (2016). Simultaneous chemical and refractive index sensing in the 1–2.5 μm near-infrared wavelength range on nanoporous gold disks. *Nano Lett.* **16**, pp. 4641–4647.
49. Johnson, P. B. and Christy, R. W. (1972). Optical constants of the noble metals. *Phys. Rev. B* **6**, pp. 4370–4379.

50. Sancho-Parramon, J., Janicki, V. and Zorc, H. (2010). On the dielectric function tuning of random metal-dielectric nanocomposites for metamaterial applications. *Opt. Express.* **18**, pp. 26915–26928.
51. Li, K., Stockman, M. I. and Bergman, D. J. (2003). Self-similar chain of metal nanospheres as an efficient nanolens. *Phys. Rev. Lett.* **91**, p. 227402.
52. Li, K., Stockman, M. I. and Bergman, D. J. (2006). Li, Stockman, and Bergman reply. *Phys. Rev. Lett.* **97**, p. 079702.
53. Li, Z., Yang, Z. and Xu, H. (2006). Comment on "Self-similar chain of metal nanospheres as an efficient nanolens". *Phys. Rev. Lett.* **97**, p. 079701.
54. Kravets, V. G., Zoriniants, G., Burrows, C. P., Schedin, F., Casiraghi, C., Klar, P., Geim, A. K., Barnes, W. L. and Grigorenko, A. N. (2010). Cascaded optical field enhancement in composite plasmonic nanostructures. *Phys. Rev. Lett.* **105**, p. 246806.
55. Jiang, R. B., Qin, F., Ruan, Q. F., Wang, J. F. and Jin C. J. (2014). Ultrasensitive plasmonic response of bimetallic Au/Pd nanostructures to hydrogen. *Adv. Funct. Mater.* **24**, pp. 7328–7337.
56. Bao, Z. Y., Lei, D. Y., Jiang, R., Liu, X., Dai, J., Wang, J., Chan, H. L. and Tsang, Y. H. (2014). Bifunctional Au@Pt core-shell nanostructures for insitu monitoring of catalytic reactions by surface-enhanced Raman scattering spectroscopy. *Nanoscale* **6**, pp. 9063–9070.
57. Xie, W., Herrmann, C., Kömpe, K., Haase, M. and Schlücker, S. (2011). Synthesis of bifunctional Au/Pt/Au Core/shell nanoraspberries for in situ SERS monitoring of platinum-catalyzed reactions. *J. Am. Chem. Soc.* **133**, pp. 19302–19305.
58. Zhang, W., Yang, J. and Lu, X. (2012). Tailoring galvanic replacement reaction for the preparation of Pt/Ag bimetallic hollow nanostructures with controlled number of voids. *ACS Nano* **6**, pp. 7397–7405.
59. Goris, B., Polavarapu, Bals, S., Van Tendeloo, G. and Liz-Marzán, L. M. (2014). Monitoring galvanic replacement through three-dimensional morphological and chemical mapping. *Nano Lett.* **14**, pp. 3220–3226.
60. Henglein, A. (2000). Preparation and optical absorption spectra of $Au_{core}Pt_{shell}$ and $Pt_{core}Au_{shell}$ colloidal nanoparticles in aqueous solution. *J. Phys. Chem. B* **104**, pp. 2201–2203.
61. Wang, Y. and Toshima, N. (1997). Preparation of Pd-Pt bimetallic colloids with controllable core/shell structures. *J. Phys. Chem. B* **101**, pp. 5301–5306.
62. Lim, B., Jiang, M., Camargo, P. H. C., Cho, E. C., Tao, J., Lu, X., Zhu, Y. and Xia, Y. (2009). Pd-Pt bimetallic nanodendrites with high activity for oxygen reduction. *Science* **324**, pp. 1302–1305.
63. Toshima, N. and Yonezawa, T. (1998). Bimetallic nanoparticles-novel materials for chemical and physical applications. *New J. Chem.* **22**, pp. 1179–1201.
64. Joseph, V., Engelbrekt, C., Zhang, J., Gernert, U., Ulstrup, J. and Kneipp, J. (2012). Characterizing the kinetics of nanoparticle-catalyzed reactions by surface-enhanced Raman scattering. *Angew. Chem., Int. Ed.* **51**, pp. 7592–7596.
65. Bao, Z. Y., Dai, J., Lei, D. Y. and Wu, Y. (2013). Maximizing surface–enhanced Raman scattering sensitivity of surfactant-free Ag-Fe_3O_4

nanocomposites through optimization of silver nanoparticle density and magnetic self-assembly. *J. Appl. Phys.* **114**, p. 124305.
66. Bao, Z. Y., Liu, X., Chen, Y., Wu, Y., Chan, H. L., Dai, J. and Lei, D. Y. (2014). Quantitative SERS detection of low-concentration aromatic polychlorinated biphenyl-77 and 2,4,6-trinitrotoluene. *J. Hazard. Mater.* **280**, pp. 706–712.
67. Sharma, B., Frontiera, R. R., Henry, A.-I., Ringe, E. and Van Duyne, R. P. (2012). SERS: Materials, applications, and the future. *Materials Today* **15**, pp. 16–25.
68. Golightly, R. S., Doering, W. E. and Natan M. J. (2009). Surface-enhanced Raman spectroscopy and homeland security: A perfect match? *ACS Nano* **3**, pp. 2859–2869.

Chapter 8

Shell-Isolated Nanoparticle-Enhanced Raman Spectroscopy

Jian-Feng Li, Rajapandiyan Panneerselvam and Zhong-Qun Tian*

State Key Laboratory of Physical Chemistry of Solid Surfaces, iChEM, and College of Chemistry and Chemical Engineering, Xiamen University, Xiamen 361005, China
**zqtian@xmu.edu.cn*

1. Introduction

Surface-enhanced Raman spectroscopy (SERS) is a powerful molecular spectroscopic technique which was discovered in 1970s (Figure 1(a)) [1–4]. The Raman signals can be enhanced by up to ~10 orders of magnitude because of a molecule's proximity to a nanostructured, free-electron metal surface such as Au, Ag, or Cu [5–13]. When the light of a resonant wavelength is incident upon this type of nanostructures, it will initiate collective oscillations in electron density called surface plasmon that gives a very narrow region of enhanced electromagnetic field strength [14–18]. The enhanced electromagnetic field can enhance the Raman signals of probe molecules nearby. This nanostructure-based vibrational spectroscopy technique can provide fingerprint chemical and molecular information of miscellaneous target molecules down to a single molecular level, comparable to single-molecule fluorescence spectroscopy [15, 19–33].

However, SERS had not been adopted as a general tool due to the lack of substrate (materials) generality and surface (morphology) generality. Only a few metals, e.g. Ag, Au, and Cu, provide a huge

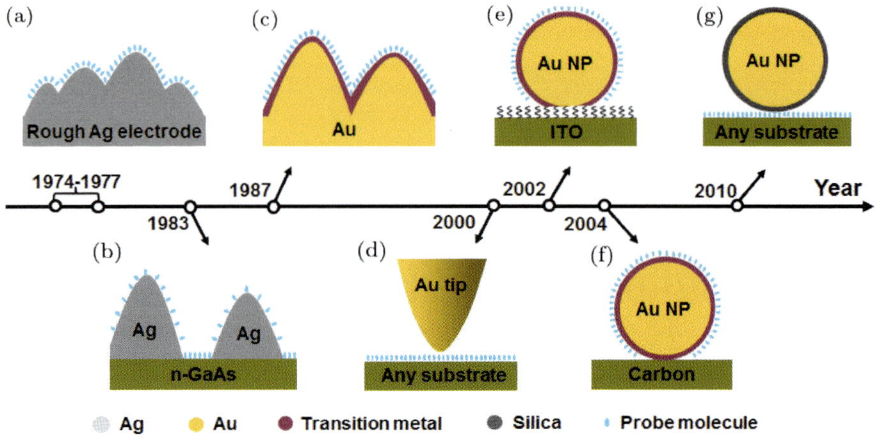

Figure 1. Illustration of the development of the "borrowing SERS activity" methods utilizing different nanostructures in the past four decades in order to expand the materials and morphology generalities.

SERS effect when the metal nanostructures are in several tens of nanometers in size [14, 34–36]. Furthermore, the signal enhancement provided by the metal nanostructures depends critically on their size, shape and spacing [14, 34–40]. Though the SERS enhancement is remarkable, only a few metals are suitable for SERS investigations. This restriction severely limits the breadth of practical applications of SERS in various fields including material science, energy, medical and life sciences. Moreover, SERS cannot be employed on atomically flat single-crystal surfaces, which are commonly used in surface science and semiconductor technology.

To overcome the above-mentioned problems, several groups have developed a strategy of "borrowing SERS activity" to employ SERS on non-traditional SERS substrates [35, 41–54]. This approach is to employ the long-range effect of the enormous electromagnetic (EM) field created by the SERS-active Au or Ag nanostructures to detect the nearby target molecules. In this strategy, probe molecules are spatially separated from the SERS-active core by a thin shell, so that the molecules located close to (within 1–10 nm) SERS-active nanoparticles or nanostructured surfaces can be enhanced/detected due to the long-range electromagnetic enhancement [55–58]. It should be noted that the strong electromagnetic field will be decreased exponentially with increasing thickness of the "spacer" film, therefore the film has to be ultra-thin, normally a few atomic layers.

In 1983, the approach of "borrowing SERS activity" was first proposed by Van Duyne and coworkers. They obtained the enhanced Raman signals of probe molecules adsorbed on non-SERS active materials such as n-GaAs electrodes by electrodepositing a discontinuous SERS-active overlayer of Ag (Figure 1(b)) [41]. The long-range effect of the strong EM field created by the SERS-active Ag islands enhanced the Raman scattering of adsorbates at the semiconductor (n-GaAs) nearby. This configuration requires that the probe molecules should selectively be adsorbed at the non-SERS-active substrate (n-GaAs) but not on the Ag islands. However, it is difficult to prevent the adsorption of probe molecules on SERS-active Ag islands.

In 1987, the Fleischmann group [42, 43] and Weaver group [44–46] developed an alternative way to form ultra-thin shells of transition metals (VIII B elements) onto SERS-active Au and Ag nanostructures (Figure 1(c)). With the aid of long-range effect of the strong EM field created by the SERS-active Au or Ag nanostructures underneath, Raman signals of probe molecules on the transition metal surfaces can be enormously enhanced. However, it is very difficult to coat the SERS-active cores with a uniform thin shell. In 2002, the Weaver group successfully coated an ultra-thin (pinhole-free) shell of Pt metal onto Au nanoparticles immobilized on ITO electrodes by underpotential deposition (UPD) and chemical redox replacement method [48], which exhibited much more intense SERS signals (Figure 1(d)). However, the preparation procedure consisted of surface functionalization, electrochemical UPD and chemical reaction. Therefore, it was highly desirable to develop a simpler and straightforward method to prepare transition metal shells on Au or Ag nanostructures for solving the substrate generality issues.

In 2004, our group developed a simpler and straightforward strategy to synthesize transition metal shells on Au nanoparticles by a seed-mediated growth method (Figure 1(e)) [35, 51]. In this approach, Au or Ag nanoparticles were first prepared and then coated with ultra-thin shells (1–10 atomic layers) of various metals. Boosted by the long-range effect of the enhanced electromagnetic field generated by the highly SERS-active Au core, Raman signal of probe molecules on Pt, Pd, Ru, Rh, Ni, and Co surfaces was substantially enhanced up to 10^4–10^5 enhancement factors (EFs) [35, 51]. However, for many other materials, such as oxides, polymers, insulators or biological membranes, it was difficult/impossible to coat them as uniform ultra-thin shells onto Au/Ag nanoparticles. Therefore, it was necessary to explore an innovative approach to extend the applications of SERS by eliminating substrate generality issues. Overall, if we analyze the above-mentioned approaches/methods carefully, target species are either adsorbed at the surface of SERS-active nanoparticles (nanostructures) or the overcoated layer (inert or transition metal shells). We can name these measurements as contact mode measurements, see Figures 2(a) and 2(b). Au/Ag nanoparticles act as Raman signal amplifiers and also as a host of probe molecules

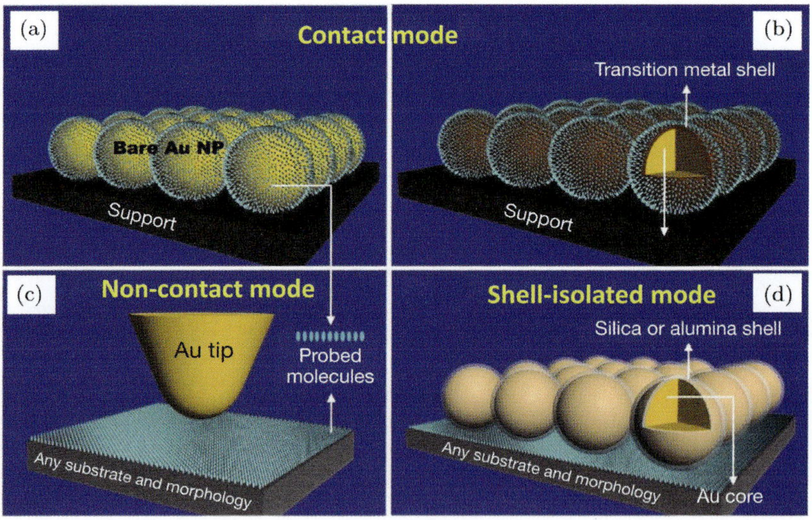

Figure 2. Three different working principles of conventional SERS, TERS, and SHINERS. (a) Bare Au nanoparticles, contact mode. (b) Au core–transition metal shell nanoparticles, contact mode. (c) Tip-enhanced Raman spectroscopy, non-contact mode. (d) Shell-isolated nanoparticle-enhanced Raman spectroscopy, shell-isolated mode. Reproduced with permission from Ref. [52] Copyright 2010, Macmillan Publishers Ltd.

simultaneously, causing the limitation of material and morphology generalities.

In 2000, tip-enhanced Raman spectroscopy (TERS) (Figure 1(f)) [47, 49, 50] was invented, leading to a breakthrough by overcoming the limitation of substrate and surface generalities of SERS. TERS is distinctly different from SERS and other previous approaches. TERS utilizes a new working mode (Figure 2(c)) which can be termed as non-contact mode that separates the target surface (molecule or materials surface) from a sharp Au/Ag tip (20–50 nm) which acts as the Raman signal amplifier. Under the excitation of a suitable laser, an enhanced electromagnetic field (hotspot) can be generated at the apex of the tip. In principle, Raman signals from any target surface regardless of material and morphology can be detected by the metallic tip with an excellent spatial resolution [49]. Surprisingly, TERS technique provides sub-nanometer spatial resolution with chemical and molecular information. However, the total Raman scattering signal from the tip area (about 20–50 nm in diameter) is rather weak, thus TERS studies have been largely limited to molecules having large Raman cross-sections. Besides, it is also a challenge for TERS to work under aqueous environments.

In 2010, we invented a new method by combining the concept of TERS and "borrowing SERS strategy" to circumvent the shortcomings described above. This approach is to replace the tip of TERS by a monolayer of SERS active nanoparticles, each nanoparticle coated with an ultra-thin inert silica shell (here denoted Au@SiO_2 nanoparticles). Thus, each nanoparticle acts as a TERS tip, and simultaneously brings thousands of TERS tips to the substrate surface to be analyzed. In this method, the enhanced Raman signals are contributed by numerous nanoparticles, which is two to three orders of magnitude higher than a single TERS tip. In addition, the SERS-active Au cores are physically and chemically separated from the area of interest/probe molecules or the environments. The main virtue of this shell-isolated mode (Figure 2(d)) is its much higher detection sensitivity, simplicity and vast practical applications to various materials with diverse morphologies. We named this technique as "shell-isolated nanoparticle-enhanced Raman spectroscopy (SHINERS)"

(Figure 1(g)) [52–54], which has been used to investigate several chemical, electrochemical biological and catalytic processes [52–54, 59–70].

In this chapter, we discuss the different modes of SERS measurements and the importance of shell-isolated mode in SERS. In the following section, we explain the electric field distribution of SHINs using 3D finite-difference time-domain (3D-FDTD) method and then we discuss the advantages of shell-isolated working mode over contact mode SERS. Obviously, the synthesis and characterization of SHINs are the most important and challenging tasks in SHINERS experiments, so we introduce the preparation and characterization procedures in detail with SHINERS principles.

From the beginning, SHINERS has been widely applied in various fields. An unequalled application of SHINERS over traditional SERS methods is its applicability on single-crystal surfaces in electrochemistry field. In the application part, we first highlight two applications of SHINERS on metal single crystal surfaces including the molecular adsorption and an electrochemical reaction. Next, we further elucidate the application of SHINERS for *in situ* monitoring of nanocatalytic chemical reactions for catalysis and then for characterizing the processes in lithium batteries for energy materials. Then we exemplify the application of SHINERS in biological systems and food safety. Overall, SHINERS can be employed on a variety of surfaces and morphologies to investigate the molecular orientations or chemical reactions to deepen our understanding in both fundamental and practical research. At last, we summarize and provide an outlook for future developments in this field. The concept of shell-isolated nanoparticle-enhancement mode can be applied to other plasmon-enhanced spectroscopic techniques, such as surface-enhanced fluorescence, TERS, and second-harmonic generation.

2. 3D-FDTD Simulations

To examine the electric field distribution and to calculate the Raman enhancement of SHINs, theoretical simulations were carried out using 3D-FDTD [71, 72]. Figure 3 shows the results obtained for a 2×2 array of Au@SiO_2 nanoparticles on a perfectly smooth gold

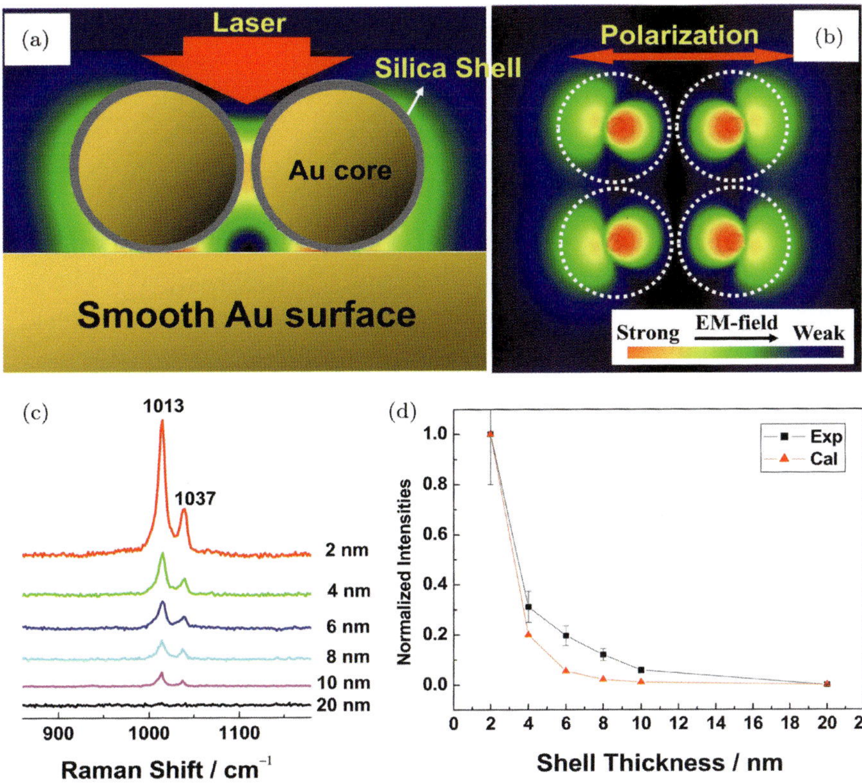

Figure 3. 3D-FDTD simulated electric field distribution for a 2×2 array of 55 nm Au@ 4 nm SiO_2 SHINs on a perfectly smooth gold surface. The shell-to-shell distance is 4 nm. Side view (a) and top view (b); the direction of incidence (a) and the polarization (b) of the 633 nm laser are also shown. (c) SHINERS spectra of pyridine (Py) molecules adsorbed on a smooth Au surface modified by 55 nm Au@SiO_2 nanoparticles with different silica shell thicknesses. (d) The shell thickness dependence of the integrated SHINERS intensity of Py (black squares) and the corresponding 3D-FDTD calculation result (red triangles). Reproduced by permission from Ref. [52]. Copyright © 2010, Macmillan Publishers Ltd.

surface [52]. When the gold cores are brought near the smooth gold surface by decreasing the thickness of the shells, localized surface plasmons are excited in the core-to-surface gaps. The electric field becomes concentrated in this gap, and Raman scattering from probe molecules adsorbed on the smooth gold surface below the nanoparticles increased dramatically. The calculations reveal that the electric field strength is enhanced 85 times for the 4 nm shell and 142 times for the 2 nm shell. The Raman scattering enhancement factors in the SHIN-to-surface junctions will be 5×10^7 and 4×10^8 for the 4 and 2 nm shell thickness, respectively. Therefore, SHINERS can be employed to characterize the adsorption behavior of a monolayer at atomically flat single-crystal gold surfaces.

Since the Raman scattering enhancement factor mainly depends on the distance between the gold core and the smooth gold surface, a combined theoretical and experimental investigation into the dependence of Raman intensity on silica shell thickness was performed [52]. The diameter of the gold core was kept constant at 55 nm, and the shell thickness was controllably changed from 2 nm to 20 nm in sequence, which was attested by high-resolution transmission electron microscopy (HR-TEM) images. Figure 3(c) presents a set of the SHINERS spectra of pyridine (Py) adsorbed on a smooth Au surface with different silica shell thicknesses. As expected, the SHINERS intensity of Py decreases significantly with the increasing shell thickness. In other words, the thinner the silica shell, the higher the SHINERS intensity. To more quantitatively evaluate the correlation of the shell thickness and the SHINERS intensity, we plotted the integrated intensity (1013 cm^{-1}, ν_1 mode) against the shell thickness in Figure 3(d) (black square). The integrated intensities were normalized by the signals obtained with the 2 nm silica thickness with the error bars of 15%. It is evident from Figure 3(d) that the SHINERS signal decreases exponentially with the increasing shell thickness.

3. Advantages of Shell-Isolated Mode over Contact Mode

By changing the working mode from the direct contact (SERS) or non-contact (TERS) to the shell-isolated mode, SHINERS has broken the long-standing limitations of SERS by investigating various systems that were previously inaccessible for Raman measurements. The thin silica shell around the SERS-active metal nanostructures has significant advantages over bare nanoparticles, for instance, the inert shell prevents aggregation, oxidation and also contamination.

Since many researchers in SERS field have been working on the contact mode for almost four decades, it is necessary to give a systematic discussion on the advantages of SHINERS over the other working modes.

In contact mode SERS, SERS-active metal nanostructures are directly deposited onto a surface that was not SERS-active, or a weak SERS-active material is deposited onto a SERS-active one [41, 42, 73]. Actually, bare Ag or Au nanoparticles are the simplest and strongest Raman signal amplifier and have been used by several groups for various applications [74, 75]. However, certain problems may come up when a chemically inert dielectric shell is absent.

(i) Contact with the chemical environment. In some cases, bare nanoparticles are not preferable for SERS measurements in aqueous environments (e.g. in biology and electrochemistry) or gaseous environments (e.g. in catalysis) that may contain several species which can adsorb onto the nanoparticle's surface (Figure 4(a)). Even if the

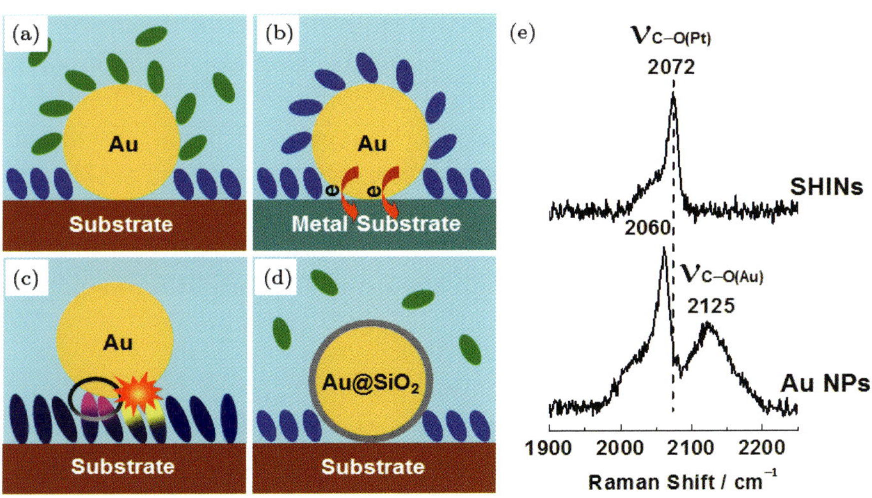

Figure 4. Schematic illustrations of four different SERS measurements. (a) Contact with a chemical environment. (b) Electrical contact with the surface. (c) Contact with probe molecules. (d) nanoparticle isolation by an inert shell. (e) SERS and SHINERS spectra showing CO adsorption on a Pt(111) electrode in 0.1 M $HClO_4$ solution saturated with CO gas. Reproduced by permission from Ref. [52]. Copyright © 2010, Macmillan Publishers Ltd.

aqueous or gaseous environment does not contain any such adsorbates, probe molecules may diffuse from the surface of interest to the surface of the bare nanoparticles. The SERS signals from molecules adsorbed on the bare nanoparticles, which is likely to be dominant, can interfere with the SERS signals from molecules adsorbed on the surface of interest [14, 55, 76].

(ii) Electrical contact with the target surface. If bare nanoparticles are used for the examination of metal or semiconductor surfaces, bare nanoparticles may have direct contact with the surface of interest, the difference in Fermi levels may result in charge transfer and creation of a contact potential that can significantly affect the electronic structure of the target system (Figure 4(b)) [77, 78].

For example, SERS and SHINERS spectra of CO adsorption on a Pt(111) surface are compared in Figure 4(e). With bare gold nanoparticles, two strong bands have been observed in the spectrum, one at 2060 cm^{-1} that is due to C–O(Pt) stretching and another at 2125 cm^{-1} that is due to C–O(Au) stretching. The latter is clearly undesirable as it could mislead the interpretation of the spectrum. More importantly, the C–O(Pt) stretching frequency is different in the SERS and SHINERS spectra. It is downshifted to 2060 cm^{-1} in the SERS spectrum because the work functions of these two metals are different (5.1 eV for polycrystalline gold, 5.7 eV for Pt(111) [79] and charge transfer occurs from gold to platinum. The true C–O(Pt) stretching frequency at 2072 cm^{-1} was obtained by SHINERS. Thus, SHINERS can be used to obtain the original information about the target systems.

(iii) Contact with probe molecules. Generally, probe molecules may interact with bare nanoparticles as well, possibly adopting a two-end adsorption rather than a one-end adsorption (pink molecules in Figure 4(c)) [80]. This direct contact would likely change the electron density distribution in the probe molecule and its adsorption behavior, resulting in significant changes to its SERS spectral features. In some cases, probe molecules may undergo photocatalytic reactions owing to a direct contact between one of its functional groups and the bare nanoparticles (yellow molecules in Figure 4(c))

[81–85]. For instance, Tian *et al.* observed the chemical transformation of para-aminothiophenol (PATP) to 4,4′-dimercaptoazobenzene (DMAB) on Ag electrodes under laser illumination.

SHINERS, with its unique shell-isolated mode, can overcome the three above-mentioned problems (Figure 4(D)). By coating the Au nanoparticles with a thin (chemically and electrically) inert shell, it completely isolates the Au/Ag nanoparticles from probe molecules/other molecules regardless of whether the measurement is in liquid or gas. Therefore, the measured Raman signal originates solely from the sample surface to be measured. If an appropriate shell material is chosen, inadvertent adsorption, charge-transfer and photocatalytic reactions can be avoided. Furthermore, the stability of metal nanoparticles are greatly improved by the inert silica shell for more than one year.

4. Synthesis and Characterization of Various Types of SHINs

Figure 5 shows an overview of the experimental procedure for the preparation of SHINs. To employ the SHINERS technique, it is essential to understand the following parameters: first, the metallic

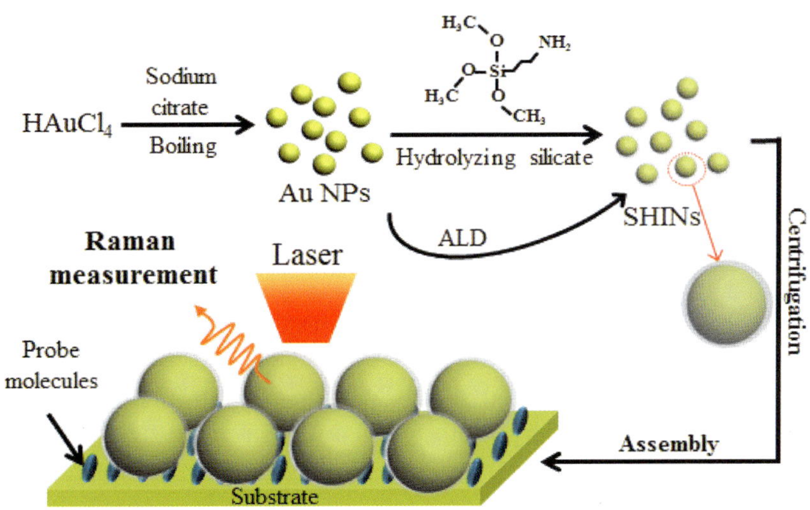

Figure 5. General overview of SHINs synthesis.

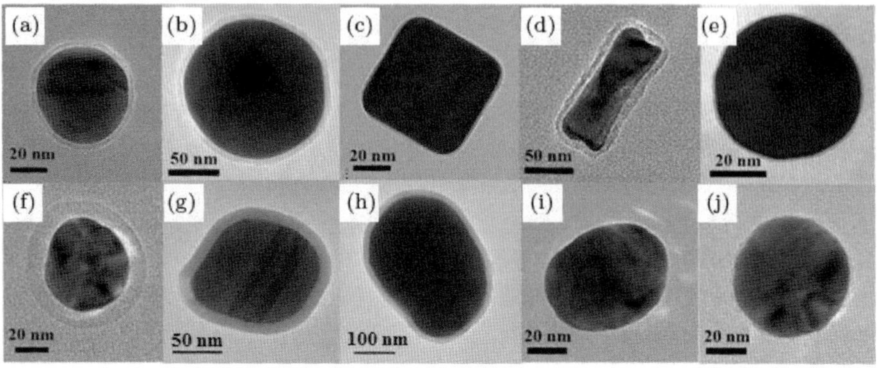

Figure 6. HR-TEM pictures of different kinds of SHINs. 55 nm Au@SiO_2 nanoparticles (a), 120 nm Au@SiO_2 nanoparticles (b), a nanocube core of Au@SiO_2 nanoparticles (c), a nanorod core of Au@SiO_2 nanoparticles (d) Ag@SiO_2 nanoparticles (e), 55 nm Au@20 nm SiO_2 nanoparticle (f), Ag@TiO_2 (g), Ag@Al_2O_3 (h), MnO_2 (i), and Ag_2S (j). Reproduced with permission [52]. Copyright 2010, Macmillan Publishers Ltd. Reproduced with permission [54]. Copyright 2013, Macmillan Publishers Ltd. Reproduced with permission [93] Copyright 2015, John Wiley & Sons, Ltd.

nanoparticle core should be SERS-active with an appropriate size. In our case, the suitable diameter of gold cores is 55 nm (Figure 6(a)) [86] for SHINERS measurements, and bigger diameter gold nanoparticles, such as 120 nm (Figure 6(b)), can be synthesized for enormous enhancements. Shell-isolated gold nanocubes and nanorods (Figures 6(c) and 6(d)) were synthesized as their maximum SPR peak can be simply tuned by adjusting the aspect ratio [51, 54, 63, 87–90]. It is well-known that the SPR property of nanoparticles will be affected by its shape and size. For Au nanorods, following its aspect ratio increase, its SPR absorption maximum can be tuned from visible to the near infrared range. Additionally, Ag@SiO_2 nanoparticles (Figure 6(e)) were synthesized because Ag nanoparticles exhibit higher enhancement compared to the Au nanoparticles under the same size, and the SPR peak position of the corresponding Ag nanoparticles can be shifted to any wavenumber at a larger scale by varying the particle size [91, 92].

Second, the inert shell should be ultra-thin as the SHINERS signal intensity will be significantly decreased with the increasing shell thickness (Figure 6(f)). For most of the target systems, the shell thickness of SHINs should be in the range of 1–5 nm to exhibit significant electromagnetic field enhancement. The shell thickness can be varied by carefully changing the synthesis conditions such as the seed concentration, reaction time, reaction temperature, pH values and so on. Third, the shell material of SHINs should be chemically and electrically inert to avoid any chemical reaction and electrical interferences. For example, the shell material of SHINs can be TiO_2 (Figure 6(g)), Al_2O_3 (Figure 6(h)), MnO_2 (Figure 6(i)), and Ag_2S (Figure 6(j)), which can protect SHINs from various chemical and physical environments [54, 67, 93].

Here, we elaborately explain the experimental procedure of SHINs consisting of 55 nm Au@SiO_2. The Au nanoparticles (55 nm) were synthesized by the classical Frens procedure [86]. In detail, 200 mL 0.01% $HAuCl_4$ solution was stirred in a round-bottom flask under heating conditions. After boiling, 1.4 mL 1% sodium citrate solution was added and the reaction was continued for 1 h, then the solution was cooled down to room temperature. The SHINs were prepared as follows [54]: 30 mL Au nanoparticles solution was added into a 100 mL round-bottom flask for stirring at room temperature and then dropped 400 μL 1 mM APTMS solution. After 15 min stirring, 3 mL 0.54% Na_2SiO_3 solution (pH $\approx$ 10.3) was added into the above solution. About 3 min later, the mixed solution was transferred to a 95°C bath and stirred for some time. By controlling the reaction time, pH values, temperature and gold nanoparticles concentration, the SiO_2 shell could be tuned with different thicknesses [54]. After certain heating time, the hot solution was quickly cooled down and then centrifuged several times. At last, the concentrated SHINs was diluted with pure water to a certain volume for further electrochemical or Raman measurements. It is important to mention that SHINs can easily form a sub-monolayer on a target surface almost without obvious aggregates. Importantly, to get reproducible Raman signals and higher enhancement, it is important to note that the target surface should be covered with an appropriate amount of SHINs.

Last, the shell of SHINs should be pinhole-free. One of the most important and challenging requirements of a SHINERS experiment is to fabricate a uniform ultra-thin shell. However, if the silica shell is too thin (<2 nm), pinholes can be easily formed and the probe/matrix molecules can easily contact the SERS-active Au core. This contact can yield false information from the target system [54]. If silica shells are filled with pinholes, its isolating function will be affected and the probe molecules or impurities from the outer environment can contact gold or silver nanoparticles core, which could yield mixture of information. Therefore, it is very important to note that the SHINs should be pinhole-free so that the obtained Raman signals only originate from the probe molecules which adsorbed on the target surface. Overall, SHINs should have ultra-thin, uniform, and pinhole-free shells to find applications in a wide range of biological, chemical, and electrochemical systems [52].

Before SHINERS measurements, it is essential to characterize the pinhole-free property of SHINs [52, 54]. To characterize this property, high-resolution transmission electron microscopy (HR-TEM) images, electrochemical cyclic voltammetry (CV) and SERS spectra from pyridine molecules were primarily used. The HR-TEM images of SHINs are shown in Figure 6. However, one cannot find/detect pinholes from TEM images. So, the electrochemical CV method was employed to obtain cyclic voltammograms of SHINs [52]. As shown in Figure 7, the blue curve (Figure 7(A-a)) shows the typical electrochemical feature of bare Au nanoparticles with a typical reduction peak around 0.9 V (vs. SCE), while the pinhole-free SHINs are represented by the black curve (Figure 7(A-c)), without the gold characteristic reduction peak, which indicates that the Au cores are isolated by the inert silica shells. Furthermore, Raman spectroscopy was used to investigate the pinhole effect of SHINs. As a standard Raman probe molecule, pyridine molecules strongly interact with metal nanoparticles, but it does not adsorb on the inert shell surface. In this way, the presence of pinholes on SHINs can be characterized using cyclic voltammetry and Raman measurements.

For Raman measurements, the procedure is as follows: first, 10 μL of SHINs solution was deposited (with pinhole and centrifuged) on an Si wafer and dried for a certain time; then 20 μL of 10 mM pyridine solution was deposited on the SHINs and covered with a thin quartz window. Finally, the sample was subjected to Raman

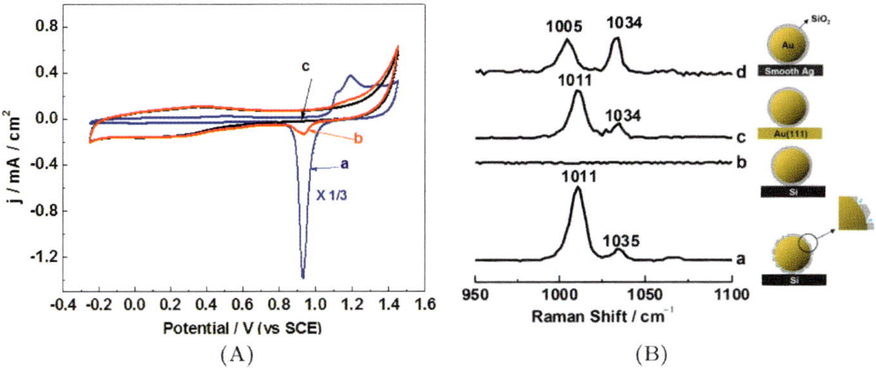

Figure 7. (A) CVs of 55 nm Au nanoparticles (a-red line) Au@SiO_2 nanoparticles without pinhole (c-black line) and Au@SiO_2 nanoparticles with pinhole (b-red line) on glassy carbon (GC) electrodes in 0.5 M H_2SO_4 solution. (B) To investigate the pinhole effect, SHINERS spectra were obtained from adsorbed pyridine molecules; SHINs with (a) and without (b) pinholes on a Si surface. SHINs without pinholes on an Au(111) electrode surface (c). SHINs without pinholes on a smooth Ag film surface (d). Reproduced with permission [52] Copyright 2010, Macmillan Publishers Ltd.

measurements. Undoubtedly, strong Raman signals from pyridine were seen at 1011 and 1035 cm^{-1} because pyridine can interact with the gold cores through pinholes (Figure 7(B-a)). When the procedure was repeated with pinhole-free SHINs, the Raman signals of pyridine were not observed for a long time (Figure 7(B-b)). When the Au(111) electrode surface was covered with pinhole-free SHINs, the typical Raman signals of pyridine at 1011 and 1034 cm^{-1} were observed (Figure 7(B-c)), which originated from the adsorbed pyridine molecules on the Au(111) single crystal surface. In order to confirm this result, pinhole-free SHINs were placed on a smooth Ag film and the typical Raman peaks of pyridine at 1005 cm^{-1} and 1034 cm^{-1} were observed (Figure 7(B-d)), which indicates that the Raman signals originated from the adsorbed pyridine molecules on the Ag film.

5. SHINERS Applications on Single-Crystal Surfaces

Single-crystal surfaces possess well-defined metal electronic levels and surface structure, which are basic parameters to unveil the chemical enhancement mechanism of SERS and other important reaction mechanism. Thus, researchers are highly interested to utilize single-crystal surfaces to obtain concrete information about the orientation and configuration of various probe molecules. However, strong SERS enhancement can only be obtained from a limited number of free-electron metals and other materials. Additionally, atomically flat single-crystal surfaces cannot support strong surface plasmon resonance effectively [94, 95], so it is difficult to obtain Raman signals from a single crystal surface using the SERS method. With the new working mode, SHINs can create a strong electromagnetic field, and SHINERS can be employed to characterize the adsorption and orientation of probe molecules on a variety of single-crystal surfaces. As a new technique, SHINERS method has been successfully applied to investigate a wide range of probe molecules and to detect the mechanism of various electrocatalytic processes on single crystal surfaces.

Pyridine (Py) is a standard probe molecule that can be utilized to understand the coordination and orientation behavior of chemical molecules at metal interfaces. Our group employed the EC-SHINERS method to examine the electrochemical adsorption behavior of Py on Au(hkl) single crystal surfaces in 0.1 M $NaClO_4$ solution [60, 96, 97]. Different factors, such as crystallographic orientation, Py concentration and applied potential, were studied systematically. Figure 8(a) shows the CVs of 1 mM Py on three basic low-index Au(hkl) single crystal electrodes in 0.1 M $NaClO_4$ solution. Figure 8(b) shows the typical EC-SHINERS spectra of Py on the Au(111) surface from -0.8 V to 0.4 V potential range. The peak at 1011 cm^{-1} was ascribed to the Py ν_1 ring breathing mode, and the peak at 1035 cm^{-1} was attributed to the Py ν_{12} symmetric triangular ring deformation mode [51, 98–101].

In principle, the Raman peak of a probe molecule is associated with adsorption configuration and surface coverage rate. By using ν_1 mode of Py on Au(111), we found that the Raman frequency and intensity of ν_1 mode around 1011 cm^{-1} increases with increasing potential (Figure 8(b)), but the enhancement factor is not the same in different potential regions as shown in Figures 8(c)–8(d). At low potential region ($E < 0.1$ V), the Raman frequency

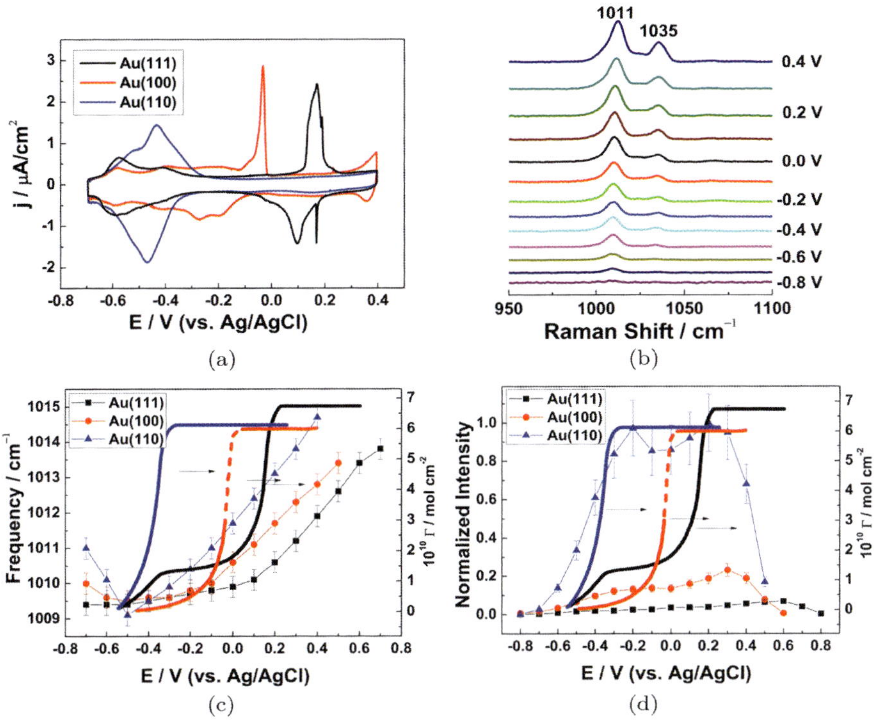

Figure 8. (a) CVs of Au(hkl) electrodes in 0.1 M $NaClO_4$ solution containing 1 mM Py. (b) SHINER spectra of Py molecules on Au(111). The relationship of Raman frequency (c) and normalized Raman intensity (d) with potentials for the ν_1 mode, and accompanied with surface concentration isotherms (bold curves). Reproduced with permission [97] Copyright 2015, American Chemical Society.

of ν_1 mode increased little, but when $E > 0.1$ V, the frequency of ν_1 mode undergoes obvious change and the Stark tuning rate is about $5.6\,\text{cm}^{-1}/\text{V}$, which indicates the adsorption configuration or orientation of Py is different compared to the negative potential region. When Py molecules were flat-adsorbed on the electrode surface, there was a weak binding interaction between π-orbital electrons and the electrode surface, thus the relevant Raman signal was weak; when Py molecules were vertically adsorbed on the electrode surface, there was a strong interaction between the nitrogen atom lone pair of electrons and the electrode surface, so the relevant Raman signal was strong. From the potential-Raman frequency relationship and potential-Raman intensity relationship results, we can conclude that Py flat-adsorbed on the Au(111) electrode surface when $E < 0.1$ V, and vertically adsorbed on the Au(111) electrode surface when $E \geq 0.1$ V, which was in good agreement with the reported articles that Py forms a full monolayer on Au(111) at higher potentials [102–104].

Additionally, we found that the orientation of Py on Au(100) and Au(110) surface would be changed from flat adsorbed to vertically adsorbed following the potential increase too [105, 106]. The trend of the potentials of Py monolayer completion decreasing in the order of Au(111) > Au(100) > Au(110) on three basic-index gold single crystal surfaces is the same as the potential of zero charge (EPZC) order obtained without Py. It means that the electrode surface charge plays an important role during the adsorption process. But the Raman peak intensity of ν_1 mode for adsorbed Py molecules on three basic-index Au(hkl) electrode surfaces increases in the order of Au(111) < Au(100) ≪ Au(110) (Figure 8(d)). Thus, EC-SHINERS can be employed to investigate the adsorption behavior of various probe molecules on single-crystal electrodes at electrochemical interfaces.

On the other side, the formation of oxide during an electrocatalytic process greatly affects the function of a catalyst and also the reaction mechanism [107, 108]. Therefore, it is important to understand the electrocatalytic processes to fabricate or design efficient new catalysts. However, the conventional analytical techniques are facing difficulties to detect the key intermediates because of the complexity in *in situ* studies. Most of the reaction mechanisms were deduced based on the electrochemical techniques and combined with theoretical calculations, but it was still a challenging task to prove these results by direct *in situ* Raman information [109, 110].

In this regard, *in situ* EC-SHINERS technique offers a novel way for real-time investigation of electrocatalytic reactions at well-defined single-crystal surfaces. Our group used *in situ* EC-SHINERS method to look into the electro-oxidation processes at Au(hkl) single crystal surfaces and rendered spectral evidences for the electro-oxidation intermediates [111]. As shown in Figure 9, there is no obvious Raman peak in the range of 300–800 cm^{-1} during the positive scan until 0.4 V

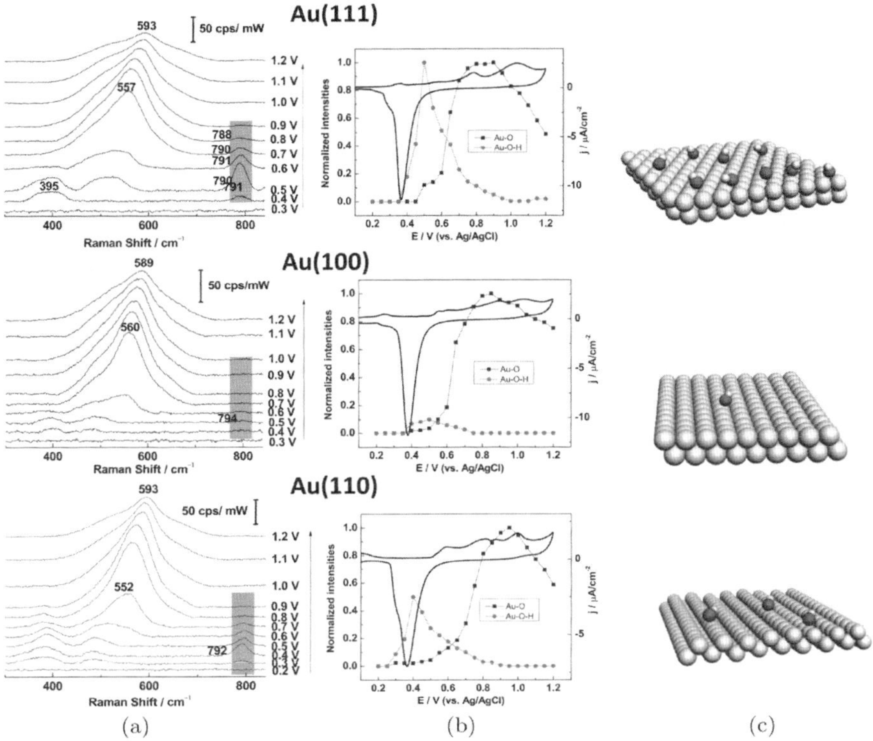

Figure 9. (a) The SHINERS spectra of electrooxidation at Au(111) Au(10) and Au(110) electrode surfaces in 0.1 M $NaClO_4$ (pH is ~9), (b) normalized SHINERS intensities of the stretching mode of AuO and the bending mode of AuOH at different potentials. CV of Au(111) electrode in 0.1 M $NaClO_4$ is presented (pH is ~9, scan rate is 2 mV/s), (c) schematic diagrams of OH species on three low-index Au(hkl) surfaces. Reproduced with permission [111]. Copyright 2015, American Chemical Society.

on Au(111) single-crystal electrode surface. There was a Raman peak around 790 cm^{-1} following the potential increase to 0.4 V, and its intensity reached maximum at 0.5 V, then decreased and almost disappeared at 0.9 V. Based on the references and the DFT calculation results, the peak at 790 cm^{-1} was assigned to the gold hydroxide bending mode δAuOH of the adsorption on top sites [112–114].

To affirm our conclusion, a deuterium isotopic experiment was also performed, and the results showed that the bending mode of AuOH at 790 cm^{-1} shifts toward the lower wavenumber 694 cm^{-1}

in deuterated water, which clearly confirms that the band around 790 cm^{-1} belongs to the gold hydroxide bending mode δAuOH. We also found that there was a broad band at 360–420 cm^{-1}, when the potential reached 0.3 V, the band blue-shifted about 20 cm^{-1} in heavy water experiment, which is also in agreement with the previous reports. When the potential was at 0.7 V, there was another broad band at 567 cm^{-1}, it shifted to 593 cm^{-1} following the potential increase, which was attributed to Au–O stretching.

We then examined the effect of crystallographic orientation, anion and pH values for the electrooxidation process on Au(hkl) single electrode surface. As shown in Figure 9, the intensities of the bending mode δ_{AuOH} at three single-crystal facets increased in the order of Au(111) > Au(110) $\gg$ Au(100), which unveiled the OH species are selectively adsorbed on three low-index gold single-crystal surfaces. However, this order is opposite to the order of activity in the oxygen reduction reaction [115–119], which may be because of the OH formation during the electrooxidation of Au surface. Along with the crystallographic orientation, pH, and anion effects, EC-SHINERS was employed to systematically characterize and monitor the electrooxidation process on gold single crystal electrodes. *In situ* EC-SHINERS offers a unique way for a real-time investigation of electrocatalytic reaction processes at single-crystal electrode surfaces, which can open the door for technological innovations in surface catalysis.

6. *In situ* Monitoring of Nanocatalytic Chemical Reactions

In plasmon-enhanced chemical reactions, metal nanostructures can play a dual function of spectroscopic signal amplifier and photocatalyst for the chemical reactions [120, 121]. Plasmonic nanostructures can generate ultra-high local electromagnetic fields for optical signals, meanwhile "hot" electrons could be excited on the metal nanostructure surface for chemical transformation or energy conversion. Recently, Liu *et al.* performed single-particle LSPR sensing experiments to probe a local reaction environment with 120 nm Au SHINs dispersed on a thin Pd film [122]. The transformation from Pd to Pd hydride upon the uptake of hydrogen changes the coupling efficiency between Au SHINs and Pd film, and leads to the resonance shift of scattering spectra.

Due to the dual function of nanostructures, it is difficult to explore the mechanism of plasmon-enhanced chemical reactions. It is truly necessary to clarify the mechanism while maintaining the higher catalytic performance. To understand the reaction mechanism, Au/Ag SHINs and Ag SITERS have been employed to alter the reaction pathways by manipulating the "hot" electrons with a gap-mode configuration [52, 123]. Under the laser illumination in the presence of bare metal nanoparticles, 4-aminothiophenol (PATP) was oxidized to 4,4-dimercaptoazobenzene (DMAB) due to the "hot" electron transfer between metal surface and surface adsorbed O_2. On the contrary, the photoinduced oxidation was distinctively prevented with the shell-isolated nanostructures which "smartly" blocked the charge transfer and other non-radiative processes.

As shown in Figure 10(a), a sub-monolayer of ~25 nm Ag nanoparticles was assembled onto the ~100 nm Ag SHIN surface to form the 3D Ag core–satellite superstructures. The "hot" local fields were located among the junction between small Ag nanoparticles and Ag SHIN surface under the illumination. Along with the gap-mode configuration, bifunctional 3D metal core–satellite superstructures exhibited an excellent surface-selective catalysis performance when combined with the shell-isolated mode [124, 125]. In contrast to small Ag nanoparticles which are incapable of catalyzing the hydride reduction reaction, the core–satellite superstructures revealed the photocatalytic activity for the reduction of 4-nitrothiophenol (PNTP) to PATP. It should be noted that both the protons and halide anions are required for the "hot" electron-induced reaction. This is evidenced by the spectral feature that the C–C stretching mode of PATP at around 1590 cm^{-1} (the peak indicated by a red dash vertical line) is only observable in the presence of both H^+ and Cl^-. The schematic illustration in Figure 10(b) demonstrates the mechanism of six-electron photocatalytic reaction where the recycling of Ag atoms was achieved between photoinduced oxidation to the electron compensation with halide anions. Interestingly, PNTP was reduced with the "hot" electrons generated upon the excitation of LSPR while the "hot" holes

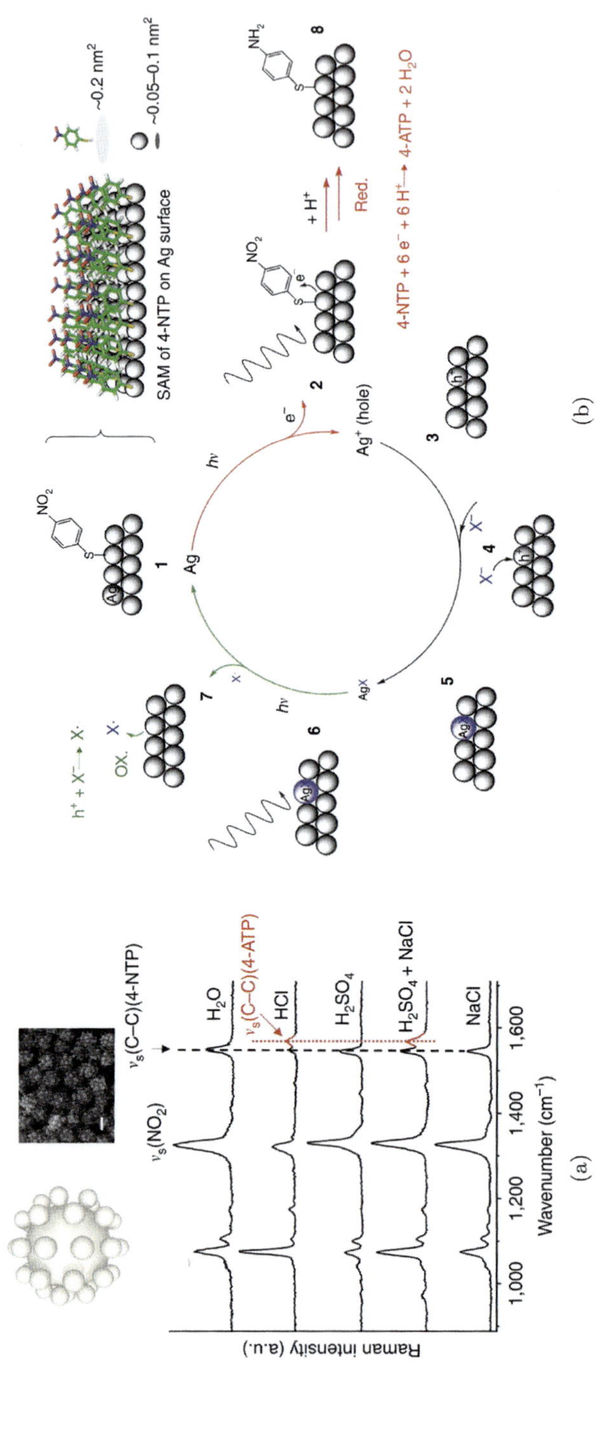

Figure 10. (a) SERS spectra of plasmon-mediated reduction of PNTP to PATP with Ag SHIN core–Ag nanoparticles satellites superstructure. The cartoon model, electric field simulation and SEM image of satellites superstructures are presented as well. (b) Schematic illustration of six-electron photoinduced reaction where the Ag atoms was recycled from photoinduced oxidation to the electron compensation with halide anions, and the photodissociation reaction of Ag halide. Reproduced with permission [124] Copyright 2015, Nature Publishing Group.

combined with the halide anions. Ag halide was then decomposed to the fresh Ag atom and X• via photodissociation. Thus, SHINERS can be used for *in situ* monitoring of heterogeneous nanocatalytic processes using 3D core–satellite superstructures, and the reaction pathways could be well distinguished with the presence of "smart" shell-isolated mode.

7. SHINERS for Batteries/Fuel Cell Characterization

In energy science, especially lithium ion batteries and fuel cells have garnered great attention among researchers for a long time. The interfacial reactions are considered as important issues that related to the reversible capacity, chemical and physical stability of batteries. Importantly, the formation of interfacial species in batteries greatly affects the battery performance. Along this direction, SHINERS technique has been employed to monitor the interfacial processes in Li-ion batteries [126]. In this study, the SHINERS signals were identified to examine the solid/electrolyte interphase (SEI) growth on Si electrode surface before reduction of ethylene carbonate during charge–discharge cycling (Figure 11(a)). In the first charge and discharge process, there were a series of SHINERS peaks that appeared. According to references, the authors considered these peaks at 356, 381, 820 and 3560 cm^{-1} are related to LiOH–H_2O, and these peaks at 204, 738, 756, and 1090 cm^{-1} are related to Li_2CO_3, respectively (Figure 11(b)) and the authors also claimed that all these peaks can be assigned to the reduction products of the electrolyte, which is the SEI composition. Meanwhile, in contrast to the normal Raman signals from the bulk Si electrode, the SHINERS signals revealed different amorphization rates at the surface and bulk of the electrode, which may indicate that the lithiation process from the surface to bulk is non-uniform. In later research, Hwang *et al.* employed SHINERS method that directly investigated the formation of Li_2O during electrochemical cycling on Li-rich cathode [127]. From the experiment results, the authors noticed a SHINERS peak at 534 cm^{-1}, which appeared when the potential reaches the oxygen-activating plateau 4.4 V, and then continued to increase until 4.52 V during the

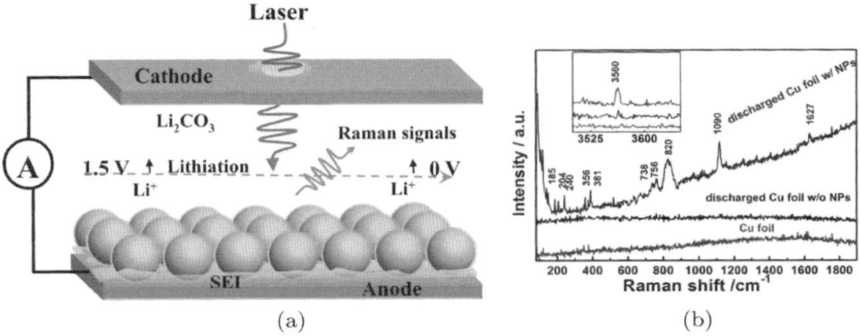

Figure 11. (a) *In situ* SHINERS study mode of SEI formation on the surface of a Si electrode in a Li-ion battery. The Au@SiO_2 nanoparticles were deposited on the electrode surface before the assembly of the coin cell. A small hole was made through the top cap, Li foil and separator for the laser to enter and the Raman signal to exit the cell. (b) Raman spectra of Cu foil, electrochemically discharged Cu foil with and without SHINs. Inset shows the higher range for the Raman shift Reproduced with permission [126]. Copyright 2014, Elsevier B.V.

charging process. The peak at 534 cm^{-1} may be related to Li_2O. The experiment results also indicated that Li_2O will reduce at the end of the oxygen-activating plateau and the formation of Li_2O was only observed with the compositions exhibiting oxygenactivation. Importantly, the formation of Li_2O on the cathode leads to the generation of H_2O and LiOH, as well as the changes within the battery electrolyte, which finally results in deteriorated performance. Though there may be a few unclear points, for example, the choice of excitation laser, it is still a very valuable attempt for SHINERS to investigate in real time the interfacial reactions in Li-ion batteries.

Similarly, Liu *et al.* employed SHINERS method for real-time investigation of solid oxide fuel cell (SOFC) materials at a high temperature [128]. With 10 nm SiO_2 shell, Ag@SiO_2 SHINs can still maintain the structural stability of the plasmonic nanostructures under high temperature. They found obvious Raman signals of doped ceria and deposited carbon from the SHINERS results, respectively, which were not found or very weak using normal Raman under a high temperature. Overall, these studies attest that SHINERS has a great potential for *in situ* investigations of energy-related devices under practical conditions. Thus, SHINERS offers a unique method to deepen our understanding of surface reactions that heavily influence the overall battery performance.

8. SHINERS Study on Corrosion Science

As a convenient method, SHINERS technique can be easily employed for various practical/industrial applications, and one of the important applications is for corrosion inhibition study. For instance, benzotriazole (BTAH) is widely used for polishing and plating purposes to prevent copper and relevant alloys from corrosion because BTAH can form a coordination polymer film on the surface to prevent metal oxidation. Gewirth's group [66] used EC-SHINERS method to investigate the very different film growth behavior between the single crystal Cu(hkl) electrode surfaces and unroughened polycrystalline Cu electrode surfaces. The authors found that the film formation is irreversible on single-crystal Cu(hkl) electrodes, while it is reversible on Cu(poly) surfaces. The reversible growth of BTA-film on Cu(poly) surface may due to the presence of grain boundaries, which also lead to the cathodic degradation of BTA-polymeric films on Cu(poly) electrode.

In the recent years, several researchers have attempted to achieve quantitative analysis, especially for practical applications [129–131]. However, quantitative analysis is a daunting challenge for plasmon-enhanced Raman spectroscopies. For example, gold–thiosulfate leaching process is inhibited by the formation of a passive layer that prevents the full dissolution of gold from ore samples. Lipkowski's group [130] employed SHINERS method for quantitative analysis of temporal changes in the passive layer at the Au electrode surface in a thiosulfate solution. In their study, the Raman band of APTES was utilized as an internal standard to compensate the fluctuations of surface enhancement. The peaks of APTES species can be effectively removed by subtracting the SHINERS spectrum in a NaF solution and the corrected SHINERS spectra was in good agreement with the relevant SERS spectra. As shown in Figure 12(a), the SHINERS spectra for long immersion times in thiosulfate electrolyte with several new bands are compared to the spectra recorded at shorter immersion times. It provides a unique

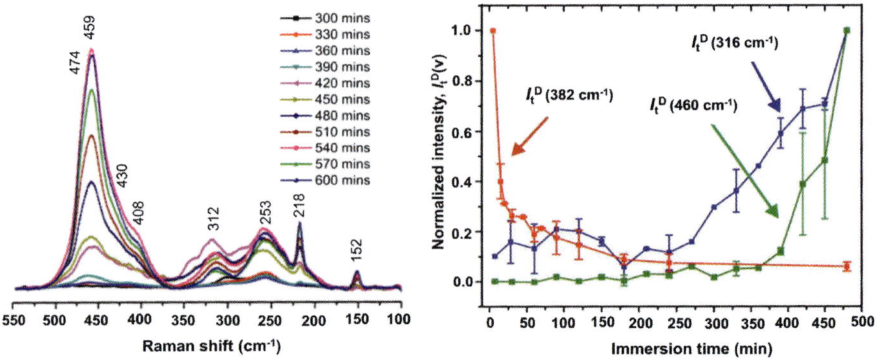

(a) (b)

Figure 12. (a) SERS spectra of the Au/SHINs electrode immersed in a 0.1 M $Na_2S_2O_3$ electrolyte (pH = 10.0) between 5 and 10 h. (b) The normalized integrated band intensity as a function of immersion time, I_t^D, of the $[Au(S_2O_3)_2]^{3-}$ complex at 382 cm^{-1} (red), adsorbed sulfide at 316 cm^{-1} (blue), and polymeric sulfur at 460 cm^{-1} (green). Reproduced by permission from Ref. [130]. Copyright © 2015, American Chemical Society.

opportunity to correlate the rate of gold leaching in thiosulfate solutions with quantitative changes in the intensity of the Raman bands. The electrochemical results show that the gold leaching rate rapidly decreases within the first 50 min of immersion and then slows down to an almost linear decay as time progresses. Figure 12(b) shows the plots of normalized and background corrected Raman intensities of 382, 316 and 460 cm^{-1}, which belong to $[Au(S_2O_3)_2]^{3-}$ complex at the surface, adsorbed sulfide and polymeric sulfur species, respectively. The peak at 382 cm^{-1} shows good correlation with leaching current as the intensity decreases quickly at early stages of time and then changes slowly at longer time scales. However, the peaks at 316 cm^{-1} (blue curve) and 460 cm^{-1} (green curve) in Figure 12(b) show a very different trend, which indicates that the adsorbed sulfide and polymeric sulfur species increase. This results provide strong quantitative evidence that gold leaching in thiosulfate solutions is inhibited by the formation of elemental sulfur species, and adsorbed sulfur blocks interactions between thiosulfate and the gold surface preventing the formation of the $[Au(S_2O_3)_2]^{3-}$ complex at the gold–solution interface. This methodological advancement allows us to monitor temporal changes of the passive layer composition and correlate Raman band intensities with the rate of gold dissolution in order to identify the nature of the passivating species. Overall, SHINERS technique can be easily applied to anticorrosion research of different substrates for qualitative and quantitative analysis, more importantly, the strategy of quantitative SHINERS analysis on various electrode surface species will be very useful for both fundamental and practical applications.

9. SHINERS for Safety Inspection and Biological Analysis

SHINERS has been practically applied for various purposes such as food inspection, explosive identification, biological analysis, public security, etc. For example, Au@SiO_2 SHINs were applied to accurately study different kinds of breast lesions, including fibroadenoma (FD), atypical ductal hyperplasia (ADH), ductal carcinoma (DCIS), invasive ductal carcinoma (IDC) and normal breast tissues (NB) [132]. Those closely related pathologies (especially ADH and DCIS) in breast lesions can be detected and distinguished by this method, which shows that the early diagnosis will be beneficial. In this way, SHINERS provides an efficient alternative method to traditional histopathological analysis.

Trinitrotoluene (TNT) is a highly explosive and environmentally dangerous substance, however, it is hard to be detected because of its small Raman cross-section. With the amplification by uniform and pinhole-free Au@poly(2-aminothiophenol) (PAT) nanoparticles, Raman signals of TNT were easily acquired. The electron-withdrawing groups –NO_2 would easily react with electron-donating amino groups, which are located outside the PAT shells, to form Meisenheimer complexes. TNT molecules can be captured by Au@PAT nanoparticles, dispersed in the environment and then detected by Raman spectrometer. With this method, high sensitivity of TNT detection can be realized with the concentration of 10^{-5} M.

With portable Raman spectrometers, SHINERS could be practically applied as a simple-to-use tool for rapid inspection of toxic poison and food products [52, 133]. Cyanide is a great threat to public security due to its extremely high toxicity and widespread usage in industry. Xie's group employed SHINs with pinholes to rapidly detect the lower concentration of cyanide. In their research, the cyanide anions interacted with the uncovered Au surface and generated a magnificent local electromagnetic field enhancement. On the other side, the inert SiO_2 shell could obviously improve the stability of SHINs and make this method more feasible for practical applications. The detection limit of cyanide down to 1 μg/L level in water was achieved.

Meanwhile, our group also used SHINs as "smart dusts" by simply spreading on the fruit (orange) surface, and then we were able to detect the pesticide residues on the fruit surface using a portable Raman spectrometer easily. As shown in Figure 13(A), two peaks at 1165 and 1526 cm^{-1}, attributed to carotenoid, were obtained on the pericarps before (curve a) and after (curve b) parathion contamination by normal Raman measurements [134]. No parathion signal could be found if SHINs are absent. However, two additional peaks at 1109 and 1340 cm^{-1} (curve c), which are attributed to parathion residues [129], could be clearly identified with the SHINERS method. As a Raman signal amplifier, SHINs have great ability to detect hazardous substances/pesticide residues on vegetables. Sulfur-fumigated ginger looks normal in color, however, sulfide may possess destructive effect on human health. In Figure 13(B), residual sulfides [7] (610 cm^{-1} on curve f) on a ginger surface were successfully detected by SHINERS (curve e). Just by spreading "smart dust" on the surface of any substance, vegetable or fruit, Raman signals could be highly amplified, and the sensitive and accurate Raman spectra can be easily recorded with a portable Raman spectrometer.

Moreover, the inert SiO_2 shell protects the plasmonic performance of chemically active Au/Ag nanoparticles so that the stability of the nanoparticles is maintained for almost one year. Furthermore, interaction between SHINs and analyte molecules could be effectively avoided, so that the result of the analysis is reliable. This method does not need any complex sample treatment and bulky instrument, and it is convenient, efficient, low cost and reliable.

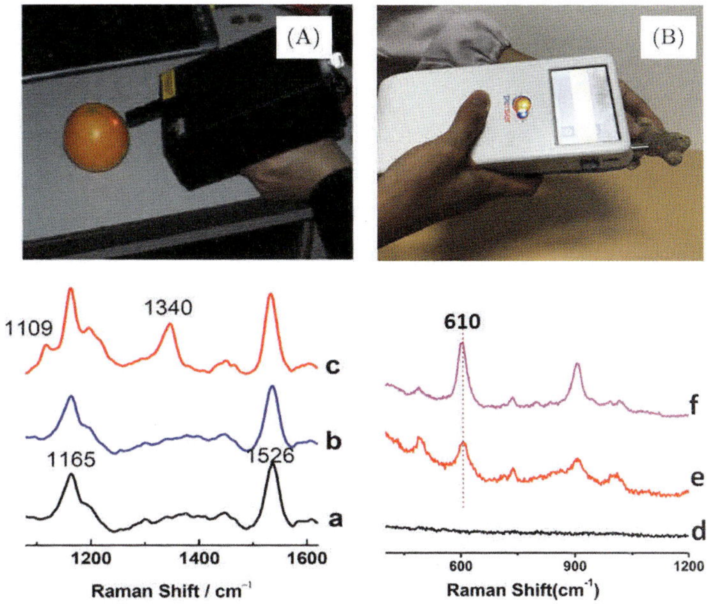

Figure 13. *In situ* detection of pesticide residues on fruit and vegetable surfaces with a portable Raman spectrometer. (A) Normal Raman spectra were recorded on fresh pericarps of orange (curve a) and parathion contaminated (curve b) pericarps of an orange. SHINERS spectrum of parathion contaminated orange (curve c). (B) SHINERS spectra were recorded on cabbage leaves contaminated by parathion (curve d), fenthion (curve e) pesticides or both two kinds of pesticides (curve f). (A) is reproduced with permission [52] Copyright 2010 Nature Publishing Group.

10. Summary and Outlook

From the beginning, SHINERS has been successfully employed in multidisciplinary fields including electrochemistry, catalysis, biology, surface science, and material science [135]. Compared with the other working modes in plasmon-enhanced vibrational spectroscopies (e.g. SERS and TERS), as an analytical technique, SHINERS is fast, inexpensive, simple and reliable. Beyond that, the shell-isolated mode is extremely flexible and of particular interest in many areas of science and technology. Particularly, SHINERS enjoys a reputation as a sensitive method for examining and monitoring various probe molecules at atomically flat single-crystal electrode surfaces.

In heterogeneous catalysis, SHINs can be easily used to examine and distinguish photoinduced reaction pathways by blocking the charge-transfer. The ultra-thin dielectric shell smartly protects the SERS-active metal core from the outer environment and also maintains the plasmonic performance for more than one year.

Most of the spectroelectrochemical studies generally focus on the analytes having strong interaction with the electrode surface because large cross-section molecules can be easily examined with spectroscopic techniques. However, one of the basic requirements of an analytical technique is its sensitivity, therefore, it is necessary to improve the detection limit of SHINERS to monitor the interaction of weakly adsorbed low cross-section species. To take full advantage of SHINERS, it is preferable to develop thinnest 2D shell materials, such as graphene [136–138] and h-NB [139–141], so that the SHINERS sensitivity will be further increased by one order of magnitude, but it is challenging to avoid pinholes. Interestingly, the sensitivity of surface-enhanced spectroscopies can be further enhanced by designing nanostructures which support plasmonic Fano-resonance [142, 143] or double resonance [144, 145] due to the near-field or far-field EM interference between nanostructures. Particularly, signals from analytes adsorbed on the plasmonic Fano-resonance supported nanostructures will be significantly augmented by the local fields of the excited plasmon modes.

The concept of SHINs enhancement is being applied to other spectroscopies such as infrared absorption spectroscopy, fluorescence, and sum frequency generation (SFG). For example, Aroca and Li' groups have applied the concept to fluorescence spectroscopy [146, 147]. Yang and Li *et al.* have extended the strategy to second-harmonic generation (SHG) [148]. Zenobi and Li groups are currently designing experiments to develop new scanning probe microscopy tip structures for shell-isolated tip-enhanced spectroscopy (SITERS) [149].

With the current developments in nanoscience and Raman spectroscopy, SHINERS plays a significant role in various fields including electrochemistry, heterogeneous catalysis, material science and biology. We believe that the achievements attained by SHINERS in the past six years are promising and have not been fully tapped.

Optimistically, we anticipate that SHINERS will provide more useful mechanistic information regarding electrochemical reactions, biological processes, and nanocatalytic processes [150]. Further, there are several efforts now afoot that are trying to implement this shell-isolated strategy in other techniques, and we believe that the potential of SHINERS will be realized and widespread acceptance of this technique can be achieved.

Acknowledgments

We are deeply grateful to Z. L. Yang, T. Wandlowski, Z. L. Wang, A. Gewirth, J. Lipkowski, B. J. Hwang, S. Schlucker and R. F. Aroca for their contribution during the development of SHINERS, and C. Y. Li, S. Y. Ding, Y. F. Huang, Q. Z. Chen and J. C. Dong for their experimental assistance. This work was supported by the NSFC (21522508 and 21427813), and the MOST (2011YQ030124 and 2015CB932301), and the Thousand Youth Talents Plan of China.

References

1. Fleischmann, M., Hendra, P. J. and McQuillan, A. J. (1974). Raman spectra of pyridine adsorbed at a silver electrode. *Chem. Phys. Lett.* **26**, pp. 163–166.
2. Jeanmaire, D. L. and Van Duyne, R. P. (1977). Surface Raman spectroelectrochemistry. Part 1: Heterocyclic, aromatic, and aliphatic-amines adsorbed on anodized silver electrode. *J. Electroanal. Chem.* **84**, pp. 1–20.
3. Albrecht, M. G. and Creighton, J. A. (1977). Anomalously intense Raman-spectra of pyridine at a silver electrode. *J. Am. Chem. Soc.* **99**, pp. 5215–5217.
4. Van Duyne, R. P. (1979). In Moore, C. B. (ed.). Chemical and Biochemical Applications of Lasers; Academic Press, New York.
5. Chang, R. K. and Furtak, T. E. (eds.) (1982). Surface Enhanced Raman Scattering (Plenum Press, New York).
6. Pockrand, I. (ed.) (1984). Surface Enhanced Raman Vibrational Studies at Solid/Gas Interfaces (Springer Tracts in Modern Physics, Springer, New York).
7. Smith, E. and Dent, G. (eds.) (2005). Modern Raman Spectroscopy: A Practical Approach. John Wiley, Chichester, UK.
8. Tian, Z. Q. (2005). Surface enhanced Raman spectroscopy. *J. Raman Spectrosc.* **36**, pp. 466–470.
9. Aroca, R. (2006). Surface-Enhanced Vibrational Spectroscopy. John Wiley, New York.
10. Brown, R. J. C., Cohen, L. F. and Milton, M. J. T. (2006). Surface enhanced Raman spectroscopy. *Faraday Discuss.* **132**, pp. 1–340.

11. Kneipp, K., Moskovits, M. and Kneipp, H. (eds.) (2006). Surface-Enhanced Raman Scattering Physics and Applications. Springer Science & Business Media, New York.
12. Schlucker, S. (ed.) (2011). Surface Enhanced Raman Spectroscopy: Analytical, Biophysical and Life Science Applications. Wiley, Weinheim, Germany.
13. Ozaki, Y., Kneipp, K. and Aroca, R. (eds.) (2014). Frontiers of Surface-Enhanced Raman Scattering: Single Nanoparticles and Single Cells. Wiley, Chichester.
14. Moskovits, M. (1985). Surface-enhanced spectroscopy, *Rev. Mod. Phys.* **57**, pp. 783–826.
15. Willets, K. A. and Van Duyne, R. P. (2007). Localized surface plasmon resonance spectroscopy and sensing, *Annu. Rev. Phys. Chem.* **58**, pp. 267–297.
16. Camden, J. P., Dieringer, J. A., Zhao, J. and Van Duyne, R. P. (2008). Controlled plasmonic nanostructures for surface-enhanced spectroscopy and sensing *Acc. Chem. Res.* **41**, pp. 1653–1661.
17. Le Ru, E. C. and Etchegoin, P. G. (2009). Principles of Surface-Enhanced Raman Spectroscopy and Related Plasmonic Effects. Elsevier, Amsterdam/Oxford.
18. Aroca, R. F. (2013). Plasmon enhanced spectroscopy. *Phys. Chem. Chem. Phys.* **15**, pp. 5355–5363.
19. Kneipp, K., Wang, Y., Kneipp, H., Perelman, L. T., Itzkan, I., Dasari, R. and Feld, M. S. (1997). Single molecule detection using surface-enhanced Raman scattering (SERS). *Phys. Rev. Lett.* **78**, pp. 1667–1670.
20. Nie, S. M. and Emery, S. R. (1997). Probing single molecules and single nanoparticles by surface-enhanced Raman scattering. *Science* **275**, pp. 1102–1106.
21. Nie, S. M. and Zare, R. N. (1997). Optical detection of single molecules. *Annu. Rev. Biophys. Biomol. Struct.* **26**, pp. 567–596.
22. Michaels, A. M., Nirmal, M. and Brus, L. E. (1999). Surface enhanced Raman spectroscopy of individual rhodamine 6G molecules on large Ag nanocrystals. *J. Am. Chem. Soc.* **121**, pp. 9932–9939.
23. Xu, H. X., Bjerneld, E. J., Kall, M. and Borjesson, L. (1999). Spectroscopy of single hemoglobin molecules by surface enhanced Raman scattering. *Phys. Rev. Lett.* **83**, pp. 4357–4360.
24. Cao, Y. W. C., Jin, R. C. and Mirkin, C. A. (2002). Nanoparticles with Raman spectroscopic fingerprints for DNA and RNA detection. *Science* **297**, pp. 1536–1540.
25. Jackson, J. B. and Halas, N. J. (2004). Surface-enhanced Raman scattering on tunable plasmonic nanoparticle substrates. *P. Natl. Acad. Sci. USA* **101**, pp. 17930–17935.
26. Qin, L., Zou, S., Xue, C., Atkinson, A., Schatz, G. C. and Mirkin, C. A. (2006). Designing, fabricating and imaging Raman hot spots. *P. Natl. Acad. Sci. USA* **103**, pp. 13300–13303.
27. Anker, J. N., Hall, W. P., Lyandres, O., Shah, N. C., Zhao, J. and Van Duyne, R. P. (2008). Biosensing with plasmonic nanosensors. *Nat. Mater.* **7**, pp. 442–453.

28. Chen, Z., Tabakman, S. M., Goodwin, A. P., Kattah, M. G., Daranciang, D., Wang, X., Zhang, G., Li, X., Liu, Z., Utz, P. J., Jiang, K., Fan, S. and Dai, H. (2008). Protein microarrays with carbon nanotubes as multicolor Raman labels. *Nat. Biotechnol.* **26**, pp. 1285–1292.
29. Graham, D., Thompson, D. G., Smith, W. E. and Faulds, K. (2008). Control of enhanced Raman scattering using a DNA-based assembly process of dye-coded nanoparticles. *Nat. Nanotechnol.* **3**, pp. 548–551.
30. Jain, P. K., Huang, X. H., El-Sayed, I. H. and El-Sayed, M. A. (2008). Noble Metals on the nanoscale: Optical and photothermal properties and some applications in imaging, sensing, biology, and medicine. *Acc. Chem. Res.* **41**, pp. 1578–1586.
31. Qian, X. M., Peng, X. H., Ansari, D. O., Yin-Goen, Q., Chen, G. Z., Shin, D. M., Yang, L., Young, A. N., Wang, M. D. and Nie, S. M. (2008). In vivo tumor targeting and spectroscopic detection with surface-enhanced Raman nanoparticle tags. *Nat. Biotechnol.* **26**, pp. 83–90.
32. Dieringer, J. A., Wustholz, K. L., Masiello, D. J., Camden, J. P., Kleinman, S. L., Schatz, G. C. and Van Duyne, R. P. (2009). Surface-enhanced Raman excitation spectroscopy of a single rhodamine 6G molecule. *J. Am. Chem. Soc.* **131**, pp. 849–854.
33. Han, X. X., Zhao, B. and Ozaki, Y. (2009). Surface-enhanced Raman scattering for protein detection. *Anal. Bioanal. Chem.* **394**, pp. 1719–1727.
34. Moskovits, M. (2005). Surface-enhanced Raman spectroscopy: A brief retrospective. *J. Raman Spectrosc.* **36**, pp. 485–496.
35. Tian, Z. Q., Ren, B., Li, J. F. and Yang, Z. L. (2007). Expanding generality of surface-enhanced Raman spectroscopy with borrowing SERS activity strategy. *Chem. Commun.* pp. 3514–3534.
36. Stiles, P. L., Dieringer, J. A., Shah, N. C. and Van Duyne, R. P. (2008). Surface-enhanced Raman spectroscopy. *Annu. Rev. Anal. Chem.* **1**, pp. 601–626.
37. Freeman, R. G., Grabar, K. C., Allison, K. J., Bright, R. M., Davis, J. A., Guthrie, A. P., Hommer, M. B., Jackson, M. A., Smith, P. C., Walter, D. G. and Natan, M. J. (1995). Self-assembled metal colloid monolayers — An approach to SERS substrates. *Science* **267**, pp. 1629–1632.
38. Li, W. Y., Camargo, P. H. C., Lu, X. M. and Xia, Y. N. (2009). Dimers of silver nanospheres: Facile synthesis and their use as hot spots for surface-enhanced Raman scattering. *Nano Lett.* **9**, pp. 485–490.
39. Greeneltch, N. G., Blaber, M. G., Henry, A. I., Schatz, G. C. and Van Duyne, R. P. (2013). Immobilized nanorod assemblies: Fabrication and understanding of large area surface-enhanced Raman spectroscopy substrates. *Anal. Chem.* **85**, pp. 2297–2303.
40. Zhang, X. Y., Zheng, Y. H., Liu, X., Lu, W., Dai, J. Y., Lei, D. Y. and MacFarlane, D. R. (2015). Hierarchical porous plasmonic metamaterials for reproducible ultrasensitive surface-enhanced Raman spectroscopy. *Adv. Mater.* **27**, pp. 1090–1096.
41. Van Duyne, R. P. and Haushalter, J. P. (1983). Surface-enhanced resonance Raman-spectroscopy of adsorbates on semiconductor electrode surfaces —

tris(bipyridine)ruthenium(ii) adsorbed on silver-modified N-GaAs(100). *J. Phys. Chem.* **87**, pp. 2999–3003.

42. Fleischmann, M. and Tian, Z. Q. (1987). The induction of SERS on smooth Ag by the deposition of Ni and Co. *J. Electroanal. Chem.* **217**, pp. 411–416.
43. Fleischmann, M., Tian, Z. Q. and Li, L. J. (1987). Raman-spectroscopy of adsorbates on thin-film electrodes deposited on silver substrates. *J. Electroanal. Chem.* **217**, pp. 397–410.
44. Leung, L. W. H. and Weaver, M. J. (1987). Extending the metal interface generality of surface-enhanced Raman-spectroscopy: Underpotential deposited layers of mercury, thallium, and lead on gold electrodes. *J. Electroanal. Chem.* **217**, pp. 367–384.
45. Leung, L. W. H. and Weaver, M. J. (1987). Extending surface-enhanced Raman spectroscopy to transition-metal surfaces: Carbon monoxide adsorption and electrooxidation on platinum-and palladium-coated gold electrodes. *J. Am. Chem. Soc.* **109**, pp. 5113–5119.
46. Zou, S. Z. and Weaver, M. J. (1998). Surface-enhanced Raman scattering an uniform transition metal films: Toward a versatile adsorbate vibrational strategy for solid-nonvacuum interfaces?. *Anal. Chem.* **70**, pp. 2387–2395.
47. Stockle, R. M., Suh, Y. D., Deckert, V. and Zenobi, R. (2000). Nanoscale chemical analysis by tip-enhanced Raman spectroscopy. *Chem. Phys. Lett.* **318**, pp. 131–136.
48. Park, S., Yang, P. X., Corredor, P. and Weaver, M. J. (2002). Transition metal-coated nanoparticle films: Vibrational characterization with surface-enhanced Raman scattering. *J. Am. Chem. Soc.* **124**, pp. 2428–2429.
49. Pettinger, B., Ren, B., Picardi, G., Schuster, R. and Ertl, G. (2004). Nanoscale probing of adsorbed species by tip-enhanced Raman spectroscopy. *Phys. Rev. Lett.* **92**, p. 096101.
50. Ren, B., Picardi, G., Pettinger, B., Schuster, R. and Ertl, G. (2005). Tip-enhanced Raman spectroscopy of benzenethiol adsorbed on Au and Pt single-crystal surfaces. *Angew. Chem. Int. Ed.* **44**, pp. 139–142.
51. Wu, D. Y., Li, J. F., Ren, B. and Tian, Z. Q. (2008). Electrochemical surface-enhanced Raman spectroscopy of nanostructures. *Chem. Soc. Rev.* **37**, pp. 1025–1041.
52. Li, J. F., Huang, Y. F., Ding, Y., Yang, Z. L., Li, S. B., Zhou, X. S., Fan, F. R., Zhang, W., Zhou, Z. Y., Wu de, Y., Ren, B., Wang, Z. L. and Tian, Z. Q. (2010). Shell-isolated nanoparticle-enhanced Raman spectroscopy. *Nature* **464**, pp. 392–395.
53. Anema, J. R., Li, J. F., Yang, Z. L., Ren, B. and Tian, Z. Q. (2011). Shell-isolated nanoparticle-enhanced Raman spectroscopy: Expanding the versatility of surface-enhanced Raman scattering. *Annu. Rev. Anal. Chem.* **4**, pp. 129–150.
54. Li, J. F., Tian, X. D., Li, S. B., Anema, J. R., Yang, Z. L., Ding, Y., Wu, Y. F., Zeng, Y. M., Chen, Q. Z., Ren, B., Wang, Z. L. and Tian, Z. Q. (2013). Surface analysis using shell-isolated nanoparticle-enhanced Raman spectroscopy. *Nat. Protoc.* **8**, pp. 52–65.

55. Gersten, J. and Nitzan, A. (1980). Electromagnetic theory of enhanced Raman scattering by molecules adsorbed on rough surfaces. *J. Chem. Phys.* **73**, pp. 3023–3037.
56. Aravind, P., Nitzan, A. and Metiu, H. (1981). The interaction between electromagnetic resonances and its role in spectroscopic studies of molecules adsorbed on colloidal particles or metal spheres. *Surf. Sci.* **110**, pp. 189–204.
57. Murray, C., Allara, D. and Rhinewine, M. (1981). Silver-molecule separation dependence of surface-enhanced Raman scattering. *Phys. Rev. Lett.* **46**, p. 57.
58. Nitzan, A. and Brus, L. (1981). Theoretical model for enhanced photochemistry on rough surfaces. *J. Chem. Phys.* **75**, pp. 2205–2214.
59. Huang, Y. F., Li, C. Y., Broadwell, I., Li, J. F., Wu, D. Y., Ren, B. and Tian, Z. Q. (2011). Shell-isolated nanoparticle-enhanced Raman spectroscopy of pyridine on smooth silver electrodes. *Electrochim. Acta* **56**, pp. 10652–10657.
60. Li, J. F., Ding, S. Y., Yang, Z. L., Bai, M. L., Anema, J. R., Wang, X., Wang, A., Wu, D. Y., Ren, B., Hou, S. M., Wandlowski, T. and Tian, Z. Q. (2011). Extraordinary enhancement of Raman scattering from pyridine on single crystal Au and Pt electrodes by shell-isolated Au nanoparticles. *J. Am. Chem. Soc.* **133**, pp. 15922–15925.
61. Li, J. F., Li, S. B., Anema, J. R., Yang, Z. L., Huang, Y. F., Ding, Y., Wu, Y. F., Zhou, X. S., Wu, D. Y., Ren, B., Wang, Z. L. and Tian, Z. Q. (2011). Synthesis and characterization of gold nanoparticles coated with ultrathin and chemically inert dielectric shells for SHINERS applications. *Appl. Spectrosc.* **65**, pp. 620–626.
62. Li, J. F., Anema, J. R., Yu, Y. C., Yang, Z. L., Huang, Y. F., Zhou, X. S., Ren, B. and Tian, Z. Q. (2011). Core-shell nanoparticle based SERS from hydrogen adsorbed on a rhodium(111) electrode. *Chem. Commun.* **47**, pp. 2023–2025.
63. Li, S. B., Li, L. M., Anema, J. R., Li, J. F., Yang, Z. L., Ren, B., Sun, J. J. and Tian, Z. Q. (2011). Shell-isolated nanoparticle-enhanced Raman spectroscopy (SHINERS) based on gold-core silica-shell nanorods. *Z. Phys. Chem.* **225**, pp. 775–783.
64. Liu, B., Blaszczyk, A., Mayor, M. and Wandlowski, T. (2011). Redox-switching in a viologen-type adlayer: An electrochemical shell-isolated nanoparticle enhanced Raman spectroscopy study on Au(111)-(1×1) single crystal electrodes. *ACS Nano* **5**, pp. 5662–5672.
65. Butcher, D. P., Boulos, S. P., Murphy, C. J., Ambrosio, R. C. and Gewirth, A. A. (2012). Face-dependent shell-isolated nanoparticle enhanced Raman spectroscopy of 2,2′-bipyridine on Au(100) and Au(111). *J. Phys. Chem. C*, **116**, pp. 5128–5140.
66. Honesty, N. R. and Gewirth, A. A. (2012). Shell-isolated nanoparticle enhanced Raman spectroscopy (SHINERS) investigation of benzotriazole film formation on Cu(100), Cu(111), and Cu(poly). *J. Raman Spectrosc.* **43**, pp. 46–50.
67. Lin, X. D., Uzayisenga, V., Li, J. F., Fang, P. P., Wu, D. Y., Ren, B. and Tian, Z. Q. (2012). Synthesis of ultrathin and compact Au@MnO2

nanoparticles for shell-isolated nanoparticle-enhanced Raman spectroscopy (SHINERS). *J. Raman Spectrosc.* **43**, pp. 40–45.

68. Qian, K., Liu, H., Yang, L. and Liu, J. (2012). Functionalized shell-isolated nanoparticle-enhanced Raman spectroscopy for selective detection of trinitrotoluene. *Analyst* **137**, pp. 4644–4646.
69. Uzayisenga, V., Lin, X. D., Li, L. M., Anema, J. R., Yang, Z. L., Huang, Y. F., Lin, H. X., Li, S. B., Li, J. F. and Tian, Z. Q. (2012). Synthesis, characterization, and 3D-FDTD simulation of Ag@SiO_2 nanoparticles for shell-isolated nanoparticle-enhanced Raman spectroscopy. *Langmuir* **28**, pp. 9140–9146.
70. Tian, X. D., Liu, B. J., Li, J. F., Yang, Z. L., Ren, B. and Tian, Z. Q. (2013). SHINERS and plasmonic properties of Au Core SiO2 shell nanoparticles with optimal core size and shell thickness. *J. Raman Spectrosc.* **44**, pp. 994–998.
71. Kunz, K. S. and Luebbers, R. J. (1993). The Finite Difference Time Domain Method for Electromagnetics. CRC Press, Boca Raton, FL.
72. Sherry, L. J., Chang, S. H., Schatz, G. C., Van Duyne, R. P., Wiley, B. J. and Xia, Y. N. (2005). Localized surface plasmon resonance spectroscopy of single silver nanocubes. *Nano Lett.* **5**, pp. 2034–2038.
73. Gui, J. and Devine, T. M. (1991). Obtaining surface-enhanced Raman-spectra from the passive film on iron. *J. Electrochem. Soc.* **138**, pp. 1376–1384.
74. Zheng, J. W., Zhou, Y. G., Li, X. W., Ji, L., Lu, T. H. and Gu, R. A. (2003). Surface-enhanced Raman scattering of 4-aminothiophenol in assemblies of nanosized particles and the macroscopic surface of silver. *Langmuir*, **19**, pp. 632–636.
75. Ikeda, K., Fujimoto, N., Uehara, H. and Uosaki, K. (2008). Raman scattering of aryl isocyanide monolayers on atomically at Au(111) single crystal surfaces enhanced by gap-mode plasmon excitation. *Chem. Phys. Lett.* **460**, pp. 205–208.
76. Mullins, D. R. and Campion, A. (1984). Unenhanced Raman scattering from pyridine chemisorbed on a stepped silver surface: Implications for proposed SERS mechanisms. *Chem. Phys. Lett.* **110**, pp. 565–570.
77. Rodriguez, J. A. and Goodman, D. W. (1992). The nature of the metal-metal bond in bimetallic surfaces. *Science* **257**, pp. 897–903.
78. Zhang, J., Sasaki, K., Sutter, E. and Adzic, R. R. (2007). Stabilization of platinum oxygen-reduction electrocatalysts using gold clusters. *Science* **315**, pp. 220–222.
79. Weast, R. C. and Astle, M. J. (eds.), (1982). CRC Handbook of Chemistry and Physics. CRC Press, Boca Raton, FL.
80. Huang, Y. F., Yin, N. N., Wang, X., Wu, D. Y., Ren, B. and Tian, Z. Q. (2010). Vibrational signature of double-end-linked molecules at Au nanojunctions probed by surface-enhanced Raman spectroscopy. *Chem-Eur, J.* **16**, pp. 1449–1453.
81. Grirrane, A., Corma, A. and Garcia, H. (2008). Gold-catalyzed synthesis of aromatic azo compounds from anilines and nitroaromatics. *Science* **322**, pp. 1661–1664.

82. Wu, D. Y., Liu, X. M., Huang, Y. F., Ren, B., Xu, X. and Tian, Z. Q. (2009). Surface catalytic coupling reaction of p-Mercaptoaniline linking to silver nanostructures responsible for abnormal SERS enhancement: A DFT study. *J. Phys. Chem. C* **113**, pp. 18212–18222.
83. Huang, Y. F., Zhu, H. P., Liu, G. K., Wu, D. Y., Ren, B. and Tian, Z. Q. (2010). When the signal is not from the original molecule to be detected: Chemical transformation of para-aminothiophenol on Ag during the SERS measurement. *J. Am. Chem. Soc.* **132**, pp. 9244–9246.
84. Van Schrojenstein Lantman, E. M., Deckert-Gaudig, T., Mank, A. J. G., Deckert, V. and Weckhuysen, B. M. (2012). Catalytic processes monitored at the nanoscale with tip-enhanced Raman spectroscopy. *Nat. Nanotechnol.* **7**, pp. 583–586.
85. Huang, Y. F., Zhang, M., Zhao, L. B., Feng, J. M., Wu, D. Y., Ren, B. and Tian, Z. Q. (2014). Activation of oxygen on gold and silver nanoparticles assisted by surface plasmon resonances. *Angew. Chem. Int. Ed.* **53**, pp. 2353–2357.
86. Frens, G. (1973). Controlled nucleation for regulation of particle-size in monodisperse gold suspensions. *Nat. Phys. Sci.* **241**, pp. 20–22.
87. Nikoobakht, B. and El-Sayed, M. A. (2003). Preparation and growth mechanism of gold nanorods (NRs) using seed-mediated growth method. *Chem. Mater.* **15**, pp. 1957–1962.
88. Pérez-Juste, J., Correa-Duarte, M. A. and Liz-Marzán, L. M. (2004). Silica gels with tailored, gold nanorod-driven optical functionalities. *Appl. Surf. Sci.* **226**, pp. 137–143.
89. Halas, N. J., Lal, S., Chang, W. S., Link, S. and Nordlander, P. (2011). Plasmons in strongly coupled metallic nanostructures. *Chem. Rev.* **111**, pp. 3913–3961.
90. Ye, X., Jin, L., Caglayan, H., Chen, J., Xing, G., Zheng, C., Doan-Nguyen, V., Kang, Y., Engheta, N. and Kagan, C. R. (2012). Improved size-tunable synthesis of monodisperse gold nanorods through the use of aromatic additives. *ACS Nano* **6**, pp. 2804–2817.
91. Zhang, Y., Luo, Y., Zhang, Y., Yu, Y. J., Kuang, Y. M., Zhang, L., Meng, Q. S., Luo, Y., Yang, J. L. and Dong, Z. C. (2016). Visualizing coherent intermolecular dipole–dipole coupling in real space. *Nature* **531**, pp. 623–627.
92. Zhao, Y., Zhang, Y. J., Meng, J. H., Chen, S., Panneerselvam, R., Li, C. Y., Jamali, S. B., Li, X., Yang, Z. L., Li, J. F. and Tian, Z. Q. (2016). A facile method for the synthesis of large-size Ag nanoparticles as efficient SERS substrates. *J. Raman Spectrosc.* **47**, pp. 662–667.
93. Zhang, W., Dong, J. C., Li, C. Y., Chen, S., Zhan, C., Panneerselvam, R., Yang, Z. L., Li, J. F. and Zhou, Y. L. (2015). Large scale synthesis of pinhole-free shell-isolated nanoparticles (SHINs) using improved atomic layer deposition (ALD) method for practical applications. *J. Raman Spectrosc.* **46**, pp. 1200–1204.
94. Campion, A. and Mullins, D. R. (1983). Normal Raman scattering from pyridine adsorbed on the low-index faces of silver. *Chem. Phys. Lett.* **94**, pp. 576–579.

95. Sakamoto, K., Hashizume, H., Nagafusa, M., Sato, H. and Ushioda, S. (1996). UV resonant Raman spectra of nitrobenzene adsorbed on Ni(111). *Surf. Sci.* **368**, pp. 292–295.
96. Li, J. F., Rudnev, A., Fu, Y. C., Bodappa, N. and Wandlowski, T. (2013). *In situ* SHINERS at electrochemical single-crystal electrode/electrolyte interfaces: Tuning preparation strategies and selected applications. *ACS Nano*, **7**, pp. 8940–8952.
97. Li, J. F., Zhang, Y. J., Rudnev, A. V., Anema, J. R., Li, S. B., Hong, W. J., Rajapandiyan, P., Lipkowski, J., Wandlowski, T. and Tian, Z. Q. (2015). Electrochemical shell-isolated nanoparticle-enhanced Raman spectroscopy: Correlating structural information and adsorption processes of pyridine at the Au(hkl) single crystal/solution interface. *J. Am. Chem. Soc.* **137**, pp. 2400–2408.
98. Creighton, J. A., Blatchford, C. G. and Albrecht, M. G. (1979). Plasma resonance enhancement of Raman-scattering by pyridine adsorbed on silver or gold sol particles of size comparable to the excitation wavelength. *J. Chem. Soc. Faraday Trans.* **2**, pp. 790–798.
99. Wu, D. Y., Ren, B., Jiang, Y. X., Xu, X. and Tian, Z. Q. (2002). Density functional study and normal-mode analysis of the bindings and vibrational frequency shifts of the pyridine-M (M = Cu, Ag, Au, Cu+, Ag+, Au+, and Pt) complexes. *J. Phys. Chem. A* **106**, pp. 9042–9052.
100. Zhao, L. L., Jensen, L. and Schatz, G. C. (2006). Pyridine-Ag-20 cluster: A model system for studying surface-enhanced Raman scattering. *J. Am. Chem. Soc.* **128**, pp. 2911–2919.
101. Wu, D. Y., Liu, X. M., Duan, S., Xu, X., Ren, B., Lin, S. H. and Tian, Z. Q. (2008). Chemical enhancement effects in SERS spectra: A quantum chemical study of pyridine interacting with copper, silver, gold and platinum metals. *J. Phys. Chem. C* **112**, pp. 4195–4204.
102. Stolberg, L., Morin, S., Lipkowski, J. and Irish, D. E. (1991). Adsorption of pyridine at the gold (111)-solution interface. *J. Electroanal. Chem.* **307**, pp. 241–242.
103. Lipkowski, J. and Stolberg, L. (1992). Adsorption of Molecules at Metal Electrodesn. In Lipkowski, J. and Ross, P. N. (eds.). VCH, New York, pp. 171–238.
104. Lipkowski, J., Stolberg, L., Yang, D. F., Pettinger, B., Mirwald, S., Henglein, F. and Kolb, D. M. (1994). Molecular adsorption at metal electrodes, Electrochim. *Acta* **39**, pp. 1045–1056.
105. Stolberg, L., Lipkowski, J. and Irish, D. E. (1987). Adsorption of pyridine at the gold (100)-solution interface. *J. Electroanal. Chem.* **238**, pp. 333–353.
106. Stolberg, L., Lipkowski, J. and Irish, D. E. (1990). Adsorption of pyridine at the gold (110)-solution interface. *J. Electroanal. Chem.* **296**, pp. 171–189.
107. Markovic, N. M. and Ross, P. N. (2002). Surface science studies of model fuel cell electrocatalysts. *Surf. Sci. Rep.* **45**, pp. 121–229.

108. Hughes, M. D., Xu, Y. J., Jenkins, P., McMorn, P., Landon, P., Enache, D. I., Carley, A. F., Attard, G. A., Hutchings, G. J., King, F., Stitt, E. H., Johnston, P., Griffin, K. and Kiely, C. J. (2005). Tunable gold catalysts for selective hydrocarbon oxidation under mild conditions. *Nature* **437**, pp. 1132–1135.
109. Conway, B. E. (1995). Electrochemical oxide film formation at noble-metals as a surface-chemical process. *Prog. Surf. Sci.* **49**, pp. 331–452.
110. Strbac, S. and Adzic, R. R. (1996). The influence of OH-chemisorption on the catalytic properties of gold single crystal surfaces for oxygen reduction in alkaline solutions. *J. Electroanal. Chem.* **403**, pp. 169–181.
111. Li, C. Y., Dong, J. C., Jin, X., Chen, S., Panneerselvam, R., Rudnev, A. V., Yang, Z. L., Li, J. F., Wandlowski, T. and Tian, Z. Q. (2015). *In situ* monitoring of electrooxidation processes at gold single crystal surfaces using shell-isolated nanoparticle-enhanced Raman spectroscopy. *J. Am. Chem. Soc.* **137**, pp. 7648–7651.
112. Desilvestro, J. and Weaver, M. J. (1986). Surface structural-changes during oxidation of gold electrodes in aqueous-media as detected using surface-enhanced Raman-spectroscopy. *J. Electroanal. Chem.* **209**, pp. 377–386.
113. Zhang, Y., Gao, X. P. and Weaver, M. J. (1993). Nature of surface bonding on voltammetrically oxidized noble-metals in aqueous-media as probed by real-time surface-enhanced Raman-spectroscopy. *J. Phys. Chem.* **97**, pp. 8656–8663.
114. Li, X. and Gewirth, A. A. (2003). Peroxide electroreduction on Bi-modified Au surfaces: Vibrational spectroscopy and density functional calculations. *J. Am. Chem. Soc.* **125**, pp. 7086–7099.
115. Hamelin, A. and Martins, A. (1996). Cyclic voltammetry at gold single-crystal surfaces. Part 2. Behaviour of high-index faces. *J. Electroanal. Chem.* **407**, pp. 13–21.
116. Schmidt, T. J., Stamenkovic, V., Arenz, M., Markovic, N. M. and Ross, P. N. (2002). Oxygen electrocatalysis in alkaline electrolyte: Pt(hkl), Au(hkl) and the effect of Pd-modification. *Electrochim. Acta* **47**, pp. 3765–3776.
117. Kim, J. W. and Gewirth, A. A. (2006). Mechanism of oxygen electroreduction on gold surfaces in basic media. *J. Phys. Chem. B* **110**, pp. 2565–2571.
118. Kondo, T., Morita, J., Hanaoka, K., Takakusagi, S., Tamura, K., Takahasi, M., Mizuki, J. I. and Uosaki, K. (2007). Structure of Au (111) and Au (100) single-crystal electrode surfaces at various potentials in sulfuric acid solution determined by in situ surface X-ray scattering. *J. Phys. Chem. C* **111**, pp. 13197–13204.
119. Min, X., Chen, Y. and Kanan, M. W. (2014). Alkaline O2 reduction on oxide-derived Au: high activity and 4e− selectivity without (100) facets. *Phys. Chem. Chem. Phys.* **16**, pp. 13601–13604.
120. Knight, M. W., Sobhani, H., Nordlander, P. and Halas, N. J. (2011). Photodetection with active optical antennas. *Science* **332**, pp. 702–704.
121. Mubeen, S., Lee, J., Singh, N., Krämer, S., Stucky, G. D. and Moskovits, M. (2013). An autonomous photosynthetic device in which all charge carriers derive from surface plasmons. *Nat. Nanotechnol.* **8**, pp. 247–251.

122. Tittl, A., Yin, X., Giessen, H., Tian, X. D., Tian, Z. Q., Kremers, C., Chigrin, D. N. and Liu, N. (2013). Plasmonic smart dust for probing local chemical reactions. *Nano Lett.* **13**, pp. 1816–1821.
123. Li, C. Y., Meng, M., Huang, S. C., Li, L., Huang, S. R., Chen, S., Meng, L. Y., Panneerselvam, R., Zhang, S. J., Ren, B., Yang, Z. L., Li, J. F. and Tian, Z. Q. (2015). "Smart" Ag Nanostructures for Plasmon-Enhanced Spectroscopies. *J. Am. Chem. Soc.* **137**, pp. 13784–13787.
124. Xie, W. and Schlücker, S. (2015). Hot electron-induced reduction of small molecules on photorecycling metal surfaces. *Nat. Commun.* **6**, pp. 7570–7575.
125. Xie, W., Walkenfort, B. and Schlücker, S. (2012). Label-free SERS monitoring of chemical reactions catalyzed by small gold nanoparticles using 3D plasmonic superstructures. *J. Am. Chem. Soc.* **135**, pp. 1657–1660.
126. Hy, S., Felix, Chen, Y. H., Liu, J. Y., Rick, J. and Hwang, B. J. (2014). *In situ* surface enhanced Raman spectroscopic studies of solid electrolyte interphase formation in lithium ion battery electrodes. *J. Power Sources* **256**, pp. 324–328.
127. Hy, S., Felix, F., Rick, J., Su, W. N. and Hwang, B. J. (2014). Direct *In situ* observation of Li2O evolution on Li-rich high-capacity cathode material, Li [Ni×Li (1–2 x)/3Mn (2–x)/3] O2 (0≤ x≤ 0.5). *J. Am. Chem. Soc.* **136**, pp. 999–1007.
128. MacFarlane, D. R., Tachikawa, N., Forsyth, M., Pringle, J. M., Howlett, P. C., Elliott, G. D., Davis, J. H., Watanabe, M., Simon, P. and Angell, C. A. (2014). Energy applications of ionic liquids. *Energ. Environ. Sci.* **7**, pp. 232–250.
129. Lee, D., Lee, S., Seong, G. H., Choo, J., Lee, E. K., Gweon, D. G. and Lee, S. (2006). Quantitative analysis of methyl parathion pesticides in a polydimethylsiloxane microfluidic channel using confocal surface-enhanced Raman spectroscopy. *Appl. Spectrosc.* **60**, pp. 373–377.
130. Smith, S. R., Leitch, J. J., Zhou, C., Mirza, J., Li, S. B., Tian, X. D., Huang, Y. F., Tian, Z. Q., Baron, J. Y., Choi, Y. and Lipkowski, J. (2015). Quantitative SHINERS analysis of temporal changes in the passive layer at a gold electrode surface in a thiosulfate solution. *Anal. Chem.* **87**, pp. 3791–3799.
131. Shen, W., Lin, X., Jiang, C., Li, C., Lin, H., Huang, J., Wang, S., Liu, G., Yan, X., Zhong, Q. and Ren, B. (2015). Reliable quantitative SERS analysis facilitated by core-shell nanoparticles with embedded internal standards. *Angew. Chem. Int. Ed.* **54**, pp. 7308–7312.
132. Zheng, C., Shao, W., Paidi, S. K., Han, B., Fu, T., Wu, D., Bi, L., Xu, W., Fan, Z. and Barman, I. (2015). Pursuing shell-isolated nanoparticle-enhanced Raman spectroscopy (SHINERS) for concomitant detection of breast lesions and microcalcifications. *Nanoscale*, **7**, pp. 16960–16968.
133. Gao, J., Guo, L., Wu, J., Feng, J., Wang, S., Lai, F., Xie, J. and Tian, Z. Q. (2014). Simple and sensitive detection of cyanide using pinhole shell-isolated nanoparticle-enhanced Raman spectroscopy. *J. Raman Spectrosc.* **45**, pp. 619–626.

134. Schulte, F., Mäder, J., Kroh, L. W., Panne, U. and Kneipp, J. (2009). Characterization of pollen carotenoids with in situ and high-performance thin-layer chromatography supported resonant Raman spectroscopy. *Anal. Chem.* **81**, pp. 8426–8433.
135. Li, J. F., Anema, J. R., Wandlowski, T. and Tian, Z. Q. (2015). Dielectric shell isolated and graphene shell isolated nanoparticle enhanced Raman spectroscopies and their applications. *Chem. Soc. Rev.* **44**, pp. 8399–8409.
136. Xu, W. G., Mao, N. N. and Zhang, J. (2013). Graphene: A platform for surface-enhanced Raman spectroscopy. *Small* **9**, pp. 1206–1224.
137. Liu, Y. M., Hu, Y. and Zhang, J. (2014). Few-layer graphene-encapsulated metal nanoparticles for surface-enhanced Raman spectroscopy. *J. Phys. Chem. C* **118**, pp. 8993–8998.
138. Song, Z. L., Chen, Z., Bian, X., Zhou, L. Y., Ding, D., Liang, H., Zou, Y. X., Wang, S. S., Chen, L., Yang, C., Zhang, X. B. and Tan, W. H. (2014). Alkyne-functionalized superstable graphitic silver nanoparticles for Raman imaging. *J. Am. Chem. Soc.* **136**, pp. 13558–13561.
139. Nagashima, A., Tejima, N., Gamou, Y., Kawai, T. and Oshima, C. (1995). Electronic-structure of monolayer hexagonal boron-nitride physisorbed on metal-surfaces. *Phys. Rev. Lett.* **75**, pp. 3918–3921.
140. Ci, L., Song, L., Jin, C. H., Jariwala, D., Wu, D. X., Li, Y. J., Srivastava, A., Wang, Z. F., Storr, K., Balicas, L., Liu, F. and Ajayan, P. M. (2010). Atomic layers of hybridized boron nitride and graphene domains. *Nat. Mater.* **9**, pp. 430–435.
141. Dean, C. R., Young, A. F., Meric, I., Lee, C., Wang, L., Sorgenfrei, S., Watanabe, K., Taniguchi, T., Kim, P., Shepard, K. L. and Hone, J. (2010). Boron nitride substrates for high-quality graphene electronics. *Nat. Nanotechnol.* **5**, pp. 722–726.
142. Ye, J., Wen, F., Sobhani, H., Lassiter, J. B., Van Dorpe, P., Nordlander, P. and Halas, N. J. (2012). Plasmonic nanoclusters: Near field properties of the Fano resonance interrogated with SERS. *Nano Lett.* **12**, pp. 1660–1667.
143. Gallinet, B., Siegfried, T., Sigg, H., Nordlander, P. and Martin, O. J. F. (2013). Plasmonic radiance: Probing structure at the angstrom scale with visible light, *Nano Lett.* **13**, pp. 497–503.
144. Chu, Y., Zhu, W., Wang, D. and Crozier, K. B. (2011). Beamed Raman: directional excitation and emission enhancement in a plasmonic crystal double resonance SERS substrate. *Opt. Express* **19**, pp. 20054–20068.
145. Chu, Y. Z., Wang, D. X., Zhu, W. Q. and Crozier, K. B. (2011). Double resonance surface enhanced Raman scattering substrates: An intuitive coupled oscillator model. *Opt. Express* **19**, pp. 14919–14928.
146. Guerrero, A. R. and Aroca, R. F. (2011). Surface-enhanced fluorescence with shell-isolated nanoparticles (SHINEF). *Angew. Chem. Int. Ed.* **50**, pp. 665–668.
147. Osorio-Roman, I. O., Guerrero, A. R., Albella, P. and Aroca, R. F. (2014). Plasmon enhanced fluorescence with aggregated shell-isolated nanoparticles. *Anal. Chem.* **86**, pp. 10246–10251.

148. Shen, S. X., Meng, L. Y., Zhang, Y., Han, J., Ma, Z., Hu, S., He, Y., Li, J. F., Ren, B., Shih, T. M., Wang, Z., Yang, Z. L. and Tian, Z. Q. (2015). Plasmon-enhanced second-harmonic generation nanorulers with ultrahigh sensitivities. *Nano Lett.* **15**, pp. 6716–6721.
149. Opilik, L., Dogan, U. Z., Li, C. Y., Stephanidis, B., Li, J.F. and Zenobi, R. (2016). Chemical production of thin protective coatings on optical nanotips for tip-enhanced Raman spectroscopy. *J. Phys. Chem. C*, doi: 10.1021/acs.jpcc.6b02147.
150. Ding, S. Y., Yi, J., Li, J. F., Ren, B., Wu, D. W., Panneerselvam, R. and Tian, Z. Q. (2016). Nanostructure-based plasmon-enhanced Raman spectroscopy for surface analysis of materials. *Nat. Rev. Mats.* doi:10.1038/natrevmats.2016.21.

Chapter 9

Ultra-High Vacuum Tip-Enhanced Raman Spectroscopy

Naihao Chiang*, Guillaume Goubert†, Eric A. Pozzi†, Michael O. McAnally†, Craig Chapman†, Nan Jiang‡, George C. Schatz*,† and Richard P. Van Duyne*,†,§

*Applied Physics Graduate Program
†Department of Chemistry, Northwestern University, Evanston, Illinois 60208, USA

‡Department of Chemistry, University of Illinois at Chicago, Illinois 60607, USA
§vanduyne@northwestern.edu

1. Introduction

Optics and photonics are recognized as technologies that have extensive impact on a broad range of fields including healthcare, communications, defense (viz., chemical and biological threats), chemical analysis and imaging, among many others. Plasmonics, a field at the intersection of optics and photonics, utilizes the collective electronic excitations in noble metal nanoparticles and nanostructures to control, manipulate, and amplify light on the nanometer length scale. These important properties translate to new spectroscopic and microscopic techniques that greatly extend the ability of scientists to "see" single molecules and small ensembles working at their most fundamental level. Molecular plasmonics investigates the interaction between surface plasmons and molecules adsorbed to nanoparticle and nanostructured surfaces. The toolbox of techniques used to study molecule–plasmon interactions includes surface-enhanced Raman spectroscopy (SERS) [1–6], localized surface plasmon resonance (LSPR) spectroscopy [1, 7, 8], tip-enhanced Raman spectroscopy (TERS) [9–11], super-resolution SERS and plasmon microscopy [12, 13], surface-enhanced femtosecond stimulated Raman spectroscopy (SE-FSRS) [14, 15], and other plasmonic spectroscopies [16]. This chapter will focus on the developments of ultrahigh vacuum tip-enhanced Raman spectroscopy (UHV-TERS) that chart the pathway to increased understanding of plasmonic spectroscopy and extending its applicability to a broader palette of molecular problems.

2. Ultra-High Vacuum Tip-Enhanced Raman Spectroscopy

There are significant motivations to look at structural information at the nanoscale and study single molecule properties. Important applications include the study of material properties that can be explained and enhanced by engineered nanostructures [17–19], as well as chemical reactivity that can be dominated by active sites positioned at undercoordinated atoms or at a metal/oxide interface on a catalyst [20–22]. Many microscopy techniques have been developed to this end, some of the most common ones include transmission electron microscopy (TEM) and scanning probe microscopy (SPM) that can access atomic resolution.

SPMs capable of being used in many experimental conditions (pressure, temperature) do not require the analytes to be optically tagged and do not involve high energy electrons that could damage organic molecules. The two most prominent techniques in SPM are scanning tunneling microscopy (STM) and atomic force microscopy (AFM). Both of these techniques have been applied to the study of interfaces at the nanoscale in many fields of science.

In STM, a sharp metallic tip is brought very close to a conducting sample, where a bias voltage is then applied between the tip and the sample to allow the tunneling current to flow from one to the other. The tunneling current (I_t) is extremely sensitive to the distance between sample and tip:

$$I_t \propto e^{-2kd} \quad (1)$$

with $k^2 = 2m(V - E)/\hbar^2$, where V is the height of the tunneling barrier (can be approximated to the vacuum level of the sample), d is the tip-sample distance and E the energy of the state in which tunneling occurs. This exponential relationship between tip current and distance gives STM its high spatial resolution. The value of k in Equation (1) is on the order of 1 Å^{-1} [23]. Thus, a change of distance between the tip and the sample (the gap) of 0.1 nm increases the current by roughly an order of magnitude.

For a very simple model of the tip and in the limit of weak coupling between tip and sample, the tunneling current is proportional to the local density of states (LDOS) at the Fermi level of the sample at the position of the tip. Thus, the tunneling current contains a mixture of topological and electronic information. If the presence of an adsorbate decreases the LDOS, this adsorbate will be captured as a depression in the image, even though it is protruding from the surface. In the most common mode of operation, the tunneling current is maintained constant in a feedback loop by changing the gap with a piezoelectric actuator. Another pair of piezoelectric actuators is used to scan the tip in the plane of the sample and acquire an image of the sample. Under feedback, the tip position changes in reaction to changes in the tunneling current: if the current decreases, the piezoelectric actuator will extend to decrease the gap and increase the current and vice versa. In mild imaging conditions, the STM tip is expected to be at a distance >3 Å from the point of contact with the sample, and the chemical interaction between the tip and the sample should be negligible. The point of contact is defined as the position where the conductance of the tunneling junction becomes independent of the tip–sample distance when the tip approaches the sample [24, 25].

However, STM alone cannot capture detailed chemical information such as adsorbate–adsorbate interactions and adsorbate–binding site interactions except for few-atom molecules using inelastic electron tunneling spectroscopy at liquid helium temperatures.

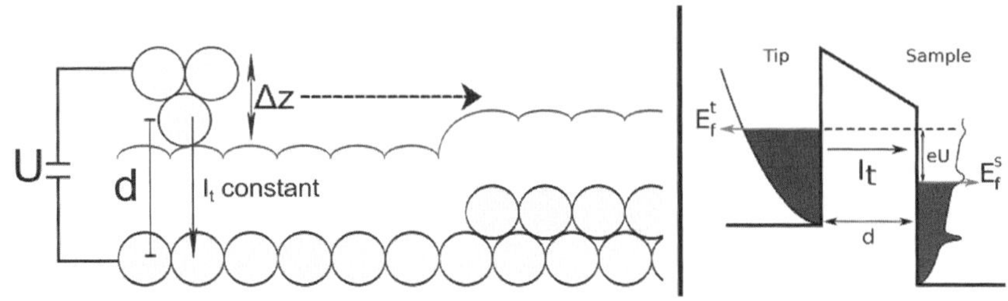

(a) (b)

Figure 1. (a) Principle of the STM showing the tip at a distance d from the sample. U is the amplitude of the bias in V. I_t is the tunneling current. The tip moves perpendicular to the sample to maintain the current constant. (b) STM principle in terms of energy, the bias allows the current to flow from tip into energy levels of the sample.

Tip-enhanced Raman spectroscopy (TERS) combines the imaging capabilities of STM with the chemical information provided by Raman spectroscopy and thereby bridges this knowledge gap. In TERS, illumination of noble metal nanostructures at the apex of a scanning probe with electromagnetic radiation causes collective electron oscillations on the metal surface, known as localized surface plasmon resonances (LSPR). LSPR induces intense local electromagnetic fields, causing a drastic (10^8) enhancement of the Raman scattering signals emanating from molecules located in close vicinity to the metal surface. Therefore, spatial resolution in TERS is determined by the spatial confinement of the electromagnetic field and not by the wavelength of the illuminating light [26–29]. In addition to continuous-wave excitation, pulsed electromagnetic radiation can be used. Pulsed excitation allows for studying molecular dynamics near plasmonic surfaces, which is highly important to understand mechanisms of plasmon-mediated chemical reactions, such as plasmonically enhanced photocatalytic or photovoltaic processes.

2.1. *Benefits of UHV environment*

As an important part of TERS technology, SPM can be used to provide atomic spatial resolution images. However, maintaining clean nanoscale structures over the time scale of typical experiments requires a level of chemical purity that cannot be obtained through normal cleaning procedures. Open-air environments provide too many opportunities for the contamination of the sample surface. Keeping the sample in an ultrahigh vacuum (UHV) chamber lowers the chance of contamination. Moreover, *in situ* preparation of samples in a clean environment prevents the introduction of contaminants during deposition of molecules of interest and ensures that the Raman signals come from the adsorbates under study. In the case of sample preparation under ambient conditions, airborne contaminants as well as traces of solvents or contaminants that were present in the liquid solution can complicate the analysis of TER spectra.

In addition, the UHV environment can protect plasmonic metal TERS tips, especially silver, from oxidation and passivation. Thus, to take full advantage of TERS, a UHV work environment is in high demand. However, the optics alignment for TERS in a UHV environment adds another level of complexity. By using *ex situ* laser focusing and Raman collection optical, optimization of the optical path can be achieved without compromising the vacuum. On the other hand, *in situ* optics can provide tighter focus and larger collection solid angle since the focal lenses could be very close to the tip–sample junction.

In 2007, Pettinger and coworkers demonstrated the design of STM for the tip-enhanced optical processes, which was employed in UHV [32]. They mounted an optical platform with all necessary optical elements to a rigid frame with STM in a UHV system. In 2008, they reported the first UHV-TERS mapping of a single brilliant cresyl blue (BCB) molecule with a lateral resolution of 15 nm combining molecular resolution with STM imaging [33]. They also demonstrated that a UHV environment can decrease the photobleaching rate of molecules, allowing longer data collection time. In 2009, J. Maultzsch and coworkers reported their development of the TERS setup in UHV [34]. An additional inverted view-port was mounted to a UHV-AFM system. Both groups prepared their samples using the solution of BCB molecules, which may significantly influence the quality of the scanning imaging.

When the sample preparation and tip cleaning is carried out under UHV conditions, atomically clean surfaces can be maintained. UHV conditions greatly enhance the stability of the tip–sample

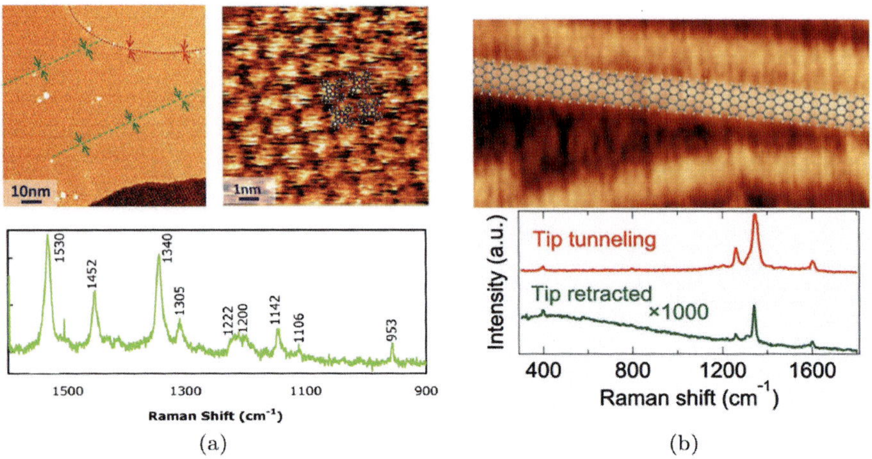

Figure 2. (a) Molecular resolution topographic STM image of a CuPc adlayer on Ag(111) with the tip illuminated by the CW laser, and UHV-TERS spectra of a CuPc adlayer on Ag(111). (b) Molecular resolution topographic STM imaging and TERS of graphene nanoribbons (GNRs) fabricated on Au(111) by the on-surface polymerization technique under UHV conditions. Reprinted with permission from Refs. [30, 31].

junction and ensure minimal contamination in the field enhancement region beneath the STM tip. Following this approach, Van Duyne and coworkers demonstrated multiple vibrational modes for large polyatomic molecular adsorbates on a well-defined surface in TER spectra obtained concurrently with molecular resolution STM image [30]. A sub-nm molecular resolution STM revealed subtle features in the images, including the molecular orientation and domain boundaries in the CuPc adlayer (Figure 2(a)). In 2014, Wolf and coworkers reported a TERS study of graphene nanoribbons prepared on an Au(111) surface under UHV conditions [31]. Molecular resolution imaging, as shown in Figure 2(b), provides detailed structural information about the polymerization process and supports the interpretation of the UHV-TER spectra. With this development, TERS has demonstrated an unprecedented level of spatial and spectroscopic resolution.

In the context of UHV-TERS, STM is the only SPM technique currently used in the literature. We anticipate that this will change.

Non-contact mode AFM (NC-AFM), a method using tuning forks to transform the force experienced by the probe into a frequency shift, has been used to achieve atomic resolution in organic molecule and inorganic solids [35], but no NC-AFM UHV-TERS system has been demonstrated to date.

2.2. *Low-temperature UHV-TERS*

UHV-TERS at cryogenic temperature is a more promising approach for unraveling the intricacies of adsorbate–substrate interactions that are inaccessible by other means (Figure 3(a)). Liquid helium temperature (19 K) TER spectra differed from those taken at room temperature due to decreased linewidths. Shifting of peaks

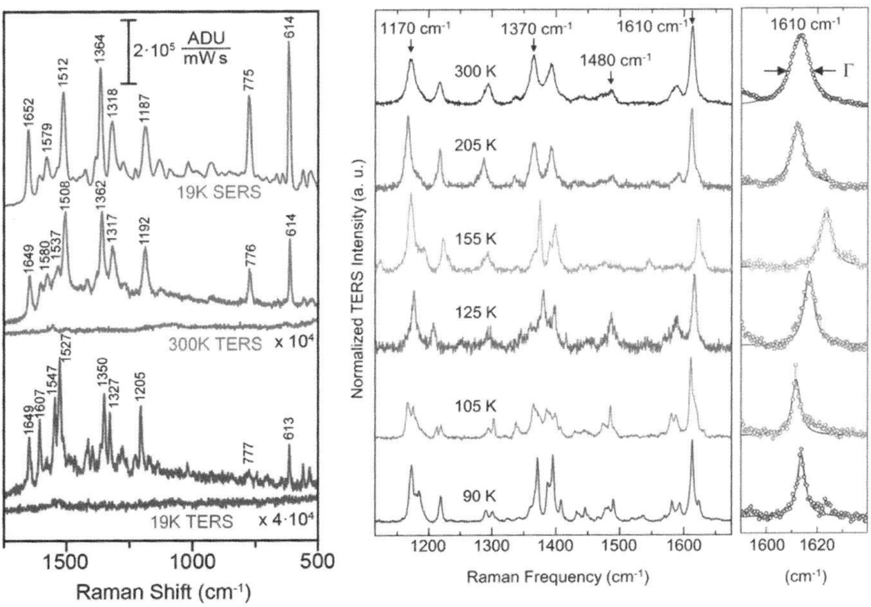

(a) (b)

Figure 3. (a) R6G surface diffusion is suppressed at 19 K. Its adsorption geometry on Ag(111) can be characterized by the differences between the 300 K and 19 K UHV-TERS. (b) Temperature-dependent TERS spectra of MG. Spectra are background subtracted and intensity normalized. Lorentzian line fit analysis of the N−C stretch mode at 1610 cm^{-1} as an example (Right graph of (b), narrowest spectra are selected from a large data set). Reprinted with permission from Refs. [36, 37].

at low temperature revealed additional chemical information about the adsorbate–substrate interactions. Specifically, the orientation of R6G on Ag(111) was determined and corroborated by time-dependent density-functional theory (TDDFT) calculations [37]. This experiment indicates that the technique is capable of determining interactions between molecules and specific binding sites, a very important feature in understanding structural, physical and chemical properties of single molecules and nanostructures in various environments.

The investigation of intramolecular coupling, adsorbate–substrate coupling and dynamic processes is possible by performing variable temperature TERS, in particular at cryogenic temperature. The Raschke group reported a temperature-dependent study of malachite green (MG), prepared *ex situ*, with shear-force AFM under high vacuum conditions. Between 90 and 300 K, TERS spectra were acquired at six different temperatures (Figure 3(b)) [36]. The width of the Raman peaks decreased with temperature, and several of them split revealing new peaks, a finding which was assigned to intramolecular coupling. Furthermore, the statistical analysis of fluctuating spectra at low temperature showed the approach measures single-molecule local dynamics with structural sensitivity. No spatial information was retrieved, most likely due to the preparation of samples being performed outside of vacuum.

2.3. *Demonstration of lateral resolution below 5 nm*

The major advantage afforded by TERS is its capability of obtaining chemical information at sub-5 nm length scale on surfaces. In contrast to the near-field scanning optical microscopy (NSOM), the spatial resolution of TERS is not limited by the size of the aperture (~20 nm).

In 2008, Pettinger and coworkers published the first single molecule chemical image obtained by UHV-TERS [33]. They spatially mapped the TERS signal while raster scanning a gold STM tip over an isolated BCB molecule on Au(111). Spatial resolution of ~15 nm was demonstrated with TERS. The original hypotheses for the lateral resolution of TERS was that the radius of curvature

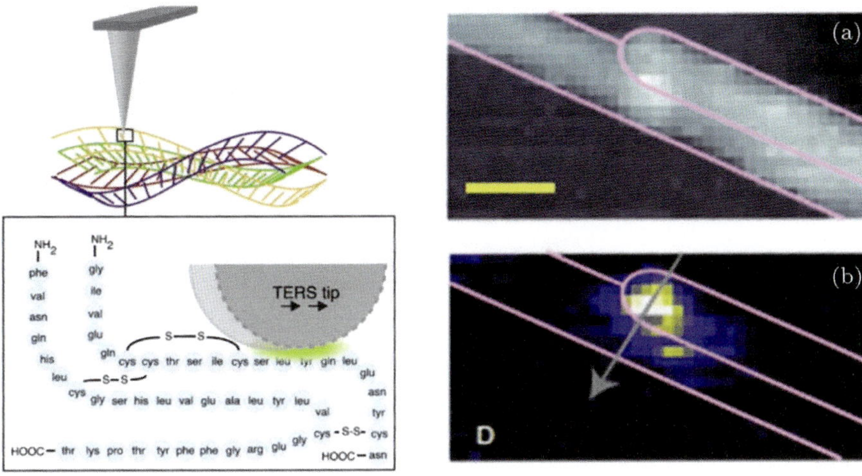

(A) (B)

Figure 4. (A) Schematic of ambient AFM-TERS on a single amyloid fibril. (B) Ambient AFM-TERS image of a single wall carbon nanotube. Reprinted with permission from Refs. [37, 38].

of the SPM tips at the apex (~15–20 nm) was the limiting factor of the localized electromagnetic field. However, recent studies by both the Deckert and Kawata Groups (Figure 4) have independently demonstrated lateral resolution below 2 nm with ambient AFM-TERS [38, 39].

Thanks to the pristine environment provided by UHV, the full chemical imaging capabilities of TERS for molecular systems have recently been achieved for three distinct surface-bound molecular systems: an isolated molecule [40], two different analytes segregated into different condensed phases [41], and a single analyte that forms different phases on a surface [42].

Dong and co-workers first demonstrated sub-nanometer lateral resolution on an isolated porphyrin molecules at liquid nitrogen temperature with a UHV-STM-based TERS system [40]. Their TER images allow the visualization of the inner structure of a single molecule by monitoring the changes in Raman intensity as the STM tip scans across the molecule (Figure 5(a)). Also demonstrated in this work was the ability to correlate the observed Raman signal with a

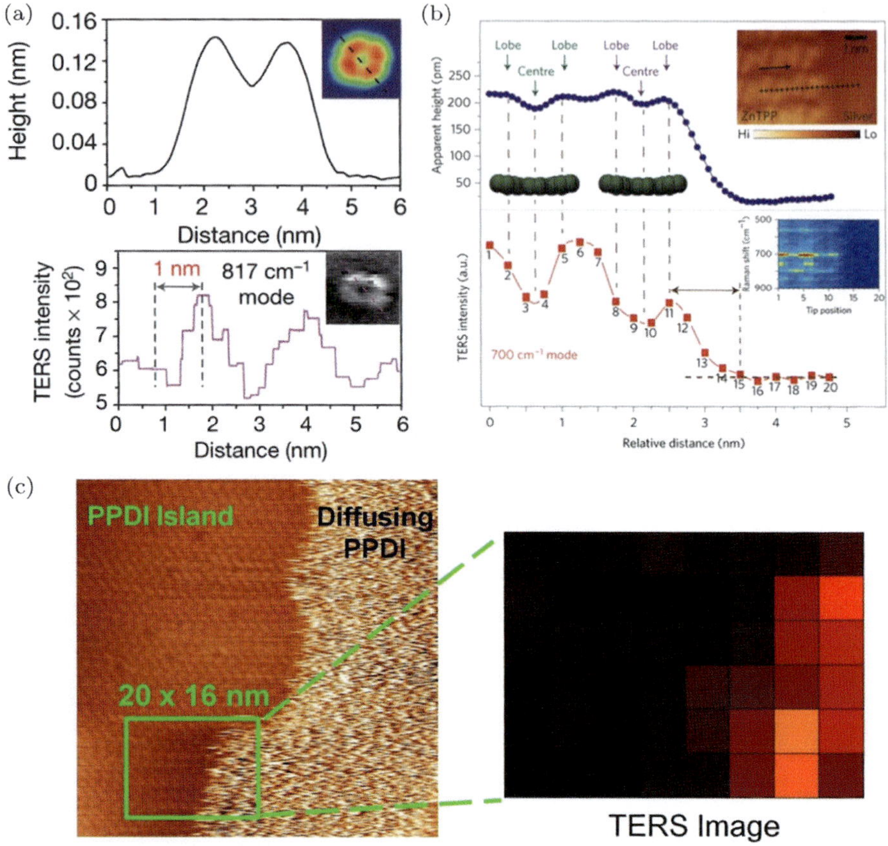

Figure 5. (a) Sub-nanometer resolution TERS image of an isolated porphyring molecule at 80 K. (b) ~0.5 nm resolution TERS mapping of adjecent porphyrin molecules at 80 K. (c) ~4 nm resolution TERS image on a dynamic 2D phase boundary of a perylene diimide derivative at room temperature. Reprinted with permission from Refs. [40–42].

molecule's local environment using TERS imaging. This work makes internal structure determination inside one molecule possible.

Utilizing the high resolution of TERS imaging under UHV conditions, Jiang *et al.* have unambiguously distinguished two similar porphyrin molecules which aggregated into separate molecular islands on a silver substrate under 80 K [41]. They have achieved ~0.5 nm lateral resolution and thus enable the ability to obtain Raman spectrum of adjacent molecules which adsorbed on the step edge of the silver single crystal (Figure 5(b)).

Also performed under UHV conditions, a dynamic 2D phase boundary has been imaged by TERS under room temperature [42]. In this study, the condensed phase (molecular island) and 2D gas phase of the perylene diimide derivative coexist on the Ag surface. Around 4 nm resolution of TERS image has been achieved by mapping out the dynamic phase boundary (Figure 5(c)). Also demonstrated in this study, TERS shows more intense signal in the 2D gas phase than in the condensed phase due to different molecular orientations. Therefore, TERS provides access to molecular adsorption geometries in the regions lacking STM topographical information.

TERS has set the stage for future high-resolution chemical imaging studies at few Angstrom levels. However, the detailed mechanistic understanding of such high lateral resolution remains unknown. Clearly, the original explanation which is derived purely from classical electrodynamics theories is insufficient to explain resolutions below 5 nm.

2.4. *Possible theories for the high spatial resolution of TERS*

Explaining the sub-nanometer lateral resolution UHV-STM-TERS from Jiang *et al.* [41] and Zhang *et al.* [40] has been of great interest to the theoretical community. Some of the approaches to explain the high resolution beyond the predicted spatial resolution due to the physical tip nanostructure described in this section include optomechanical cavities [43], nonlinear Raman scattering effects [44, 45], and previously formulated SERS theories including EM mechanism, image field effect, and electric field gradients [46, 47].

The optomechanical cavity effects in surface-enhanced Raman scattering were explored by Roelli *et al.*, who observe nonlinear plasmonic enhancement effects in certain physical limits [43]. The theory of the optomechanical cavity applied to SERS and TERS relies on a coherent coupling between the molecular vibrations (mechanical oscillator) and the localized surface plasmon (electromagnetic cavity). The molecular vibration perturbs the plasmon such that the

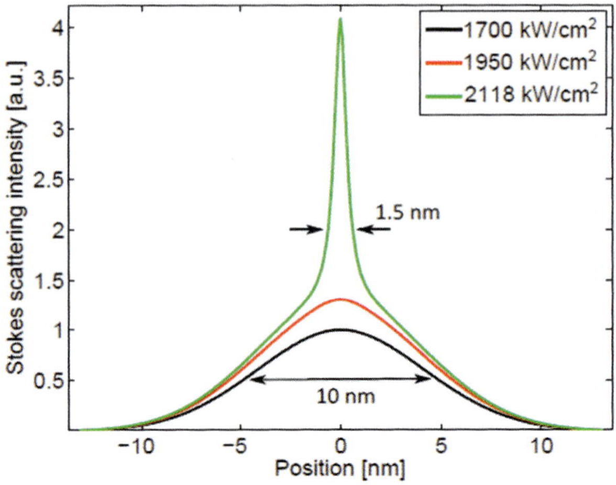

Figure 6. Comparison of theoretical Stokes scattering intensity as a function of lateral position from a point-source scatter at the origin and a Gaussian plasmonic field of 10 nm FWHM. At high field strengths, a strong nonlinear response is observed leading to an increased spatial resolution. Reprinted with permission from Ref. [43].

instantaneous plasmon occupancy changes which then acts back on the molecular vibration.

By pushing the optomechanical cavity to limits exceeding experimental capabilities, a strong nonlinear plasmonically enhanced Raman response is observed (Figure 6). The nonlinear Raman response can be interpreted as a possible explanation to the high lateral resolution TERS of Zhang *et al.* [40]. However, we note that typical UHV-STM-TERS experiments rarely exceed 1 mW/cm^2, thus the nonlinear region observed in the optomechanical cavity theory is unlikely to be accessed by UHV-STM-TERS.

Other justifications for the increased lateral resolution of Zhang *et al.* [40] by nonlinear Raman scattering effects have suggested that TERS can be effectively described by stimulated Raman scattering (SRS) [44–46]. These reports develop a theory of SRS where the two fields, Raman pump and Raman probe, are given by the incident laser and the optical plasmon response to the incident laser. However, the SRS theory developed is not fully representative of the UHV-STM-TERS observations. Particularly, the experiments done were on molecular electronic resonance which when observed in resonant femtosecond stimulated Raman scattering (FSRS) experiments leads to dispersive lineshapes due to the interference of multiple Feynman pathways for the light–matter interaction [48], which is not observed in TERS experiments. In addition, the theory in question was derived in a limit that is applicable to all surface-enhanced Raman experiments suggesting an alternative to the established EM mechanism of SERS. Finally, the SRS theory developed gives a higher lateral resolution of the TERS signal by deriving a higher order function for the spatial confinement of the plasmonic field than that expected by the EM mechanism. Using the higher order confinement from SRS, Duan *et al.* [45] showed higher lateral resolution in TERS by the nonlinear Raman effects compared to normal SERS effects (Figure 7).

Finally we discuss two theoretical approaches that while previously applied to SERS, are applied to TERS to attempt to explain the higher lateral resolution observed by Zhang *et al.* [40]: image field

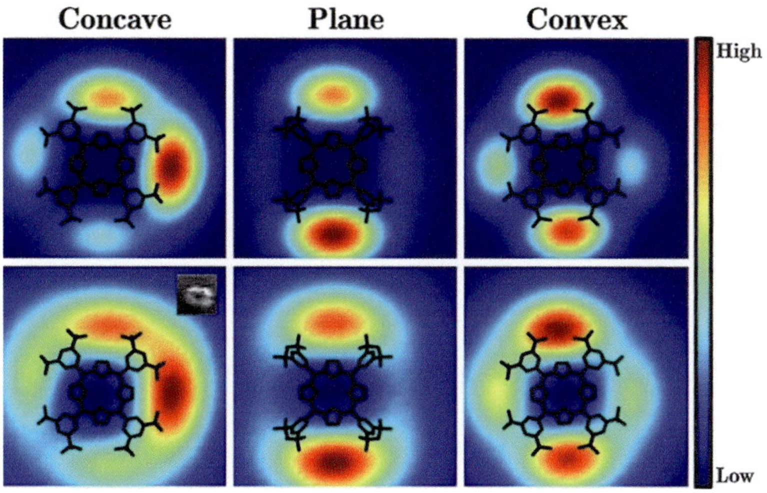

Figure 7. Comparison of nonlinear (top) and total (bottom) tip-enhanced Raman images of a H_2TBPP molecule under concave, planar, and convex configurations by a plasmonic field with a FWHM of 20 Å. Inset in the bottom left image is the experimental data from Ref. [40]. As seen, the nonlinear Raman response can lead to tighter confinement for the Raman image and higher lateral resolution. Reprinted with permission from Ref. [45].

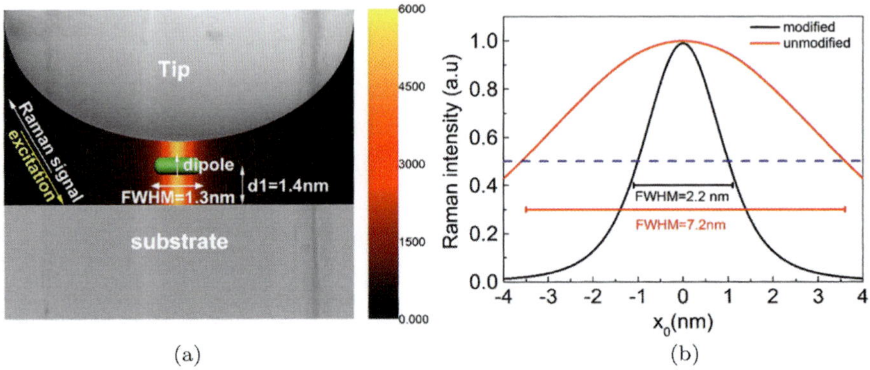

Figure 8. (a) Diagram of coupled molecular dipole–nanogap TERS system. (b) Comparison of TERS lateral resolution in the cases of no molecular self-interaction (red) and inclusion of molecular self-interaction (black). Including the self-interaction leads to higher lateral resolution in TERS imaging. Reprinted with permission from Ref. [47].

effect [47] and electric field gradient effects [46]. The image field effect of SERS [49] was considered by the Li group as a self-interaction of the molecular dipole with an image dipole in the substrate. The stronger spatial dependence of self-interaction leads to higher lateral resolution in TERS imaging as shown in Figure 8. The rationale of the higher lateral resolution comes from the self-interaction relying on multiple elastic scattering events within the Ag tip–substrate gap. While this theory does provide a higher lateral resolution than that predicted by conventional EM mechanisms, it does not reach the sub-nanometer resolution observed. In addition, work from Marinica *et al.* suggests that the spatial regime where image field and self-interaction effects would strongly contribute is overtaken by quantum plasmonic effects [50]. The effect of electric field gradients on lateral resolution in TERS was explored by Meng *et al.* where they looked at a system of H_2TBPP molecules with plasmonic tips [46]. The authors show that electric field gradients can lead to higher TERS lateral resolution than predicted by traditional EM mechanisms; however, these effects occur in very small tip–substrate distances where, again, quantum plasmonic effects are expected to contribute greatly [50, 51].

In conclusion, the source of high lateral resolution TERS imaging in the sub-nanometer limit is still unresolved. Mechanisms that rely on optomechanical cavities require the use of sufficiently high laser power regimes that they are not realistic, SRS mechanisms lead to unobserved experimental signatures, and other mechanisms rely on distance regimes that have stronger quantum plasmonic effects. Hence, more theoretical studies on high resolution TERS are necessary to fully address the experimental observables from Zhang *et al.* [40]

3. Toward Ultra-fast TERS

Processes occurring at ultra-fast timescales are critical to the function of chemical systems, both natural and man-made. Vibrational motion, proton transfer, and transition state formation and destruction, to name a few, take place on timescales of 10 ps or faster [52]. In order to spectroscopically investigate such transient processes and species, nonlinear techniques are required. Ultra-fast Raman techniques typically involve pulses compressed to a few picoseconds or shorter. Extending these studies to the nanoscale may become possible by incorporating ultra-fast spectroscopic techniques with TERS. A great challenge in doing so lies in the perturbation of the probed system by the high local EM fields experienced at the tip–sample junction under irradiation by ultrashort pulses. These pulses are produced in a pulse train operated at a fixed frequency (commonly 100 kHz to 100 MHz). As a result, duty cycles, which refer to the fraction of time the system is active, in ultrafast spectroscopy are of order 10^{-9} to 10^{-4}. Peak powers within ultrashort pulses are then orders of magnitude higher than the average power of the laser. However, in an effort to keep the signal above the noise in ultrafast spectroscopy, modest average laser power must be used, resulting in transiently intense fields within each individual pulse.

It is well known that energy flow within ultrashort pulses can be strong enough to damage irradiated media. In fact, this characteristic has led to the use of laser pulses for material removal with micron or better precision in areas including micromachining and ophthalmic surgery. When combined with plasmonic enhancement, near-field intensity can become prohibitively high. As previously discussed, the electric field is enhanced of order 100× in a TERS junction. The effects of plasmonic enhancement and pulsed excitation then compound, resulting in instantaneously intense near-field strengths. Multiple groups have demonstrated the morphological changes, including complete melting, that can occur when plasmonic structures are irradiated by ultrafast pulses [53, 54]. Even more sensitive to ultrashort pulses are the organic analytes studied in plasmonically enhanced ultrafast experiments, which can readily undergo physical and chemical modifications in these environments [55]. In a previous report of plasmonically enhanced ultrafast Raman spectroscopy, in this case surface-enhanced femtosecond stimulated Raman spectroscopy (SE-FSRS), a clear plateau in signal gain was observed with increasing power of the Raman pump (1 ps pulses), which the authors attribute to a damage mechanism involving either the probed molecules or the plasmonic substrates [14].

The Van Duyne group previously conducted experiments to assess the feasibility of integrating pulsed laser excitation with TERS. Sonntag *et al.* collected TERS of R6G and malachite green isothiocyanate (MGITC) in atmosphere using 1 ps pulses from an optical parametric oscillator (OPO) with a repetition rate of 80 MHz [57]. Spontaneous TERS was successfully collected, but the signal was observed to decay irreversibly on the timescale of tens of seconds. If the tip was moved to a different surface location within the laser focus, however, signal would transiently reappear. Signal decay was then attributed to a molecular damage mechanism as opposed to one involving damage to the plasmonic activity of the tip. Analysis of the signal decay allowed the authors to postulate a mechanism involving reactive decay chemistry of the photoexcited molecules with small molecules present in ambient (e.g. O_2).

The pristine environments achievable in UHV provide a means to mitigate photodissociation in pulsed-excitation TERS. Pozzi *et al.* subsequently used the same OPO described above to collect UHV-TERS of R6G using picosecond excitation [56]. Multimodal TERS was observed only with the tip in close proximity to the Ag(111) surface (Figure 9). Importantly, the signal was not observed to decay

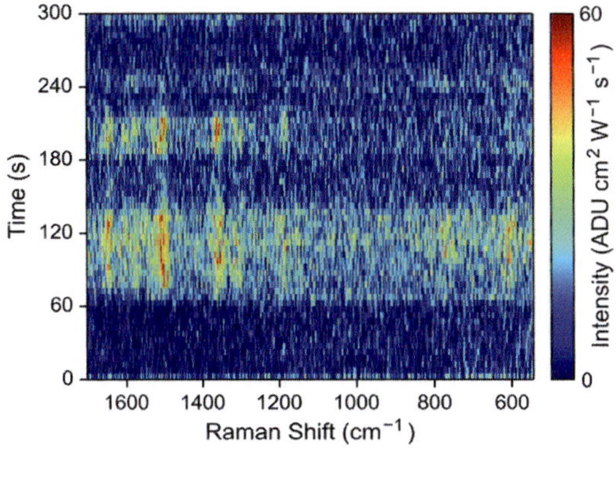

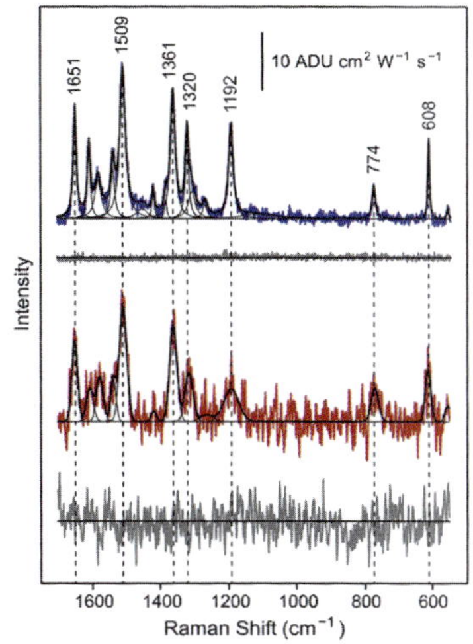

Figure 9. (a) Waterfall plot of sixty 5 s TER spectra of R6G collected using ps irradiation ($\lambda = 532\,\text{nm}$, $P_{\text{ex}} = 0.05\,\text{W/cm}^2$). The initially retracted Ag tip was brought into tunneling range at $t = 60\,\text{s}$. (b) CW TER engaged (blue) and retracted (gray) spectra ($\lambda = 532\,\text{nm}$, $P_{\text{ex}} = 1.3\,\text{W/cm}^2$, $t_{\text{acq}} = 30\,\text{s}$) plotted above averages of 12 ps TER engaged spectra with signal/noise ratios above 12 (red) and 12 retracted spectra (gray) from (a). Black lines represent individual and composite peak fits. The included scale bar applies to all plotted spectra. Reprinted with permission from Ref. [56].

irreversibly as in ambient conditions, but fluctuation in the signal intensity prevented kinetic analysis of the UHV-TER signal with continued irradiation. The authors determined the effects of UHV on signal persistence using SERS, wherein averaging over many hot spots reduces intensity fluctuations. Decay was observed to be slowed by a factor of 5 in UHV compared to that in atmosphere. Amplifying signal decay using higher excitation intensity enabled a critical analysis of the decay mechanism, which was determined to be a result of surface diffusion of the R6G across the silver surface. While signal decay in pulsed excitation TERS can arise from photochemistry as well as surface diffusion, a UHV environment is clearly an asset.

Advances in ultrafast TERS have also been made in ambient. The Wickramasinghe group reported stimulated TERS of a thiolated azobenzene derivative on an evaporated Au film [58]. Here, the authors reported an estimated stimulated gain of 1.0×10^9, thus increasing sensitivity compared to spontaneous TERS. Using this increased sensitivity, TER images were collected with the imaging rates limited not by TER signal strength but instead by the STM scanning rate. Using the ideas presented in this chapter, a pulsed Raman beam overlapped in time with the stimulating beam would allow stimulated TERS to resolve ultrafast dynamics of surface-bound molecules.

The Kawata group has reported integration of TERS with a well-established nonlinear Raman technique, coherent anti-Stokes Raman spectroscopy (CARS). In initial experiments, individual modes of robust molecular systems, namely, adenine encased in DNA nanocrystals [59, 60] and single-wall carbon nanotubes (SWNTs) [61], were probed with narrow band excitation from mode-locked Ti:sapphire lasers. Tip enhancement of the signal-enabled samples to be imaged and resulting optical images exhibited higher spatial resolution than topographic imaging, which was argued to be a result of the nonlinearity of the CARS process. These reports touch upon the inevitability of an electronic CARS signal arising from the third-order nonlinear susceptibility of Ag. This issue was not explored in depth, but is an important consideration in any surface- or tip-enhanced CARS experiment. Later, a so-called broadband variant of the technique applied to semiconducting carbon nanotubes (s-CNTs) was reported [62]. In this report, a delay between pulses allowed the kinetics of near-field signal decay to be discerned. While only one mode of the s-CNTs was visible in the collected spectra, the study serves as a proof of concept for future TE-CARS experiments using complete molecular vibrational fingerprints.

4. Summary and Outlook

UHV-TERS has developed into a powerful tool for chemical interrogation at sub-molecular resolution. Interest in the technique has grown considerably, and multiple companies have commercialized UHV-TERS instruments. Using either commercial [40] or homebuilt [30, 32] instruments, chemical investigation at single-molecule level has been routinely achieved. The observed TER signal not only identifies the surface-bounded analytes but also interrogates their internal analytes but also interrogates their internal structures and relevant interactions

Moving forward, successful combination of multiplex stimulated Raman techniques with TERS will enable investigations of transient behavior in molecular systems with nanoscale spatial resolution. Subpopulations and potentially single molecules may be interrogated, allowing surface sites and adsorption configurations to be differentiated and correlated with transient kinetics. Resulting experiments have the potential to characterize ultrafast processes at the molecular level with unprecedented detail.

References

1. Haes, A. J., Haynes, C. L., McFarland, A. D., Schatz, G. C., Van Duyne, R. P. and Zou, S. (2005). Plasmonic materials for surface-enhanced sensing and spectroscopy. *MRS Bull.* **30**(5), pp. 368–375.
2. Haynes, C. L., McFarland, A. D. and Van Duyne, R. P. (2005). Surface-enhanced Raman spectroscopy. *Anal. Chem.* **77**(17), pp. 338A–346A.
3. Haynes, C. L., Yonzon, C. R., Zhang, X. Y. and Van Duyne, R. P. (2005). Surface-enhanced Raman sensors: early history and the development of sensors for quantitative biowarfare agent and glucose detection. *J. Raman Spectrosc.* **36**(6–7), pp. 471–484.
4. Lal, S., Grady, N. K., Kundu, J., Levin, C. S., Lassiter, J. B. and Halas, N. J. (2008). Tailoring plasmonic substrates for surface enhanced spectroscopies. *Chem. Soc. Rev.* **37**(5), pp. 898–911.
5. Schatz, G. C. and Van Duyne, R. P. (2006). Electromagnetic Mechanism of Surface-Enhanced Spectroscopy. In *Handbook of Vibrational Spectroscopy.* John Wiley & Sons, Ltd., Hoboken, NJ.
6. Stiles, P. L., Dieringer, J. A., Shah, N. C. and Van Duyne, R. R. (2008). Surface-Enhanced Raman Spectroscopy. In *Annual Review of Analytical Chemistry.* Annual Reviews: Palo Alto, pp. 601–626.
7. Anker, J. N., Hall, W. P., Lyandres, O., Shah, N. C., Zhao, J. and Van Duyne, R. P. (2008). Biosensing with plasmonic nanosensors. *Nature Mater.* **7**(6), pp. 442–453.
8. Willets, K. A. and Van Duyne, R. P. (2007). Localized surface plasmon resonance spectroscopy and sensing. In *Annual Review of Physical Chemistry.* Annual Reviews: Palo Alto, pp. 267–297.
9. Pettinger, B. (2006). Tip-enhanced Raman spectroscopy (TERS). Surface-Enhanced Raman Scattering, *Phys. Appl.* **103**, pp. 217–240.
10. Pettinger, B. (2010). Single-molecule surface- and tip-enhanced Raman spectroscopy. *Mole. Phys.* **108**(16), pp. 2039–2059.

11. Pettinger, B., Schambach, P. and Ojeda, C. V. (2011). Advances in tip-enhanced Raman spectroscopy. Abstracts of Papers of the *Am. Chem. Soc.* **242**, p. 1.
12. Sagle, L. B., Ruvuna, L. K., Bingham, J. M., Liu, C. M., Cremer, P. S. and Van Duyne, R. P. (2012). Single plasmonic nanoparticle tracking studies of solid supported bilayers with ganglioside lipids. *J. Am. Chem. Soc.* **134**(38), pp. 15832–15839.
13. Stranahan, S. M. and Willets, K. A. (2010). Super-resolution optical imaging of single-molecule SERS hot spots. *Nano Lett.* **10**(9), pp. 3777–3784.
14. Frontiera, R. R., Henry, A.-I., Gruenke, N. L. and Van Duyne, R. P. (2011). Surface-enhanced femtosecond stimulated Raman spectroscopy. *J. Phys. Chem. Lett.* **2**(10), pp. 1199–1203.
15. Frontiera, R. R., Gruenke, N. L. and Van Duyne, R. P. (2012). Fano-like resonances arising from long-lived molecule-plasmon interactions in colloidal nanoantennas. *Nano Lett.* **12**(11), pp. 5989–5994.
16. Grubisic, A., Ringe, E., Cobley, C. M., Xia, Y. N., Marks, L. D., Van Duyne, R. P. and Nesbitt, D. J. (2012). Plasmonic near-electric field enhancement effects in ultrafast photoelectron emission: Correlated spatial and laser polarization microscopy studies of individual Ag nanocubes. *Nano Lett.* **12**(9), pp. 4823–4829.
17. Poudel, B., Hao, Q., Ma, Y., Lan, Y., Minnich, A., Yu, B., Yan, X., Wang, D., Muto, A., Vashaee, D., Chen, X., Liu, J., Dresselhaus, M. S., Chen, G. and Ren, Z. (2008). High-thermoelectric performance of nanostructured bismuth antimony telluride bulk alloys. *Science* **320**(5876), pp. 634–638.
18. Lu, K., Lu, L. and Suresh, S. (2009). Strengthening materials by engineering coherent internal boundaries at the nanoscale. *Science* **324**(5925), pp. 349–352.
19. Gleiter, H. (2000). Nanostructured materials: Basic concepts and microstructure. *Acta Mater.* **48**(1), pp. 1–29.
20. Bell, A. T. (2003). The impact of nanoscience on heterogeneous catalysis. *Science* **299**(5613), pp. 1688–1691.
21. Lauritsen, J. V., Kibsgaard, J., Helveg, S., Topsoe, H., Clausen, B. S., Laegsgaard, E. and Besenbacher, F. (2007). Size-dependent structure of MoS2 nanocrystals. *Nat. Nano.* **2**(1), pp. 53–58.
22. Lai, X. and Goodman, D. W. (2000). Structure–reactivity correlations for oxide-supported metal catalysts: new perspectives from STM. *J. Mole. Catal. A, Chem.* **162**(1–2), pp. 33–50.
23. Hansma, P. K. and Tersoff, J. (1987). Scanning tunneling microscopy. *J. Appl. Phys.* **61**(2), pp. R1–R24.
24. Gimzewski, J. K. and Möller, R. (1987). Transition from the tunneling regime to point contact studied using scanning tunneling microscopy. *Phys. Rev. B* **36**(2), pp. 1284–1287.
25. Kim, H. and Hasegawa, Y. (2015). Site-dependent evolution of electrical conductance from tunneling to atomic point contact. *Phys. Rev. Lett.* **114**(20), p. 206801.

26. Stockle, R. M., Suh, Y. D., Deckert, V. and Zenobi, R. (2000). Nanoscale chemical analysis by tip-enhanced Raman spectroscopy. *Chem. Phys. Lett.* **318**(1–3), pp. 131–136.
27. Pettinger, B., Picardi, G., Schuster, R. and Ertl, G. (2000). Surface enhanced Raman spectroscopy: Towards single molecular spectroscopy. *Electrochemistry* **68**(12), pp. 942–949.
28. Hayazawa, N., Inouye, Y., Sekkat, Z. and Kawata, S. (2000). Metallized tip amplification of near-field Raman scattering. *Opt. Commun.* **183**(1–4), pp. 333–336.
29. Anderson, M. S. (2000). Locally enhanced Raman spectroscopy with an atomic force microscope. *Appl. Phys. Lett.* **76**(21), pp. 3130–3132.
30. Jiang, N., Foley, E. T., Klingsporn, J. M., Sonntag, M. D., Valley, N. A., Dieringer, J. A., Seideman, T., Schatz, G. C., Hersam, M. C. and Van Duyne, R. P. (2011). Observation of multiple vibrational modes in ultrahigh vacuum tip-enhanced Raman spectroscopy combined with molecular-resolution scanning tunneling microscopy. *Nano Lett.* **12**(10), pp. 5061–5067.
31. Shiotari, A., Kumagai, T. and Wolf, M. (2014). Tip-enhanced Raman spectroscopy of graphene nanoribbons on Au(111). *J. Phys. Chem. C* **118**(22), pp. 11806–11812.
32. Steidtner, J. and Pettinger, B. (2007). High-resolution microscope for tip-enhanced optical processes in ultrahigh vacuum. *Rev. Sci. Instrum.* **78**(10), p. 103104.
33. Steidtner, J. and Pettinger, B. (2008). Tip-enhanced Raman spectroscopy and microscopy on single dye molecules with 15 nm Resolution. *Phys. Rev. Lett.* **100**(23), p. 236101.
34. Peica, N., Rohrig, S., Rudiger, A., Brose, K., Thomsen, C. and Maultzsch, J. (2009). Characterization of dye molecules and carbon nanostructures by tip-enhanced Raman spectroscopy. *Phys. Status Solidi B, Basic Solid State Phys.* **246**(11–12), pp. 2708–2712.
35. Altman, E. I., Baykara, M. Z. and Schwarz, U. D. (2015). Noncontact atomic force microscopy: An emerging tool for fundamental catalysis research. *Accounts Chem. Res.* **48**(9), pp. 2640–2648.
36. Park, K. D., Muller, E. A., Kravtsov, V., Sass, P. M., Dreyer, J., Atkin, J. M. and Raschke, M. B. (2016). Variable-temperature tip-enhanced Raman spectroscopy of single-molecule fluctuations and dynamics. *Nano Lett.* **16**(1), pp. 479–487.
37. Klingsporn, J. M., Jiang, N., Pozzi, E. A., Sonntag, M. D., Chulhai, D., Seideman, T., Jensen, L., Hersam, M. C. and Van Duyne, R. P. (2014). Intramolecular insight into adsorbate-substrate interactions via low-temperature, ultrahigh-vacuum tip-enhanced Raman spectroscopy. *J. Am. Chem. Soc.* **136**(10), pp. 3881–3887.
38. Chen, C., Hayazawa, N. and Kawata, S. (2014). A 1.7 nm resolution chemical analysis of carbon nanotubes by tip-enhanced Raman imaging in the ambient. *Nat. Commun.* **5**, p. 3312.

39. Deckert-Gaudig, T., Kämmer, E. and Deckert, V. (2012). Tracking of nanoscale structural variations on a single amyloid fibril with tip-enhanced Raman scattering. *J. Biophoton.* **5**(3), pp. 215–219.
40. Zhang, R., Zhang, Y., Dong, Z. C., Jiang, S., Zhang, C., Chen, L. G., Zhang, L., Liao, Y., Aizpurua, J., Luo, Y., Yang, J. L. and Hou, J. G. (2013). Chemical mapping of a single molecule by plasmon-enhanced Raman scattering. *Nature* **498**(7452), pp. 82–86.
41. Jiang, S., Zhang, Y., Zhang, R., Hu, C., Liao, M., Luo, Y., Yang, J., Dong, Z. and Hou, J. G. (2015). Distinguishing adjacent molecules on a surface using plasmon-enhanced Raman scattering. *Nat. Nano.* **10**(10), pp. 865–869.
42. Jiang, N., Chiang, N., Madison, L. R., Pozzi, E. A., Wasielewski, M. R., Seideman, T., Ratner, M. A., Hersam, M. C., Schatz, G. C. and Van Duyne, R. P. (2016). Nanoscale chemical imaging of a dynamic molecular phase boundary with ultrahigh vacuum tip-enhanced Raman spectroscopy. *Nano Lett.* **16**(6), pp. 3898–3904.
43. Roelli, P., Galland, C., Piro, N. and Kippenberg, T. J. (2016). Molecular cavity optomechanics as a theory of plasmon-enhanced Raman scattering. *Nat Nano.* **11**(2), pp. 164–169.
44. Duan, S., Tian, G. and Luo, Y. (2016). Visualization of vibrational modes in real space by tip-enhanced non-resonant Raman spectroscopy. *Angew. Chem. Inter. Ed.* **55**(3), pp. 1041–1045.
45. Duan, S., Tian, G., Ji, Y., Shao, J., Dong, Z. and Luo, Y. (2015). Theoretical modeling of plasmon-enhanced Raman images of a single molecule with subnanometer resolution. *J. Am. Chem. Soc.* **137**(30), pp. 9515–9518.
46. Meng, L., Yang, Z., Chen, J. and Sun, M. (2015). Effect of electric field gradient on sub-nanometer spatial resolution of tip-enhanced Raman spectroscopy. *Sci. Rep.* **5**, p. 9240.
47. Zhang, C., Chen, B.-Q. and Li, Z.-Y. (2015). Optical origin of subnanometer resolution in tip-enhanced Raman mapping. *J. Phys. Chem. C* **119**(21), pp. 11858–11871.
48. Frontiera, R. R., Shim, S. and Mathies, R. A. (2008). Origin of negative and dispersive features in anti-Stokes and resonance femtosecond stimulated Raman spectroscopy. *J. Chem. Phy.* **129**(6), p. 064507.
49. King, F. W., Van Duyne, R. P. and Schatz, G. C. (1978). Theory of Raman scattering by molecules adsorbed on electrode surfaces. *J. Chem. Phys.* **69**(10), pp. 4472–4481.
50. Marinica, D. C., Kazansky, A. K., Nordlander, P., Aizpurua, J. and Borisov, A. G. (2012). Quantum plasmonics: Nonlinear effects in the field enhancement of a plasmonic nanoparticle dimer. *Nano Lett.* **12**(3), pp. 1333–1339.
51. Zuloaga, J., Prodan, E. and Nordlander, P. (2009). Quantum description of the plasmon resonances of a nanoparticle dimer. *Nano Lett.* **9**(2), pp. 887–891.
52. Zewail, A. H. (2000). Femtochemistry: Atomic-scale dynamics of the chemical bond. *J. Phys. Chem. A* **104**(24), pp. 5660–5694.
53. Fujiwara, H., Yanagida, S. and Kamat, P. V. (1999). Visible laser induced fusion and fragmentation of thionicotinamide-capped gold nanoparticles. *J. Phys. Chem. B* **103**(14), pp. 2589–2591.

54. Link, S., Burda, C., Mohamed, M., Nikoobakht, B. and El-Sayed, M. A. (1999). Laser photothermal melting and fragmentation of gold nanorods: energy and laser pulse-width dependence. *J. Phys. Chem. A* **103**(9), pp. 1165–1170.
55. Littleford, R. E., Cunningham, D., Matousek, P., Towrie, M., Parker, A. W., Khan, I., McComb, D. and Ewen Smith, W. (2005). Surface-enhanced resonance Raman scattering using pulsed and continuous-wave laser excitation. *J. Raman Spectrosc.* **36**(6–7), pp. 600–605.
56. Pozzi, E. A., Sonntag, M. D., Jiang, N., Chiang, N., Seideman, T., Hersam, M. C. and Van Duyne, R. P. (2014). Ultra-high vacuum tip-Enhanced Raman spectroscopy with picosecond excitation. *J. Phys. Chem. Lett.* **5**(15), pp. 2657–2661.
57. Klingsporn, J. M., Sonntag, M. D., Seideman, T. and Van Duyne, R. P. (2013). Tip-enhanced Raman spectroscopy with picosecond pulses. *J. Phys. Chem. Lett.* **5**(1), pp. 106–110.
58. Wickramasinghe, H. K., Chaigneau, M., Yasukuni, R., Picardi, G. and Ossikovski, R. (2014). Billion-fold increase in tip-enhanced Raman signal. *ACS Nano* **8** (4), pp. 3421–3426.
59. Ichimura, T., Hayazawa, N., Hashimoto, M., Inouye, Y. and Kawata, S. (2004). Application of tip-enhanced microscopy for nonlinear Raman spectroscopy. *Appl. Phys. Lett.* **84**(10), pp. 1768–1770.
60. Ichimura, T., Hayazawa, N., Hashimoto, M., Inouye, Y. and Kawata, S. (2004). Tip-enhanced coherent anti-Stokes Raman scattering for vibrational nanoimaging. *Phys. Rev. Lett.* **92**(22), p. 220801.
61. Hayazawa, N., Ichimura, T., Hashimoto, M., Inouye, Y. and Kawata, S. (2004). Amplification of coherent anti-Stokes Raman scattering by a metallic nanostructure for a high resolution vibration microscopy. *J. Appl. Phys.* **95**(5), pp. 2676–2681.
62. Furusawa, K., Hayazawa, N., Catalan, F. C., Okamoto, T. and Kawata, S. (2012). Tip-enhanced broadband CARS spectroscopy and imaging using a photonic crystal fiber based broadband light source. *J. Raman Spectrosc.* **43**(5), pp. 656–661.

Chapter 10

Tip-Enhanced Raman Spectroscopy for Surface and Interface Analysis

Jin-Hui Zhong, Xiang Wang, Teng-Xiang Huang,
Sheng-Chao Huang and Bin Ren*

*State Key Laboratory of Physical Chemistry of Solid Surfaces,
The MOE Key Laboratory of Spectral Analysis and Instrumentation,
iChEM, and College of Chemistry and Chemical Engineering,
Xiamen University, Xiamen 361005, China*
**bren@xmu.edu.cn*

1. Introduction

Surface-enhanced Raman spectroscopy (SERS) is a technique that utilizes the localized surface plasmon resonance (SPR) effect of metal nanostructures (mainly Au, Ag, or Cu) under proper light illumination to enhance the electromagnetic field in the near field of metal and thus the Raman signal of species in close vicinity to the metal surface [1–4]. The discovery of SERS, which can enhance the surface Raman signal by over a million times, has significantly pushed Raman spectroscopy in becoming a powerful and versatile spectroscopic tool. Such a high enhancement overcomes the inherent low sensitivity of Raman spectroscopy (as a result of the extremely small Raman cross-section of molecules, usually in the order of $10^{-29}\,\text{cm}^2\,\text{sr}^{-1}$). In the past four decades, different strategies have been developed to extend SERS from coinage metals to other metal or semiconductor surfaces, such as "borrowing SERS", SERS from pure transition metals, and shell-isolated nanoparticle-enhanced Raman spectroscopy (SHINERS) (Chapter 8, Jian-Feng Li, ZQ Tian). Therefore, SERS can now be considered as a tool with material generality and morphology generality.

However, both SERS (regardless of different working modes) and SHINERS belong to the conventional optical method that suffers a limited spatial resolution of hundreds of nanometers (roughly half of the wavelength of light) due to the diffraction of light. Therefore, the obtained information is an average of large amounts of different surface sites, making it highly challenging to correlate specific surface structure to its properties. Such site-specific information is, however, vital for the understanding of various surface/interfacial processes that are surface structure sensitive, especially on heterogeneous material surfaces. Therefore, it is crucial to develop new (enhanced) Raman techniques with spatial resolution beyond the diffraction limit of light. Indeed, pushing the spatial resolution down to nanometer or even atomic scale has been a longstanding target in electron and optical microscopies. In 1928, Synge proposed to pass light through an aperture with a dimension smaller than the wavelength of light to create a sub-diffraction limit light source [5]. This idea was realized to be the aperture scanning near-field microscopy (SNOM) [6]. When the aperture fiber is brought into close distance (within near-field regime, i.e. a few nanometers) to the sample surface, it can achieve an optical resolution determined by the dimension of the aperture. Although the aperture SNOM is an elegant way to break the diffraction limit, the throughput of light from the aperture is pretty low, severely limiting its vast applications.

In 1982, scanning tunneling microscopy (STM) was invented, allowing atomic scale resolution imaging of the morphology of a surface [7]. In 1985, Wessel proposed a surface-enhanced optical microscopy by illuminating laser on a single metal nanoparticle [8]. The nanoparticle has two functions: (i) to excite surface plasmon for the enhancement of electromagnetic field and thus the optical signal; and (ii) to confine the optical field to the adjacent space (near-field) of the nanoparticle to achieve a spatial resolution comparable or even smaller than the size of the nanoparticle. The nanoparticle can be controlled by a scanning probe microscope (SPM) to do a raster scan of a sample. In principle, optical imaging with a spatial resolution of a few tens of nanometers or even several nanometers can be obtained. Wessel's idea is a prototype of aperture-less SNOM. After that, many groups have attempted to realize such ideas. The Kawata group demonstrated the detection of scattered light from the near field of a tip made of stainless steel in 1994 [9]. They pointed out a future direction of combining surface-enhanced spectroscopies in their system. In 1995, the Tian group combined STM and SERS aiming to simultaneously obtain the morphology and Raman signal of an electrode surface [10]. Indeed, different approaches to combine SPM with SERS have been growing in 1990s. In 1999, Xie *et al.* realized tip-enhanced two-photon fluorescence measurements [11].

In 2000, the groups of Zenobi [12], Anderson [13], and Kawata [14] independently realized tip-enhanced Raman spectroscopy (TERS) for the first time, using plasmonic silver or gold tips with proper laser excitation. In these pioneering experiments, a clear contrast in the detected signals was observed when the tip was engaged and retracted. Zenobi group studied C_{60} molecules and BCB molecules deposited as a thin layer on a glass substrate using silver-coated AFM tip [12]. Anderson group used gold-coated AFM tip to enhance the Raman signal of thin films of sulfur deposited on a quartz substrate [13]. They both used dielectric substrates and the tip solely acted as a nanoantenna to support the surface plasmon. When the tip was retracted, they can hardly detect Raman signal from the molecules on the substrate. Kawata group also used Ag-coated AFM tips and instead of dielectric substrate, they used rough Ag films as substrate modified with rhodamine 6G (Rh6G) [14]. Such a substrate was SERS active and the far-field spectra can be easily detected even when the tip was retracted. However, they observed a clear enhancement when the tip was brought into contact with the sample, indicating a strong tip-enhanced effect. Later in 2000, Pettinger group reported TERS results using STM and sharp metallic Ag tips electrochemically etched from Ag wire [15]. They obtained TERS signal for monolayer BCB molecules adsorbed on smooth gold films. Different from the method of Kawata group, the smooth gold substrate did not generate SERS signal and all the signals obtained were from near-field TERS. Note that in both experiments of Kawata and Pettinger

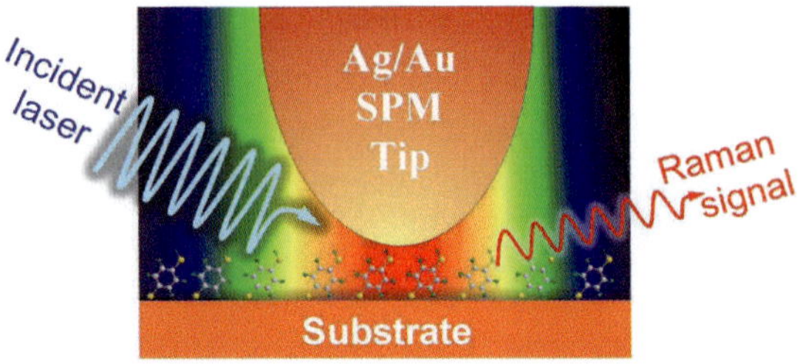

Figure 1. Schematic illustration of principle of TERS. An SPM controlled silver or gold tip is placed above the target substrate at a distance of few nanometers. Laser illumination induces localized SPR that can enhance the electromagnetic field and the Raman signal of species in close vicinity of the tip apex. The red zone indicates the highly localized enhanced region that enables the nanometer spatial resolution of TERS.

groups, there was a strong plasmonic coupling between the tip and the substrate (later known as "gap mode") and a well-defined hot spot was formed in the gap between tip and metallic substrate. Such a gap mode leads to much stronger plasmonic enhancement than the dielectric substrate case, as will be discussed later.

The invention of TERS is truly a breakthrough in the field of plasmon-enhanced spectroscopies. Figure 1 shows a scheme of TERS in which a sharp plasmonic tip (usually made of gold or silver) is brought into close distance to the sample surface and used as the Raman signal amplifier. The spatial resolution of TERS is generally comparable or even smaller than the radius of tip apex ($\sim$20 nm). Impressively, sub-nanometer spatial resolution has been achieved very recently [16, 17]. The most striking feature of TERS that differs with SERS and SHINERS is its high spatial resolution, making it promising to obtain the site specific information. In principle, TERS can be performed on any material surface regardless of morphology, which significantly expands the application of plasmon-enhanced Raman spectroscopy (PERS). TERS is an organic integration of SPM techniques and plasmon-enhanced effect. It can not only image the morphology of a surface at the nanometer/atomic resolution as normal SPM, but also offers chemical identification of the probe materials with Raman fingerprint information, which allows to explore a clear structure–function relationship of various materials. TERS has been growing into an important surface analysis tool in the past 16 years.

In this chapter, we first discuss about the electromagnetic enhancement mechanism in TERS. We then introduce the development of various methods to fabricate tips suitable for different SPM systems as the tip is the most important part of TERS, determining its sensitivity and spatial resolution. Following the discussion of determination of enhancement factor of TERS, we discuss the application of TERS for surface and interface analysis (including surface processes on single-crystal surfaces molecular electronics) in ambient air ultrahigh vacuum, aqueous solution and electrochemical environment. Finally, we provide a perspective of future developments of TERS in surface and interface analysis.

2. Enhancement Mechanism of TERS

2.1. *Electromagnetic enhancement in TERS*

Plasmon-enhanced spectroscopies such as SERS and TERS have been developed into powerful surface and interface analysis tools with the contribution of SPR effect of metal nanostructures. The collective oscillation of free electrons in metal nanostructures under the excitation of light, which is called the SPR effect, is able to efficiently increase the surface charge density of metals, resulting in a strongly enhanced electric-field on metal surfaces. As a result, the optical processes such as Raman and fluorescence will be enhanced. The field intensity decreases exponentially with the increased distance from the surface, which makes SERS/TERS a surface sensitive technique. The efficiency of SPR effect depends on the material, geometry, size, shape, and environment of metal nanostructures. Among all the materials, coinage metals such as gold and silver are still the best choices to achieve high SPR efficiency in the visible region due to their large density and mobility of free electrons.

Note that the field enhancement of simple isolated nanostructure is still quite limited, usually in the order of 10–30. Fortunately, when two isolated metal structures approach each other, and under the proper light excitation, their electric fields start to interact. Such an interaction increases dramatically with the decreasing separation between the nanoparticles, leading to huge field enhancement in the gap. When the separation is below 2–5 nm (forming the so-called nanogap), the field enhancement can be extremely high, giving SERS and TERS techniques the single molecule sensitivity. The position that can produce extremely large enhancement is called the hotspot.

2.2. *Optimizing TERS enhancement*

In TERS, a sharp metallic tip approaches the surface of a sample controlled by SPM. When a beam of light with appropriate wavelength and polarization is focused at the tip apex, the surface charge will accumulate at the tip apex due to SPR and lightning rod effect, resulting in an electric field enhancement in the vicinity of the tip apex. It is well recognized that TERS enhancement is dominantly contributed by the electromagnetic effect. Therefore, the factors determining the SPR effect may significantly influence the TERS enhancement, including the excitation laser, the tip, and substrate, as given in Figure 2. As a rule of thumb, the excitation laser should match the SPR response of the TERS system (which should

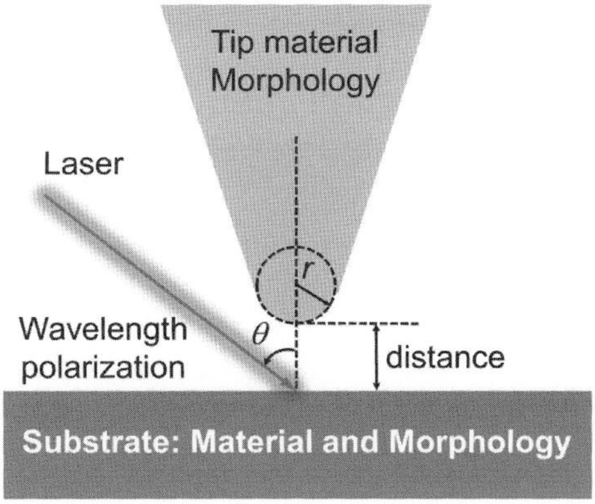

Figure 2. Scheme to illustrate the factors that influence the TERS enhancement, including (i) the wavelength and polarization of the laser, and the incident angle (θ) of the laser with respect to tip axis. (ii) Tip: the material and morphology of the tip, and the radius (r) of tip apex. (iii) Substrate: the material and morphology of the substrate. (iv) The distance between tip and substrate Redrawn with permission from Ref. [18], Copyright Springer-Verlag, Berlin, Heidelberg, 2015.

consider the influence of the substrate beneath) in each individual experiment to obtain optimized signals.

In the following, we discuss the effects of different factors determining the enhancement of TERS.

2.2.1. *Incident laser*

Polarization: In TERS, the enhanced electric field needs to be confined to the very end of the tip apex to achieve a high TERS enhancement. Consequently, a p-polarized incident beam may produce a stronger near-field enhancement than that of s-polarized beam, independent of the illumination configuration. The radially polarized beam has been widely used in the inverted illumination configuration, due to a large component of longitudinal field, which can effectively excite the localized SPR at the tip apex. As a result, it can produce a 5–10 times stronger signal compared with that excited by the linear polarized laser [19, 20].

Incident angle (θ): For the top and inverted illumination mode, the incident angle is simply determined by the numerical aperture (NA) of the objective. An objective with large NA will not only provide a large electric field component of the excitation laser along the tip axis, but also ensure a high collection efficiency. For the side illumination mode, the field enhancement is not very sensitive (within one order of magnitude) to the incident angle. The optimized incident angle is in the range of 40–60° from the tip axis for Au tip–Au substrate system excited at 633 nm [18].

Wavelength: Maximum field enhancement can be achieved when the wavelength of excitation laser matches the localized SPR of the tip. However, it is impractical for most of the labs to scan the wavelength of the laser. In comparison, it is much more feasible to change the SPR frequency to match the excitation laser to achieve a high-field enhancement.

2.2.2. *Tip and substrate*

The tip plays a crucial role in TERS. It should possess not only a proper shape to provide the high electric field enhancement at a specific excitation wavelength, but also an atomic sharp end to probe the surface morphology at a high spatial resolution. Due to their strong plasmonic effect in the visible region, silver and gold are the two most commonly employed materials to fabricate TERS tips. The imaging quality has to be compromised due to the high mobility of the gold and silver atom under ambient conditions. The tip usually is brought to a distance of about 1–2 nm to the sample surface, as the field enhancement decreases dramatically with the increased distance. Field enhancement and spatial resolution are the two key points to evaluate the performance of TERS, which strongly rely on the choice of substrates, geometry and morphology of tips.

Substrate: The TERS working modes can be categorized into two types according to the material of substrates: uncoupled and coupled types. The uncoupled-type TERS involves the use of dielectric substrates (mica, SiO_2, polymer, etc.) that lack free electrons. As a result, there is no efficient electromagnetic coupling between the dielectric material and the tip, resulting in a low enhancement. The coupled-type TERS involves the use of coinage metal substrates, such as Au or Ag. The high free electron density and mobility of these metals allows an effective electromagnetic coupling with the tip, leading to the coupled plasmon resonance. Such a coupled plasmon resonance results in a strong-field enhancement, which is two to four orders of magnitude higher than that of the uncoupled type, and thus enables the single-molecule detection sensitivity. The coupled plasmon resonance also leads to a tightly confined electric field and a higher spatial resolution. Other transition metals, such as Pt, Pd, which are important in surface science and heterogeneous catalysis, have a lower density of free conduction electrons compared with that of coinage metals, leading to relatively weak enhancement. Anyhow, the enhancement from transition metal substrates is still about 100 times stronger than that of dielectric ones, ensuring a reasonable TERS intensity from these surfaces for studying catalytic systems.

Tip radius (r): The size effect of the tip on enhancement is relatively complicated to analyze, because the size of both the tip and the substrate now is a substantial fraction of or even exceeds the excitation wavelength, where the electrostatic approximation fails. However, there are still some empirical trends. The radius of tip apex determines the electric field distribution (more precisely, the extending range of the electric field) in the vicinity of the tip. Most of the theoretical studies reveal that the spatial resolution of TERS is comparable with or smaller than the tip radius and usually increases with the decreased radius, both for uncoupled and coupled types of TERS. In terms of the field enhancement, it also increases with decreased radius for the uncoupled-type TERS because of the lightning rod effect and the decreased radiation loss of SPR. Regarding the coupled-type TERS, there is an optimal tip radius. We have found in our AFM TERS study that the maximum enhancement for coupled-type TERS can be achieved when the tip radius is in the range of 50–80 nm, for Au tip and Au substrate system under 633 nm excitation [21].

Tip morphology: From a basic understanding, when the gold/silver tip is excited by a light polarized along the tip axis, the surface charge will oscillate at the same frequency as the exciting field. The surface charges form an oscillating standing wave and support the surface plasmon. Therefore, the induced surface charges have a large chance to accumulate at the apex of the tip and give strongly enhanced electric field. In this regard, the tip should possess a smooth surface to avoid the scattering loss of plasmon. Recently, there are reports on the further engineering of the electrochemical etched tip and the use of AFM TERS tip with discrete nanoparticles [22].

2.3. *New development in the understanding of TERS*

In the framework of the classic electromagnetic theory, most of the phenomena in TERS can be fully explained. However, recent works of TERS in ultra-high vacuum (UHV) environment show a sub-nanometer resolution, as shown in Figure 3(a) [16]. The extremely high spatial resolution and the ultrahigh confinement of the electric-field enhancement cannot be fully understood with classic electromagnetic theory. They described the TERS process as an analogue to the broadband femtosecond stimulated Raman scattering, in which the laser acts as the Raman pump and the surface plasmon acts as the Raman probe (Figure 3(b)). The sub-nanometer resolution inspired quite a few theoreticians to understand the TERS and

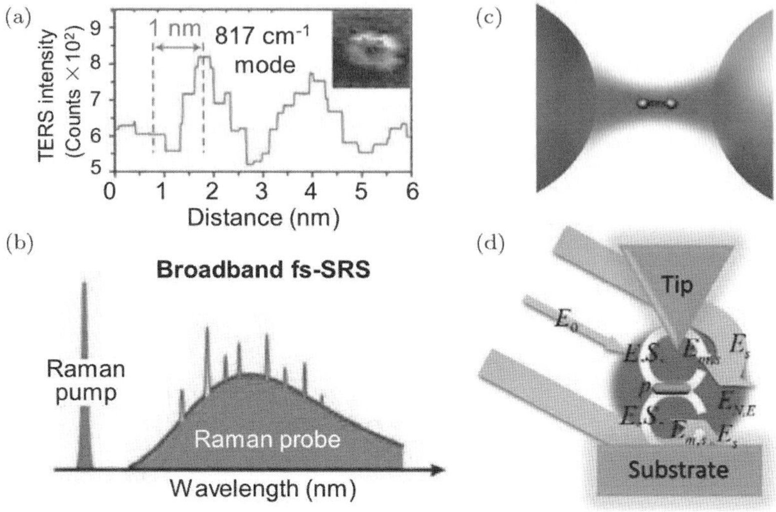

Figure 3. (a) TERS intensity profile of the line trace in the inset Raman map. (b) A schematic draw of the broadband femtosecond stimulated Raman scattering. Reproduced with permission from Ref. [16], copyright 2013 Macmillan Publishers Limited. (c) Schematic mapping of a plasmonic hot spot hosting a molecule with a Raman-active internal vibrational mode sketched as two masses connected by a spring. Reproduced with permission from Ref. [23], copyright 2016 Macmillan Publishers Limited. (d) Excitation process of spontaneous Raman enhancement when considering the molecule self-interaction with the Ag tip–substrate gap through multiple elastic scattering (white arrows, abbreviated by E.S.) for the incident light. $E_{m,s}$ is the modified excitation field due to the E.S. mechanism and p is the induced dipole moment describing the molecule response to the excitation light. Reproduced with permission from Ref. [24], copyright 2015 American Chemistry Society.

SERS processes taking into account the active role of molecules. A canonical model of cavity optomechanics was proposed by assuming a dynamic interaction between the molecular vibration and the localized plasmon as two parametrically coupled and non-resonant harmonic oscillators, as shown in Figure 3(c) [23]. The theory predicted that an excitation laser blue-detuned from the plasmon resonance is able to parametrically amplify the molecular vibration, leading to a nonlinear enhancement of Raman emission. Multiple elastic scattering of light by the molecules located at the nanogap was also proposed to induce a strong near-field self-interaction of molecule with the plasmonic structures (Figure 3(d)) [24]. Consequently, the self-interaction effect dramatically modulates the Raman excitation and radiation in the signal intensity and spatial resolution, resulting in a super hotspot and subnanometer lateral resolution of Raman mapping. Though there are still some controversies about the understanding of the subnanometer resolution and its related ultra-high-field enhancement, such exciting results point to a new era not only for experimentalists but also for the theoreticians.

3. Tip Issues and Enhancement Factor

As discussed above, the tip not only determines the enhancement of the signal but also the spatial resolution of TERS and SPM imaging quality. In addition, the reliability of TERS results also relies on the stability of the TERS tips. As a decisive component of the TERS technique, the development of reliable and highly enhanced TERS tips is a prerequisite for further development and wide application of TERS.

STM, AFM, and shear-force microscopy (SFM) are widely employed to control a TERS system. Accordingly, there are three types of tips: STM TERS tips, AFM TERS tips, and SFM TERS tips. In STM TERS, a tip with a high conductivity is required for the flow of the tunneling current. In AFM TERS, the AFM feedback relies on either the changed reflection of the AFM laser diode or the change in the resonance frequency of the tip. It requires a cantilever-based tip with different resonance frequencies in the range of 10–400 KHz. In SFM TERS, almost all kinds of tips can be attached to a quartz crystal tuning fork to work in the SFM feedback. Up to now, considerable efforts have been devoted to the fabrication of TERS tips with an improved enhancement and reproducibility. In this section, we will make a short overview of the existing methods for the fabrication of TERS tips.

3.1. *STM TERS tips*

Sharp tips are essential for the high resolution STM imaging. Although cutting with scissors may produce very nice STM tips, it does not produce reliable tips for TERS due to the irregular

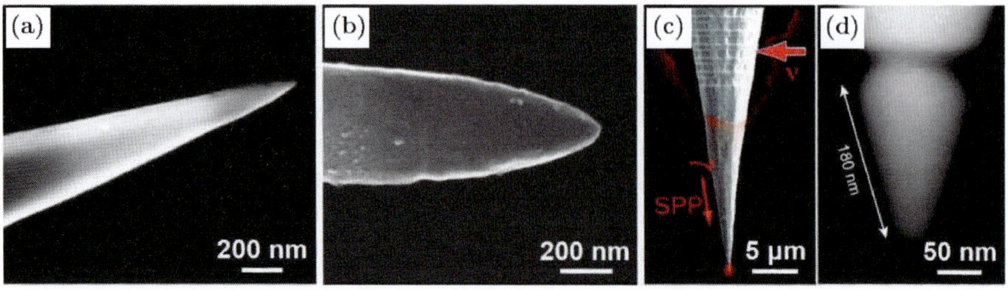

Figure 4. SEM image of the STM TERS tips. (a) EC etched Au tip; Redrawn with permission from Ref. [26], copyright 2007 American Institute of Physics. (b) EC etched Ag tip; Redrawn with permission from Ref. [18], copyright 2015 Springer-Verlag Berlin Heidelberg. (c) FIB milled EC etched Au tip superimposed with optical image of grating illumination and apex emission; Reprinted with permission from Ref. [40], copyright 2010 American Chemical Society. (d) FIB milled single groove near the apex of EC etched Au tip. Reprinted with permission from Ref. [41], copyright 2010 American Chemical Society.

shape of tip apex. A more common and reproducible way to produce STM TERS tips is using the electrochemical (EC) etching, which is to anodically etch a metal wire in an electrolyte. Gold [25–27] (Figure 4(a)) and silver [28–33] (Figure 4(b)) tips are produced through the "drop off" mechanism [34, 35]. The quality of the tip fabricated by the EC etching depends on the composition of the etchant, the etching voltage, the method to cut off the circuit, etc. [18] However, up to now, it is still very complicated to discuss about the detailed etching mechanism and determining parameters because it was found that the etching temperature and the concentration of the original chemicals also play key roles in determining the reproducibility.

Due to the reproducibility issue of the EC-etched Ag tips and the requirement for tips with specialized structures for special purposes, other methods for fabricating STM TERS tips have been developed. Coating of Ag nanostructures on the sharp tungsten tip apex is one of the most commonly use a methods [36]. However, the tip apex becomes blunt after Ag sputtering. Another option is the use of Ag nanowires [37, 38], but it is still time-consuming to finish the attachment processes. Focused ion beam (FIB) milling has been used for further sharpening the tip [39]. More interestingly,

Raschke group used FIB milling to fabricate gratings on the tip shaft (Figure 4(c)) [40]. When the laser is focused onto the shaft, surface plasmon polaritons (SPP) can be excited due to the coupling effect of the grating, which can subsequently propagate to the tip apex for the nanoscale excitation. Compared with the normal tip by direct illumination at the apex, this grating coupling scheme could eliminate the far-field background. More recently, FIB method was also used to engineer a neck on the tip shaft to produce a structure with right resonance with the laser wavelength (Figure 4(d)) [41]. Although the FIB method is ideal to obtain special type of tips, it is not commonly available in normal labs.

3.2. *AFM TERS tips*

Different from STM, most AFM studies used the commercial tips for measurement. Therefore, it is also a common practice to fabricate AFM TERS tips by deposition of metal onto a commercially available silicon cantilever. The most prevalent method of preparing the AFM TERS tip is vacuum evaporation, benefiting from its ease of deposition (Figure 5(a)). However, many factors, including evaporation rate, temperature, vapor pressure, and the placement of

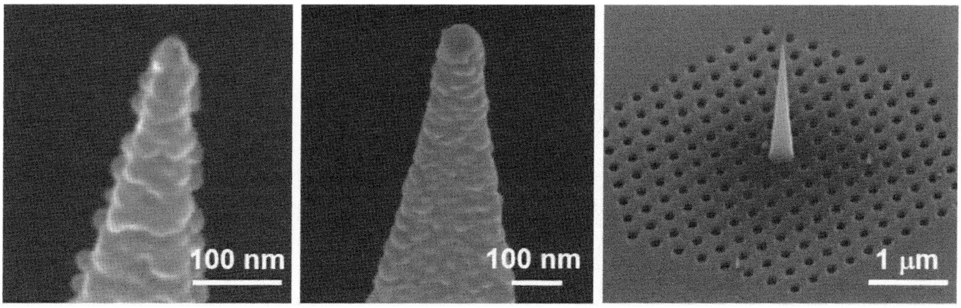

(a) (b) (c)

Figure 5. SEM image of AFM TERS tips. (a) Tip produced by vacuum evaporation; Redrawn with permission from Ref. [68], copyright 2009 The Optical Society. (b) Tip produced by pulse electrodeposition; Reprinted with permission from Ref. [21], copyright 2015 The Royal Society of Chemistry. (c) Photonic crystal cavity on a Si_3N_4 cantilever with attached tapered waveguide. Redrawn with permission from Ref. [62], copyright 2010 Macmillan Publishers Limited.

the heater, source and substrate, may affect the quality of the as-deposited film [18, 39, 42–45]. There are still some issues needed to be addressed concerning the right morphology for high enhancement [44–47], the reproducibility, and the mechanical stability [46–48] (i.e. the wear-off of the coating layer [49]). The random arrangement of the metal nanoparticles onto the tip apex during the physical deposition process may lead to a great variation of the absolute enhancement [49].

Compared with the vacuum deposition, chemical deposition has emerged as a more facile and cost-effective method to prepare AFM TERS tip [50–53] such as mirror reaction and Galvanic reaction, especially when the tip is non-conductive. The quality of the coating layer can be controlled by the reductant (or the deposition potential/current) and deposition time. More recently, we developed a pulsed electrodeposition method to conveniently fabricate Au-coated AFM TERS tips with a high reproducibility (Figure 5(b)) [21]. The surface morphology of the coated gold layer can be controlled by the deposition potential and the tip radius can be tuned by the deposition time.

In addition to the metal-coated tips, different kinds of specialized tips have been developed and used in AFM TERS studies. In general, non-standard tips have special metal nanostructures at the tip apex to confine the laser energy and produce a strong field enhancement. The tip can be obtained either by the attachment of nanoparticles to an AFM tip [54–58], or by the microfabrication method (Figure 5(c)) [59–65], or bending of the EC etched metal tip in the lab [66, 67]. Attaching nanoparticles to the AFM tip apex is one of the most commonly used methods to fabricated specialized tips. However, it suffers from either low efficiency or poor tip shape. Microfabrication appears to be a very promising way to fabricate TERS tips with a high enhancement and reproducibility because it provides a precise control of the size and shape of the nanomaterials as well as a much higher reproducibility. However, the method is not accessible for many TERS labs. Bending the EC etched metal tip can be done in the lab and shows excellent TERS activity. However, it is tedious and time consuming to implement.

3.3. *SFM TERS tips*

There are several advantages of using the tuning fork-based SFM [69]. First, it is not necessary to use a feedback laser to control the tip–sample distance. Therefore, the structure of the SFM head can be significantly simplified compared to AFM. Second, SFM can be applied to all different types of samples as in AFM. Third, almost all kinds of tips can be attached to the quartz tuning forks for further use.

The most widely used tip for SFM-based TERS was made by gluing a silver or gold tip prepared by EC etching to a quartz tuning fork [39, 70–72]. The mass of the glued tip and the amount of glue have a great effect on the quality factor of the quartz tuning fork. They should be as small as possible to maintain a high quality factor to achieve sufficiently high force sensitivity. SFM TERS tips can also be made by coating a metal layer or attaching nanoparticles to the sharp end of the fiber apex [73–75], which appears to have a low reproducibility and efficiency. In 2012, Johnson *et al.* [76] proposed a promising method to fabricate over one million of gold pyramids in a single piece of wafer by using a template-stripping fabrication method. These pyramids are highly reproducible and can be stored for a long time. However, the pyramid should be glued to a thin wire and then glued to the tuning fork for TERS studies. This method is very promising for commercialization if the procedure can be further simplified.

3.4. *Enhancement factor*

The absolute signal intensity is not necessarily a good indicator of the field enhancement because for tips with different radii the number of molecules detected is different. In the literature, several methods have been proposed to calculate the TERS enhancement factors [77–80], which make it difficult to directly compare the results. It should be noted that two factors must be considered when calculating the enhancement factor [49, 79]. First is the contrast between the near-field and far-field signals. Second is the geometry factor, which will affect the number of molecules to be detected during the near-field and far-field measurement. In fact, the contrast factor (C) and

TERS enhancement factor (EF_{TERS}) are the two accepted important parameters to reflect the electric field enhancement of the tip [79].

C is of more practical relevance because it is a direct measure for the signal increase in the presence of the tip and determines the imaging contrast in TERS. By definition, C can be obtained by

$$C = \frac{I_{tip}}{I_0}, \tag{1}$$

where I_{tip} is the TERS signal when the tip is engaged, and I_0 is the background signal when the tip is retracted. Considering the far-field contribution, a more precise definition is often used:

$$C = \frac{I_{NF}}{I_{FF}} = \frac{I_{tip} - I_0}{I_0} = \frac{I_{tip}}{I_0} - 1, \tag{2}$$

where I_{NF} is the near-field signal, which is the difference signal between the tip-engaged and the tip-retracted signals. I_{FF} is far-field signal when the tip is retracted. The difference between the two definitions is 1, thus it only affects the weak enhancement case.

Considering the tips with different radii, the number of molecules detected may be different. Therefore it is better to calculate the TERS enhancement factor considering the difference in the sampling area in the far-field and near-field configurations [79]:

$$EF_{TERS} = C \times \frac{A_{FF}}{A_{NF}} = C\frac{R_{laser}^2}{R_{NF}^2}, \tag{3}$$

where A_{FF} and A_{NF} are the sampling areas in far-field and near-field configurations, respectively. R_{laser} is the radius of the laser focus, which is determined by the laser wavelength and the lens. Signals enhanced by the near field created in the vicinity of the tip contribute to majority of the total signals measured in TERS. Therefore, TERS signal depends on the radius of the near-field zone, R_{NF}. The area of the near field under the tip apex is difficult to estimate. Various equations for estimating R_{NF} on the basis of the radius of the tip apex (R_{tip}) have been proposed [80–83]. It can be summarized that R_{NF} ranges from $(2R_{tip}d)^{1/2}$ to $1/2R_{tip}$, where d is the distance between the tip and the substrate, R_{tip} is the radius of the tip apex which can be determined by SEM. It is also commonly accepted that the near-field zone is of the same scale with the tip radius. In fact, a more precise way to determine the radius of the near-field zone is to determine the lateral resolution of the TERS images. However, it is tedious and time consuming.

In the case of the bulk samples, the signals should be normalized to the corresponding illuminated volumes due to the optical penetration depth. The TERS enhancement factor can be calculated according to Equation (4) [49, 83]:

$$\mathrm{EF}_{\mathrm{TERS}} = C \times \frac{V_{\mathrm{FF}}}{V_{\mathrm{NF}}} = C \times \frac{R_{\mathrm{laser}}^2}{R_{\mathrm{NF}}^2} \times \frac{h_{\mathrm{laser}}}{h_{\mathrm{NF}}}, \tag{4}$$

where V_{FF} and V_{NF} are the volume of the far-field zone and near-field zone, respectively, which are determined by the area and height of the far-field (h_{laser}) and near-field zone (h_{NF}). The TERS enhancement factor is affected by the laser wavelength, sample and the tip shape.

4. TERS Study on Single Crystal Surfaces

We have introduced the basic concept, enhancement mechanism, and tip issues above. In the following sections, we will focus on the application of TERS in surface and interface analysis.

One of the most important classes of materials studied in surface science are single-crystal surfaces, which possess well-defined atomic arrangements and energy levels, offering an ideal model to understand surface and interfacial physical and chemical processes at the atomic level. TERS is ideally suitable for the study of single-crystal surfaces because of the introduction of coupled plasmon resonance when a tip is placed above single-crystal surfaces.

Pettinger and Ren were the first to perform TERS study on single–crystal surfaces [77, 84]. In 2004, they studied the adsorbed CN^- on Au(111) and malachite green isothiocyanate (MGITC) molecules on Au(111) and Pt(110) surfaces using electrochemically etched gold tips in a STM-TERS system [77]. In the case of CN^- on Au(111), no signal can be detected when the tip is retracted because of the low Raman cross-section of CN^-. A clear enhancement was observed when the tip is in tunneling condition (~1 nm above the surface). They estimated an enhancement factor of 4×10^5. This is the first time that Raman spectra of an optically non-resonant species on a flat, single crystalline surface were obtained. For a more reliable quantification of the enhancement factor of TERS, they further studied a dye molecule MGITC, which can give detectable signal in the absence of the tip because of its electronic resonance under 633 nm laser excitation (Figure 6(a)). By comparing the quenching rate under normal Raman and TERS conditions, a local, field enhancement factor of about 30–50 was determined, corresponding to the TERS enhancement factor in the order of 10^6.

They further extended the study to benzenethiol adsorbed on Pt(110) and Au(110) surfaces (Figure 6(b)) [84]. It was found that the TERS signal on Au(110) surface is about 20 times stronger than that on Pt(110) surface when the same gold tip was used, which may be due to the stronger plasmonic coupling between Au tip and Au substrate than Pt substrate. Note that in their first report [77], an Ir tip was also used for TERS measurements on Au(111) surface. A contrast of two was observed between tip-engaged and tip-retracted conditions, which is small but still observable. In these two reports [77, 84], they used different materials of tips and substrates (Au tip–Au substrate, Ir tip–Au substrate, Au tip–Pt substrate). The results suggested that the strong plasmonic coupling between Au tip and Au substrate leads to much stronger enhancement than either Ir tip or Pt substrate used. However, it should be pointed out the Au tip–Pt substrate system can produce fairly good signal that allows one to analyze the vibrational features of sub(monolayer) molecules adsorbed on atomically flat transition metal surfaces, which indicates the possibility to study the surface properties of these catalytically active metals.

One major concern in TERS is whether the signal is from the contamination on the tip or the desired signal of probed materials. The use of single-crystal surfaces effectively avoids SERS background signal from the substrate. However, it is still important to exclude the possible contribution from the tip contamination. Ren *et al.* demonstrated this issue by performing TERS on two different single-crystal surfaces, Au(110) and Pt(110) [84]. Due to the different chemical interaction between benzenethiol and metal substrates, the Raman spectra showed characteristic peak shifts and relative

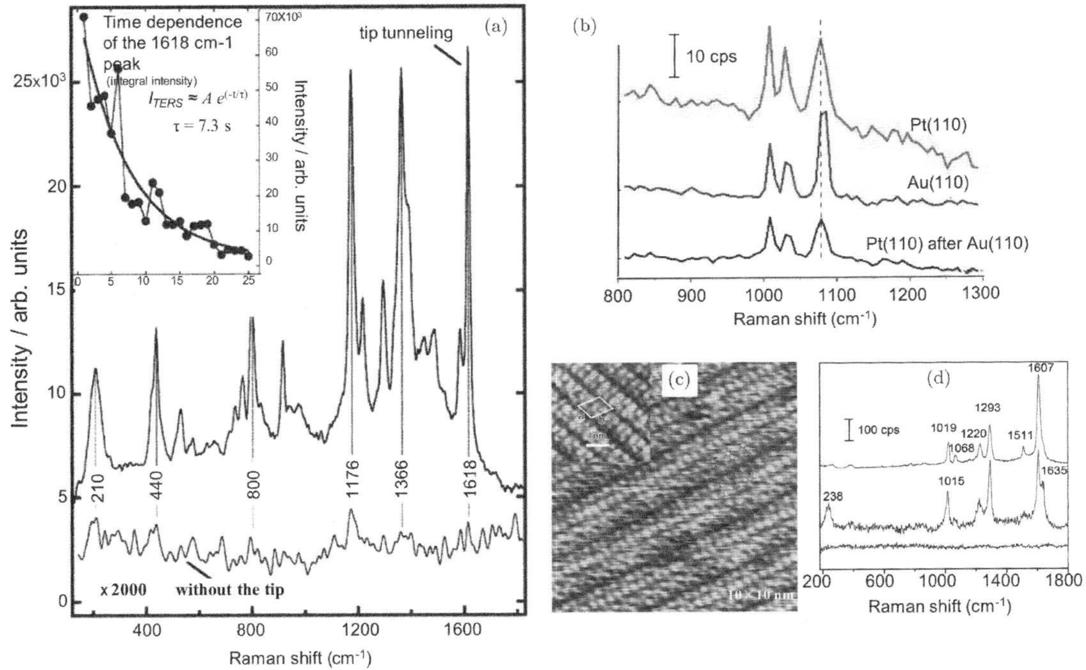

Figure 6. (a) Comparison of resonance Raman (bottom spectrum) and TERS (top spectrum) for MGITC adsorbed at a Au(111) surface. Inset: Time dependence of the integral intensity of the 1618 cm^{-1} band in TERS. Reproduced with permission from Ref. [77]. Copyright 2004 The American Physical Society. (b) Sequential TERS experiments for benzenethiol obtained first on Pt(110), then on Au(110), and finally on a Pt(110) surface again. Reproduced with permission from Ref. [84]. Copyright 2005 Wiley-VCH Verlag GmbH & Co. KGaA, Weinheim. (c) STM image of 4,4′-bipyridine adlayer on a Au(111) surface obtained with a Au tip, inset shows the high resolution STM image obtained with a Pt–Ir tip. (d) TERS spectra for the 4,4′-bipyridine adlayer on the Au(111) surface (middle spectrum), the bottom spectrum for the tip-retracted case, and the top spectrum for the SERS spectrum. Reproduced with permission from Ref. [85]. Copyright 2007 American Institute of Physics.

intensity change on these two single–crystal surfaces. The $1070\,cm^{-1}$ peak having an a1-type symmetry shifts from $1068\,cm^{-1}$ on Pt to $1075\,cm^{-1}$ on the Au surface when using the same tip for measurement (Figure 6(b)). To make sure the shift is reliable, they again carried out TERS on the Pt surface after measurement on the Au surface. Then the peak shifts back to $1068\,cm^{-1}$ characteristic of Pt (Figure 6(b)). The experiments suggested that the signal is truly from the substrate (with distinct spectral features) but not the tip. Nevertheless, they also found that the tip can be easily contaminated in some cases, for instance, when 4-pyridine-4-thiol was used as the probe. This is possibly due to the special chemical structure of the molecule, which has N and S atoms on the two ends. In this case, the tip may pick up some molecules from the substrate because of the relatively strong bonding between the molecule and the tip [84]. Therefore, special attention should be paid when detecting molecules with multiple binding groups.

It is therefore vitally important to check the possible contamination of the tip by the sample molecule after the TERS measurement. A simple way is to put the tip above a clean region (free of the target molecule and any contamination) near the interested region and perform TERS. This is often done when the probe molecule is only of sub-monolayer on a surface. In this way, the control experiment is performed under identical conditions. If no signal is detected, we can basically assume that the tip is clean. If one cannot find such a clean region in the sample, an alternative way is to put the tip on another clean substrate to carry out a control TERS experiment. Note that in certain cases, even though there may be some contamination on the tip, the signal from it is much weaker than the probe molecule and thus will not alter the signal from the substrate.

Single-crystal surfaces are ideal substrates for studying molecule-substrate and molecule–molecule interactions [85–88]. Figure 6(c) shows the STM image with molecular resolution of 4,4′-bipyridine monolayer on Au(111) surface; the corresponding TERS spectrum is present in Figure 6(d) [85]. The results suggest that the atomically flat surface allows a clear correlation between topography and Raman spectrum, especially for single molecule study (as will be discussed later), where a well-defined surface structure is highly appreciated.

5. TERS for Molecular Electronics

As a newly developed technique, TERS has significantly improved our understanding of various physical and chemical properties of a material at nanometer scale. However, in the first 10 years of development, since most TERS studies were carried out on relatively "stable" samples, much less dynamic information has been obtained by using TERS. In this section, we will describe our efforts in providing dynamic TERS information when molecular conductance was simultaneously measured during the formation of a metal–molecule–metal junction.

Molecular electronics is a promising next-generation circuit. To better understand the function of molecular electronics, optical spectroscopic methods, such as SERS [89, 90], have been developed to correlate the structure (with fingerprint spectrum) and conductance of molecules. However, these methods lack sufficient spatial resolution, and the contribution from many molecules other than the one that forms the junction would be large.

Our group developed a fishing-mode STM-TERS by lowering the feedback of the STM system to increase the probability in the formation of molecular junctions [86]. We simultaneously measured the conductance and TERS of 4,4′-bipyridine (4bipy) molecule in the junction. One very important aspect of TERS is that the highly confined electromagnetic field ensures that the spectrum mainly contributed from a single molecule forming the electrical junction (Figure 7(a)). Impressively, we found that the TERS spectra dynamically changed with the formation (ON state) and break (OFF state) of a molecular junction. A clear difference of the TERS spectra between ON and OFF states can be observed. As can be seen in Figure 7(b), in the OFF state, only one peak at $1609\,\mathrm{cm}^{-1}$ can be observed, while this peak splits to $1609\,\mathrm{cm}^{-1}$ and $1631\,\mathrm{cm}^{-1}$ in the ON state. We found that in the ON state, a stronger chemical interaction is formed between the upper pyridine ring and the tip (Figure 7(c)). This strengthens the C–N/C–C dephasing stretch

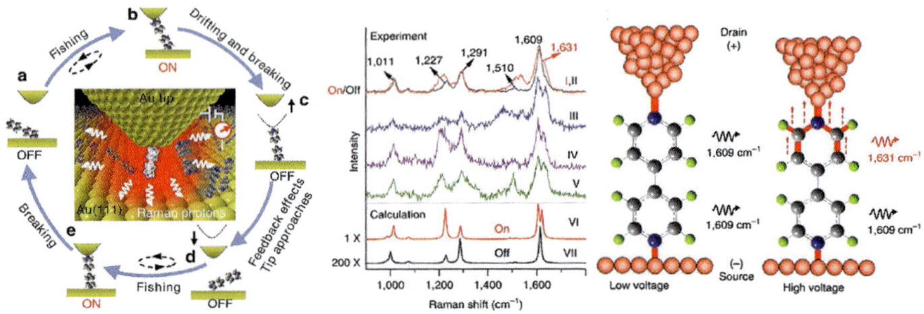

(a) (b) (c)

Figure 7. (a) An illustration of the fishing-mode STM-TERS system for simultaneous conductance and TERS measurement of single-molecule junctions. (b) TERS spectra: I is an average of all 190 ON-state TERS spectra and II is a typical OFF-state TERS spectrum. III, IV and V correspond to an average of $\geq$20 ON-state spectra. VI and VII are used to calculate ON- and OFF-state spectra, respectively. (c) A diagram illustrates why 1,609 cm^{-1} band splitting is proportional to the applied bias voltage. Reproduced with permission from Ref. [86]. Copyright 2011, Macmillan Publishers Ltd.

of the upper pyridine ring, leading to the observed splitting of the peak. The correlated conductance and TERS measurements are very important in elucidating the structural change of a molecule during the electron's transport, which may deepen our understanding of the function of molecular electronics.

6. Nanoscale Characterization of Plasmon-Enhanced Reactions

In recent years, plasmon-enhanced reactions have become an increasingly important topic in both plasmonics and heterogeneous catalysis. This is because plasmon-enhanced reactions offer a promising way to make use of light for chemical reactions. One of the most well-studied plasmon-enhanced catalytic reactions is the dimerization of either p-aminothiophenol (PATP) or p-nitrothiophenol (PNTP) to 4,4′-dimercaptoazobenzene (DMAB). Therefore, we first briefly review these reactions, which were discovered in SERS study. PATP has long been a model molecule in SERS study because it can strongly bond to metal surface and produce strong Raman signal. In an early very important electrochemical SERS study of PATP,

Osawa and co-workers found that, besides the normal "a1 mode" of PATP, some new peaks can be observed under certain electrode potentials and laser excitation [91]. They ascribed these new peaks to the "b2 mode" of PATP, resulting from the photon-driven charge transfer between PATP and metal surface. For a long time, these new peaks have been accepted as an indication of charge-transfer mechanism in SERS. However, we found that the intensity of the so-called "b2 mode" is laser power- and illumination time-dependent [92, 93]. This finding is in contrast to the charge-transfer mechanism in SERS in which the intensity of the "b2 mode" should be wavelength-, or electrode potential-dependent. We have rigorously studied the SERS of PATP adsorbed on Ag nanoparticle surface and in metal–PATP–metal junctions. With more evidences from surface mass spectroscopy and the Raman spectra from pure DMAB sample, we convincingly concluded that it is the chemical transformation of PATP to DMAB rather than the charge-transfer model that accounts for the appearance of the new peaks (Figure 8) [93, 94]. The new peaks (marked as red in Figure 8(c)) observed in SERS should be assigned to the vibrational modes of the product, DMAB, as proved experimentally and theoretically [93, 94]. Our work renewed the SERS study of PATP and PNTP, which nowadays has been well accepted as a model system in plasmon-enhanced chemical reactions and extensively studied in the past several years.

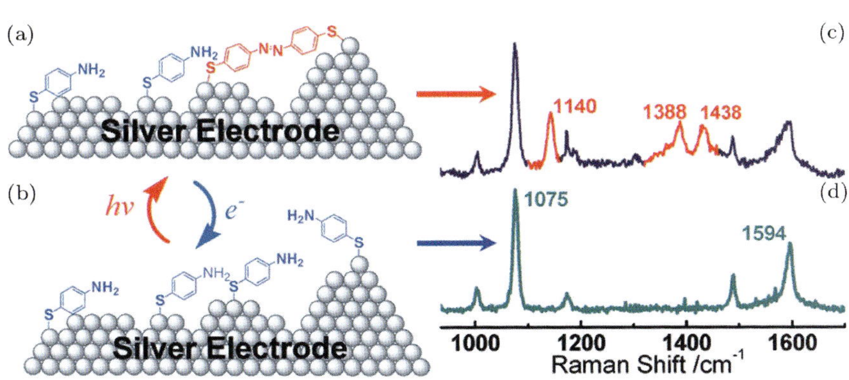

Figure 8. Transformation of DMAB (a) and (c) from PATP (b) and (d) adsorbed on Ag during the electrochemical SERS measurement. Reprinted with permission from Ref. [93], Copyright (2010) American Chemical Society.

With the successful employment of SERS for monitoring the reaction of PATP or PNTP to DMAB *in situ*, it would be very interesting to apply TERS to study these reactions by taking advantage of the high spatial resolution of TERS for studying chemical (catalytic) reactions at the nanometer scale. Similar to SERS, TERS can provide Raman fingerprint information that may offer molecular insight on the breaking and formation of specific chemical bonds. In 2012, Deckert, Weckhuysen and their co-workers utilized TERS to monitor the transformation of PNTP to DMAB [95]. They used Ag-coated AFM cantilever as TERS tip, and PNTP was adsorbed on atomically flat Au nanoplate (Figure 9(a)). They found that the reaction can only occur under 532 nm excitation in the presence of an Ag-coated tip. Therefore, the Ag-coated tip also worked as a catalyst, which means that only the molecules in close vicinity (underneath) of the tip can be catalyzed. In this regard, the reaction is spatially confined at the nanoscale (the radius of the tip is usually tens of nanometers). In their experiments, 532 nm laser was used to trigger the reaction and 633 nm laser was used to monitor the reaction. As seen in Figure 9(b), the characteristic 1440 cm^{-1} peak ($\nu_{NN} + \nu_{CC} + \beta_{C-H}$) of the product DMAB only appears after 532 nm illumination, indicating the occurrence of the reaction (Figure 9(c)). This work demonstrated that TERS is able to monitor nanoscale catalytic reaction with molecular fingerprints. Xu and co-workers also studied this reaction by using STM-TERS in a high-vacuum chamber, in which Au tip and Ag film substrate were used [96]. They found that a sufficient plasmon intensity is necessary to trigger the reaction. It would be interesting if the effect of substrate can be explored and nanoscale imaging can be done, which is one of the most advantageous functions of TERS in comparison with SERS. Roy *et al.* used Ag nanoparticle as substrate and catalyst (Figure 9(d)), while the Ag-coated AFM tip was further coated by alumina to suppress its catalytic effect while maintaining the plasmon enhancement effect [97]. In this way, the catalyst (substrate) effect will be more pronounced. They found that the catalytic activity of the assembled Ag nanoparticles for the transformation of PATP to DMAB varied at different locations (Figure 9(e)). Using the 1142 cm^{-1} peak of DMAB, they obtained a spatial resolution of about 20 nm (Figure 9(f)). This work demonstrated that it is promising to

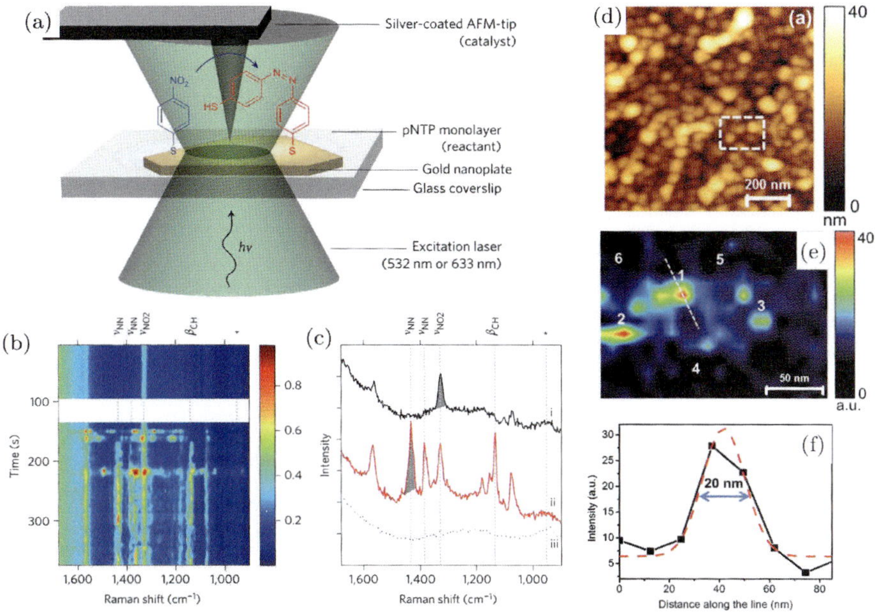

Figure 9. (a) Schematic of the experimental setup for TERS monitoring of the photocatalytic conversion of PNTP to DMAB. (b) Time-dependent TERS spectra at 633 nm excitation shown before (top) and after (below white band) 532 nm illumination. (c) Representative spectra before (i) and after (ii) the initiation of the catalytic reaction. Spectrum (ii) indicates the formation of N=N bond. Reprinted with permission from Ref. [95], Copyright 2011 Macmillan Publishers Ltd. (d) AFM topography image of an Ag substrate with Ag nanoparticles on a glass substrate. (e) TERS map from the dashed rectangle marked in (d), showing the variation of the 1142 cm^{-1} Raman peak intensity of DMAB. (f) Intensity profile along the dotted line marked in (e) showing the spatial resolution of the TERS map. Reprinted with permission from Ref. [97], Copyright 2015 The Royal Society of Chemistry.

use TERS for nanoscale imaging of the catalytic activity of a heterogeneous catalyst.

7. Environment Control of TERS: From Air to Ultra-High Vacuum and Solution

7.1. *Ultra-high vacuum TERS*

In the early stage of TERS, all the experiments were carried out under ambient air condition. Although TERS has shown its promise in nanoscale characterization of various kinds of materials, there still remain open questions on how far we can push the spatial resolution and the detection sensitivity of TERS. For instance, can we detect or image a single molecule? Under ambient air condition, it is almost impossible to image single molecules even by normal SPM because the surface can easily be contaminated by impurities from air and the high mobility of the molecules on the surface. The UHV condition has been widely used in surface science because it offers a very clean and controlled environment. As such, monolayer or sub-monolayer molecules or even a single molecule can be imaged in UHV. Therefore, it is highly significant to develop UHV-TERS.

In 2007, the Pettinger group first realized UHV-TERS by integrating optics into the UHV chamber [98]. They adopted parabolic mirror with high numerical aperture for a high detection sensitivity. As a demonstration, they detected brilliant cresyl blue (BCB) molecules on Au(111) surface. In a follow-up experiment, they decreased the surface coverage of BCB molecules on the surface to obtain isolated single molecule (Figures 10(a) and 10(b)) [99]. In UHV environment, the photo bleaching of the dye molecule is substantially suppressed compared with ambient air condition, which allows sufficiently long time TERS measurements. They achieved TERS mapping of a single BCB molecule at a resolution of ~15 nm. Pettinger's work made a great step in extending TERS to UHV environment and pushed TERS to be single molecule technique [99].

In 2012, Van Duyne group reported UHV-TERS study of copper phthalocyanine (CuPc) adlayers on Ag(111) surface (Figures 10(c)–10(e)) [100]. In their system, the optics were outside the UHV chamber. All the sample preparation procedures were performed in UHV chamber, avoiding any surface impurities. They were able to obtain sub-nanometer molecular resolution STM imaging using Ag tips under laser illumination, and simultaneously multiple vibrational modes of CuPc in TERS.

Note that in Pettinger's experiment, the spatial resolution of 15 nm is apparently larger than the size of a single molecule. This is possibly due to the thermal drift of the SPM system and the thermal movement (diffusion) of the molecules on the surface. Therefore, a

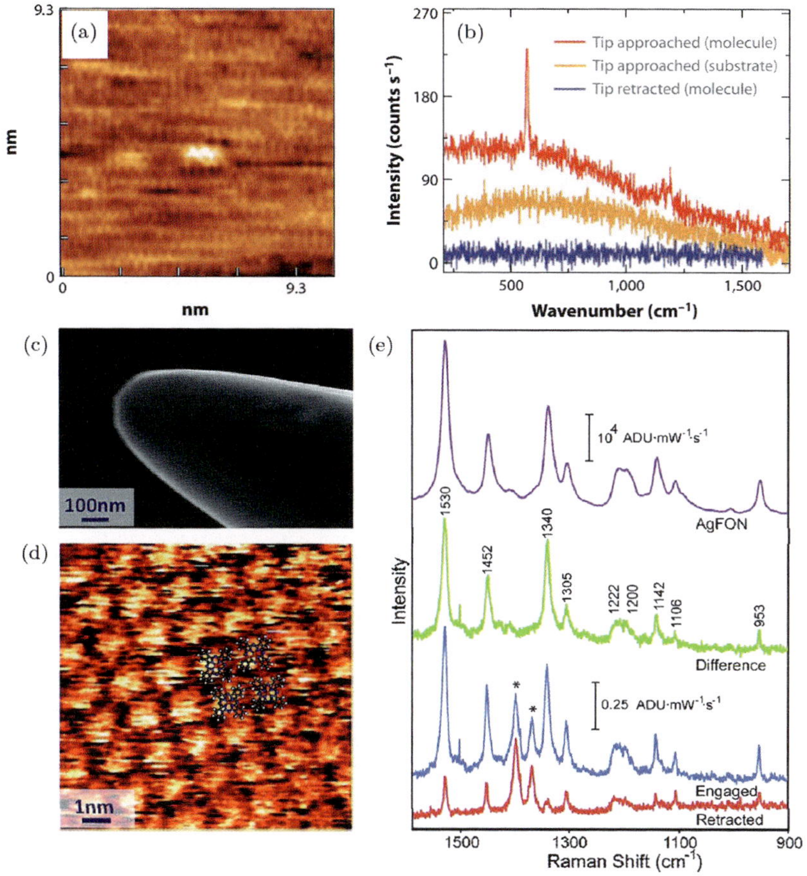

Figure 10. STM image of a single BCB molecule adsorbed on Au(111) (a), and the resonance Raman and TER spectra (b). Reproduced with permission from Ref. [99]. Copyright 2008 The American Physical Society. (c) SEM image of a silver tip. (d) Molecular resolution topographic STM image (10 nm × 10 nm) of a CuPc adlayer on Ag(111) with the tip illuminated by the laser. (e) UHV-TERS spectra of a CuPc adlayer on Ag(111) and SERS spectrum of a CuPc adlayer on AgFON. The asterisks (*) indicate peaks at 1374 and 1403 cm^{-1} that result from the sapphire windows. Reproduced with permission from Ref. [100]. Copyright © 2011 American Chemical Society.

cryogenic temperature environment is desirable to obtain a higher spatial resolution.

In 2013, the Dong group reported TERS imaging of a single porphyrin molecule on Ag(111) surface [16]. They achieved an unprecedented spatial resolution of 0.5 nm, that was able to resolve the vibrational features of different fractions of a single molecule. The experiments were performed at a temperature of 80 K and UHV conditions, which make their systems highly stable. Their work is a breakthrough in pushing the spatial resolution of TERS to sub-nanometer scale, which is not achievable by any other spectroscopic methods. This high spatial resolution makes them spatially distinguish adjacent molecules within van der Waals interaction [17].

7.2. *TERS in solution*

As a nanospectroscopic tool, TERS has been developed to be an important tool for the nanoscale analysis of solid–vacuum and solid–gas interface. However, the ambient air or UHV conditions cannot meet the requirement for the *in situ* characterization of samples in wet environment including living biological systems, solid–liquid interface and electrochemical systems. Therefore, there is an increasing interest in developing TERS capable of operating in solution [48, 101–105].

The main challenge to extend TERS to solution lies in the optical distortion of the optical path. The distorted optical path when light passes through the air and the solution interface of different refractive index will severely affect the excitation and collection efficiency of TERS signal. As shown in Figure 11, different optical geometries

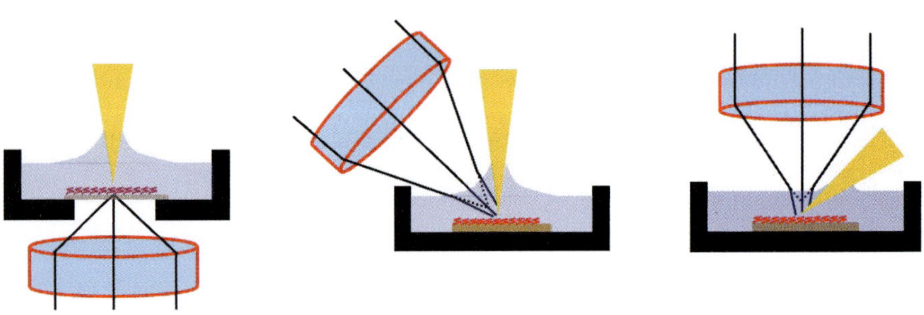

(a) (b) (c)

Figure 11. Three optical geometries of TERS setup can be considered for the extension of TERS from air to solution. (a) Bottom illumination geometry. (b) Side illumination geometry. (c) Top illumination geometry.

applied in ambient TERS experiments can be considered for the extension of TERS from air to solution. Among the three geometries, the bottom illumination geometry is the most convenient one. In this geometry, the laser is focused via an inverted microscopy objective without passing through the solution, avoiding the optical distortion. However, this backscattering geometry is limited to the transparent sample. For the side illumination and top illumination geometries, the light has to pass through the air and the solution, and the optical path will be severely distorted. Therefore, it is difficult to form a good focus, especially when the solution layer becomes thicker.

The first report of solution TERS was published in 2009 by Zenobi and his co-workers [48]. Using the AFM-TERS setup based on the bottom illumination configuration, the authors obtained the TERS spectrum from a monolayer of thiophenol on a thin transparent Au film in water environment. In TERS experiments, both the tip and the sample were immersed in water. They found that a pre-coating of SiO_x layer can improve the adhesion of the Ag coating and reduce the peeling-off of the coating layer when the AFM-TERS tip is soaking in solution. Additionally, a strategy to prevent the adsorption of analyte molecules and other contamination in solution to the tip by coating a self-assembly monolayer of ethanethiolate on the tip was present in this work. With a similar bottom illumination setup, in 2013, Fujinami *et al.* obtained the TERS spectrum from lipid bilayers on glass substrate in water [101].

However, the transmission configuration did not meet the request for the study of the opaque sample in solution. In 2016, Domke and coworkers developed a solid/liquid TERS setup based on a side-illumination design [104]. The laser beam was introduced in a tilted way and focused on the tip apex after passing through the liquid layer (Figure 12). To reduce the effect of optical distortion on TERS measurements, they carefully controlled the water layer thickness within several millimeters. It allowed them to measure a monolayer of resonant dye molecules and non-resonant thiophenol molecules on Au(111) surface in water.

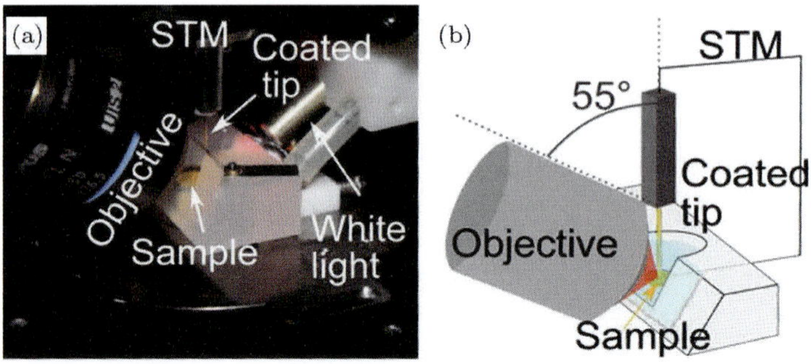

Figure 12. A solid/liquid TERS setup based on a side-illumination design (a) A photo of the setup. (b) Schematic illustration of the side-illumination geometry. Reproduced with permission from Ref. [104]. Copyright © 2016 American Chemical Society.

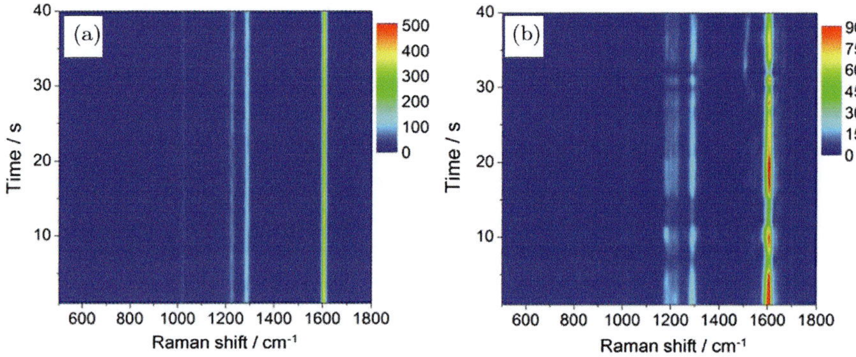

Figure 13. Time series TERS signal of 4PBT adsorbed on Au(111) surface. (a) In the solution of 0.1 M $NaClO_4$ (pH=3). (b) In air. Reproduced with permission from Ref. [102]. Copyright 2016 American Chemical Society.

In 2015, we found that the intensity and frequency of the TERS signal of 4′-(pyridin-4-yl)biphenyl-4-yl)methanethiol (4-PBT) adsorbed on Au(111) surface obtained under the solution environment was much more stable compared with that obtained in air (Figure 13) [102]. The electrolyte could be a buffer to prevent the exposure of the tip–substrate gap to air and impurities. On the other hand, the high heat capacity of water can effectively prevent the abrupt rise of the local temperature. In this regard, performing TERS in solution is advantageous compared with that in air. These important technical developments allow the use of TERS in more complex solution systems, for instance, nanoscale pH characterization (Chapter 11, Prompong Pienpinijtham and Yukihiro Ozaki) and electrochemical study.

8. Electrochemical Tip-Enhanced Raman Spectroscopy

Electrochemistry plays an important role in surface science. The electrochemical interfacial process lies in a broad range of important research disciplines including batteries, fuel cells, material corrosion, electric synthesis and semiconductor processing. To reveal the nature of electrochemical interfacial properties and unravel the related interfacial process at nanoscale, highly sensitive *in situ* techniques that can provide fingerprint molecular information at the nanometer resolution are required. In this regard, TERS integrates the advantages of nanometer spatial resolution, single molecule sensitivity, and rich vibrational information, and is a promising tool for the molecular level and nanoscale analysis of the electrochemical interface.

There are several challenges to extend TERS to electrochemistry. Besides, the challenge lies in minimizing or eliminating the effect of optical distortion produced in solution. Electrochemical TERS (EC-TERS) requires the potential control of the working electrode under investigation in an electrochemical system, thus rendering the setup even more complex. The implementation of EC-TERS has been a dream for researchers in TERS and SERS communities [105–107].

In 2015, our group first successfully demonstrated EC-TERS [102]. In the EC-STM-TERS setup (Figure 14), the potentials of the tip (WE1) and the substrate (WE2) were controlled by a bipotentiostat relative to a Pt wire. The laser beam was introduced horizontally via an objective to the electrochemical cell and focused onto the tip apex. To ensure the high throughout, we modified the EC cell by tilting the single-crystal electrode to 10° to minimize the block of the laser light by the electrode. In this horizontal illumination design, the water layer thickness in optical path can be decreased to about 1 mm to minimize the optical distortion. Additionally, the influence of water evaporation and curved liquid surface caused by the capillary

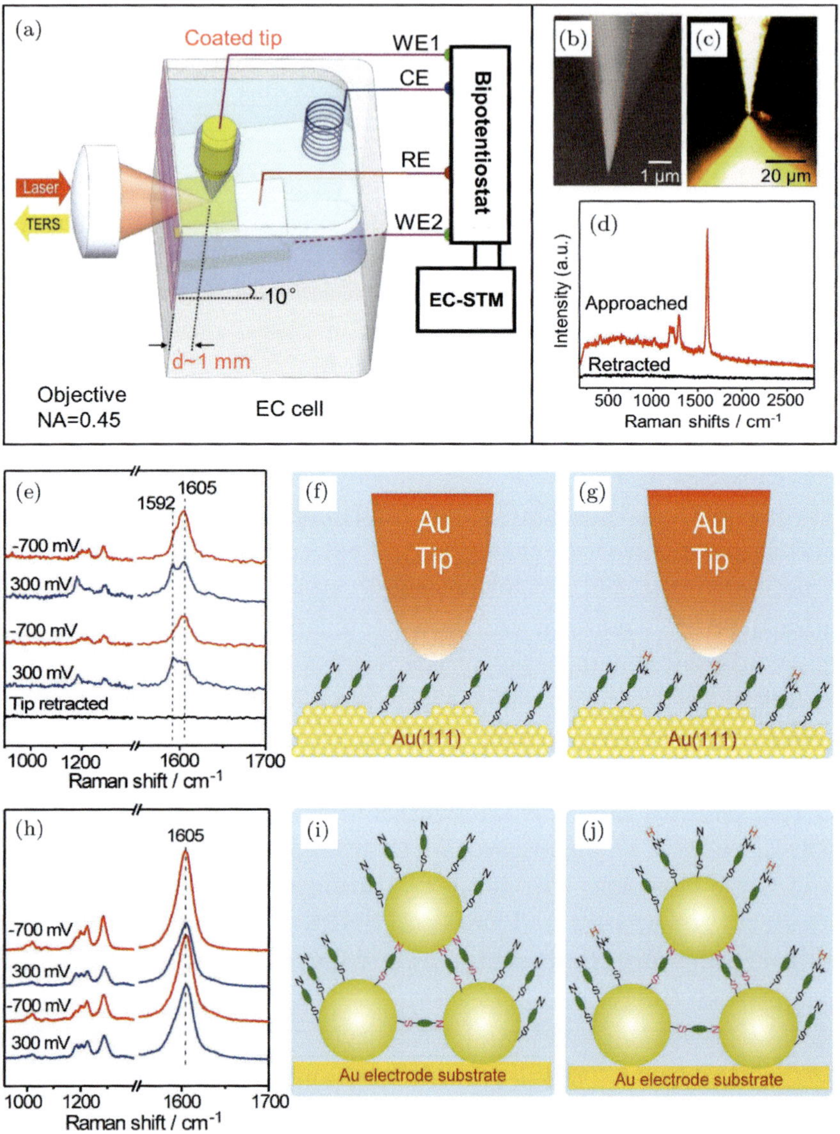

Figure 14. (a) Schematic illustration of the EC-TERS setup. (b) SEM image of an insulated gold tip. (c) Microscopic image of the tip, single–crystal substrate, and laser spot in a EC-TERS system. (d) TERS of 4-PBT adsorbed on Au(111) surface obtained while the tip was approached (top) and retracted (bottom); (e) Potential-dependent EC-TERS of 4-PBT adsorbed on Au(111) surface

effect on the focused optical path can be avoided. Benefitting from this design, the laser can be well focused and the Raman signal can be collected with a high efficiency. In EC-STM-TERS, another key issue is to minimize the Faradaic currents. We chose polyethylene glue as an insulating material. The proper insulation coating (Figure 14(b)) is key to preventing the interference from Faradaic currents and ensuring the tip to be properly approached in feedback, thus a strong and stable TERS signal can be obtained. With the insulated Au tip and the designed EC-TERS setup, we obtained EC-TERS signals (Figure 14(e)) of monolayer 4-PBT adsorbed on an Au(111) surface as a result of the protonation and deprotonation of the molecules with a change of electrode potential (Figures 14(f) and 14(g)). The peak at $1592\,cm^{-1}$ contributed by the pyridine ring shifts to a high frequency to overlap with the vibration band of benzene ring at $1605\,cm^{-1}$ and appears as a single peak after the molecule protonated at $-700\,mV$. However, we found that this potential-dependent spectral change is not observable in EC-SERS (Figure 14(h)). In SERS configuration, 4-PBT molecules may be trapped between the nanoparticles to form double-end-bonded 4-PBT molecules. These double-end-bonded 4-PBT molecules have a frequency close to that of the protonated 4-PBT molecules, thus contributing to the majority of the Raman signal, and overwhelming the potential-dependent signal of 4-PBT (Figures 14(i) and 14(j)). This work demonstrates that EC-TERS can more sensitively reflect the interfacial structure of an electrochemical system due to the simple and well-defined configuration and much less molecules detected (<600).

Immediately after the first EC-TERS demonstration, Van Duyne group demonstrated EC-AFM-TERS based on a bottom illumination

Figure 14. (*continued*) in 0.1 M $NaClO_4$ (pH=10); (f) and (g) Models for the TERS system at positive (f) and negative potentials (g); (h) Potential-dependent SERS of a 4-PBT-modified Au nanoparticle surface in 0.1 M $NaClO_4$ (pH=10); (i, j) Models for the SERS system at positive (i) and negative potentials (j). Reproduced with permission from Ref. [102]. Copyright 2015 American Chemical Society.

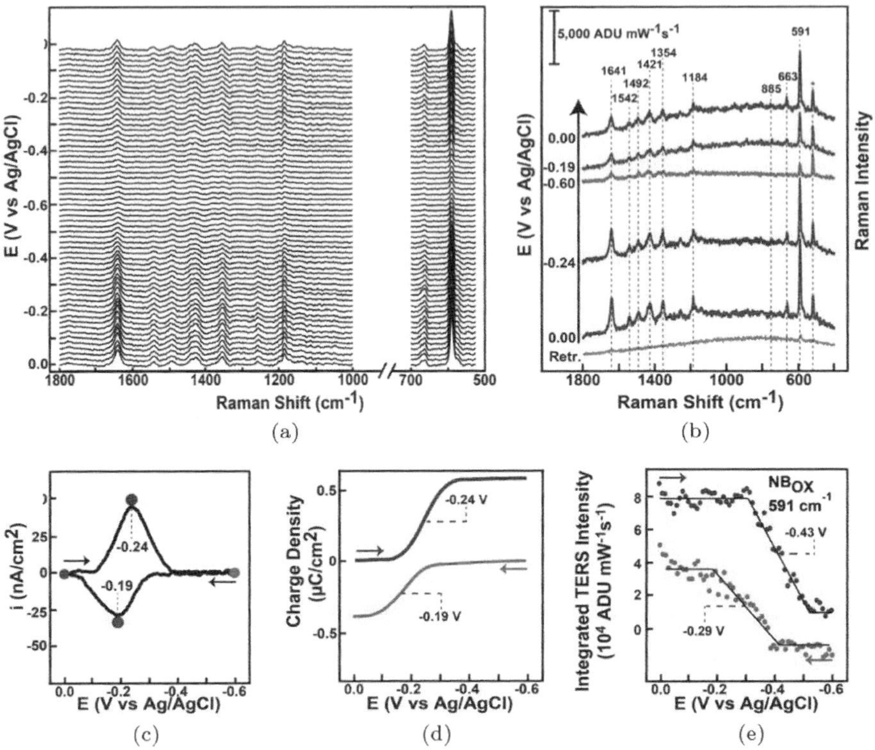

Figure 15. (a) Stack plot EC-TER spectra of NB acquired at different potentials during cyclic voltammetry (CV). (b) individual EC-TER spectra acquired at select potentials; (c) Background subtracted CV of NB on ITO. (d) Charge acquired by integrating the CV in (c), the reverse sweep is offset by 0.6 $\mu C/cm^2$ for clarity. (e) Potential dependence of the integrated TERS intensity of the 591 cm^{-1} band of NB molecules. Reproduced with permission from Ref. [103]. Copyright 2015 American Chemical Society.

to probe the nanoscale redox behavior of Nile Blue (NB) on a transparent ITO electrode (Figure 15) [103]. They monitored the reversible spectral changes of NB upon its reduction and oxidation during cyclic voltammetry (CV) (Figure 15(a)). Interestingly, a negative shift of more than 100 mV in the potential response of the TERS intensity (Figure 15(e)) relative to the conventional CV (Figures 15(c) and 15(d)) was observed. The authors found that although the tip locally perturbed the structure of the electrical double-layer, leading to a negative shift in TERS voltammogram, it had no effect on the electron transfer kinetics. Additionally, a step-like feature in some TERS voltammograms corresponding to the reduction and oxidation of single or few molecules was observed. This work indicates that EC-TERS is able to probe a redox reaction at the extreme of single or few molecules.

These works demonstrate that EC-TERS can be successfully used to monitor electrochemical processes at the nanoscale. Though EC-TERS is still at an early stage, it shows impressive potential to address more interesting and challenging issues in electro(photo)catalysis and energy-related electrochemical systems.

9. Summary and Outlook

With over 16 years' development, TERS has now grown into a very important nanospectroscopic tool for surface and interface analysis. Particularly, the sub-nanometer spatial resolution of TERS has stimulated both the theoreticians and experimentalists to reconsider the mechanism of tip-enhanced processes. The active participation of probe molecules may play a crucial role to access such a highly enhanced and tightly confined electric field. The issues of whether it is only a specific phenomenon of resonant molecules and whether nonlinear processes or other optical processed are involved in TERS will have a significant impact for TERS and plasmon-enhanced spectroscopies. In addition to classic electromagnetic theory, a unified quantum-mechanical theory of TERS/SERS should be developed to describe the enhancement process.

With respect to TERS tips, it will be very important to engineer the tip apex of both silver and gold to a specific shape and/or LSPR wavelength. To commercialize TERS tips, it requires the tip to have a good reproducibility of the signal enhancement and be resistant to mechanical damage during measurement. We expect further developments may lower the price of the TERS tips to be comparable with commercial AFM tips, thus a wide application of TERS may eventually come.

Following the understanding of the physics behind the ultra-high spatial resolution and the improvement in tip fabrication, the future of TERS is to find even better applications to advance our understanding of chemical and physical processes in surface and interface. Up to now, although TERS has been utilized to monitor nanoscale reactions, it still remains a challenge to provide an atomic and molecular level understanding of heterogeneous catalysis. TERS has the promise to correlate nanoscale surface structure with its properties in real space, which is crucial to identify catalytic active sites and explore catalytic mechanisms, and will help the rational design of new catalysts with better performance.

EC-TERS further provides the flexibility to control the potential to tune the plasmon resonance frequency as well as the interaction of the molecule with the substrate. With further improvement of the detection sensitivity, EC-TERS will become a promising tool to study the synergistic effect of potential and light at nanoscale in the electro(photo)catalysis. Additionally, with the extremely high spatial resolution, it is possible to reveal the structure–function correlation at the electrochemical interfaces.

Acknowledgments

We would like to thank all previous group members who contributed to the development of the TERS technique (X. Wang, Z. Liu and Z. C. Zeng), and collaborators (Z. Q. Tian, Z. L. Yang, D. Y. Wu) for helpful discussions. We acknowledge support from MOST (2016YFA0200601, 2011YQ03012400, and 2013CB933703), NSFC (21633005, 21227004, 21321062, and J1310024), and MOE (IRT13036).

References

1. Fleischmann, M., Hendra, P. J. and McQuillan, A. J. (1974). Raman spectra of pyridine adsorbed at a silver electrode. *Chem. Phys. Lett.* **26**, pp. 163–166.
2. Albrecht, M. G. and Creighton, J. A. (1977). Anomalously intense Raman spectra of pyridine at a silver electrode. *J. Am. Chem. Soc.* **99**, pp. 5215–5217.
3. Jeanmaire, D. L. and Van Duyne, R. P. (1977). Heterocyclic, aromatic, and aliphatic amines adsorbed on the anodized silver electrode. *J. Electroanal. Chem.* **84**, pp. 1–20.

4. Ding, S.-Y., Yi, J., Li, J.-F., Ren, B., Wu, D.-Y., Panneerselvam, R. and Tian, Z.-Q. (2016). Nanostructure-based plasmon-enhanced Raman spectroscopy for surface analysis of materials. *Nat. Rev. Mater.* **1**, Article 16021.
5. Synge, E. H. (1928). A suggested method for extending microscopic resolution into the ultra-microscopic region. *Philos. Mag.* **6**(35), pp. 356–362.
6. Dunn, R. C. (1999). Near-field scanning optical microscopy. *Chem. Rev.* **99**, pp. 2891–2928.
7. Binnig, G., Rohrer, H., Gerber, C. and Weibel, E. (1982). Surface studies by scanning tunneling microscopy. *Phys. Rev. Lett.* **49**, pp. 57–61.
8. Wessel, J. (1985). Surface-enhanced optical microscopy. *J. Opt. Soc. Am. B* **2**, pp. 1538–1541.
9. Inouye, Y. and Kawata, S. (1994). Near-field scanning optical microscope with a metallic probe tip. *Opt. Lett.* **19**, p. 159.
10. Tian, Z. Q., Li, W. H., Ren, B., Mao, B. W., Chen, J. G., Mu, J. Q., Zhuo, X. D. and Wang, D. (1996). Simultaneous STM and Raman measurements on electrochemical interfaces. *J. Electroanal. Chem.* **401**, pp. 247–251.
11. Sánchez, E. J., Novotny, L. and Xie, X. S. (1999). Near-field fluorescence microscopy based on two-photon excitation with metal tips. *Phys. Rev. Lett.* **82**, pp. 4014–4017.
12. Stockle, R. M., Suh, Y. D., Deckert, V. and Zenobi, R. (2000). Nanoscale chemical analysis by tip-enhanced Raman spectroscopy. *Chem. Phys. Lett.* **318**, pp. 131–136.
13. Anderson, M. S. (2000). Locally enhanced Raman spectroscopy with an atomic force microscope. *Appl. Phys. Lett.* **76**, pp. 3130–3132.
14. Hayazawa, N., Inouye, Y., Sekkat, Z. and Kawata, S. (2000). Metallized tip amplification of near-field Raman scattering. *Opt. Commun.* **183**, pp. 333–336.
15. Pettinger, B., Picardi, G., Schuster, R. and Ertl, G. (2000). Surface enhanced Raman spectroscopy: Towards single molecular spectroscopy. *Electrochemistry* **68**, pp. 942–949.
16. Zhang, R., Zhang, Y., Dong, Z. C., Jiang, S., Zhang, C., Chen, L. G., Zhang, L., Liao, Y., Aizpurua, J., Luo, Y., Yang, J. L. and Hou, J. G. (2013). Chemical mapping of a single molecule by plasmon-enhanced Raman scattering. *Nature* **498**, pp. 82–86.
17. Jiang, S., Zhang, Y., Zhang, R., Hu, C., Liao, M., Luo, Y., Yang, J., Dong, Z. and Hou, J. G. (2015). Distinguishing adjacent molecules on a surface using plasmon-enhanced Raman scattering. *Nat. Nanotechnol.* **10**, pp. 865–869.
18. Huang, T.-X., Huang, S.-C., Li, M.-H., Zeng, Z.-C., Wang, X. and Ren, B. (2015). Tip-enhanced Raman spectroscopy: Tip-related issues. *Anal. Bioanal. Chem.* **407**, pp. 8177–8195.
19. Hayazawa, N., Saito, Y. and Kawata, S. (2004). Detection and characterization of longitudinal field for tip-enhanced Raman spectroscopy. *Appl. Phys. Lett.* **85**, pp. 6239–6241.
20. Schultz, Z. D., Stranick, S. J. and Levin, I. W. (2008). Tip-enhanced Raman spectroscopy and imaging: An apical illumination geometry. *Appl. Spectrosc.* **62**, pp. 1173–1179.

21. Yang, L.-K., Huang, T.-X., Zeng, Z.-C., Li, M.-H., Wang, X., Yang, F.-Z. and Ren, B. (2015). Rational fabrication of a gold-coated AFM TERS tip by pulsed electrodeposition. *Nanoscale* **7**, pp. 18225–18231.
22. Taguchi, A., Yu, J., Verma, P. and Kawata, S. (2015). Optical antennas with multiple plasmonic nanoparticles for tip-enhanced Raman microscopy. *Nanoscale* **7**, pp. 17424–17433.
23. Roelli, P., Galland, C., Piro, N. and Kippenberg, T. J. (2016). Molecular cavity optomechanics as a theory of plasmon-enhanced Raman scattering. *Nat. Nanotechnol.* **11**, pp. 164–169.
24. Zhang, C., Chen, B.-Q. and Li, Z.-Y. (2015). Optical origin of subnanometer resolution in tip-enhanced Raman mapping. *J. Phys. Chem. C* **119**, pp. 11858–11871.
25. Ren, B., Picardi, G. and Pettinger, B. (2004). Preparation of gold tips suitable for tip-enhanced Raman spectroscopy and light emission by electrochemical etching. *Rev. Sci. Instrum.* **75**, pp. 837–841.
26. Wang, X., Cui, Y. and Ren, B. (2007). Fabrication of Au tips for tip-enhanced Raman spectroscopy. *Chem. J. Chin. Univ.-Chin.* **28**, pp. 522–525.
27. Boyle, M. G., Feng, L. and Dawson, P. (2008). Safe fabrication of sharp gold tips for light emission in scanning tunnelling microscopy. *Ultramicroscopy* **108**, pp. 558–566.
28. Dickmann, K., Demming, F. and Jersch, J. (1996). New etching procedure for silver scanning tunneling microscopy tips. *Rev. Sci. Instrum.* **67**, pp. 845–846.
29. Iwami, M., Uehara, Y. and Ushioda, S. (1998). Preparation of silver tips for scanning tunneling microscopy imaging. *Rev. Sci. Instrum.* **69**, pp. 4010–4011.
30. Stadler, J., Schmid, T. and Zenobi, R. (2010). Nanoscale chemical imaging using top-illumination tip-enhanced Raman spectroscopy. *Nano Lett.* **10**, pp. 4514–4520.
31. Lloyd, J. S., Williams, A., Rickman, R. H., McCowen, A. and Dunstan, P. R. (2011). Reproducible electrochemical etching of silver probes with a radius of curvature of 20 nm for tip-enhanced Raman applications. *Appl. Phys. Lett.* **99**, p. 143108.
32. Hodgson, P. A., Wang, Y., Mohammad, A. A. and Kruse, P. (2013). Note: Electrochemical etching of silver tips in concentrated sulfuric acid. *Rev. Sci. Instrum.* **84**, p. 026109.
33. Li, M., Lv, R., Huang, S., Dai, Y., Zeng, Z., Wang, L. and Ren, B. (2016). Electrochemical fabrication of silver tips for tip-enhanced Raman spectroscopy assisted by a machine vision system. *J. Raman Spectrosc.* **47**, pp. 808–812.
34. Bryant, P. J., Kim, H. S., Zheng, Y. C. and Yang, R. (1987). Technique for shaping scanning tunneling microscope tips. *Rev. Sci. Instrum.* **58**, p. 1115.
35. Melmed, A. J. (1991). The art and science and other aspects of making sharp tips. *J. Vac. Sci. Technol. B* **9**, pp. 601–608.

36. Yi, K., He, X., Zhou, Y., Xiong, W. and Lu, Y. (2008). Tip-enhanced near-field Raman spectroscopy with a scanning tunneling microscope and side-illumination optics. *Rev. Sci. Instrum.* **79**, p. 073706.
37. You, Y. M., Purnawirman, N. A., Hu, H. L., Kasim, J., Yang, H. P., Du, C. L., Yu, T. and Shen, Z. X. (2010). Tip-enhanced Raman spectroscopy using single-crystalline Ag nanowire as tip. *J. Raman Spectrosc.* **41**, pp. 1156–1162.
38. Fujita, Y., Chiba, R., Lu, G., Horimoto, N. N., Kajimoto, S., Fukumura, H. and Uji-i, H. (2014). A silver nanowire-based tip suitable for STM tip-enhanced Raman scattering. *Chem. Commun.* **50**, pp. 9839–9841.
39. Hartschuh, A., Sanchez, E. J., Xie, X. S. and Novotny, L. (2003). High-resolution near-field Raman microscopy of single-walled carbon nanotubes, *Phys. Rev. Lett.* **90**, p. 095503.
40. Berweger, S., Atkin, J. M., Olmon, R. L. and Raschke, M. B. (2010). Adiabatic tip-plasmon focusing for nano-Raman spectroscopy. *J. Phys. Chem. Lett.* **1**, pp. 3427–3432.
41. Vasconcelos, T. L., Archanjo, B. S., Fragneaud, B., Oliveira, B. S., Riikonen, J., Li, C., Ribeiro, D. S., Rabelo, C., Rodrigues, W. N., Jorio, A., Achete, C. A. and Cançado, L. G. (2015). Tuning localized surface plasmon resonance in scanning near-field optical microscopy probes. *ACS Nano* **9**, pp. 6297–6304.
42. DeRose, J. A., Thundat, T., Nagahara, L. A. and Lindsay, S. M. (1991). Gold grown epitaxially on mica: Conditions for large area flat faces. *Surf. Sci.* **256**, pp. 102–108.
43. Lüssem, B., Karthäuser, S., Haselier, H. and Waser, R. (2005). The origin of faceting of ultraflat gold films epitaxially grown on mica. *Appl. Surf. Sci.* **249**, pp. 197–202.
44. Schlegel, V. L. and Cotton, T. M. (1991). Silver-island films as substrates for enhanced Raman scattering: Effect of deposition rate on intensity. *Anal. Chem.* **63**, pp. 241–247.
45. Zhang, J., Matveeva, E., Gryczynski, I., Leonenko, Z. and Lakowicz, J. R. (2005). Metal-enhanced fluoroimmunoassay on a silver film by vapor deposition. *J. Phys. Chem. B* **109**, pp. 7969–7975.
46. Barrios, C. A., Malkovskiy, A. V., Kisliuk, A. M., Sokolov, A. P. and Foster, M. D. (2009). Highly stable, protected plasmonic nanostructures for tip enhanced Raman spectroscopy. *J. Phys. Chem. C* **113**, pp. 8158–8161.
47. Peng, L., Lee, H., Teizer, W. and Liang, H. (2009). Nanowear of gold and silver against silicon. *Wear* **267**, pp. 1177–1180.
48. Schmid, T., Yeo, B. S., Leong, G., Stadler, J. and Zenobi, R. (2009). Performing tip-enhanced Raman spectroscopy in liquids. *J. Raman Spectrosc.* **40**, pp. 1392–1399.
49. Kawata, S. and Shalaev, V. M. (2011). *Tip Enhancement.* Elsevier.
50. Brejna, P. R. and Griffiths, P. R. (2010). Electroless deposition of silver onto silicon as a method of preparation of reproducible surface-enhanced Raman spectroscopy substrates and tip-enhanced Raman spectroscopy tips. *Appl. Spectrosc.* **64**, pp. 493–499.

51. Saito, Y., Murakami, T., Inouye, Y. and Kawata, S. (2005). Fabrication of silver probes for localized plasmon excitation in near-field Raman spectroscopy. *Chem. Lett.* **34**, pp. 920–921.
52. Saito, Y., Wang, J. J., Batchelder, D. N. and Smith, D. A. (2003). Simple chemical method for forming silver surfaces with controlled grain sizes for surface plasmon experiments. *Langmuir* **19**, pp. 6857–6861.
53. Wang, J. J., Saito, Y., Batchelder, D. N., Kirkham, J., Robinson, C. and Smith, D. A. (2005). Controllable method for the preparation of metalized probes for efficient scanning near-field optical Raman microscopy. *Appl. Phys. Lett.* **86**, p. 263111.
54. Umakoshi, T., Yano, T. A., Saito, Y. and Verma, P. (2012). Fabrication of near-field plasmonic tip by photoreduction for strong enhancement in tip-enhanced Raman spectroscopy. *Appl. Phys. Express* **5**, pp. 052001–052003.
55. Wang, H. T., Tian, T., Zhang, Y., Pan, Z. Q., Wang, Y. and Xiao, Z. D. (2008). Sequential electrochemical oxidation and site-selective growth of nanoparticles onto AFM probes. *Langmuir* **24**, pp. 8918–8922.
56. Vakarelski, I. U. and Higashitani, K. (2006). Single-nanoparticle-terminated tips for scanning probe microscopy, *Langmuir* **22**, pp. 2931–2934.
57. Okamoto, T. and Yamaguchi, I. (2001). Photocatalytic deposition of a gold nanoparticle onto the top of a SiN cantilever tip. *J. Microsc.* **202**, pp. 100–103.
58. Sqalli, O., Bernal, M. P., Hoffmann, P. and Marquis-Weible, F. (2000). Improved tip performance for scanning near-field optical microscopy by the attachment of a single gold nanoparticle. *Appl. Phys. Lett.* **76**, pp. 2134–2136.
59. Comstock, D. J., Elam, J. W., Pellin, M. J. and Hersam, M. C. (2012). High aspect ratio nanoneedle probes with an integrated electrode at the tip apex. *Rev. Sci. Instrum.* **83**, pp. 113704–113704.
60. Weber-Bargioni, A., Schwartzberg, A., Cornaglia, M., Ismach, A., Urban, J. J., Pang, Y. J., Gordon, R., Bokor, J., Salmeron, M. B., Ogletree, D. F., Ashby, P., Cabrini, S. and Schuck, P. J. (2011). Hyperspectral nanoscale imaging on dielectric substrates with coaxial optical antenna scan probes. *Nano Lett.* **11**, pp. 1201–1207.
61. Fleischer, M., Weber-Bargioni, A., Altoe, M. V. P., Schwartzberg, A. M., Schuck, P. J., Cabrini, S. and Kern, D. P. (2011). Gold nanocone near-field scanning optical microscopy Probes. *ACS Nano* **5**, pp. 2570–2579.
62. De Angelis, F., Das, G., Candeloro, P., Patrini, M., Galli, M., Bek, A., Lazzarino, M., Maksymov, I., Liberale, C., Andreani, L. C. and Di Fabrizio, E. (2010). Nanoscale chemical mapping using three-dimensional adiabatic compression of surface plasmon polaritons. *Nat. Nanotechnol.* **5**, pp. 67–72.
63. Zou, Y., Steinvurzel, P., Yang, T. and Crozier, K. B. (2009). Surface plasmon resonances of optical antenna atomic force microscope tips. *Appl. Phys. Lett.* **94**, p. 171107.
64. Christiansen, S. H., Becker, M., Fahlbusch, S., Michler, J., Sivakov, V., Andra, G. and Geiger, R. (2007). Signal enhancement in nano-Raman spectroscopy by gold caps on silicon nanowires obtained by vapour-liquid-solid growth. *Nanotechnology* **18**, p. 035503.

65. Farahani, J. N., Pohl, D. W., Eisler, H. J. and Hecht, B. (2005). Single quantum dot coupled to a scanning optical antenna: A tunable superemitter. *Phys. Rev. Lett.* **95**, p. 017402.
66. Park, J., Hong, T., Lee, N., Kim, K. and Seo, Y. (2011). Viscosity dependence of electrochemical etching for gold tip fabrication. *Curr. Appl. Phys.* **11**, pp. 1332–1336.
67. Macpherson, J. V. and Unwin, P. R. (2000). Combined scanning electrochemical-atomic force microscopy. *Anal. Chem.* **72**, pp. 276–285.
68. Taguchi, A., Hayazawa, N., Saito, Y., Ishitobi, H., Tarun, A. and Kawata, S. (2009). Controlling the plasmon resonance wavelength in metal-coated probe using refractive index modification. *Opt. Express* **17**, pp. 6509–6518.
69. Karral, K. and Grober, R. D. (1995). Piezoelectric tip-sample distance control for near-field optical microscopes. *Appl. Phys. Lett.* **66**, pp. 1842–1844.
70. Neacsu, C. C., Dreyer, J., Behr, N. and Raschke, M. B. (2006). Scanning-probe Raman spectroscopy with single-molecule sensitivity. *Phys. Rev. B* **73**, p. 193406.
71. Roy, D., Wang, J. and Welland, M. E. (2006). Nanoscale imaging of carbon nanotubes using tip enhanced Raman spectroscopy in reflection mode. *Faraday Discuss.* **132**, pp. 215–225.
72. Zhang, D., Wang, X., Braun, K., Egelhaaf, H. J., Fleischer, M., Hennemann, L., Hintz, H., Stanciu, C., Brabec, C. J., Kern, D. P. and Meixner, A. J. (2009). Parabolic mirror-assisted tip-enhanced spectroscopic imaging for non-transparent materials. *J. Raman Spectrosc.* **40**, pp. 1371–1376.
73. Christiane, H. and Lukas, N. (2008). Imaging of membrane proteins using antenna-based optical microscopy. *Nanotechnology* **19**, p. 384012.
74. Kalkbrenner, T., Ramstein, M., Mlynek, J. and Sandoghdar, V. (2001). A single gold particle as a probe for apertureless scanning near-field optical microscopy. *J. Microsc.-Oxf.* **202**, pp. 72–76.
75. Le, N. V., Mevellec, J. Y., Minea, T. and Louarn, G. (2012). Gold nanoparticles as probes for nano-Raman spectroscopy: Preliminary experimental results and modeling. *Int. J. Opt.* **2012**, p. 591083.
76. Johnson, T. W., Lapin, Z. J., Beams, R., Lindquist, N. C., Rodrigo, S. G., Novotny, L. and Oh, S. H. (2012). Highly reproducible near-field optical imaging with sub-20-nm resolution based on template-stripped gold pyramids. *ACS Nano* **6**, pp. 9168–9174.
77. Pettinger, B., Ren, B., Picardi, G., Schuster, R. and Ertl, G. (2004). Nanoscale probing of adsorbed species by tip-enhanced Raman spectroscopy. *Phys. Rev. Lett.* **92**, p. 096101.
78. Pettinger, B., Ren, B., Picardi, G., Schuster, R. and Ertl, G. (2005). Tip-enhanced Raman spectroscopy (TERS) of malachite green isothiocyanate at Au(111): bleaching behavior under the influence of high electromagnetic fields. *J. Raman Spectrosc.* **36**, pp. 541–550.
79. Stadler, J., Schmid, T. and Zenobi, R. (2012). Developments in and practical guidelines for tip-enhanced Raman spectroscopy. *Nanoscale* **4**, pp. 1856–1870.

80. Domke, K. F., Zhang, D. and Pettinger, B. (2006). Toward Raman fingerprints of single dye molecules at atomically smooth Au(111). *J. Am. Chem. Soc.* **128**, pp. 14721–14727.
81. Rendell, R. W., Scalapino, D. J. and Mühlschlegel, B. (1978). Role of local plasmon modes in light emission from small-particle tunnel junctions. *Phys. Rev. Lett.* **41**, pp. 1746–1750.
82. Zhang, W. H., Yeo, B. S., Schmid, T. and Zenobi, R. (2007). Single molecule tip-enhanced Raman spectroscopy with silver tips. *J. Phys. Chem. C* **111**, pp. 1733–1738.
83. Pettinger, B., Schambach, P., Villagómez, C. J. and Scott, N. (2012). Tip-enhanced Raman spectroscopy: Near-fields acting on a few molecules. *Annu. Rev. Phys. Chem.* **63**, pp. 379–399.
84. Ren, B., Picardi, G., Pettinger, B., Schuster, R. and Ertl, G. (2005). Tip-enhanced Raman spectroscopy of benzenethiol adsorbed on Au and Pt single-crystal surfaces. *Angew. Chem. Int. Ed.* **44**, pp. 139–142.
85. Wang, X., Liu, Z., Zhuang, M. D., Zhang, H. M., Xie, Z. X., Wu, D. Y., Ren, B. and Tian, Z. Q. (2007). Tip-enhanced Raman spectroscopy for investigating adsorbed species on a single-crystal surface using electrochemically prepared Au tips. *Appl. Phys. Lett.* **91**, p. 101105.
86. Liu, Z., Ding, S. Y., Chen, Z. B., Wang, X., Tian, J. H., Anema, J. R., Zhou, X. S., Wu, D. Y., Mao, B. W., Xu, X., Ren, B. and Tian, Z. Q. (2011). Revealing the molecular structure of single-molecule junctions in different conductance states by fishing-mode tip-enhanced Raman spectroscopy. *Nat. Commun.* **2**, p. 305.
87. Wang, X., Zhong, J.-H., Zhang, M., Liu, Z., Wu, D.-Y. and Ren, B. (2016). Revealing intermolecular interaction and surface restructuring of an aromatic thiol assembling on Au(111) by tip-enhanced Raman spectroscopy. *Anal. Chem.* **88**, pp. 915–921.
88. Liu, Z., Wang, X., Dai, K., Jin, S., Zeng, Z. C., Zhuang, M. D., Yang, Z. L., Wu, D. Y., Ren, B. and Tian, Z. Q. (2009). Tip-enhanced Raman spectroscopy for investigating adsorbed nonresonant molecules on single-crystal surfaces: tip regeneration, probe molecule, and enhancement effect. *J. Raman Spectrosc.* **40**, pp. 1400–1406.
89. Tian, J.-H., Liu, B., Li, Yang, Z.-L., Ren, B., Wu, S.-T., Tao and Tian, Z.-Q. (2006). Study of molecular junctions with a combined surface-enhanced Raman and mechanically controllable break junction method. *J. Am. Chem. Soc.* **128**, pp. 14748–14749.
90. Ward, D. R., Halas, N. J., Ciszek, J. W., Tour, J. M., Wu, Y., Nordlander, P. and Natelson, D. (2008). Simultaneous measurements of electronic conduction and Raman response in molecular junctions. *Nano Lett.* **8**, pp. 919–924.
91. Osawa, M., Matsuda, N., Yoshii, K. and Uchida, I. (1994). Charge-transfer resonance Raman process in surface-enhanced Raman-scattering from P-aminothiophenol adsorbed on Silver–Herzberg–Teller Contribution. *J. Phys. Chem.* **98**, pp. 12702–12707.

92. Liu, G.-K., Hu, J., Zheng, P.-C., Shen, G.-L., Jiang, J.-H., Yu, R.-Q., Cui, Y. and Ren, B. (2008). Laser-induced formation of metal–molecule–metal junctions between Au nanoparticles as probed by surface-enhanced Raman spectroscopy. *J. Phys. Chem. C* **112**, pp. 6499–6508.
93. Huang, Y.-F., Zhu, H.-P., Liu, G.-K., Wu, D.-Y., Ren, B. and Tian, Z.-Q. (2010). When the signal is not from the original molecule to be detected: Chemical transformation of para-aminothiophenol on Ag during the SERS measurement. *J. Am. Chem. Soc.* **132**, pp. 9244–9246.
94. Huang, Y. F., Wu, D. Y., Zhu, H. P., Zhao, L. B., Liu, G. K., Ren, B. and Tian, Z. Q. (2012). Surface-enhanced Raman spectroscopic study of p-aminothiophenol. *Phys. Chem. Chem. Phys.* **14**, pp. 8485–8497.
95. Lantman, E. M. V., Deckert-Gaudig, T., Mank, A. J. G., Deckert, V. and Weckhuysen, B. M. (2012). Catalytic processes monitored at the nanoscale with tip-enhanced Raman spectroscopy. *Nat. Nanotechnol.* **7**, pp. 583–586.
96. Sun, M., Zhang, Z., Zheng, H. and Xu, H. (2012). *In situ* plasmon-driven chemical reactions revealed by high vacuum tip-enhanced Raman spectroscopy. *Sci. Rep.* **2**, p. 647.
97. Kumar, N., Stephanidis, B., Zenobi, R., Wain, A. J. and Roy, D. (2015). Nanoscale mapping of catalytic activity using tip-enhanced Raman spectroscopy. *Nanoscale* **7**, pp. 7133–7137.
98. Steidtner, J. and Pettinger, B. (2007). High-resolution microscope for tip-enhanced optical processes in ultrahigh vacuum. *Rev. Sci. Instrum.* **78**(10), p. 103104.
99. Steidtner, J. and Pettinger, B. (2008). Tip-enhanced Raman spectroscopy and microscopy on single dye molecules with 15 nm resolution. *Phys. Rev. Lett.* **100**(23), p. 236101.
100. Jiang, N., Foley, E. T., Klingsporn, J. M., Sonntag, M. D., Valley, N. A., Dieringer, J. A., Seideman, T., Schatz, G. C., Hersam, M. C. and Van Duyne, R. P. (2012). Observation of multiple vibrational modes in ultrahigh vacuum tip-enhanced Raman spectroscopy combined with molecular-resolution scanning tunneling microscopy. *Nano Lett.* **12**, pp. 5061–5067.
101. Nakata, A., Nomoto, T., Toyota, T. and Fujinami, M. (2013). Tip-enhanced Raman spectroscopy of lipid bilayers in water with an alumina- and silver-coated tungsten tip. *Anal. Sci.* **29**, pp. 865–869.
102. Zeng, Z.-C., Huang, S.-C., Wu, D.-Y., Meng, L.-Y., Li, M.-H., Huang, T.-X., Zhong, J.-H., Wang, X., Yang, Z.-L. and Ren, B. (2015). Electrochemical tip-enhanced Raman spectroscopy. *J. Am. Chem. Soc.* **137**, pp. 11928–11931.
103. Kurouski, D., Mattei, M. and Van Duyne, R. P. (2015). Probing redox reactions at the nanoscale with electrochemical tip-enhanced Raman spectroscopy. *Nano Lett.* **15**, pp. 7956–7962.
104. Martin Sabanes, N., Driessen, L. M. and Domke, K. F. (2016). Versatile side-illumination geometry for tip-enhanced Raman spectroscopy at solid/liquid interfaces. *Anal. Chem.* **88**, pp. 7108–7114.

105. Tian, Z.-Q. and Ren, B. (2007). Raman spectroscopy of electrode surfaces. In Bard, A. J., Stratmann, M. and Unwin, P. (Eds.). *Encyclopedia of Electrochemistry.* Wiley-VCH Verlag GmbH & Co. KGaA, pp. 572–659.
106. Tian, Z. Q. and Ren, B. (2004). Adsorption and reaction at electrochemical interfaces as probed by surface-enhanced Raman spectroscopy. *Annu. Rev. Phys. Chem.* **55**, pp. 197–229.
107. Pettinger, B. (2008). Tip-enhanced Raman spectroscopy — Recent developments and future prospects. In Alkire, R. C., Kolb, D. M., Lipkowski, J. and Ross, P. N. (eds.), *Advances in Electrochemical Sciences and Engineering.* Wiley-VCH Verlag GmbH, New York, pp. 377–418.

Chapter 11

Tip-Enhanced Raman Scattering in Liquid/Solution

Prompong Pienpinijtham*,‡ and Yukihiro Ozaki†,§

*Sensor Research Unit (SRU),
Department of Chemistry, Faculty of Science,
Chulalongkorn University, Thailand
†Department of Chemistry,
School of Science and Technology,
Kwansei Gakuin University, Japan
‡prompong.p@chula.ac.th
§ozaki@kwansei.ac.jp

1. Introduction

The use of tip-enhanced Raman scattering (TERS) in solutions is very important, especially in the study of biological samples. These biological samples normally function in aqueous solutions, and *in situ* measurements are required to explore their biological activities in wet environments. Some properties, such as pH, appear only in aqueous systems. Living cells and biological molecules also behave differently in different environments of acidity or alkalinity. Moreover, a huge number of chemical reactions in nature commonly occur in aqueous environments. Understanding the fundamental of these reactions at nanoscale can also solve problems in various fields, e.g. energy storages, catalysts, electronic devices, medicines, alternative energy sources.

In this chapter, we introduce the rise of TERS, the development of TERS in solution including a theoretical point of view and propose experimental setups, and some promising potential applications of TERS in liquids and solutions.

2. Rise of Tip-Enhanced Raman Scattering

To investigate the nanoscale world nowadays, scanning probe microscopy (SPM), e.g. atomic force microscopy (AFM), scanning tunneling microscopy (STM), and near-field scanning optical microscopy (NSOM/SNOM), is one of the most popular and suitable techniques. This technique is based on the principle of a tiny nanoneedle scanning and touching the sample surface and providing topological information by plotting the height or response (e.g. force, current) of a tip/probe at each specific (x, y) coordinates. However, the molecular information of sample cannot be obtained by using only this technique.

It has been over 40 years since the first surface-enhanced Raman scattering (SERS) was fortuitously observed [1]. The report presented an extraordinarily high intensity of Raman signals from the monolayer of pyridine adsorbed on an electrochemically rough surface of silver electrode. From that time, SERS is well known to yield molecular information of trace molecules attached to a rough metal surface, including a surface of metal nanoparticles [2, 3]. Among the intense argument on SERS-enhancement mechanism, two proposed mechanisms are widely accepted, which are "electromagnetic mechanism" and "chemical or charge-transfer mechanism". The electromagnetic mechanism plays an important role in tremendous electric field enhancement of 10^2–10^{14} [4] while the chemical mechanism is always necessarily required with the enhancement up to 10^2 times [5]. Nevertheless, obtaining molecular information of trace chemicals on a specifically small area along with viewing their surface topology was still unimaginable at that time.

By gathering the advantages of SPM and SERS, tip-enhanced Raman scattering (TERS) was formally operated for the first time in 2000. That year, many research groups coincidentally reported enhanced Raman signals from the combination of SPM (including AFM, STM, NSOM/SNOM, etc.) and SERS [6–8]. TERS utilizes a metal tip (particularly gold and silver) of SPM not only to scan the topology of the sample surface, but also to enhance Raman signals of molecules closed to the tip. When the tip is irradiated by an excitation laser with a frequency corresponding to that of electron oscillation on a tip apex, plasmonic coupling on the surface of the metal tip produces a large electric field. The trace molecules closed to the tip apex interact with this enhanced electric field, and then enhanced Raman signals arising from these molecules are obtained [9–12]. This way, the nanoscale topology of the sample and molecular information at a given position can be obtained simultaneously.

Extensive developments in TERS have greatly benefited studies on molecular structures and interactions in the world of nanotechnology [6–8, 13–15]. Moreover, computational calculations and simulations of electric field enhancement have been carried out to understand the TERS phenomenon [10, 12]. TERS is widely used in various fields, e.g. few- or single-molecule studies [16, 17], RNA sequencing [18, 19], mechanistic studies of biological molecules, [20, 21] carbon nanotube and graphene characterization [22–24], and TERS imaging/mapping [25–28]. For example, Bailo and Deckert (2008) proposed a new direct method for sequencing single RNA strands using the TERS technique [18]. Thus, using this method, both the AFM images of RNA strands, and the TERS spectra at a specific point on these strands can be obtained to provide a sequence of RNA strands. This sequencing method was further developed for higher lateral resolution and single-molecule sensitivity by Treffer *et al.* (2012) [19]. They demonstrated that Raman signals with respect to the four nucleobases can be remarkably recorded and determined. Furthermore, Yeo *et al.* (2008) applied the TERS technique to investigate the membrane heme protein cytochrome c (cyt c). This technique gave more structural information on large biomolecules when compared to SERS [21]. Recently, Vantasin *et al.* used the TERS technique to determine submicro/nanostructures of epitaxial graphene, formed by using an annealing SiC method to perceive the growth mechanism of local nanostructures [24].

Thus far, most of the TERS measurements were carried out to investigate dry samples. Due to a number of limitations including problems with laser focusing, sample preparation, the feedback system of SPM, and adsorbed contaminants on the TERS tip, it is practically difficult to operate TERS measurement in a liquid/solution.

3. Simulation of TERS in an Aqueous Solution

Before the first TERS measurement in a liquid/solution was set up practically, the theoretical calculations had already been done. Several methods such as discrete dipole approximation (DDA), finite element, or finite-difference time-domain (FDTD) were used to calculate the electric field around the apex of TERS tip in order to gain a deep understanding about TERS phenomenon and its enhancement.

The TERS enhancement is described by the TERS enhancement factor (EF). Experimentally, it is determined by EF $= (I_{\mathrm{Tip-in}}/I_{\mathrm{Tip-out}} - 1)(A_{\mathrm{FF}}/A_{\mathrm{NF}})$, where $I_{\mathrm{Tip-in}}$ and $I_{\mathrm{Tip-out}}$ are the Raman peak intensities measured with the tip in contact and retracted from the sample; A_{FF} is the area of the far-field laser probe; A_{NF} is the effective area of TERS probe, which is usually estimated from the geometric diameter of the tip apex. The term $(I_{\mathrm{Tip-in}}/I_{\mathrm{Tip-out}} - 1)$ is also described as "contrast" [29]. Theoretically, *EF* can be calculated from the fourth power of the ratio between near-field and far-field electric-field intensities around the tip under the irradiation of excitation laser, which will be discussed in detail in this section.

In 2006, Downes *et al.* aimed to employ a finite element model to theoretically calculate the optical enhancement as a function of wavelength or Raman shift using atomically flat surfaces of gold, mica, and silicon as substrates, which are materials commonly used as SPM substrates [10]. Aqueous environments were modeled for "the first time" comparing with findings for scanning in air.

ANSYS Multiphysics software with a high-frequency electromagnetic option was used to create finite element models by solving Maxwell's equations across a defined region containing the tip, substrate, and air or water gap. In the calculations, a p-polarized plane wave interacts with the simulated tip with a hemisphere of 20-nm radius, which represents a tip radius obtained from electrochemical etching in hydrochloric acid [30], placed at the end of a cone of half angle 10°. The amplitude of wave is set to 1 in order to make the calculated value of electric field amplitude in the tip apex region equal to the enhancement of the incident electric field amplitude, E_{elec}.

The simulation was calculated based on the electromagnetic mechanism of SERS enhancement, which is a two-stage process of scattering from the tip region. The tip region must first be excited, then the region radiates. The overall enhancement of the scattered light intensity, E_{scat}, is known as two-fold electromagnetic enhancement and is given by the product of the incident electric-field intensity enhancement and the radiated electric-field intensity enhancement. The expression of this enhancement is $E_{\text{scat}} = (E_{\text{elec}}(\lambda_{\text{inc}}))^2 \cdot (E_{\text{elec}}(\lambda_{\text{rad}}))^2$, where λ_{inc} and λ_{rad} are the incident and radiated wavelengths, respectively. Due to the symmetry of incident and radiative processes, when the incident and radiated wavelengths are the same (i.e. no losses), the intensity enhancement (E_{scat}) equals to $(E_{\text{elec}}(\lambda_{\text{inc}}))^4$ in this work.

The discrepancy between theoretically simulated and experimentally measured TERS enhancement factors can be found because of some assumptions. In experiments, EF is measured from all signals scattered back from the tip where the intensity of the electric field is not homogeneously distributed over the tip surface. On the other hand, EF in the simulation is considered from the position under the tip with the strongest electric field intensity. TERS signals in experiments also depend on a number of molecules in the effective area of TERS probe under the tip, which needs to be calculated using any other assumption. Moreover, TERS signals are practically lost due to scattering in an optic path in a Raman system. Last but not least, it is difficult to fabricate a completely perfect TERS tip similar to a model in the simulation.

In real biological samples, some ions can be dissolved and form an aqueous solution. However, the ions are in relatively small quantities. Thus, it can be assumed in this work that the aqueous environment is pure water.

When the gap between tip and substrate is water instead of air, significant differences can be observed. There is a red-shift in plasmon band with a decrease in the gap separation. A mica substrate

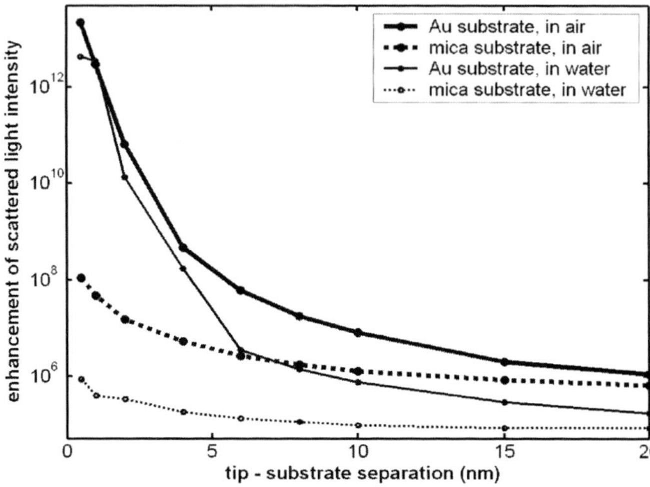

Figure 1. The enhancement of the scattered light intensity for a 20-nm radius gold tip and substrates of a gold or mica in air or water. Reprint with permission from Ref. [10].

shows the spectra with lower peak mobility than the gold substrate. It suggests that mica is more stable for imaging. The intensity enhancement in aqueous environments is lower than that in air, as shown in Figure 1. With small separation, Au substrates provide extremely high enhancements of 10^{12} while the enhancement from mica substrate does not exceed 10^8. The intensity enhancement rapidly decays with an order of magnitude when the tip–substrate separation increases for only 0.5 nm. At very large separation, the tip behaves like an isolated tip with the enhancements of 10^5–10^6. This work also demonstrates that the lateral resolution under water is similar to those in air, except at small separation (slightly less than 2 nm). However, for large separation on mica substrates, the lateral resolution is closed to the tip radius.

4. The First TERS Measurement in a Solution

Almost 3 years after the report from Downes *et al.*, the first TERS measurement in a solution was proposed by Schmid *et al.* (2009) [31]. The motivation for the first TERS also comes from many outstanding questions in biology and medicine, which require analytical tools providing image and chemical information with high spatial resolution. At that time, TERS had been performed only in air or ultra-high vacuum but the active state of most biological samples such as cells and tissues can only be studied in aqueous buffers. Therefore, the first TERS setup for aqueous conditions was introduced. Briefly, the AFM tip was mounted on a Teflon plate with a window for the AFM feedback laser. The tip and sample were immersed in a droplet of water situated between the Teflon plate and the transparent sample stage. The excitation laser was illuminated through the transparent sample stage (bottom illumination), as shown in Figure 2. A diode-pumped solid-state laser with a wavelength of 532 nm was employed for excitation. AlF_3/Ag-coated and SiO_x/Ag-coated tips were used as a TERS tip with a good enhancement in the blue to green range of excitation laser.

Self-assembled monolayers (SAMs) of thiophenol (PhS) on a gold substrate were used in a proof-of-principle study. The results show the peeling off of coating layer for AlF_3/Ag-coated tips. Hence, this kind of tip cannot be used in TERS measurements in water or aqueous buffers. However, a SiO_x/Ag-coated TERS tip can stay almost intact and give a good enhancement in water. The tips were also

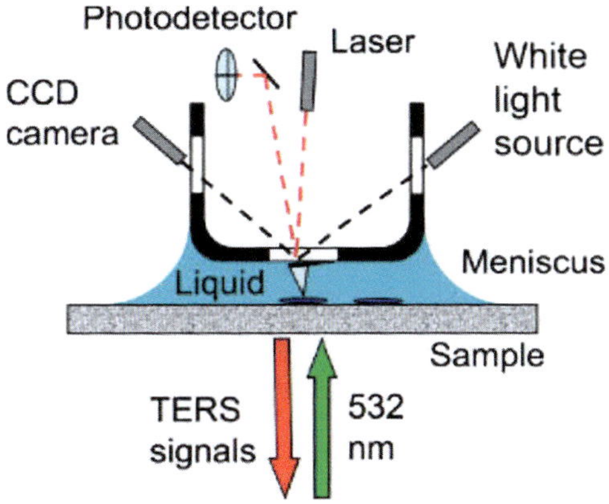

Figure 2. The first TERS setup for aqueous conditions. Reprint with permission from Ref. [31].

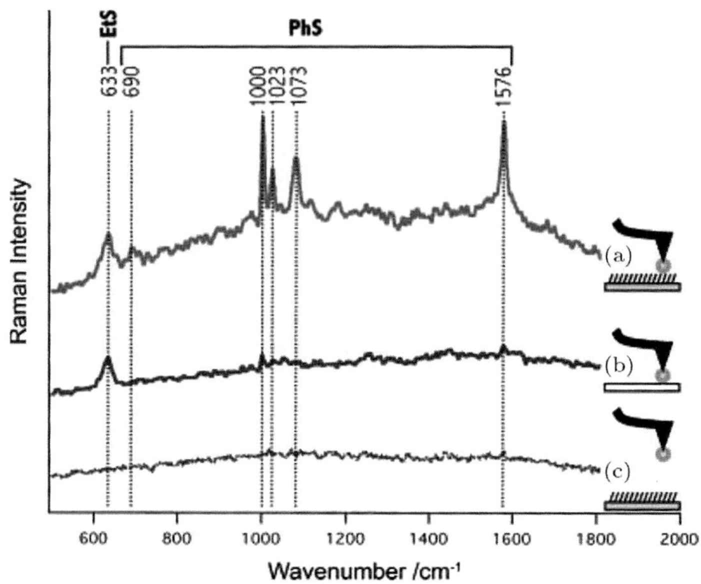

Figure 3. Proof-of-principle experiment demonstrating the feasibility of performing TERS to detect PhS on Au in water using EtS-protected tips. Reprint with permission from Ref. [31].

prevented from unwanted binding of molecules during the experiment or even remove/replace the contaminant due to the protective layer of EtS SAM. TERS of PhS on Au in water using EtS-protected tips can be successfully obatained, as shown in Figure 3. TER spectrum of a PhS SAM on Au collected in water shows characteristic peaks of PhS at 1576, 1073, 1023, 1000 and 690 cm^{-1}. On the other hand, the control experiment performed after measurement with the same tip in contact with a clean Au surface and far-field measurement shows no peak from PhS. The contrast between near-field and far-field signal indicates an enhancement factor of $\sim 3 \times 10^3$. Again, this is the first TERS study at liquid/solid interfaces, and this study greatly opens the door to use TERS for the investigation of biological cells, cell membranes, or supported lipid bilayers in aqueous environments.

5. TERS for Biological Application in Water

After the first proposal of TERS in solution, studies about biological systems using TERS have been reported. The real behaviors in nature (in aqueous conditions or wet environments) of biological molecules that we cannot observe in dry systems are gradually revealed.

Biomembranes (lipid bilayers) play crucial roles in various activities of living cells. The structure of biomembranes is dynamic and heterogeneous. There is an incessant movement inside the biomembranes due to the formation of lipid domains or lipid rafts from groups of lipids that segregate from other regions [32–36]. The size of these domains is generally less than 100 nm, which must be determined by electron microscopy and scanning probe microscopy. To perform an *in situ* characterization of biomembrane structures, chemical reactions and dynamics, TERS measurement in water is a good choice.

Nakata *et al.* mimicked biomembranes using two kinds of supported phospholipid bilayers, i.e. 1,2-dipalmitoyl-*sn*-glycero-3-phosphocholine (DPPC) and 1,2-dioleoyl-*sn*-glycero-3-phosphocholine (DOPC) and monitored them in an aqueous environment [37]. The TERS setup in this work is a bottom illumination using a 532-nm excitation wavelength generated by a diode-pumped solid-state laser. The beam was radially polarized to increase the z-polarized electric field at the sample position. The distance between the tips and samples was controlled by shear–force feedback. The TERS tip was attached to a tuning fork from a crystal oscillator. To protect tuning fork from the aqueous environment, it was covered by a glass tube filled with air. Only then the TERS tip was immersed in the water. The TERS tips were made from electropolished tungsten coated with alumina and then silver particles. Alumina was coated here under the silver nanoparticle layer as a dielectric layer to control the localized surface plasmon resonance wavelength (a change in dielectric environments of silver nanoparticles shifts their localized surface plasmon resonance wavelength) and give stability to the tip (low-solubility

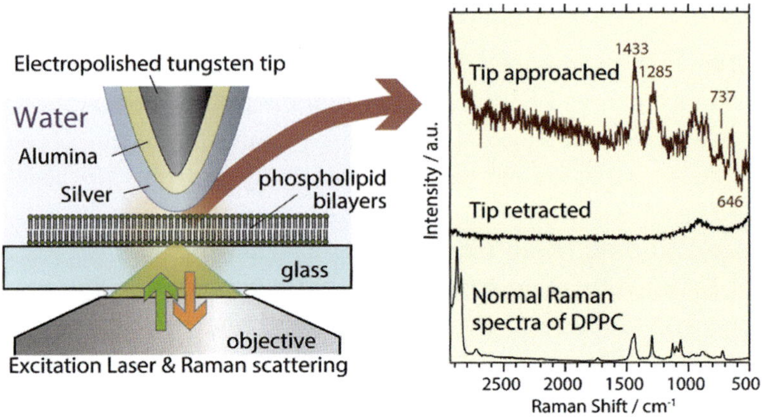

Figure 4. The schematic setup of TERS for the study of biomembranes. The beam was radially polarized to increase the z-polarized electric field at the sample position. Reprint with permission from Ref. [37].

alumina layer from the anodization of aluminum prevents detachment of the silver in water). Lipid bilayers supported on a glass substrate were prepared by a vesicle fusion method [38]. The resulting membrane bilayer was immersed in water for the experiments at room temperature (20–23°C). The schematic setup of this system is displayed in Figure 4.

The results show that the TERS spectrum from the fabricated tips provides the signals 2.5 times stronger compared to normal Raman spectrum and the tips have a good stability even after being immersed in water for 1 h. A small enhancement of TERS signal is observed. It is because the tip is not a pure silver tip, but a tungsten tip modified with alumina and silver layers. The small enhancement is compensated by good stability of the tip in water. Spectral changes can be observed in time-lapse TERS spectra of supported DOPC bilayers (see Figure 5), which indicate inhomogeneity and mobility of fluid bilayer membranes near the tip. However, for supported DPPC bilayers, there is no spectral change during the measurements and no transient peaks appeared in the spectra. It might be explained that at room temperatures, DOPC forms a lamellar liquid-crystalline phase while DPPC is in a lamellar-gel phase with low fluidity [39]. The lateral diffusion of lipid molecules in a lamellar liquid-crystalline phase

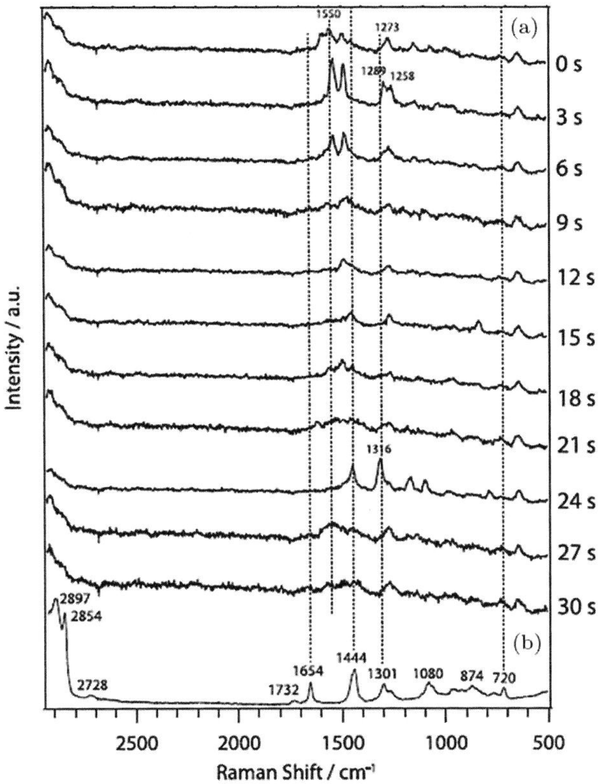

Figure 5. (a) Time-lapse TERS spectra of supported DOPC bilayers. (b) A normal Raman spectrum of multilayer DOPC. Reprint with permission from Ref. [37].

is relatively faster than that of a lamellar-gel phase. Therefore, spectral changes can be observed in a DOPC case with this time interval. These results show a great potential of TERS in solution to monitor fluidity and fluctuations of structurally different components coexisting in biomembranes. Furthermore, conformational or structural differences in biomembranes at length scales below the diffraction limit can be examined.

Moreover, carbon nanotubes, which are currently a great topic of interest, are generally studied in dry conditions. It is possible that their behaviors and properties might be different when they are in liquid or water. Nagata *et al.* showed that behavior by using TERS to investigate single-walled carbon nanotubes in a liquid [40].

6. Electrochemical TERS

Electrochemical (EC) or redox reaction is very useful nowadays and will remain so in future as well. There are many related applications such as energy storage, catalysts, fuel cells, and alternative energy sources. Moreover, several biochemical processes in plants, animals, or even in humans are also related to electrochemical reaction, for example, photosynthesis process, biological electron transfer, and cellular respiratory process. Therefore, a deep understanding about electrochemical processes at the nanoscale is very important. However, it is difficult and there is no capable technique to directly acquire chemical information from molecules undergoing heterogeneous charge-transfer at the nanoscale. TERS could provide local chemical information, while traditional cyclic voltammetry (CV) could provide average electrochemical properties on a bulk working electrode. To investigate electrochemical processes at the nanoscale, it is necessary to couple TERS machine with scanning electrochemical microscope (SECM), which provides local electrochemical information. This idea leads to the development of electrochemical TERS (EC-TERS) [41, 42].

The very first EC-TERS results and the design of EC-TERS setup have been presented in the Third International Conference on Tip-Enhanced Raman Spectroscopy (TERS III) in Zurich in August 2013, which is well-known to the TERS community. Later, Zeng *et al.* published the first EC-TERS setup in a solution based on an EC-STM system using horizontal illumination mode in 2015 (see Figure 6) [41]. (It is noted that the results of this work have been reported earlier in Heraeus Seminar "Electrochemical Surface Science" in Bad Honnef in January 2015.) The tip in STM-based TERS commonly plays two important roles. The first is to produce the enhanced electromagnetic field and the second is to allow the tunneling current to pass. In EC-TERS, the tip must additionally be coated with an insulating material to avoid the interference of Faradaic current from electrochemical process with the tunneling current. In this work, the

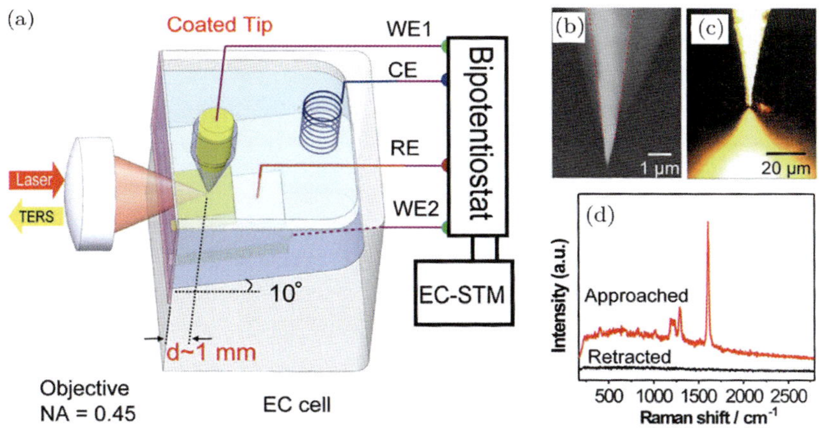

Figure 6. (a) Schematic of the EC-TERS setup by Zeng *et al.* (b) SEM image of an insulated gold tip. (c) Microscopic image of the tip, single-crystal substrate, and laser spot in an EC-TERS system. (d) TERS of 4-PBT adsorbed on the Au(111) surface obtained while the tip was approached (top) and retracted (bottom). Reprint with permission from Ref. [41].

polyethylene glue-coated Au TERS tips were employed to obtain potential-dependent EC-TERS of 4-PBT (4′-(pyridin-4-yl)biphenyl-4-yl)methanethiol) adsorbed on the Au(111) surface. The potentials of the tip (WE1) and the substrate (WE2) were controlled by a bipotentiostat to obtain local electrochemical information. This function came from an EC-STM system. The laser was introduced horizontally via a microscope objective to the electrochemical cell and illuminated in the gap between the tip and Au(111) surface. Then, the produced TERS signal was collected by the same objective to obtain local chemical information. Therefore, local electrochemical information and local chemical information of 4-PBT can be obtained in the same time by this EC-TERS setup. The results present a substantial change in the 4-PBT molecule configuration with potential, which results in the protonation and deprotonation of the 4-PBT molecule.

At about the same time in 2015, Kurouski *et al.* demonstrated a different idea of EC-TERS setup for studying the nanoscale redox behavior [42]. EC-TERS of Nile Blue (NB) was studied and the results were compared with that of conventional CV. The setup of EC-TERS is shown in Figure 7. Briefly, NB molecules were first

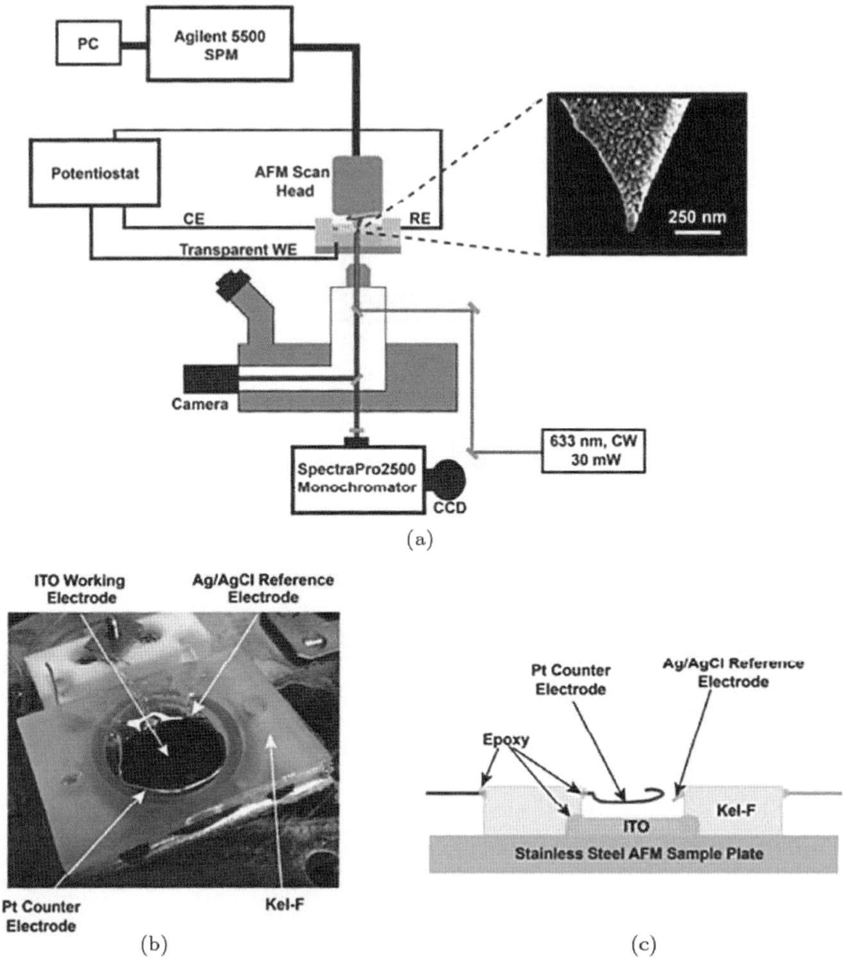

Figure 7. (a) Schematic diagram of the EC-AFM-TERS instrument by Kurouski *et al.* An inset shows SEM image of a Au coated contact mode AFM-TERS tip. (b) The electrochemical AFM-TERS cell. (c) Schematic cross-section of the cell in (b). Reprint with permission from Ref. [42].

deposited on ITO coverslips. The NB-deposited ITO coverslip was fixed at the bottom of the Kel-F wet-cell and sealed by epoxy to act as a working electrode. To make external electrical contact, a silver wire was attached to the top of the ITO film by silver epoxy. Epoxy was used to insulate both the silver epoxy and the Ag wire from the working solution. Pt and Ag/AgCl wires were used as the counter electrode and reference electrode, respectively. The Ag/AgCl wire reference electrode was directly immersed in the supporting electrolyte. TERS tips were prepared by thermal evaporation of a 70 nm Au film onto contact mode AFM cantilevers. A 633-nm continuous wave excitation laser was employed. After the TERS tip approached the ITO surface, the laser was focused on the tip apex using bottom illumination. The feedback laser of AFM scanner was an IR diode laser to prevent spectral interference in TERS measurements. The potential of the ITO surface was controlled by a bipotentiostat.

From this setup, EC-TERS can be operated to study the nanoscale redox behavior of NB. The disappearance of the 591 cm^{-1} band of NB upon reduction and its reversible reappearance upon oxidation during the CV can be observed. The observation of a negative shift in the onset of the potential response of the TERS intensity of the 591 cm^{-1} band suggests that there is perturbation of the electrical double-layer by the TERS tip. It locally alters the effective potential for NB molecules in the tip–sample junction. Furthermore, the results also suggest that the tip has no effect on the local charge-transfer kinetics. In addition, step-like behavior in some TERS voltammograms corresponding to reduction and oxidation of single or few NB molecules can be observed. This work also shows that the coverage of NB is not uniform across the surface of ITO coverslips.

7. Nanoscale pH Measurement Using TERS

Acidity or alkalinity (pH) exists only in aqueous system and its properties cannot be determined from dry samples. The pH involves a lot of biological processes. Under different pHs, living cells, organelles, and biomolecules also behave differently. Furthermore, different regions in living cells have different activities which produce different amounts of protons or hydroxide ions leading to different local pH in cells. This local pH can indicate the pathological changes or other biological processes at a single cellular level. Therefore, the determination of local pH at the nanoscale level is one of the crucial keys to investigate and understand cellular behaviors including cancers and diseases. Some works determined the local pH using SERS technique with metal nanoparticles [43–46]. However, metal nanoparticles might disturb samples.

Pienpinijtham *et al.* presented the nanoscale pH profile at a solution/solid interface using chemically modified tip-enhanced Raman scattering [47]. In this study, a TERS tip was chemically modified with pH-sensitive molecules for use in an aqueous system. The selected pH-sensitive molecules were *para*-mercaptobenzoic acid (*p*MBA) and *para*-aminothiophenol (*p*ATP), which are topics of interest in several papers on pH/chemical probes using SERS techniques [43–46]. The TERS setup was top-illumination and top-collection using the same objective lens. A 514.5-nm diode-pumped solid-state (DPSS) laser was employed for the excitation. The angle between the tip and sample surface was 45° (see TERS measurement setup in Figure 8).

The *p*MBA-modified tip can be used to determine pH in the range of 7–9 by following a peak at 1422 cm^{-1} due to the –COO$^-$ stretching vibration from a transformation of a –COOH group to a –COO$^-$ moiety with an increase in alkalinity (Figure 9). The peak at 1586 cm^{-1}, which is assigned to ring-breathing vibration mode, was selected as the reference because it slightly involves the proton dissociation process.

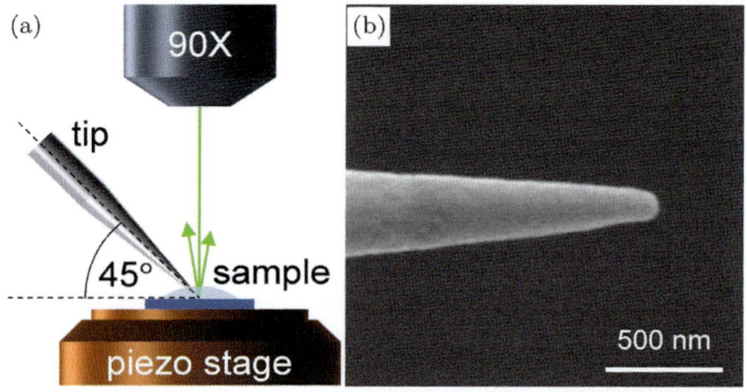

Figure 8. (a) TERS measurement setup used by Pienpinijtham *et al.* (b) An SEM image of TERS tip. Reprint with permission from Ref. [47].

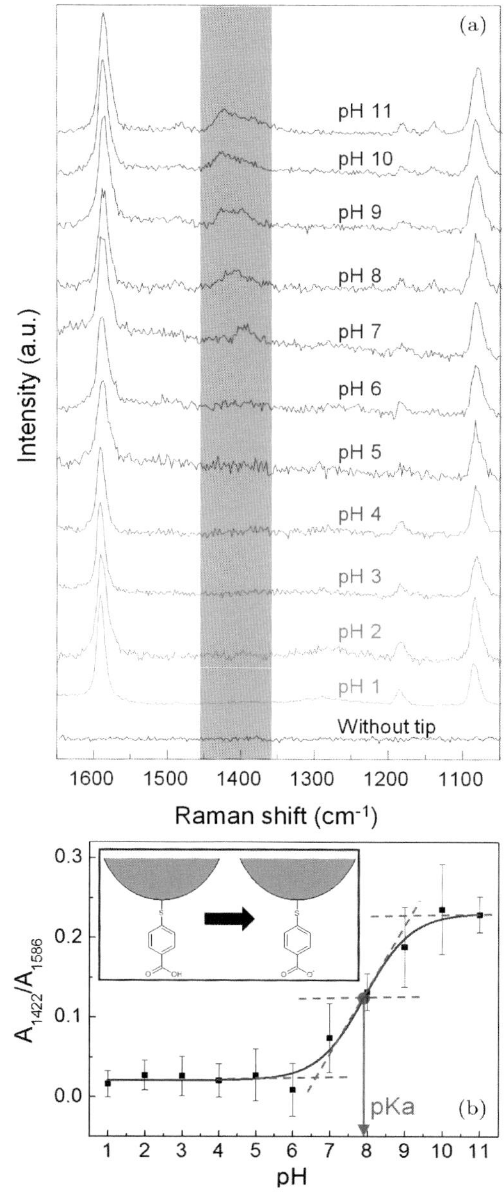

Figure 9. (a) TERS spectra collected from a *p*MBA-modified TERS tip in different pH solutions. (b) A plot of the ratio of two peak areas at 1422 cm^{-1} and 1586 cm^{-1} versus pH of solutions. The inset shows a change in molecular structure of pMBA attached on the TERS tip. Reprint with permission from Ref. [47].

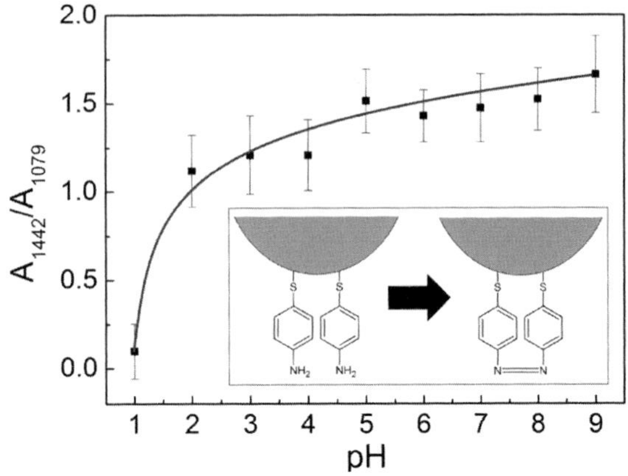

Figure 10. A plot of the ratio of two peak areas at 1442 cm^{-1} and 1079 cm^{-1} versus the pH of solutions. The inset shows a change in molecular structure of *p*ATP attached on TERS tip. Reprint with permission from Ref. [47].

On the other hand, the *p*ATP-modified tip was used to determine pH in the range of 1–2 by detecting a peak at 1442 cm^{-1} due to the N=N stretching vibration arising from the formation of 4,4′-dimercaptoazobenzene (DMAB), as shown in Figure 10. The peak at 1079 cm^{-1}, which is ascribed to C–S stretching vibration mode, is used as the reference peak. When *p*ATP is placed in an alkaline solution, it reacts with an adjacent molecule to form an azo (N=N) dimer, i.e. DMAB, via two Ph–NH_2 moieties, which is stable in acidic environments.

Moreover, these two kinds of molecules also help protect the tip surface from oxidation (from O_2 in the atmosphere) and adsorbed contaminants.

To construct a nanoscale pH profile, a $4 \times 4 \mu m^2$ area of NH_2-anchored glass slide in an aqueous solution is determined using *p*MBA-modified tip. Only half of the glass slide was modified. Therefore, the constructed pH profile shows two different regions, which indicate the different pH over the areas (Figure 11). The results also show the spatial resolution of ∼400 nm and the measured pH becomes the pH of the bulk solution when the tip is ∼200 nm from the surface. This proposed method has a large number of potential applications, including intracellular pH sensing in biological samples such as cells

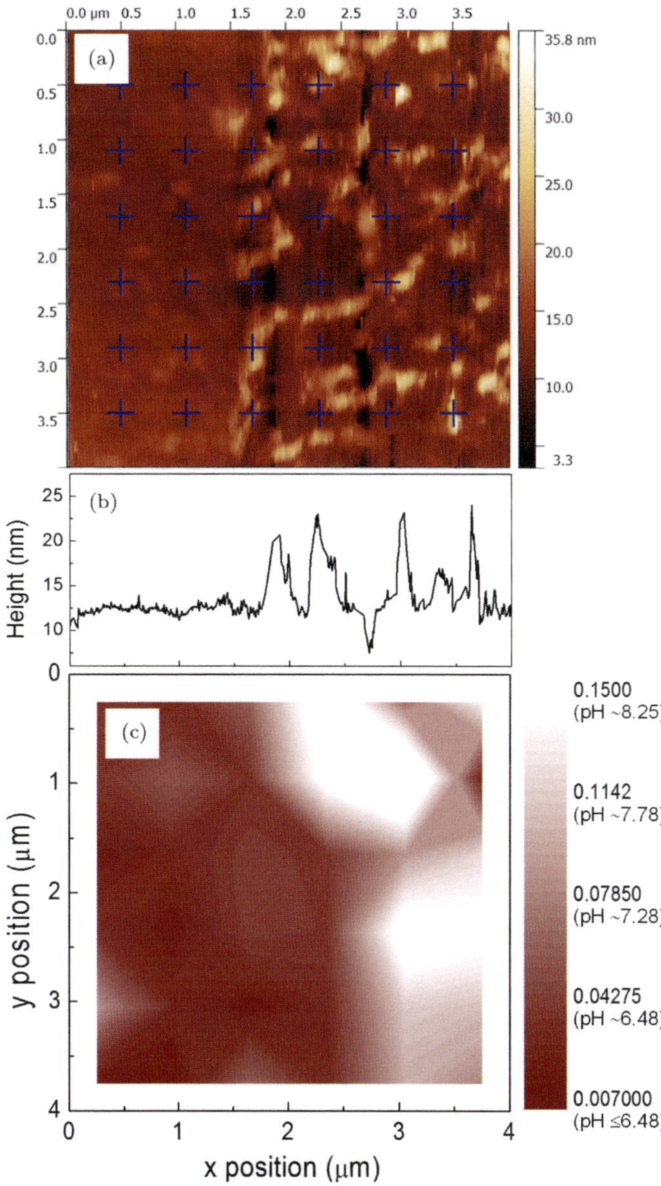

Figure 11. (a) An AFM image of the glass slide where a half of surface was anchored by $-NH_2$ group. (b) The height of the surface along selected line scan on x-axis ($y = 0.5\mu$m) in (a). (c) A contour map constructed by plotting the ratio of two peak areas at 1422 and 1586 cm^{-1} measured from TERS spectra of pMBA-modified tip versus corresponding positions (+) in (a). Reprint with permission from Ref. [47].

and organisms, and *in situ* measurements of other properties in the solution by using tips modified with other molecules.

8. Conclusion

This chapter shows the impact of applications of TERS in liquids/solutions, which are widely used in various fields. If the nanoscale topology of samples is required with the molecular information, there will be not so many methods or techniques that be employed. TERS is one of the really good choices for this case. Especially, if the sample is required to be under liquid/aqueous environments, other methods or techniques cannot provide researchers with useful information. TERS measurement in liquids/solutions opens an essential door to explore a small world, which is a basic unit for macroscopic systems, with molecular information from the real situation in the nature. The knowledge from TERS in liquids/solutions will fulfill our understanding and be used to solve many problems.

References

1. Fleischmann, M., Hendra, P. J. and McQuillan, A. J. (1974). Raman spectra of pyridine adsorbed at a silver electrode, *Chem. Phys. Lett.*, **26**, pp. 163–166.
2. Aroca, R. (2006). *Surface-Enhanced Raman Scattering, in Surface-Enhanced Vibrational Spectroscopy.* John Wiley & Sons, Ltd., UK pp. 73–106.
3. Ozaki, Y., Kneipp, K. and Aroca, R. (2014). *Frontiers of Surface-Enhanced Raman Scattering: Single Nanoparticles and Single Cells.* John Wiley & Sons, Ltd., UK.
4. Jeanmaire, D. L. and Van Duyne, R. P. (1977). Surface Raman electrochemistry part I. Heterocyclic, aromatic and aliphatic amines adsorbed on the anodized silver electrode. *J. Electroanal. Chem.* **84**, pp. 1–20.
5. Albrecht, M. G. and Creighton, J. A. (1977). Anomalously intense Raman spectra of pyridine at a silver electrode. *J. Am. Chem. Soc.* **99**, pp. 5215–5217.
6. Anderson, M. S. (2000). Locally enhanced Raman spectroscopy with an atomic force microscope. *Appl. Phys. Lett.* **76**, pp. 3130–3132.
7. Hayazawa, N., Inouye, Y., Sekkat, Z. and Kawata, S. (2000). Metallized tip amplification of near-field Raman scattering. *Opt. Commun.* **183**, pp. 333–336.
8. Stöckle, R. M., Suh, Y. D. Deckert, V. and Zenobi, R. (2000). Nanoscale chemical analysis by tip-enhanced Raman spectroscopy. *Chem. Phys. Lett.* **318**, pp. 131–136.

9. Bailo, E. and Deckert, V. (2008). Tip-enhanced Raman scattering. *Chem. Soc. Rev.* **37**, pp. 921–930.
10. Downes, A., Salter, D. and Elfick, A. (2006). Finite element simulations of tip-enhanced Raman and fluorescence spectroscopy. *J. Phys. Chem. B* **110**, pp. 6692–6698.
11. Pettinger, B. *et al.* (2004). Nanoscale probing of adsorbed species by tip-enhanced Raman spectroscopy. *Phys. Rev. Lett.* **92**, p. 096101.
12. Watanabe, H. *et al.* (2005). DFT vibrational calculations of rhodamine 6G adsorbed on silver:? Analysis of tip-enhanced Raman spectroscopy. *J. Phys. Chem. B* **109**, pp. 5012–5020.
13. Deckert, V. (2009) Tip-enhanced Raman spectroscopy. *J. Raman Spectrosc.* **40**, pp. 1336–1337.
14. Hecht, B. *et al.* (2000). Scanning near-field optical microscopy with aperture probes: Fundamentals and applications. *J. Chem. Phys.* **112**, pp. 7761–7774.
15. Zenobi, R. and Deckert, V. (2000) Scanning near-field optical microscopy and spectroscopy as a tool for chemical analysis. *Angew. Chem. Int. Ed.* **39**, pp. 1746–1756.
16. Pettinger, B. *et al.* (2012). Tip-enhanced Raman spectroscopy: Near-fields acting on a few molecules. *Annu. Rev. Phys. Chem.* **63**, pp. 379–399.
17. Zhang, W. *et al.* (2007). Single molecule tip-enhanced Raman spectroscopy with silver tips. *J. Phys. Chem. C* **111**, pp. 1733–1738.
18. Bailo, E. and Deckert, V. (2008) Tip-enhanced Raman spectroscopy of single RNA strands: Towards a novel direct-sequencing method. *Angew. Chem. Int. Ed.* **47**, pp. 1658–1661.
19. Treffer, R. *et al.* (2011) Distinction of nucleobases — a tip-enhanced Raman approach. *Beilstein J. Nanotechnol.* **2**, pp. 628–637.
20. Treffer, R. *et al.* (2012). Advances in TERS (tip-enhanced Raman scattering) for biochemical applications. *Biochem. Soc. Trans.* **40**, pp. 609–614.
21. Yeo, B.-S. *et al.* (2008). Tip-enhanced Raman spectroscopy can see more:? The case of cytochrome c. *J. Phys. Chem. C* **112**, pp. 4867–4873.
22. Saito, Y. *et al.* (2009). Nano-scale analysis of graphene layers by tip-enhanced near-field Raman spectroscopy. *J. Raman Spectrosc.* **40**, pp. 1434–1440.
23. Suzuki, T. *et al.* (2014). Tip-enhanced Raman spectroscopic measurement of stress change in the local domain of epitaxial graphene on the carbon face of 4H-SiC(000-1). *Phys. Chem. Chem. Phys.* **16**, pp. 20236–20240.
24. Vantasin, S. *et al.* (2014). Tip-enhanced Raman scattering of the local nanostructure of epitaxial graphene grown on 4H-SiC (0001^{-}). *J. Phys. Chem. C* **118**, pp. 25809–25815.
25. El-Khoury, P. Z. *et al.* (2015). Tip-enhanced Raman nanographs: Mapping topography and local electric fields. *Nano Lett.* **15**, pp. 2385–2390.
26. Ichimura, T. *et al.* (2004). Tip-enhanced coherent anti-stokes Raman scattering for vibrational nanoimaging. *Phys. Rev. Lett.* **92**, p. 220801.
27. Saito, Y. *et al.* (2008). Stress imagining of semiconductor surface by tip-enhanced Raman spectroscopy. *J. Microsc.* **229**, pp. 217–222.

28. Schmid, T. *et al.* (2013). Nanoscale chemical imaging using tip-enhanced Raman spectroscopy: A critical review. *Angew. Chem. Int. Ed.* **52**, pp. 5940–5954.
29. Kumar, N., Rae, A. and Roy, D. (2014). Accurate measurement of enhancement factor in tip-enhanced Raman spectroscopy through elimination of far-field artefacts. *Appl. Phys. Lett.* **104**, p. 123106.
30. Billot, L., Berguiga, L., de la Chapelle, M. L., Gilbert, Y. and Bachelot, R. (2005). Production of gold tips for tip-enhanced near-field optical microscopy and spectroscopy: Analysis of the etching parameters. *Eur. Phys. J. Appl. Phys.* **31**, pp. 139–145.
31. Schmid, T. *et al.* (2009). Performing tip-enhanced Raman spectroscopy in liquids. *J. Raman Spectrosc.* **40**, pp. 1392–1399.
32. Brown, D. A. and London, E. (1998). Functions of lipid rafts in biological membranes. *Annu. Rev. Cell Dev. Biol.* **14**, pp. 111–136.
33. Harder, T., Scheiffele, P., Verkade, P. and Simons, K. (1998). Lipid domain structure of the plasma membrane revealed by patching of membrane components. *J. Cell Biol.* **141**, pp. 929–942.
34. Munro, S. (2003). Lipid rafts: Elusive or illusive? *Cell* **115**, pp. 377–388.
35. Simons, K. and Ikonen, E. (1997). Functional rafts in cell membranes. *Nature* **387**, pp. 569–572.
36. Simons, K. and Toomre, D. (2000). Lipid rafts and signal transduction. *Nat. Rev. Mol. Cell Biol.* **1**, pp. 31–39.
37. Nakata, A. *et al.* (2013). Tip-enhanced Raman spectroscopy of lipid bilayers in water with an alumina- and silver-coated tungsten tip. *Anal. Sci.* **29**, pp. 865–869.
38. Leonenko, Z. V., Carnini, A. and Cramb, D. T. (2000). Supported planar bilayer formation by vesicle fusion: The interaction of phospholipid vesicles with surfaces and the effect of gramicidin on bilayer properties using atomic force microscopy. *Biochim. Biophys. Acta* **1509**, pp. 131–147.
39. Ladbrooke, B. D. and Chapman, D. (1969). Thermal analysis of lipids, proteins and biological membranes. A review and summary of some recent studies. *Chem. Phys. Lipids* **3**, pp. 304–356.
40. Nagata, Y., Saito, Y. and Verma. P. (2013). Raman investigation of single-walled carbon nanotubes in liquid by tip-enhanced Raman spectroscopy, JSAP-OSA Joint Symposia 2013 Abstracts, Optical Society of America: 16p_D5_6.
41. Zeng, Z.-C. *et al.* (2015). Electrochemical tip-enhanced Raman spectroscopy. *J. Am. Chem. Soc.* **137**, pp. 11928–11931.
42. Kurouski, D., Mattei, M. and Van Duyne, R. P. (2015). Probing redox reactions at the nanoscale with electrochemical tip-enhanced Raman spectroscopy. *Nano Lett.* **15**, pp. 7956–7962.
43. Guerrini, L. *et al.* (2015). SERS detection of amyloid oligomers on metallorganic-decorated plasmonic beads. *ACS Appl. Mater. Interfaces* **7**, pp. 9420–9428.
44. Talley, C. E. *et al.* (2004). Intracellular pH sensors based on surface-enhanced Raman scattering, *Anal. Chem.* **76**, pp. 7064–7068.

45. Wang, F. *et al.* (2012). Surface-enhanced Raman scattering detection of pH with silica-encapsulated 4-mercaptobenzoic acid-functionalized silver nanoparticles, *Anal. Chem.* **84**, pp. 8013–8019.
46. Zong, S. *et al.* (2011). Intracellular pH sensing using *p*-aminothiophenol functionalized gold nanorods with low cytotoxicity. *Anal. Chem.* **83**, pp. 4178–4183.
47. Pienpinijtham, P. *et al.* (2016). Nanoscale pH profile at a solution/solid interface by chemically modified tip-enhanced Raman scattering. *J. Phys. Chem. C* **120**, pp. 14663–14668.

Chapter 12

Tip-Enhanced Raman Scattering of Nanocarbons

Sanpon Vantasin*, Yoshito Okuno†, Yuika Saito‡
and Yukihiro Ozaki*

**Department of Chemistry,*
School of Science and Technology,
Kwansei Gakuin University,
Sanda, Hyogo 669-1337, Japan

†Horiba Ltd. 2 Miyanohigashi,
Kisshoin, Minami-ku,
Kyoto 601-8510, Japan

‡Department of Chemistry,
School of Science, Gakushuin University,
Mejiro, Toshima,
Tokyo 171-0031, Japan

Nanomaterials such as fullerenes, carbon nanotubes, and graphenes have been matters of great interest in a wide variety of research fields because of their unique material properties, for example, charge carrier mobility, optical transparency, superconductivity, and elasticity. This review outlines tip-enhanced Raman scattering (TERS) studies of fullerenes, carbon nanotubes, graphenes, and polymer nanocomposites containing carbon nanomaterials. TERS is an emerging spectroscopic tool that combines scanning probe microscopy and Raman spectroscopy. Near-field enhancement from a metallic nanotip enables one to obtain Raman spectra with spatial resolution surpassing the diffraction limit. Compared with normal Raman spectroscopy, TERS has much better spatial resolution (several tens of nanometers) and sensitivity. The purpose of this review is to demonstrate the potential of TERS spectroscopy in exploring the structure and interactions of nanomaterials.

1. Introduction

Recently, a number of carbon nanomaterials have been hot topics in the field of research because besides their curious optical and electronic properties, they exhibit novel nanomechanical properties [1–4] Fullerenes, carbon nanotubes (CNTs), and graphene are among the representative materials. A variety of novel effects occur in materials when their structures have sizes comparable to any one of the many possible length scales such as the optical wavelengths of high-energy photons and the de Broglie wavelength electrons [5–7]. In these cases, the properties of the materials are greatly affected by quantum mechanics. Due to the nanoscale origin, these properties can be altered by interfacial interactions with substrates and distortions of the local structure. Therefore, it is of particular importance to explore the interaction and distortion in very small areas of the nanomaterials.

Raman spectroscopy has widely been used for characterization of carbon nanomaterials because Raman spectra can be obtained for a single sheet of sp^2-hybridized carbon atoms, the 2D graphene sheet and a narrow strip of graphene sheet rolled up into a 1 nm diameter cylinder to form the 1D single-wall CNTs (SWCNTs) [8–11]. Raman spectra of CNTs and graphenes change with their structure, defects, and strain. The numbers of layers, structure, and defects of graphenes can be determined by using peak shifts, intensity variations, and changes in bandwidth [10–15]. Moreover, Raman imaging measurement allows visualization of local structure distribution in CNTs and graphenes [16–18]. A Raman spectrum of graphene is mainly characterized using three representative signals called G, G′, and D bands, which are observed at about 1600, 2700 and 1350 cm^{-1}, respectively [8–11]. G band is a simple Raman band originating from doubly degenerate (E_{2g}) LO and iTO phonons around Γ point, while G′ band is a result of double resonance process consisting of two scatterings of iTO phonons around K point. The D band is due to defects in graphene, allowing the scattering of single iTO phonon around K point to be observed despite being normally forbidden. However, Raman cross-section is extremely small, requiring the use of large aperture leading to poor resolution and long acquisition time.

Usually, the special resolution of Raman spectroscopy is in the scale of 400 nm to 1 μm.

On the other hand, the spatial resolution of tip-enhanced Raman scattering (TERS) is determined by the radius of a tip and can exceed the diffraction limit of light [19, 20]. TERS is a technique that employs near-field enhancement from a metallic nanotip to obtain Raman spectra with spatial resolution surpassing the diffraction limit [21–26]. It combines information-rich Raman spectroscopy and nanoscole resolution of STM/AFM. The enhancement field from plasmon resonance is confined to a small area around the tip, resulting in the improvement of the spatial resolution. A typical interesting feature of TERS is that it allows molecular-scale view of small ensembles of molecules or even single molecules. Thus, it has been proved that TERS is a unique technique for probing very small portions of nanomaterials such as CNTs and graphenes [19, 20, 27–41, 41–43]. Recent progress on TERS of buckministerfullerene, CNTs, graphenes, and carbon nanocomposites will be discussed in the following section.

2. TERS on Buckminsterfullerene

2.1. *Non-encapsulated buckminsterfullerene*

Fullerenes are allotropes of carbon which basically are graphene sheets rolled into tubes or spheres. Fullerenes show various attractive physical and chemical properties, among which heat resistance and superconductivity are particularly attractive. Buckminsterfullerene (C_{60}) is a ball type fullerene which consists of pentagonal and hexagonal carbon rings folded into football-like envelope. Due to its cage-like structure and dispersible properties in solvents, while still possessing the inertness of fullerene, it has been considered to hold potential medicinal use in drug delivery [44].

TERS studies on buckminsterfullerene have employed both the capability of TERS in improving the spatial resolution and in tip pressure control. Stockle *et al.* [45] demonstrated TERS on buckminsterfullerene with good resolution and reported that the probing area is around ~55 nm. The signal enhancement by tip is very high, even though the far-field Raman measurement (without tip) shows

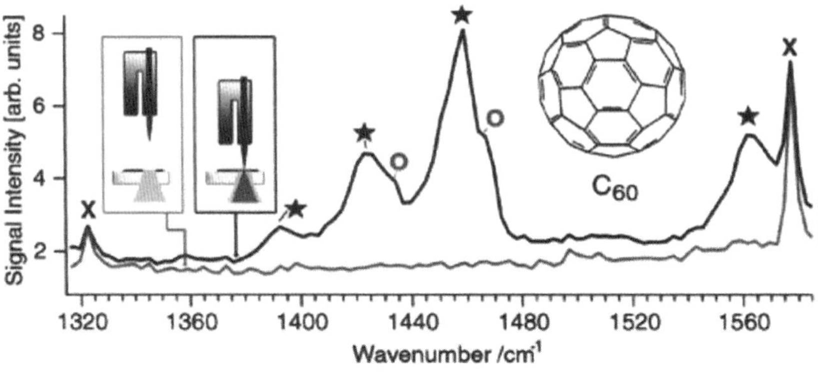

Figure 1. TERS spectra of C_{60}. (a) Tip-attached position and (b) tip-retracted position. Peaks labeled with circle, star, and cross are normal C_{60} bands, SERS bands, and immersion oil peak, respectively (From Ref. [45]).

no peak at all, all characteristic peaks of C_{60} are clearly presented. Signal-to-noise ratio is improved by at least 4000% (40 times). The TERS spectra show the shift of primary bands in a similar way to SERS spectra of C_{60} enhanced by gold (Figure 1). This indicates that C_{60} interacts with TERS tips in a similar manner as the conventional SERS substrate. The exceeding spatial resolution of TERS is useful for the study in distribution and arrangement of C_{60} on surfaces. For example, C_{60} dispersed on Au(111) surface can be visualized by STM to reveal the formation of many small islands (~40 nm in size). However, only TERS technique can probe the diffusion of C_{60} in the islands. Kharintsev *et al.* visualized this dispersion of fullerene by performing TERS imaging on fullerene film [46]. TERS can also detect symmetry reduction from substrate adsorption or from lateral interaction between C_{60} [47]. Verma *et al.* [48] investigated the effect of tip pressure on TERS spectra. Upon increasing of tip pressure, C_{60} deforms and the symmetry is reduced. The Hg mode normally becomes different in five-fold degenerate. Thus, the Raman band responsible for this mode becomes broader and its intensity decreases. Both bands for Hg mode, 272 cm^{-1} and 710 cm^{-1}, experience this broadening but the proportion of compression is a little bit different. The Ag mode at 496 cm^{-1}, which is the radial breathing mode, gets compressed and thus this Raman band shifts

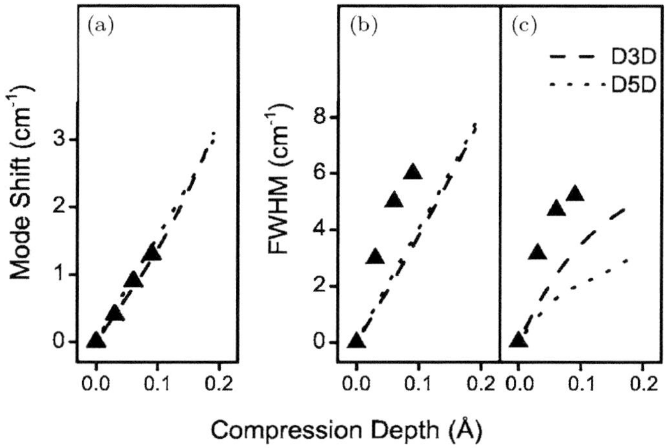

Figure 2. Plot of (a) 496 cm^{-1} Ag mode shift, (b) 272 cm^{-1} Hg mode FWHM, and (c) 710 cm^{-1} Hg mode FWHM against compression depth from tip. Triangle points are experimental values while dotted lines are theoretical calculations using D3D and D5D compression model (From Ref. [48]).

to higher frequency as the tip pressure increases (Figure 2). These shifts closely trend with the data from the theoretical model of compressed C_{60}. It is interesting that the dispersion of C_{60} on substrate also affects the band shift by the pressure. In the sparsely distributed C_{60}, the tip force is distributed to smaller number of molecules, thus each molecule experiences more pressure from the same tip force. This subsequently leads to higher shift rate against the tip force applied. This study of pressure effects by TERS tip provides unique information because the pressure effect on C_{60} has previously been studied using hydraulic pressure. However, the hydraulic pressure is isotropic and thus the symmetry in C_{60} molecules has still persisted. Therefore, the effects on Raman band are different between the two schemes of pressure application.

2.2. *Encapsulated fullerene*

The temporal fluctuation from a nanometric area of the sample was also studied by TERS using C_{60} molecules encapsulated inside SWNT (en-C_{60}) as the sample [49]. Figure 3(a) is the TEM image of C_{60} encapsulated SWCNT. In general, a moderately intense

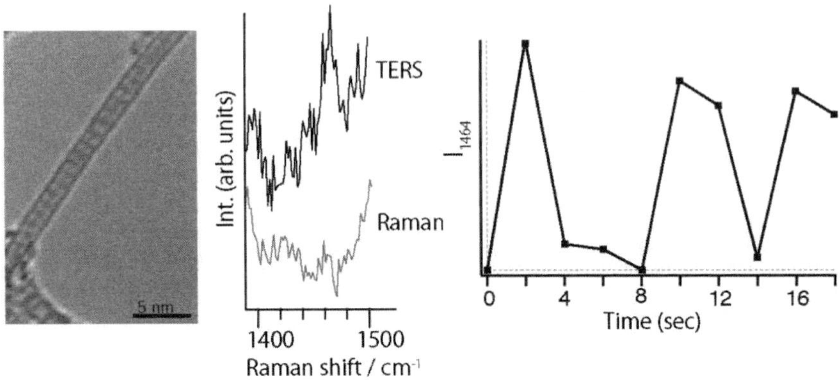

(a) (b) (c)

Figure 3. Temporal fluctuation of en-C60. (a) TEM image of the sample, (b) TERS and Raman spectra, and (c) temporal fluctuation of polymerization and depolymerization process observed in nanometric area (from Ref. [49]).

532 nm laser irradiation invokes complete photo-polymerization of non-encapsulated bulk C_{60} molecules, which was confirmed by the extinction of Ag(2) mode from a monomer. However, in contrast, the en-C_{60} molecules never get completely polymerized, instead, simultaneous occurrence of polymerization and depolymerization during the process of irradiation was observed. Figure 3(b) shows a shot of TERS spectra from the en-C_{60}, where the red spectrum was measured with tip in close contact with the sample, while the reference far-field spectrum in blue was measured by taking the tip off the sample. The enhanced peak coming from Ag(2) mode of C_{60} monomer can be recognized at 1467 cm^{-1}. Figure 3(c) shows the intensity fluctuation of the 1467 cm^{-1} peak in TERS, measured from the same area of the sample at intervals of 2 min. The temporal fluctuation in Figure 3(c) clearly indicates that the C_{60} molecules encapsulated in SWCNT jump back and forth between the monomer and polymer phase under light irradiation. The origin of this movement was not clarified. However, one possibility is that that the interaction between C_{60} molecules with the inner wall of the nanotube in en-C_{60} sample generates frictional heat, responsible for the scission of polymeric bonds between neighboring C_{60} molecules. The frictional scission competes with the photo-polymerization process, resulting

in a temporally dynamic situation where the sample can never get completely polymerized even under strong irradiation. This kind of fluctuation is more clearly seen from a smaller volume of the sample, where there are lesser possibilities of averaged observations. Also, TERS results show a faster fluctuation than the far-field measurements because owing to the field enhancement TERS has the capabilities of faster measurements.

3. TERS on Carbon Nanotubes

CNTs are members of the fullerene structural family. These cylindrical carbon molecules provide remarkable electronic, thermal, optical, and mechanical properties. In particular, owing to their extraordinary thermal conductivity and mechanical and electrical properties, CNTs have found various applications as, for example, additives. The unique tubular structure of CNTs prevents the propagation of cracks and dislocations, which provides CNTs with great potential as materials with high Young's moduli. The partial $sp^2 - sp^3$ hybridization of the C–C bonds of CNTs can induce high flexibility. CNTs are categorized as single-walled nanotubes (SWNTs) and multi-walled nanotubes (MWNTs).

3.1. *TERS studies on SWNTs and MWNTs*

Manipulation of electronic properties of SWCNTs, such as chemical functionalization [50–52], encapsulation of molecules inside SWCNTs and structural deformation [53, 54], has been a great interest for researchers because of their potential applicability in device development. Especially, electronic band modulation by structural deformation formed with bending SWCNTs [55, 56] and bridging SWCNTs over electrodes or other SWCNTs [57–59] has attracted researchers. In this chapter, we discuss TERS study of deformation effect on their electronic properties in a nanometer scale.

Yano *et al.* [60] demonstrated nanovisualization of strain distribution in SWCNTs with TERS microscopy. Within single image, nanolocalized strain distributions and their types were recorded, representing a contrast for strain degree and colors for types of strains. This image revealed how structural deformation relates to strains induced upon SWCNTs in nanometer scale. Okuno *et al.* [61] demonstrated how electronic properties of semiconducting SWCNTs change by structural deformation as visualization of the localized semiconductor-to-metal (S-to-M) transition, which is important to understand a relation between strain and electronic properties. In their study, "X-CNT", which is formed by laying one semiconducting SWCNT over another, was investigated to reveal how electronic band modulation occurs at a point where strong localized deformation was induced. An inverted optical microscope combined with a contact-mode AFM was utilized to perform a TERS microscopic investigation on "X-CNT" based on the knowledge that it provides high success rate on imaging.

3D AFM image and TERS intensity image of "X-CNT" are shown in Figures 4(a) and 4(b). TERS image was constructed by the intensity of the G band. Interestingly, depending the on spatial position of "X-CNT", there was slight difference in a shape of G-band as shown in Figure 4(c). In the contact point, where semiconducting SWCNTs crossed each other, G-band was distorted and decomposed to 3 Lorentzian mode and 1 Fano mode (Figure 4 (c-iii, iv)). Fano mode in G band, which is one of bands in G-band, is known to be a coupling mode of conducting plasmon with LO phonon. A degree of this interaction was estimated by coupling factor of the plasmon with the phonon represented as "$1/q$" in the following formula of Raman intensity.

$$I(\omega) = I_0 \frac{(1 + (\omega - \omega_{\mathrm{BWF}})/q\Gamma)^2}{1 + [(\omega - \omega_{\mathrm{BWF}})/\Gamma]^2}, \tag{1}$$

where $I_0, \omega_{\mathrm{BWF}}$, Γ, and $1/q$ are intensity, renormalized frequency, broadening parameter and the asymmetric parameter, respectively.

TERS image constructed by the absolute values of $1/q$ reveals distribution of S-to-M transitions as shown in Figure 5. Figure 5(a) shows a 3D TERS image of "X-CNT" constructed by peak intensity of G-band for its contrast and the value of $1/q$ for z-axis. A strong localization of $1/q$ was seen as a sharp peak in the Z-direction. In addition, a line profile along yellow dashed line marked in Figure 5(a) reveals a strong localization of the S-to-M transition in the sample as shown in Figure 5(b). Figures 5(c) and 5(d) show

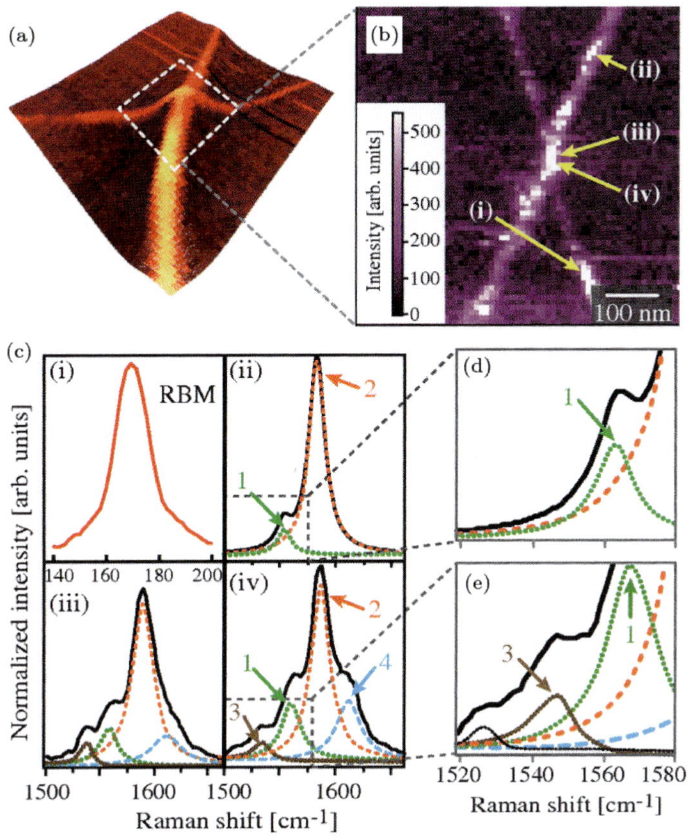

Figure 4. (a) A 3D AFM image of X-CNT sample (b) G+ mode TERS image of sample are enclosed by the white dashed line in (a). (c) TERS spectra obtained from points (i) through (iv) indicated in (b). Spectrum at point (i) shows the RBM region with the presence of only one peak. The G-band spectrum obtained far from the junction at point (ii) could be decomposed into two symmetric Lorentzian peaks (peak 1 and peak 2), while the spectra obtained near the junction at points (iii) and (iv) required two additional peaks, one asymmetric Fano peak (peak 3) and one symmetric Lorentzian peak (peak 4) for the best fitting. (d) and (e) show magnified Raman spectra obtained far from the junction and near the junction, respectively. Reproduced with permission from Ref. [61].

the spatial confinement of the new mode, which corresponds to the Lorentzian mode represented as "4" (from now on, we call it "peak 4") in decomposed spectra shown in Figure 4(c), related to radial symmetry breaking. The dashed white line in Figure 5(c) indicates

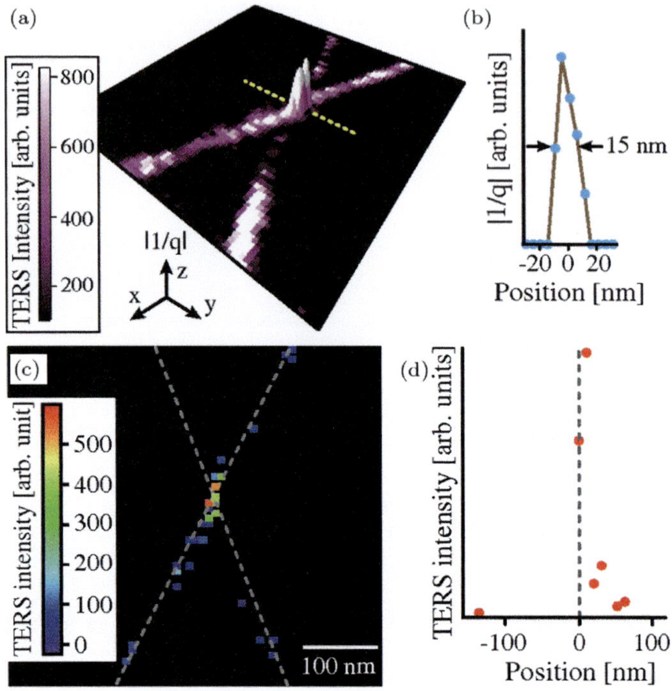

Figure 5. (a) A TERS image of our X-CNT sample, where x and y axes show the sample plane, the color represents TERS intensity of the G mode in accordance with the color bar shown on the left, and the z-axis shows the value of $j_1 = qj$. (b) A line profile of $j_1 = qj$ along the yellow dashed line in (a). (c) White dashed lines depict the location of the nanotubes in the X-CNT sample, whereas the colored spots indicate the TERS intensity of peak 4, in accordance with the color bar on the left. (d) TERS intensity plotted along one branch of the nanotube across the junction showing sudden increase in intensity at the junction. Reproduced with permission from Ref. [61].

the location of the sample and the colored dots indicate the intensity of "peak 4". There was a strong localization of "peak 4" in the very close vicinity of the contact point. This result is displayed in Figure 5(d), where the intensity of "peak 4" is plotted with respect to the distance measured from the contact. Therefore, Figures 5(c) and 5(d) show the distortion of its structure and the localized strong symmetry reduction near the contact, which finally indicates a close relation with its S-to-M transition.

In summary, by utilizing TERS spectroscopy and microscopy techniques, metal transition of semiconducting SWCNTs was revealed as nanoscale optical image, which shows the great ability of imaging analysis for properties of SWCNTs.

3.2. *TERS on CNT-polymer nanocomposite*

Polymer nanocomposites consist of a polymer and filler(s). As fillers, various kinds of nanomaterials such as titania, layered silicate, clay filler, carbon blacks, and carbon nanotubes have been used. These fillers reinforce the polymer, resulting in novel materials with improved functionality over that of conventional polymer. Among polymer nanocomposites, polymer/carbon nanotube (CNT) materials have attracted great interest because the addition of even small amount of CNTs into a polymer can improve the mechanical, physical, and electrical properties of polymers significantly [62, 63]. A number of studies have been reported for the physical properties and structure of CNT-based polymer nanocomposites. For example, Bokobza *et al.* [63] found a 45% increase in modulus and a 70% increase in tensile length by incorporating 1% w/w of MWCNT into SBR. In the study, MWCNT/poly(dimethylsiloxane) (PDMS) composites were also investigated. The results indicated that well-dispersed CNT forms conductive interconnecting networks even at small CNT concentration (5 wt.%). It is known that physical and chemical properties of polymer nanocomposites are improved via the interactions between a polymer and a nanofiller and an alteration of the polymer structure [64, 65]. However, the mechanism has not been fully clarified. Since the interactions occur at the nanoscale polymer–nanofiller interface, the investigation with spatial resolution of <100 nm is required for such study.

Raman spectroscopy is a powerful tool for the characterization of CNT-based materials, but normal Raman experiments can determine only average information from nanocomposites. Thus, TERS is expected to play an important role in exploring the interfacial structure and interaction of polymer composites including CNT-based nanocomposites. The local molecular interactions on the interface of styrene-butadiene rubber (SBR)-multiwalled carbon nanotube

(MWCNT) nanocomposites were investigated by Suzuki *et al.* [65] by using TERS. Yan *et al.* [66] carried out a TERS study on the interaction of single-walled carbon nanotubes (SWCNTs) and polystyrene (PS) at the interface in the SWCNT-PS nanocomposites.

Suzuki *et al.* investigated the local molecular interaction between the polymer and the nanofiller for SBR and MWCNTs nanocomposite films by using TERS [65]. They found that TERS bands due to SBR phenyl groups were strong when MWCNT bands were strong, whereas vinyl-group TERS bands were strong when the MWCNT bands were weak. These observations led them to conclude that the local distribution of polymer chains is modified with changes in the orientation of the phenyl rings by $\pi - \pi$ interactions between the polymer chains and the MWCNTs [65].

Raman spectra of pure SBR, pure MWCNTs, and 1 phr SBR/MWCNT nanocomposites are shown in Figures 6(a)–6(c), respectively [65]. TERS spectra of the composite are presented in Figure 7. The assignments of Raman bands in the three spectra are summarized in Table 1 [65]. It is noted that the G band in the SBR/MWCNT spectrum shifted to a higher wavenumber compared

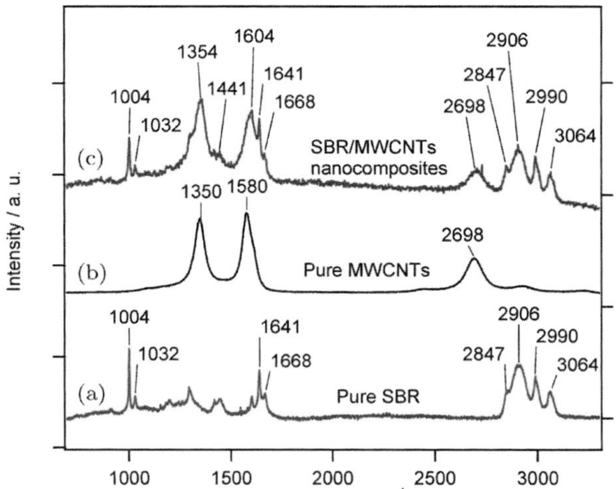

Figure 6. Raman spectra of (a) pure SBR, (b) pure MWCNTs, and (c) 1 phr SBR/MWCNT nanocomposites. (from Ref. [65]).

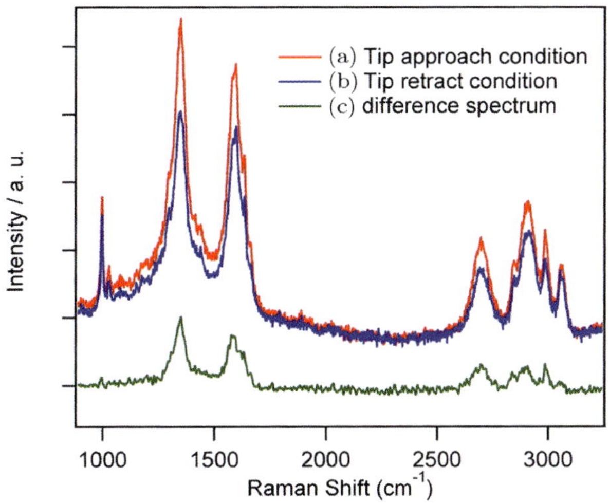

Figure 7. Raman spectra of SBR/MWCNT nanocomposites measured under (a) tip approaching conditions, and (b) tip retracted conditions. (c) The difference spectrum calculated by subtracting spectrum (b) from spectrum (a) (TERS spectrum) (from Ref. [65]).

Table 1. Assignment of Raman Bands of 1 phr SBR/MWCNT Nanocomposites (From Ref. [54]).

Wavenumber (cm^{-1})	Assignment
1004	phenyl ring breathing SBR (phenyl)
1032	phenyl ring stretching SBR (phenyl)
1354	D band MWCNTs
1604	G band MWCNTs
1641	cis C=C stretching SBR (C=C)
1668	trans C=C stretching SBR (C=C)
2698	G′ band MWCNTs
2847	aliphatic CH symmetric stretching SBR (main chain)
2906	aliphatic CH asymmetric stretching SBR (main chain)
2990	vinyl C–H stretching SBR (C=C)
3064	aromatic C–H stretching SBR (phenyl)

with the pure MWCNT spectrum, suggesting the shrinkage of the CNTs induced by the pressure from the matrix of the polymer [67, 68]. Figure 8 displays Raman spectra of 1 phr SBR/MWCNT nanocomposites measured at eight points (Points 1–8) [65]. Note

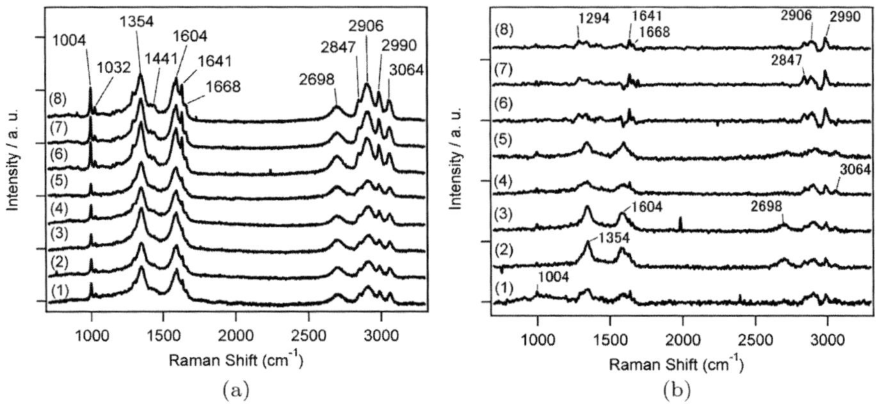

Figure 8. (a) Raman and (b) TERS spectra of 1 phr SBR/MWCNT nanocomposites measured at the eight points (from Ref. [65]).

that the Raman spectra in Figure 8 show point-dependent spectral changes. For example, the Raman signals assigned to SBR from Point 7 were relatively strong, while the spectrum obtained at Point 3 yielded relatively strong MWCNT signals. These results demonstrated that TERS is capable in exploring position-dependent structural variations of polymer/MWCNT composite.

The 1800 – 800 and 3200 – 2500 cm^{-1} regions of TERS spectra of Points 7 and 3 in Figure 8 are shown in Figures 9(a) and 9(b), respectively [65]. In contrast to Raman spectra for Points 7 and 3 in Figure 7(a), where the Raman band at 1004 cm^{-1} was more intense than those at 1668 and 1641 cm^{-1} in the TERS spectra (Figure 9(a)), the bands at 1668 cm^{-1} and 1641 cm^{-1} were stronger than the 1004 cm^{-1} band. In addition, the Raman spectrum in the higher wavenumber region gives rise to four SBR bands at 3064, 2990, 2906, and 2847 cm^{-1} (Figure 6(a)). However, in the TERS spectrum in Figure 9, the 3064 cm^{-1} band was nearly missing, and the relative intensity of the signal at 2990 cm^{-1} became stronger than that of the corresponding signal in the Raman spectrum. The relatively weak TERS bands at 3064 cm^{-1} and 1004 cm^{-1} were assigned to the aromatic C–H groups and the phenyl rings of the polymer side chains, respectively.

Figure 9(b) clearly shows a different spectral pattern from Figure 9(a); the TERS signals due to MWCNTs at 2698, 1604, and

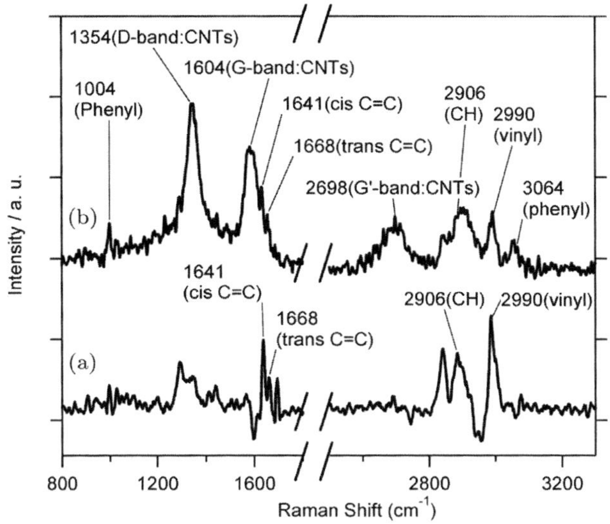

Figure 9. Enlarged TERS spectra of 1 phr SBR/MWCNTs: (a) spectrum 7 and (b) spectrum 3 (from Ref. [65]).

1354 cm^{-1} were clearly observed, and a peak from the aromatic C–H appeared at 3064 cm^{-1}. The 2990 cm^{-1} vinyl peak weakens while the 1604 cm^{-1} strengthens. This result indicates that the local distribution of the nanocomposites at the interface between the polymer and the filler was markedly different from that of polymer areas without MWCNTs.

Figure 10 illustrates the structure and interaction of the surface of the SBR/MWCNT nanocomposites proposed by Suzuki *et al.* In areas of polymer nanocomposites without MWCNTs, the C=C groups and the phenyl rings seem to be parallel to the polymer surface. In this orientation, the C=C signal would be much more enhanced than that of the phenyl ring. On the other hand, at the interface of the polymer and the filler (Figure 10(b)), MWCNTs are likely to have a significant interaction with the phenyl rings of the polymer through $\pi-\pi$ interactions, inducing different orientations of the polymer chain and its side residue. The phenyl rings at the interface may be oriented to be perpendicular to the surface because of the $\pi-\pi$ interaction between the CNTs and the phenyl rings shown in Figure 10(b).

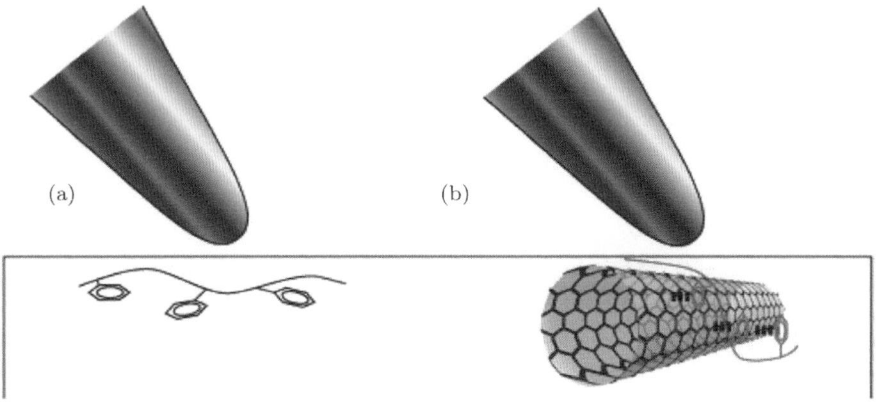

Figure 10. Schematic diagram of the surface structure of SBR/MWCNT nanocomposites: (a) far from and (b) near the CNT area. At the interface between the polymer and the filler, the phenyl ring should be oriented perpendicular to the surface through a $\pi - \pi$ interaction between the CNT and the phenyl ring (from Ref. [65]).

These studies have revealed that TERS spectroscopy is useful in investigating the interactions at interfaces in polymer nanocomposites. TERS is much more powerful than Raman spectroscopy in disclosing local position-dependent spectral variations. TERS enables one to observe not only the relative intensity change of MWCNT and SBR bands by location but also the relative intensity variation of the vinyl and phenyl bands.

4. TERS on Graphene

Graphene, a 2D honeycomb structure of carbon atoms is a novel material with unique and promising properties. Studies show that graphene could provide zero bandgap [69, 70], tunable bandgap [71, 72] quantum hall effects [73, 74], ultra-high charge mobility [75, 76], exceptional high tensile strength and Young's modulus [1], Raman enhancement capability [77–80], transparency [81, 82] etc. These impressive properties allows graphene to be a promising material in a broad range of applications, including but not limited to extremely high frequency transistor [83] ultra-fast photodetector [84] high-speed electronics [85], and supercapacitor [86–88].

Epitaxial graphene is a graphene synthesized by silicon sublimation method from the SiC substrate [89, 90]. This method is more suitable to produce graphene for electronics application compared to the original exfoliation method because it provides graphene with extremely low defects and large sheet area on the substrate which is transparent and electronically insulating [76, 82, 90–92]. In-depth studies suggest that epitaxial graphene sample is not perfectly flat and contains nanostructures such as nanosteps, nanoridges, and nanocracks [39, 92–96]. There are more occurrences of these structures in graphene grown on the C-face of SiC than on the Si-face because graphene on C-face has weaker interaction with SiC substrate [97, 98]. These structures do affect the properties of the graphene sheet [93]. However, they are too small to be resolved by conventional confocal Raman spectroscopy, and thus TERS is a crucial tool to study these structures [38, 39, 92].

4.1. *Introduction to TERS of graphene*

TERS has been used in the characterization of the important properties of graphene in the nanoscale. Number of layers in the small local areas [33, 99–102], nanoscale defects [35, 36], strain and stress variation [38, 39, 92], graphene edges [103–105], and chemical contamination [11, 37, 106] have been studied with exceeding spatial resolution. Therefore, the true nature of graphene in the nanoscale can be revealed.

Due to its significant role in nanoscale Raman spectroscopic characterization, the nature of graphene properties have been revealed or confirmed by TERS. Saito *et al.* [33] measured TERS spectra of graphene layers on glass, which were fabricated by scotch-tape exfoliation process. The measurement provided spatial resolution of 30 nm. They investigated position-dependent TERS spectra and found that G band intensity closely follows the number of graphene layers and D band intensity fluctuates due to defects on the edges between layers (Figure 11). Although graphene edge, in some configuration, has been known long before to exhibit D band [71], this study showed that this defect is definitely confined near the edge. The nanoscale correlations between G band number of layers and D band edge/defect

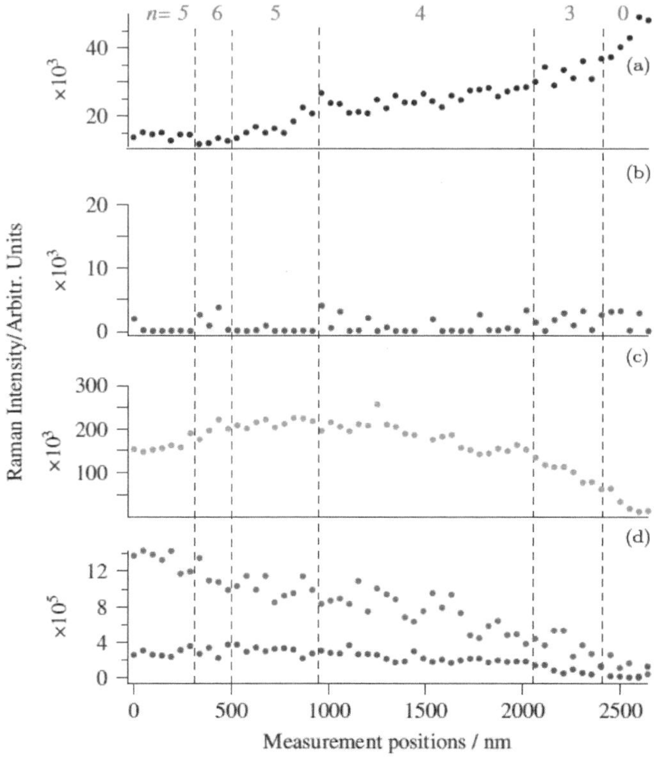

Figure 11. Intensities of (a) silicon tip and graphene (b) G band, (c) D band, and (d) two components of G′ band in TERS spectra along various numbers of graphene layers (from Ref. [33]).

were later repeated, confirmed, and then visualized by TERS mapping by Hoffman *et al.* [34], Ghislandi *et al.* [101] (Figure 12), and Su and Roy [104].

Wang *et al.* [35] performed a TERS study of exfoliated graphene using a TERS probe consisting of gold nanoparticles. They induced reversible artificial defects and controlled them using pressure from the TERS probe. The emergence of D band in the process provides more understanding of the relationship between structural deformation and D band (Figure 13). This effect on D band was then further studied by Rickman and Dunstan [36] by following TERS signal from overlapping areas and folds on the graphene sheet. On the point with structural deformation, TERS enhancement for the D band (1.2)

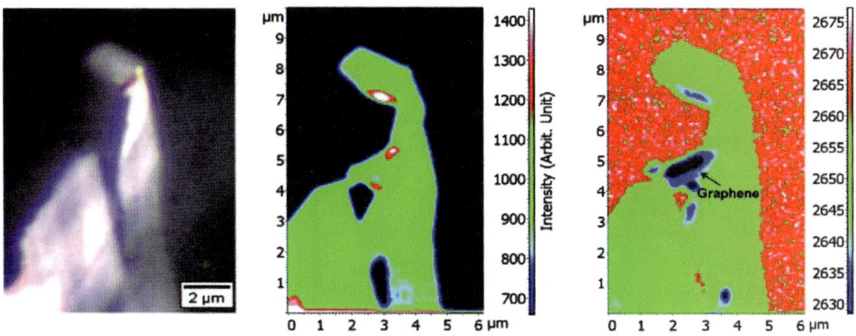

Figure 12. Optical image and TERS images of G′ band intensities and positions (from Ref. [101]).

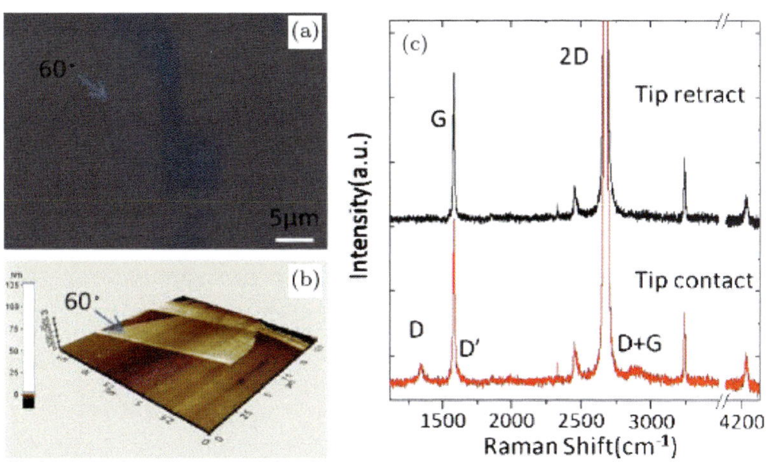

Figure 13. (a) Optical image, (b) AFM image, and (c) TERS spectra of artificial defect-inducing experiment (from Ref. [35]).

is significantly higher than the G band (0.16). This indicates that the defect/deformation of exfoliated graphene is accumulated in very small domains on its sheet rather than loosely distributed. Pashaee *et al.* [99] employed a gap-mode trick using gold nanoplates to provide very high TERS signal of exfoliated graphene (Figure 14), then reinvestigated the discoveries of previously mentioned papers, but with much better signal quality. Even though gap-mode TERS is a common technique in TERS field and has been used to measure

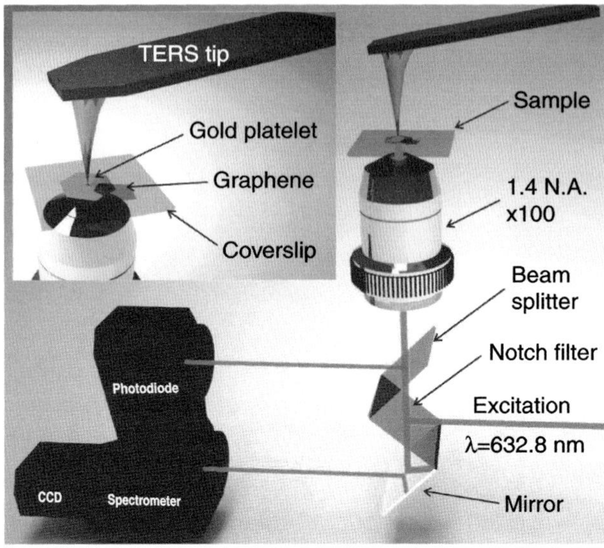

Figure 14. Apparatus of gap-mode TERS using gold nanoplate (from Ref. [99]).

graphene before by Poliani *et al.* [37], gold nanoplates are very useful because their transparency allows gap mode for any TERS setup.

However, even the exfoliated graphene is sufficient for the theoretical investigation. It is not practical for electronics and optics application due to a very small sheet area. Thus, graphene with better quality from other synthesis methods is needed to be investigated to provide more real-life nanoscale information. A few groups reported TERS studies of graphenes prepared by a silicon carbide (SiC) thermal decomposition method [101] and a chemical vapor position (CVD) method [107]. These methods can be used to prepare homogeneous graphene with higher quality in a larger area [62]. Stadler *et al.* [11] investigated CVD graphene by TERS and acquired spatial resolution of around 12 nm (Figure 15). The number of layers, small defects, and molecular contamination have been studied in great detail. Shiotari *et al.* [103] performed TERS on CVD graphene nanoribbon, which is a truly interesting material because its narrow width allows quantization. Graphene nanoribbon with zigzag edge has long edges available for spin polarization and thus has a future in spintronics.

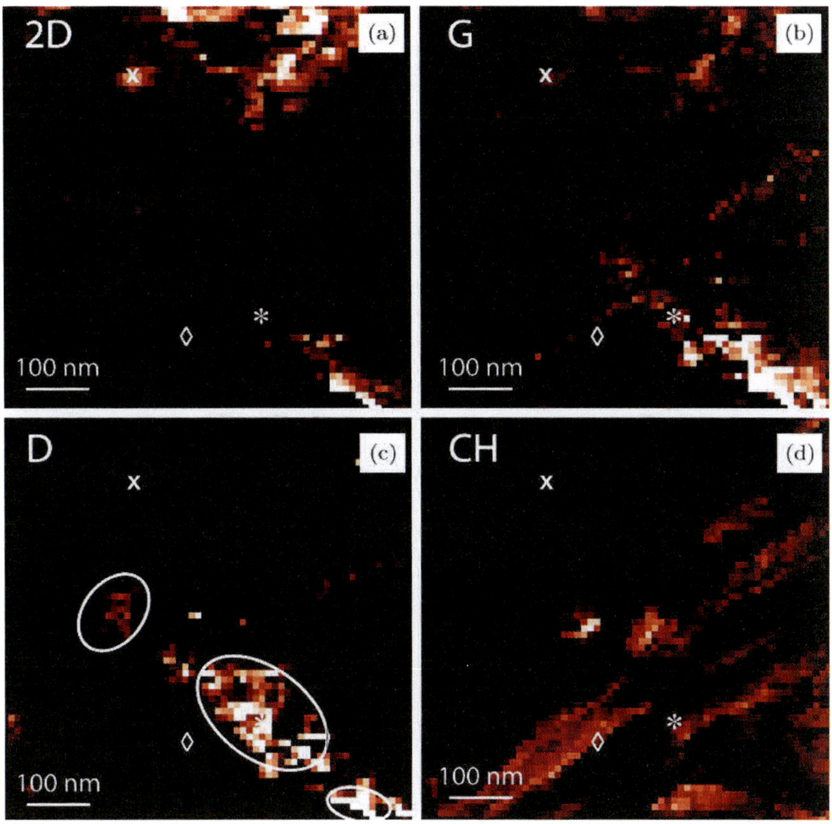

Figure 15. TERS imaging showing 2D, G, and D band intensities of CVS graphene and CH vibration intensities from contamination (from Ref. [11]).

4.2. *TERS of epitaxial graphene*

Epitaxial graphene is graphene grown by the silicon vaporization from SiC. The carbon atoms left on the SiC then form into graphene. Graphene synthesized by this method generally has much larger sheet area than exfoliated graphene. Together with its low defect and electrical insulating substrate, epitaxial graphene is highly interesting in a wide range of applications, including electronics [76, 89, 90, 108]. Graphene grown on C-face ($000\bar{1}$) of SiC has a weaker interaction to the substrate than graphene grown on Si-face (0001), and therefore has some different properties, such as more curvatures [39, 92].

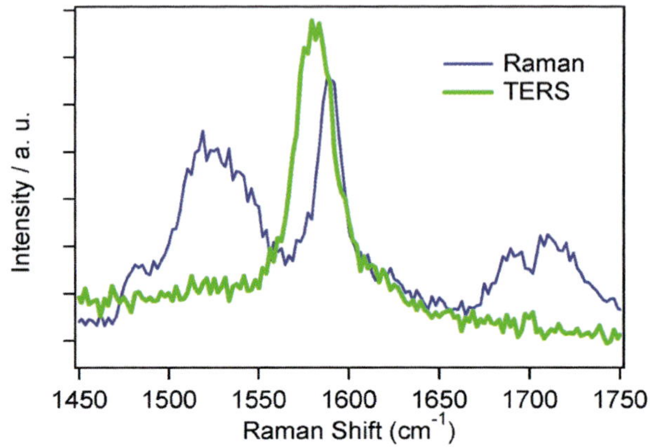

Figure 16. Raman and TERS spectra in the 1750–1450 cm^{-1} region of epitaxial graphene on the carbon surface of 4H-SiC (000$\bar{1}$) (from Ref. [38]).

Domke and Pettinger [31] demonstrate a suppression of E1 modes of SiC in the graphene on 6H-SiC system. Suzuki *et al.* [38] reported a TERS study on strain change in the local domain of epitaxial graphene on the carbon face of 4H-SiC (000$\bar{1}$) by following the change of G band positions and FWHMs in the small spatial location. The local strain is an important aspect of graphene grown on SiC since the synthesis involves heating to high temperature (>1000°C) and cooling down to room temperature. Thus, SiC-grown graphene (and also CVD graphene) always contains inherent strain. The study also shows the relationship between the shift of G band position and ΔFWHM (Figures 16 and 17) although this investigation is quite qualitative, and the relationship of such changes to the graphene topology is just a speculation. This study provides nanoscale informations and confirms that SiC-grown graphene has many structural distinctions compared with exfoliated graphene, down to the nanoscale level. Vantasin *et al.* [39] later proved that the shift of Raman bands of SiC-grown graphene has really originated from graphene topology. TERS spectra on step, ridge, crack sub-micro/nanostructres of SiC-grown graphene were measured and then correlated with the topology at the corresponding points. This study provides evidence for the hypothesis that graphene nanoridges are a result of relaxation against strain that occurred during the cooling down

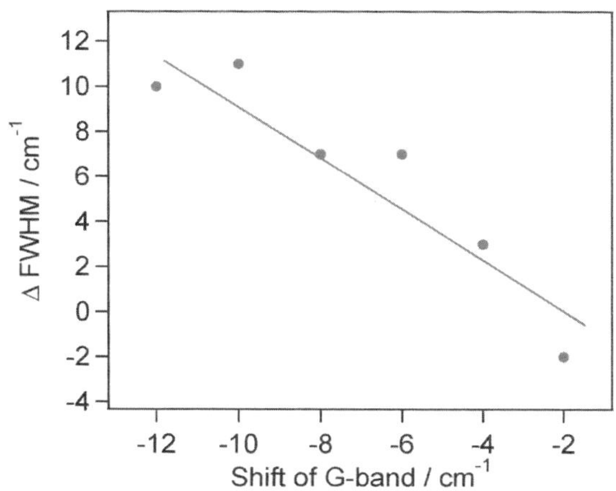

Figure 17. The plot of a correlation between the peak shift and the ΔFWHM of the G band at different positions (from Ref. [38]).

from synthesis temperature, since the thermal expansion coefficients between graphene and SiC are mismatched. The evidence was apparent by the direct quantitative measurement of strain on graphene nanoridge. The positions of G′ band in TERS spectra confirm that the ridge center has smaller compressive strain compared to neighboring area (Figure 18). Another paper by Vantasin *et al.* [92] further investigates this strain relaxation effect in nanoridges on SiC-grown graphene microisland. The shift G′ band positions in TERS spectra from each nanoridge display a similar degree of strain relaxation with those nanoridges on large sheet graphene in the previous paper (Figure 19). Since the microisland is small, very regular, and undoubtedly single crystal and single layer, it is an ideal sample for a mechanism check. Nanoridges which are connected to the edge of island are open-ended and thus cannot collect Si vapor. TERS result on this sample rules out the possibility of proposed alternative mechanism in which Si vapor accumulates into nanoridges. TERS measurement on ridge junction also hints at a complicated strain level, due to the non-uniform relaxation in various directions from the different nanoridges. These papers about TERS on SiC-grown graphene will be discussed here in some detail.

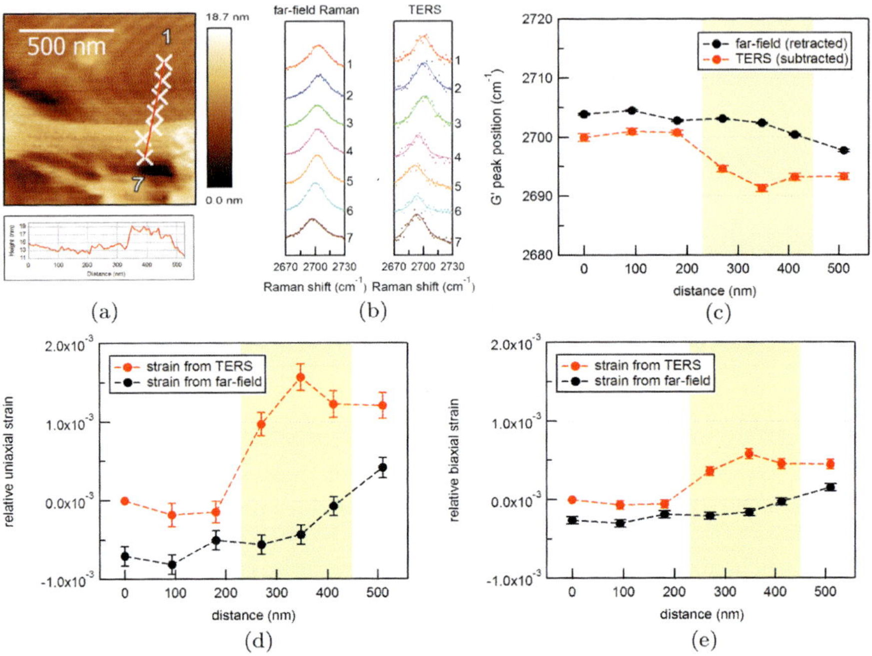

Figure 18. TERS measurement on nanoridge structure. (a) AFM topology image and line height profile. (b) G′ band in far-field Raman and TERS spectra from seven measurement points shown in (a). (c) Peak position of each G′ band in (b), plotted against distances from point 1 along the line profile in (a). (d) and (e) Calculated relative uniaxial and biaxial strains on each point, plot in the same fashion as (c), respectively. Error bars in (c), (d), and (e) were calculated from uncertainties in Lorentzian fitting and Grüneisen parameter (from Ref. [39]).

4.3. *Stress change in the local domain of epitaxial graphene on the carbon face of 4H-SiC* ($000\bar{1}$) *studied by TERS*

Suzuki *et al.* [38] investigated the stress change in the local domain of epitaxial graphene on the carbon face of 4H-SiC ($000\bar{1}$) by using TERS. They found that the G band in the TERS spectra of epitaxial graphene on the C-face of 4H-SiC shows position-dependent variations in both lower wavenumber shifts and spectral broadening. It was found from the analysis of the variations that the shifts and broadenings are linearly correlated to each other. Thus, it is very likely that the position-dependent changes in both wavenumber

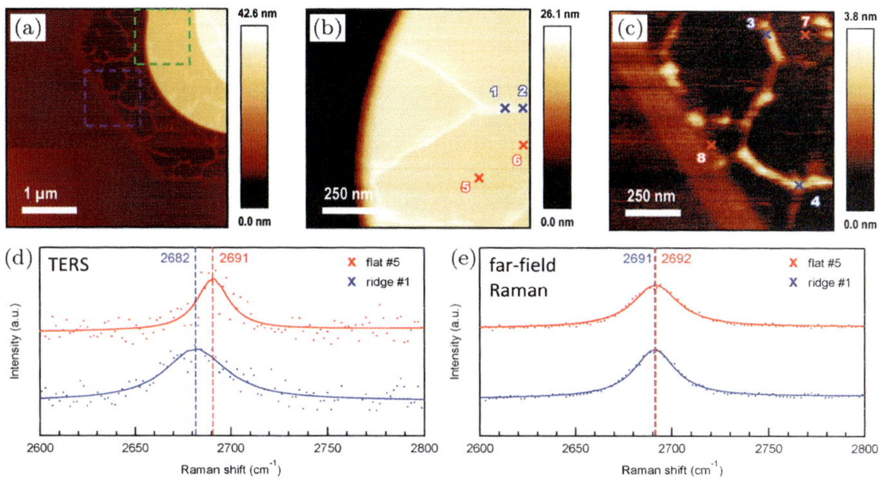

Figure 19. AFM images of graphene nanoridges, TERS, and far-field Raman at corresponding points (from Ref. [92]).

shifts and broadenings are induced by the position-dependent local stress on graphene based on a uniaxial strain model. Figures 16(a) and 16(b) depict Raman spectra of a particular spot in epitaxial graphene on the 4H-SiC $(000\bar{1})$ surface collected under tip-approaching and tip-retracted conditions, respectively.

The difference between normal Raman and the TERS spectra in the G band position and in the G band width are correlated. This is shown in Figure 17 It was reported that the G band shift happens upon applying the stress to graphene because of the distortion of the graphene lattice [109]. In the SiC thermal decomposition method, graphene undergoes thermal expansion. The coefficient of thermal expansion of graphene layer (9×10^{-6} K^{-1}) is different from that of the SiC substrate (4×10^{-6} K^{-1}) [110–112], inducing the difference that generates thermal stress between SiC and graphene during a cooling process. For graphene with unidirectional stress, the G band is split into two directions where the stress is applied and not applied, yielding the G^+ and G^- bands [113]. For a very weak stress small splitting appears as broadening instead.

In the graphene used by Suzuki *et al.* [38] the graphene surface consists of a wide flat area and a very small ridge structure area.

Therefore, normal Raman spectrum of graphene shows mainly signals from the flat area, while the signal from the ridge structure is only clearly observed in the TERS spectra. Thus, Suzuki *et al.* [38] hypothesized that variations in the local stress at ridge structures are the origin of the observed variations in the G band. This band shift due to the presence of ridge structure was later proofed by Vantasin *et al.* [39].

4.4. *A TERS study of the local nanostructure of epitaxial graphene grown on 4H-SiC* (000$\bar{1}$)

Vantasin *et al.* [39] explored step, ridge, and crack sub-micro/nanostructures of epitaxial graphene on 4H-SiC (000$\bar{1}$) by TERS and AFM. This TERS study investigated the effect of nanoridge on the local strain of graphene by analyzing the G′ band position. The strain on the ridge was calculated to be 1.6×10^{-3} assuming uniaxial strain model. The graphene content in the sub-micrometer crack was also found to be much lower compared to nearby areas.

Figure 1 of Ref. [39] shows tip-attached, tip-retracted, and subtracted (TERS) spectra of single-layer epitaxial graphene on the C-face of SiC [39]. It can be seen from the figure that G′ band near 2700 cm^{-1} is more intense than G band near 1590 cm^{-1}, and FWHM of the G′ band is lower than 27 cm^{-1} (in this case, 23 cm^{-1} for TERS spectra) [39]. Note that the actual sample has 1.7 layers of graphene on average, but every spectrum in the study was collected from single-layer areas to eliminate the layer effect. Moreover, it is noted that this sample has extremely low defect, as evidenced by the missing of the D band in the 1350 cm^{-1} region.

As already mentioned, during the cooling after the synthesis, SiC shrank more than graphene due to the difference in thermal expansion coefficient. This exerted compressive strain on graphene. However, graphene can be pushed out of plane into a ridge when the generated strain exceeds a limit called critical buckling strain. Figure 3(b) of Ref. [39] reveals that G′ band position in the TERS spectra from the ridge center shows a low-frequency shift by 8.7 cm^{-1} from the TERS spectra from flat area. This shift does not arise from the tip interaction because the tip was controlled by the non-contact mode. Therefore, the shift must be ascribed to the deformation cause by the ridge.

The downshift of G′ band usually indicates a stretching (tensile) strain. However, the G′ band position at 2691 cm^{-1} in the TERS spectra from the ridge is still far higher than the band position at 2674 cm^{-1} from unstrained and non-doped graphene [114]. This indicates that the shift of G′ band from the ridge is actually due to a smaller compressive strain compared to neighboring area, not due to a stretching strain.

Data in Figures 3(c) and 3(d) of Ref. [39] yield another interesting information in comparison between TERS and the normal Raman spectra. TERS monitors a small area around the tip while far-field Raman spectroscopy probes a much larger area, and in every point the G′ band positions in TERS spectra are lower, indicating that the even the flat areas around the ridge also have weaker compressive strain relative to areas far from the ridge. The ridge itself, without a doubt, has weakest compressive strain since the difference in G′ band between near-field and far-field spectra on the ridge is as high as 11.1 cm^{-1}.

Vantasin *et al.* [39] also explored crack sub-microstructure using TERS. The TERS spectra of the crack structure showed a reduced graphene/SiC signal ratio, which cannot be probed by far-field Raman spectroscopy. These results of step, ridge, and crack structures of epitaxial graphene suggested that TERS is an excellent tool in the characterization of nanostructures on epitaxial graphene.

4.5. *Side-illuminated TERS study of edge phonon in graphene at the electrical breakdown limit*

There are scientific curiosities in how a single-layer graphene (SLG) is damaged by an application of high current density, which is controversial when graphene-based electronic devices are scaled down in a nanometer scale, on the graphene [115]. Okuno *et al.* [105] explored the effects of high current density ($\sim 3.0 \times 10^8$ A/cm^2) on a graphene on Si/SiO_2 substrate using non-gap mode side-illumination TERS.

An AFM image of the graphene indicated that island structures appeared after an application of the high current density

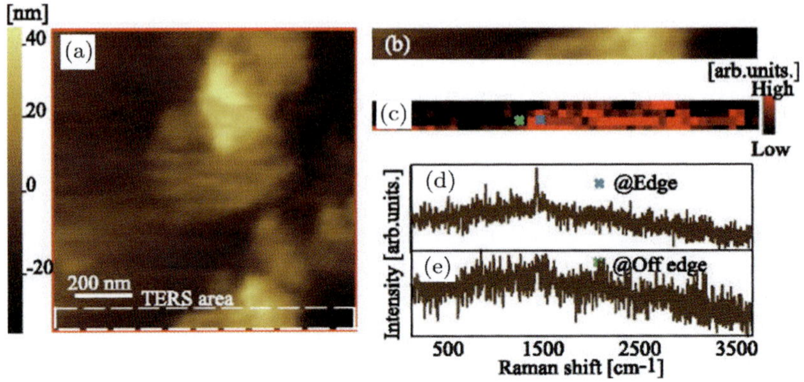

Figure 20. (a) AFM image of the graphene. (b) Region of panel (a) enclosed by the dashed rectangle. (c) Corresponding TERS image. (d) and (e) TERS spectra measured at on-island and off-island locations, respectively. Reproduced with permission from Ref. [105].

(Figure 20(a)) [105]. AFM and TERS images of the target area in Figure 20(a) are shown in Figures 20(b) and 20(c), respectively. The TERS image using integration of 1530 cm^{-1} peak shows a correspondence of their image with nanometer scale. The strong Raman band at 1530 cm^{-1} (Figure 20(d)), which was reported by Ren *et al.* [116] as a localized vibration coming from armchair edge, leads them to conclude that the distribution of an edge structure was induced by an application of high current density. It is worth noting that the TERS image revealed that the phonons at 1530 cm^{-1} were uniformly distributed over the islands, which was not confirmed by conventional far-field Raman spectra of graphene measured just before the TERS measurements.

In TERS spectra measured in the areas out of island and on island (Figures 21(a) and 21(b), respectively), the differences between the two spectra indicate that the spectrum from out of islands contained peaks at 3341 and 1530 cm^{-1}, whereas the spectrum from on island had peaks at 1245, 1530, and 1586 (G-band) cm^{-1}. The bands at 1245 and 1530 cm^{-1}, which originate from the edge structures, imply the coexistence of crystalline and defects in islands [115]. Outside islands, TERS spectra give 2D′ band at 3341 cm^{-1} due to a double resonance mode of D′ band and edge phonon at 1530 cm^{-1} without G-band as

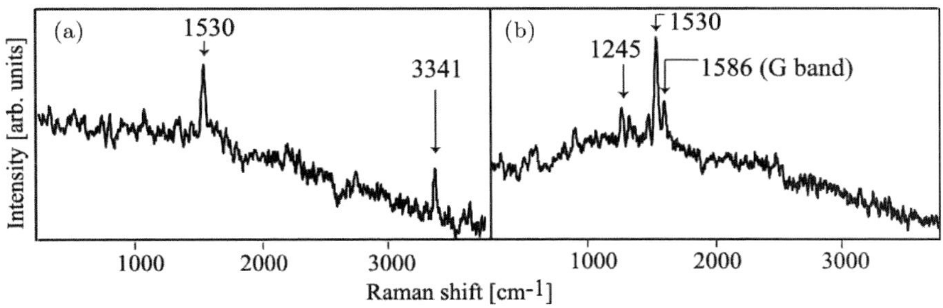

Figure 21. (a) TERS spectrameasured in the areas out of island and (b) on island Reproduced with permission from Ref. [105].

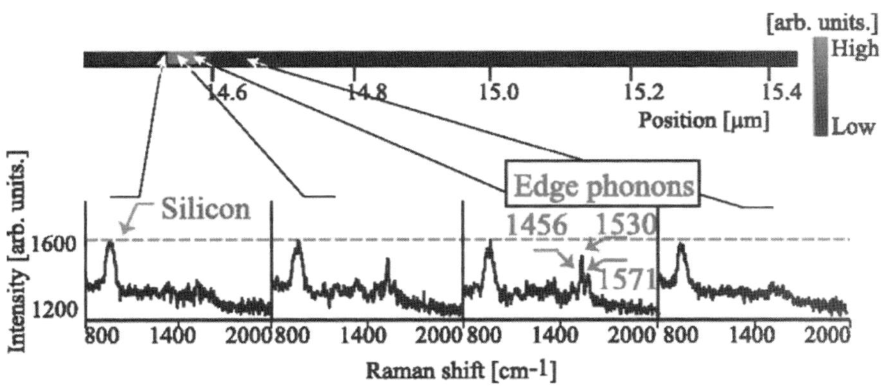

Figure 22. TERS line analysis of graphene in a 20 nm × 1 μm area (top). The TERS line was constructed by plotting peak intensity at 1530 cm^{-1}. TERS spectra of graphene extracted from the line profile above (bottom). The intensity of the edge phonons varied at the nanometer scale despite the presence of a Raman band from the Si substrate with approximately the same intensity Reproduced with permission from Ref. [105].

an indication of the existence of crystallinity. This is an evidence that defects in the SLG were induced by high current density and crystallinity increase on islands.

TERS line analysis on the islands showed an origin of edge phonon. Intensity variance of edge phonon was revealed as line analysis proceeded (Figure 22). Phonons due to an H-terminated armchair structure have been theoretically predicted to appear at 1575 cm^{-1}, but in this study they are located at 1530 cm^{-1}. As Ren *et al.*

reported, an edge phonon at 1530 cm^{-1} is expected to be derived from H-terminated C=C double bonds in the armchair edge because of an increased mass from H atom giving blue-shift in phonon frequency from 1580 cm^{-1} to 1530 cm^{-1} [116]. In the same manner, the peak at 1456 cm^{-1} was be cause of H-terminated C=C double bonds in the zigzag structure [116]. Both results of H-termination with the armchair and zigzag structure reveal a similarity in their reactivity of C≡C triple bonds with H atoms.

In this way, Okuno *et al.* revealed and discussed the edge structures with phonon frequency analysis [105]. The edge phonons at 1456 and 1530 cm^{-1} were presented as an evidence of the existence of edge structures. The reactivity of C≡C triple bonds with H atoms in armchair and zigzag edge is also discussed. This is, of course, important for the production of nanoelectronic devices with graphene.

5. Conclusions

This chapter reported recent progress in TERS studies of polymer nanocomposites and graphene. TERS was demonstrated to be powerful in the investigation of the local structure and interaction of nanomaterials. TERS reveled insights into the nature of the intermolecular interactions and interface structure of polymer nanocomposites. This may open a new avenue for investigating polymer nanomaterials interfaces.

To further explore graphene nature with TERS, one needs a high-quality standard graphene sample with few defects and a definite number of layers. Recently, in general, the quality of graphene has become much better than before, and thus previously hidden properties of graphene can be revealed by TERS.

New techniques such as TERS imaging and low-temperature low-pressure TERS have recently been topics of keen interest. The former is widely used and still continuously developed. TERS imaging would become much easier and faster, providing a strong spectroscopic analysis method for nanoscience and technology. TERS instruments with ultra-low temperature and ultra-high vacuum are also under progress. Under these environments, samples are very stable, so high quality TERS spectra/imaging would become more feasible.

References

1. Lee, C., Wei, X., Kysar, J. W. and Hone, J. (2008). Measurement of the elastic properties and intrinsic strength of monolayer graphene. *Science* **321**, pp. 385–388.
2. Geim, A. K. (2009). Graphene: Status and prospects. *Science* **324**, pp. 1530–1534.
3. Loh, K. P., Bao, Q., Ang, P. K., and Yang, J. (2010). The chemistry of graphene. *J. Mater. Chem.* **20**(12), pp. 2277–2289.
4. Bae, S., Kim, H., Lee, Y., Xu, X., J.-Park, S., Zheng, Y., Balakrishnan, J., Lei, T., H. Ri Kim, Song, Y. I., *et al.* (2010). Roll-to-roll production of 30-Inch graphene films for transparent electrodes. *Nat. Nanotechnol.* **5**(8), pp. 574–578.
5. Popov, V. N. (2004). Carbon nanotubes: Properties and application. *Mater. Sci. Eng. R Rep.* **43**(3), pp. 61–102.
6. Yan, X., Itoh, T., Kitahama, Y., Suzuki, T., Sato, H., Miyake, T. and Ozaki, Y. (2012). A raman spectroscopy study on single-wall carbon nanotube/polystyrene nanocomposites: Mechanical compression transferred from the polymer to single-wall carbon nanotubes. *J. Phys. Chem. C* **116**(33), pp. 17897–17903.
7. Sgobba, V. and Guldi, D. M. (2008). Carbon nanotubes — electronic/electrochemical properties and application for nanoelectronics and photonics. *Chem. Soc. Rev.* **38**(1), pp. 165–184.
8. Malard, L. M., Pimenta, M. A., Dresselhaus, G. and Dresselhaus, M. S. (2009). Raman spectroscopy in graphene. *Phys. Rep.* **473**(5–6), pp. 51–87.
9. Saito, R., Hofmann, M., Dresselhaus, G., Jorio, A. and Dresselhaus, M. S. (2011). Raman spectroscopy of graphene and carbon nanotubes. *Adv. Phys.* **60**(3), pp. 413–550.
10. Jorio, A. (Ed.) (2011). *Raman Spectroscopy in Graphene Related Systems.* Wiley-VCH, Weinheim, Germany.
11. Stadler, J., Schmid, T. and Zenobi, R. (2011). Nanoscale chemical imaging of single-layer graphene. *ACS Nano* **5**(10), pp. 8442–8448.
12. Casiraghi, C., Pisana, S., Novoselov, K. S., Geim, A. K. and Ferrari, A. C. (2007). Raman fingerprint of charged impurities in graphene. *Appl. Phys. Lett.* **91**(23), p. 233108.
13. Yan, J., Zhang, Y., Kim, P. and Pinczuk, A. (2007). Electric field effect tuning of electron-phonon coupling in graphene. *Phys. Rev. Lett.* **98**(16), p. 166802.
14. Pisana, S., Lazzeri, M., Casiraghi, C., Novoselov, K. S.,Geim, A. K., A. C. Ferrari and Mauri, F. (2007). Breakdown of the adiabatic born–oppenheimer approximation in graphene. *Nat. Mater.* **6**(3), pp. 198–201.
15. Graf, D., Molitor, F., Ensslin, K., Stampfer, C., Jungen, A., Hierold, C. and L. Wirtz (2007). Spatially resolved raman spectroscopy of single- and few-layer graphene. *Nano Lett.* **7**(2), pp. 238–242.
16. Kataria, S., Patsha, A., Dhara, S., Tyagi, A. K. and Barshilia, H., C. (2012). Raman imaging on high-quality graphene grown by hot-filament chemical vapor deposition. *J. Raman Spectrosc.* **43**(12), pp. 1864–1867.

17. R.-O, S., Iwamoto, A., Nishi, Y., Funase, Y., Yuasa, T., Tomita, T., Nagase, M., Hibino, H. and Yamaguchi, H. (2012). Microscopic raman mapping of epitaxial graphene on 4H-SiC(0001). *Jpn. J. Appl. Phys.* **51**(6S), p. 06FD06.
18. Stampfer, C., Molitor, F., Graf, D., Ensslin, K., Jungen, A., Hierold, C. and Wirtz, L. (2007). Raman imaging of doping domains in graphene on SiO_2. *Appl. Phys. Lett.* **91**(24), p. 241907.
19. Domke, K. F., Pettinger, B. (2010). Studying surface chemistry beyond the diffraction limit: 10 years of TERS. *Chem. Phys. Chem.* **11**(7), pp. 1365–1373.
20. Verma, P., Ichimura, T., Yano, T., Saito, Y. and Kawata, S. (2009). Nano-imaging through tip-enhanced raman spectroscopy: Stepping beyond the classical limits. *Laser Photonics Rev.* **4**(4), pp. 548–561.
21. Hartschuh, A. (2008). Tip-enhanced near-field optical microscopy. *Angew. Chem. Int. Ed.* **47**(43), pp. 8178–8191.
22. Pettinger, B., Ren, B., Picardi, G., Schuster, R. and Ertl, G. (2004). Nanoscale probing of adsorbed species by tip-enhanced raman spectroscopy. *Phys. Rev. Lett.* **92**(9), p. 096101.
23. Pettinger, B. (2006). Kneipp, K., Moskovits, M. and Kneipp, H. Springer, Berlin, Heidelberg, pp. 217–240.
24. Saito, Y., Motohashi, M., Hayazawa, N., Iyoki, M. and Kawata, S. (2006). Nanoscale characterization of strained silicon by tip-enhanced raman spectroscope in reflection mode. *Appl. Phys. Lett.* **88**(14), p. 143109.
25. Ichimura, T., Watanabe, H., Morita, Y., Verma, P., Kawata, S. and Inouye, Y. (2007). Temporal fluctuation of tip-enhanced raman spectra of adenine molecules. *J. Phys. Chem. C* **111**(26), pp. 9460–9464.
26. Bailo, E. Deckert, V. (2008). Tip-enhanced raman scattering. *Chem. Soc. Rev.* **37**(5), pp. 921–930.
27. Deckert, V. (2009). Tip-enhanced raman spectroscopy. *J. Raman Spectrosc.* **40**(10), pp. 1336–1337.
28. Hayazawa, N., Inouye, Y., Sekkat, Z. and Kawata, S. (2000). Metallized tip amplification of near-field raman scattering. *Opt. Commun.* **183**(1–4), pp. 333–336.
29. Yano, T., Kawata, S. (2014). Frontiers in Surface Enhanced Raman Scattering. *Single-Nanoparticles Single Cells*, John Wiley & Sons, Inc, Chichester, West Sussex.
30. Kawata, S., Inouye, Y. and Verma, P. (2009). Plasmonics for near-field nano-imaging and superlensing. *Nat. Photonics* **3**(7), pp. 388–394.
31. Domke, K. F., Pettinger, B. (2009). Tip-enhanced raman spectroscopy of 6H-SiC with graphene adlayers: Selective suppression of E1 modes. *J. Raman Spectrosc.* **40**(10), pp. 1427–1433.
32. Ren, B., Picardi, G. and Pettinger, B. (2004). Preparation of gold tips suitable for tip-enhanced raman spectroscopy and light emission by electrochemical etching. *Rev. Sci. Instrum.* **75**(4), pp. 837–841.
33. Saito, Y., Verma, P., Masui, K., Inouye, Y. and Kawata, S. (2009). Nanoscale analysis of graphene layers by tip-enhanced near-field raman spectroscopy. *J. Raman Spectrosc.* **40**(10), pp. 1434–1440.

34. Hoffmann, G. G., de With, G., Loos, J. (2008). Micro-raman and tip-enhanced raman spectroscopy of carbon allotropes. *Macromol. Symp.* **265**(1), pp. 1–11.
35. Wang, P., Zhang, D., Li, L., Li, Z., Zhang, L. and Fang, Y. (2012). Reversible defect in graphene investigated by tip-enhanced raman spectroscopy. *Plasmonics* **7**(3), pp. 555–561.
36. Rickman, R. H. and Dunstan, P. R. (2014). Enhancement of lattice defect signatures in graphene and ultrathin graphite using tip-enhanced raman spectroscopy: Enhancement of defect signatures in graphene using tip-enhanced raman. *J. Raman Spectrosc.* **45**(1), pp. 15–21.
37. Poliani, E., Nippert, F. and Maultzsch, J. (2012). Effect of gap modes on graphene and multilayer graphene in tip-enhanced raman spectroscopy. *Phys. Status Solidi B* **249**(12), pp. 2511–2514.
38. Suzuki, T., Itoh, T., Vantasin, S., Minami, S., Kutsuma, Y., Ashida, K., Kaneko, T., Morisawa, Y., Miura, T. and Ozaki, Y. (2014). Tip-enhanced raman spectroscopic measurement of stress change in the local domain of epitaxial graphene on the carbon face of 4H-SiC(000–1). *Phys. Chem. Chem. Phys.* **16**(37), pp. 20236–20240.
39. Vantasin, S., Tanabe, I., Tanaka, Y., Itoh, T., Suzuki, T., Kutsuma, Y., Ashida, K., Kaneko, T. and Ozaki, Y. (2014). Tip-enhanced raman scattering of the local nanostructure of epitaxial graphene grown on 4H-SiC $(000\bar{1})$. *J. Phys. Chem. C* **118**(44), pp. 25809–25815.
40. Peica, N., Thomsen, C. and Maultzsch, J. (2011). Studying the local character of raman features of single-walled carbon nanotubes along a bundle using TERS. *Nanoscale Res. Lett.* **6**(1), p. 174.
41. Wang, R., Wang, J., Hao, F., Zhang, M. and Tian, Q. (2010). Tip-enhanced raman spectroscopy with silver-coated optical fiber probe in reflection mode for investigating multiwall carbon nanotubes. *Appl. Opt.* **49**(10), p. 1845.
42. Hayazawa, N., Yano, T., Watanabe, H., Inouye, Y. and Kawata, S. (2003). Detection of an individual single-wall carbon nanotube by tip-enhanced near-field raman spectroscopy. *Chem. Phys. Lett.* **376**(1–2), pp. 174–180.
43. Wu, X., Zhang, M., Wang, J. and Tian, Q. (2010). Experimental research on double-walled carbon nanotubes by using tip-enhanced raman spectroscopy. *J. Korean Phys. Soc.* **56**(4), pp. 1103–1108.
44. Shi, J., Zhang, H., Wang, L., Li, L., Wang, H., Wang, Z., Li, Z., Chen, C., Hou, L., Zhang, C. *et al.* (2013). PEI-derivatized fullerene drug delivery using folate as a homing device targeting to tumor. *Biomaterials* **34**(1), pp. 251–261.
45. Stöckle, R. M., Suh, Y. D., Deckert, V. and Zenobi, R. (2000). Nanoscale chemical analysis by tip-enhanced raman spectroscopy. *Chem. Phys. Lett.* **318**(1–3), pp. 131–136.
46. Kharintsev, S. S., Noskov, A. I., Hoffmann, G. G. and Loos, J. (2011). Near-field optical taper antennas fabricated with a highly replicable Ac electrochemical etching method. *Nanotechnology* **22**(2), p. 025202.
47. Pettinger, B., Schambach, P., Villagómez, C. J. and Scott, N. (2012). Tip-enhanced raman spectroscopy: Near-fields acting on a few molecules. *Annu. Rev. Phys. Chem.* **63**(1), pp. 379–399.

48. Verma, P., Yamada, K., Watanabe, H., Inouye, Y. and Kawata, S. (2006). Near-field raman scattering investigation of tip effects on C60 molecules. *Phys. Rev. B* **73**(4), p. 045416.
49. Saito, Y., Honda, M., Moriguchi, Y. and Verma, P. (2010). Temporally dynamic photopolymerization of C60 molecules encapsulated in single-walled carbon nanotubes. *Phys. Rev. B* **81**(24), p. 245416.
50. Liu, Z., Davis, C., Cai, W., He, L., Chen, X. and Dai, H. (2008). Circulation and long-term fate of functionalized, biocompatible single-walled carbon nanotubes in mice probed by raman spectroscopy. *Proc. Natl. Acad. Sci.* **105**(5), pp. 1410–1415.
51. Strano, M. S., Dyke, C. A., Usrey, M. L., Barone, P. W., Allen, M. J., Shan, H., Kittrell, C., Hauge, R. H., Tour, J. M. and Smalley, R. E. (2003). Electronic structure control of single-walled carbon nanotube functionalization. *Science* **301**(5639), pp. 1519–1522.
52. Kostarelos, K., Lacerda, L., Pastorin, G., Wu, W., Wieckowski, S., Luangsivilay, J., Godefroy, S., Pantarotto, D., J.-Briand, P., Muller, S. *et al.* (2007). Cellular uptake of functionalized carbon nanotubes is independent of functional group and cell type. *Nat. Nanotechnol.* **2**(2), pp. 108–113.
53. Postma, H. W. C., Sellmeijer, A., Dekker, C. (2000). Manipulation and imaging of individual single-walled carbon nanotubes with an atomic force microscope. *Adv. Mater.* **12**(17), pp. 1299–1302.
54. Tombler, T. W., Zhou, C., Alexseyev, L., Kong, J., Dai, H., Liu, L., Jayanthi, C. S., Tang, M. and Wu, S.-Y. (2000). Reversible electromechanical characteristics of carbon nanotubes under local-probe manipulation. *Nature* **405**(6788), pp. 769–772.
55. Duan, X., Son, H., Gao, B., Zhang, J., Wu, T., Samsonidze, G. G., Dresselhaus, M. S., Liu, Z. and Kong, J. (2007). Resonant raman spectroscopy of individual strained single-wall carbon nanotubes. *Nano Lett.* **7**(7), pp. 2116–2121.
56. Cronin, S. B., Swan, A. K., Ünlü, M. S., Goldberg, B. B., Dresselhaus, M. S. and Tinkham, M. (2004). Measuring the uniaxial strain of individual single-wall carbon nanotubes: Resonance raman spectra of atomic-force-microscope modified single-wall nanotubes. *Phys. Rev. Lett.* **93**(16), p. 167401.
57. Hertel, T., Walkup, R. E. and Avouris, P. (1998). Deformation of carbon nanotubes by surface van der waals forces. *Phys. Rev. B* **58**(20), pp. 13870–13873.
58. Postma, H. W. C., de Jonge, M., Yao, Z. and Dekker, C. (2000). Electrical transport through carbon nanotube junctions created by mechanical manipulation. *Phys. Rev. B* **62**(16), pp. R10653–R10656.
59. Fuhrer, M. S., Nygård, J., Shih, L., Forero, M., Yoon, Y.-G., Mazzoni, M. S. C., Choi, H. J., Ihm, J., Louie, S. G., Zettl, A. *et al.* (2000). Crossed nanotube junctions. *Science* **288**(5465), pp. 494–497.
60. Yano, T., Ichimura, T., Kuwahara, S., F. H'Dhili, Uetsuki, K., Okuno, Y., Verma, P. and Kawata, S. (2013). Tip-enhanced nano-raman analytical imaging of locally induced strain distribution in carbon nanotubes. *Nat. Commun.* **4**, p. 2592.

61. Okuno, Y., Saito, Y., Kawata, S. and Verma, P. (2013). Tip-enhanced raman investigation of extremely localized semiconductor-to-metal transition of a carbon nanotube. *Phys. Rev. Lett.* **111**(21), p. 216101.
62. Frogley, M. D., Ravich, D. and Wagner, H. D. (2003). Mechanical properties of carbon nanoparticle-reinforced elastomers. *Compos. Sci. Technol.* **63**(11), pp. 1647–1654.
63. Bokobza, L. (2009). Mechanical, electrical and spectroscopic investigations of carbon nanotube-reinforced elastomers. *Vib. Spectrosc.* **51**(1), pp. 52–59.
64. Barraza, H. J., Pompeo, F., O'Rea, E. A. and Resasco, D. E. (2002). SWNT-filled thermoplastic and elastomeric composites prepared by miniemulsion polymerization. *Nano Lett.* **2**(8), pp. 797–802.
65. Suzuki, T., Yan, X., Kitahama, Y., Sato, H., Itoh, T., Miura, T. and Ozaki, Y. (2013). Tip-enhanced raman spectroscopy study of local interactions at the interface of styrene–butadiene rubber/multiwalled carbon nanotube nanocomposites. *J. Phys. Chem. C* **117**(3), pp. 1436–1440.
66. Yan, X., Suzuki, T., Kitahama, Y., Sato, H., Itoh, T. and Ozaki, Y. (2013). A study on the interaction of Single-Walled Carbon Nanotubes (SWC-NTs) and Polystyrene (PS) at the interface in SWCNT–PS nanocomposites using tip-enhanced raman spectroscopy. *Phys. Chem. Chem. Phys.* **15**(47), pp. 20618–20624.
67. Yan, X., Kitahama, Y., Sato, H., Suzuki, T., Han, X., Itoh, T., Bokobza, L. and Ozaki, Y. (2012). Laser heating effect on raman spectra of styrene–butadiene rubber/multiwalled carbon nanotube nanocomposites. *Chem. Phys. Lett.* **523**, pp. 87–91.
68. Wood, J. R., Frogley, M. D., Meurs, E. R., Prins, A. D., Peijs, T., Dunstan, D. J. and Wagner, H. D. (1999). Mechanical response of carbon nanotubes under molecular and macroscopic pressures. *J. Phys. Chem. B* **103**(47), pp. 10388–10392.
69. Gui, G., Li, J. and Zhong, J. (2008). Band structure engineering of graphene by strain: First-principles calculations. *Phys. Rev. B* **78**(7), p. 075435.
70. Duplock, E. J., Scheffler, M. and Lindan, P. J. D. (2004). Hallmark of perfect graphene. *Phys. Rev. Lett.* **92**(22), p. 225502.
71. Low, T., Guinea, F. and Katsnelson, M. I. (2011). Gaps tunable by electrostatic gates in strained graphene. *Phys. Rev. B* **83**(19), pp. 195436.
72. Ni, Z. H., Yu, T., Lu, Y. H., Wang, Y. Y., Feng, Y. P. and Shen, Z. X. (2008). Uniaxial strain on graphene: Raman spectroscopy study and band-gap opening. *ACS Nano* **2**(11), pp. 2301–2305.
73. Jiang, Z., Zhang, Y., Tan, Y.-W., Stormer, H. L. and Kim, P. (2007). Quantum hall effect in graphene. *Solid State Commun.* **143**(1–2), pp. 14–19.
74. Jobst, J., Waldmann, D., Speck, F., Hirner, R., Maude, D. K., Seyller, T. and Weber, H. B. (2010). Quantum oscillations and quantum hall effect in epitaxial graphene. *Phys. Rev. B* **81**(19), p. 195434.
75. de Heer, W. A., Berger, C., Wu, X., First, P. N., Conrad, E. H., Li, X., Li, T., Sprinkle, M., Hass, J., Sadowski, M. L. *et al.* (2007). Epitaxial graphene. *Solid State Commun.* **143**(1–2), pp. 92–100.

76. Emtsev, K. V., Bostwick, A., Horn, K., Jobst, J., Kellogg, G. L., Ley, L., McChesney, J. L., Ohta, T., Reshanov, S. A., Röhrl, J. *et al.* (2009). Towards wafer-size graphene layers by atmospheric pressure graphitization of silicon carbide. *Nat. Mater.* **8**(3), pp. 203–207.
77. Xu, W., Mao, N. and Zhang, J. (2013). Graphene: A platform for surface-enhanced raman spectroscopy. *Small* **9**(8), pp. 1206–1224.
78. Ling, X., Moura, L. G., Pimenta, M. A. and Zhang, J. (2012). Charge-transfer mechanism in graphene-enhanced raman scattering. *J. Phys. Chem. C* **116**(47), pp. 25112–25118.
79. Ling, X., Xie, L., Fang, Y., Xu, H., Zhang, H., Kong, J., Dresselhaus, M. S., Zhang, J. and Liu, Z. (2010). Can graphene be used as a substrate for raman enhancement? *Nano Lett.* **10**(2), pp. 553–561.
80. Ling, X. and Zhang, J. (2010). First-layer effect in graphene-enhanced raman scattering. *Small* **6**(18), pp. 2020–2025.
81. Blake, P., Hill, E. W., Castro Neto, A. H., Novoselov, K. S., Jiang, D., Yang, R., Booth, T. J. and Geim, A. K. (2007). Making graphene visible. *Appl. Phys. Lett.* **91**(6), p. 063124.
82. Velez-Fort, E., Mathieu, C., Pallecchi, E., Pigneur, M., Silly, M. G., Belkhou, R., Marangolo, M., Shukla, A., Sirotti, F. and Ouerghi, A. (2012). Epitaxial graphene on 4H-SiC(0001) grown under nitrogen flux: Evidence of low nitrogen doping and high charge transfer. *ACS Nano* **6**(12), pp. 10893–10900.
83. Lin, Y.-M., Dimitrakopoulos, C., Jenkins, K. A., Farmer, D. B., Chiu, H.-Y., Grill, A. and Avouris, P. (2010). 100-GHz transistors from wafer-scale epitaxial graphene. *Science* **327**(5966), pp. 662–662.
84. Xia, F., Mueller, T., Lin, Y., Valdes-Garcia, A. and Avouris, P. (2009). Ultrafast graphene photodetector. *Nat. Immunol.* **4**(12), pp. 839–843.
85. Allen, M. J., Tung, V. C. and Kaner, R. B. (2010). Honeycomb carbon: A review of graphene. *Chem. Rev.* **110**(1), pp. 132–145.
86. Liu, C., Yu, Z., Neff, D., Zhamu, A. and Jang, B. Z. (2010). Graphene-based supercapacitor with an ultrahigh energy density. *Nano Lett.* **10**(12), pp. 4863–4868.
87. Wang, Y., Shi, Z., Huang, Y., Ma, Y., Wang, C., Chen, M. and Chen, Y. (2009). Supercapacitor devices based on graphene materials. *J. Phys. Chem. C* **113**(30), pp. 13103–13107.
88. Le, L. T., Ervin, M. H., Qiu, H., Fuchs, B. E. and Lee, W. Y. (2011). Graphene supercapacitor electrodes fabricated by inkjet printing and thermal reduction of graphene oxide. *Electrochem. Commun.* **13**(4), pp. 355–358.
89. Ushio, S., Yoshii, A., Tamai, N., Ohtani, N. and Kaneko, T. (2011). Epitaxial graphene growth on 4H-SiC (0001). with precisely controlled step-terrace surface by high temperature annealing above 2000°C in UHV. *Phys. Status Solidi C* **8**(2), pp. 580–582.
90. Ushio, S., Kutsuma, Y., Yoshii, A., Tamai, N., Ohtani, N. and Kaneko, T. (2011). 4H-SiC(0001) basal plane stability during the growth of epitaxial graphene on inverted-mesa structures. *Jpn. J. Appl. Phys.* **50**(7R), pp. 070104.

91. Ouerghi, A., Silly, M. G., Marangolo, M., Mathieu, C., Eddrief, M., Picher, M., Sirotti, F., El Moussaoui, S. and Belkhou, R. (2012). Large-area and high-quality epitaxial graphene on off-axis SiC wafers. *ACS Nano* **6**(7), pp. 6075–6082.
92. Vantasin, S., Tanaka, Y., Uemura, S., Suzuki, T., Kutsuma, Y., Doujima, D., Kaneko, T. and Ozaki, Y. (2015). Characterization of SiC-grown epitaxial graphene microislands using tip-enhanced raman spectroscopy. *Phys. Chem. Chem. Phys.* **17**(43), pp. 28993–28999.
93. Wang, F., Liu, G., Rothwell, S., Nevius, M., Tejeda, A., Taleb-Ibrahimi, A., Feldman, L. C., Cohen, P. I. and Conrad, E. H. (2013). Wide-gap semiconducting graphene from nitrogen-seeded SiC. *Nano Lett.* **13**(10), pp. 4827–4832.
94. Xu, P., Barber, S. D., Schoelz, J. K., Ackerman, M. L., Qi, D., Thibado, P. M., Wheeler, V. D., Nyakiti, L. O., Myers-Ward, R. L. and Jr, C. R. E. *et al.* (2013). Atomic-scale movement induced in nanoridges by scanning tunneling microscopy on epitaxial graphene grown on 4H-SiC(0001). *J. Vac. Sci. Technol. B* **31**(4), p. 04D101.
95. Prakash, G., Capano, M. A., Bolen, M. L., Zemlyanov, D. and Reifenberger, R. G. (2010). AFM study of ridges in few-layer epitaxial graphene grown on the carbon-face of 4H–SiC. *Carbon* **48**(9), pp. 2383–2393.
96. Sun, G. F., Jia, J. F., Xue, Q. K. and Li, L. (2009). Atomic-scale imaging and manipulation of ridges on epitaxial graphene on 6H-SiC(0001). *Nanotechnology* **20**(35), p. 355701.
97. Srivastava, N., He, Luxmi, G., Mende, P. C., Feenstra, R. M. and Sun, Y. (2012). Graphene formed on SiC under various environments: Comparison of Si-Face and C-Face. *J. Phys. Appl. Phys.* **45**(15), p. 154001.
98. Luxmi, P. J., Fisher, N., Srivastava, R. M., Feenstra, Y. Sun, J. Kedzierski, P. and Healey, G. Gu (2009). Morphology of graphene on SiC($000\bar{1}$) surfaces. *Appl. Phys. Lett.* **95**(7), p. 073101.
99. Pashaee, F., Sharifi, F., Fanchini, G. and Lagugné-Labarthet, F. (2015). Tip-enhanced raman spectroscopy of graphene-like and graphitic platelets on ultraflat gold nanoplates. *Phys. Chem. Chem. Phys.* **17**(33), pp. 21315–21322.
100. Hoffmann, G. G., Xue, L., Loos, J. and de With, G. (2011). High-resolution tip-enhanced raman mapping. *Macromol. Symp.* **305**(1), pp. 26–42.
101. Ghislandi, M., Hoffmann, G. G., Tkalya, E., Xue, L. and With, G. D. (2012). Tip-enhanced raman spectroscopy and mapping of graphene sheets. *Appl. Spectrosc. Rev.* **47**(5), pp. 371–381.
102. Snitka, V., Rodrigues, R. D. and Lendraitis, V. (2011). Novel gold cantilever for nano-raman spectroscopy of graphene. *Microelectron. Eng.* **88**(8), pp. 2759–2762.
103. Shiotari, A., Kumagai, T. and Wolf, M. (2014). Tip-enhanced raman spectroscopy of graphene nanoribbons on Au(111). *J. Phys. Chem. C* **118**(22), pp. 11806–11812.
104. Su, W. and D. Roy (2013). Visualizing graphene edges using tip-enhanced raman spectroscopy. *J. Vac. Sci. Technol. B* **31**(4), p. 041808.

105. Okuno, Y., Vantasin, S., Yang, I.-S., Son, J., Hong, J., Tanaka, Y. Y., Nakata, Y., Ozaki, Y. and Naka, N. (2016). Side-illuminated tip-enhanced raman study of edge phonon in graphene at the electrical breakdown limit. *Appl. Phys. Lett.* **108**(16), p. 163110.
106. Xu, G., Liu, Z., Xu, K., Zhang, Y., Zhong, H., Fan, Y. and Huang, Z. (2012). Constant current etching of gold tips suitable for tip-enhanced raman spectroscopy. *Rev. Sci. Instrum.* **83**(10), p. 103708.
107. Hibino, H., Kageshima, H. and Nagase, M. (2010). Epitaxial few-layer graphene: Towards single crystal growth. *J. Phys. Appl. Phys.* **43**(37), p. 374005.
108. Emtsev, K. V., Speck, F., Seyller, T., Ley, L. and Riley, J. D. (2008). Interaction, growth, and ordering of epitaxial graphene on SiC{0001} surfaces: A comparative photoelectron spectroscopy study. *Phys. Rev. B* **77**(15), p. 155303.
109. Ferralis, N. (2010). Probing mechanical properties of graphene with Raman spectroscopy. *J. Mater. Sci.* **45**(19), pp. 5135–5149.
110. Harris, G. L. (1995). *Properties of Silicon Carbide.* IET.
111. Yue, Y., Zhang, J. and Wang, X. (2011). Micro/nanoscale spatial resolution temperature probing for the interfacial thermal characterization of epitaxial graphene on 4H-SiC. *Small* **7**(23), pp. 3324–3333.
112. Mounet, N. and Marzari, N. (2005). First-principles determination of the structural, vibrational and thermodynamic properties of diamond, graphite, and derivatives. *Phys. Rev. B* **71**(20), p. 205214.
113. Mohiuddin, T. M. G., Lombardo, A., Nair, R. R., Bonetti, A., Savini, G., Jalil, R., Bonini, N., Basko, D. M., Galiotis, C., Marzari, N. *et al.* (2009). Uniaxial strain in graphene by raman spectroscopy: G peak splitting, grüneisen parameters, and sample orientation. *Phys. Rev. B* **79**(20), p. 205433.
114. Berciaud, S., Ryu, S., Brus, L. E. and Heinz, T. F. (2009). Probing the intrinsic properties of exfoliated graphene: Raman spectroscopy of free-standing monolayers. *Nano Lett.* **9**(1), pp. 346–352.
115. Son, J., Choi, M., Choi, H., Kim, S. J.,Kim, S., Lee, K.-R., Vantasin, S., Tanabe, I., Cha, J., Ozaki, Y. *et al.* (2016). Structural evolution of graphene in air at the electrical breakdown limit. *Carbon* **99**, pp. 466–471.
116. Ren, W., Saito, R., Gao, L., Zheng, F., Wu, Z., Liu, B., Furukawa, M., Zhao, J., Chen, Z. and Cheng, H.-M. (2010). Edge phonon state of mono- and few-layer graphene nanoribbons observed by surface and interference co-enhanced raman spectroscopy. *Phys. Rev. B* **81**(3), p. 035412.

Chapter 13

Chemical Identification by Sub-Nanometer Resolved Single-Molecule Raman Scattering

Yao Zhang, Yang Zhang, Zhenchao Dong* and Jianguo Hou

Hefei National Laboratory for Physical Sciences at the Microscale and Synergetic Innovation Center of Quantum Information & Quantum Physics, University of Science and Technology of China, Hefei, Anhui 230026, China

**zcdong@ustc.edu.cn*

1. Introduction

Visualizing individual molecules with chemical recognition by optical means is a longstanding target in catalysis, material science, molecular nanotechnology and biotechnology. One way to achieve this is to use fluorescence-based techniques such as stimulated emission depletion microscopy (STED) [1, 2], stochastic optical reconstruction microscopy (STORM) [3, 4] and photo-activated localization microscopy [5, 6], which can provide super-resolution at far field down to ~10 nm by statistically controlling the optical states of the fluorescence-labeled molecules. The pioneering works in this field won the Nobel Prize of Chemistry in 2014 for the development of super-resolved fluorescence microscopy. The other way to achieve chemical identification at the nanoscale is to use molecular-vibration-based techniques, such as infrared nanospectroscopy [7, 8] and tip-enhanced Raman scattering (TERS) spectroscopy [9–12], which are labeling-free and much richer in chemical fingerprints. The chemical sensitivity and spatial resolution of these vibration-based techniques are dramatically improved because of the confinement and enhancement of local plasmonic near field produced by the metallic nanostructures acting as optical nanoantennas [13, 14]. In particular, the TERS technique has shown great promise in this aspect due to its combined advantage of ultrahigh spatial resolution in scanning probe microscopy (SPM) and sensitive chemical specificity in Raman spectroscopy. During the recent decades, extensive TERS research has been carried out mainly along two directions. One is to demonstrate its wide applications for nanoscale chemical identification ranging from organic molecules [15–20] to carbon nanotubes [21–23], and even to bio-materials [24–29]. The other is to improve the spatial resolution, which has been improved from over 10 nm [17, 21] down to a few nanometers [22, 23, 29, 30].

In this chapter, after a brief introduction (Section 1), we will give a detailed account of our recent development of the TERS technique in a well-controlled condition based on the ultrahigh-vacuum (UHV) and low-temperature scanning tunneling microscope (STM). Using such STM-controlled TERS technique, the spatial resolution has been further driven down to ~0.5 nm (Section 2) [31] which could even enable the chemical distinguishing of two adjacent different molecules that are within van der Waals contact and of very similar chemical structures (Section 3) [32]. By taking into account the Raman selection rules during the density functional theory (DFT) simulations, the adsorption configurations and orientations of single molecules can be derived by fitting to the experimental spectra (Section 4). The intramolecular resolved TERS mapping image of a single molecule can also be simulated by considering the different contributions from different atoms with respect to the local electric field. The origin of the ultrahigh spatial resolution is believed to somehow correlate with the nonlinear optical process in Raman scattering, which will be discussed in Section 5.

2. Sub-Nanometer Resolved Single-Molecule TERS Mapping

Since the first development of the TERS technique in 2000 [9–12], it has provided a unique way not only to increase the intensity of the Raman photon signals via the near-field enhancement at the tip apex, but also to obtain the spatial properties through tip scanning, thanks to the combination of Raman spectroscopy with microscopic capabilities in the SPM technique [16, 17, 19, 21, 22, 33–36]. The high spatial resolution below the diffraction limit originates from the plasmonic field confinement into a localized region. However, considering that the spatial extent of the local plasmonic field is usually 5–10 nm, it imposes challenges on the further improvement of the spatial resolution down to the sub-nanometer scale [37, 38]. Moreover, conventional TERS usually requires the use of strong incident laser fields, which could result in undesired diffusion, desorption and even damage to the molecule, thus affecting the sustainability and stability of Raman spectral mapping. In this section, we describe our experimental TERS study on a single porphyrin molecule located at the STM nanocavity under UHV and low temperature conditions.

2.1. *Experimental setup of the TERS measurement*

Our STM-controlled TERS experiments were performed on a custom low-temperature UHV-STM (Unisoku) in a confocal-type side-illumination configuration (Figure 1(a)) [39] at about 80 K under a base pressure of around 10^{-10} torr. Free-base *meso*-tetrakis(3,5-ditertiarybutylphenyl)-porphyrin (H_2TBPP) molecules were thermally evaporated onto the Ag(111) surface that was previously cleaned by cycles of argon ion sputtering and annealing. Silver (Ag) is also used as a tip material because the nanogap defined by the Ag tip and the Ag substrate can offer very strong plasmonic resonance [40]. Upon fabrication via electrochemical etching, Ag tips were cleaned in UHV via outgassing and ion sputtering, with the tip status often further modified by high-voltage pulses or tip indentation in order to achieve desired plasmonic resonance modes and emission intensities. STM imaging and spectral measurements were taken in a constant-current mode with the sampled biased.

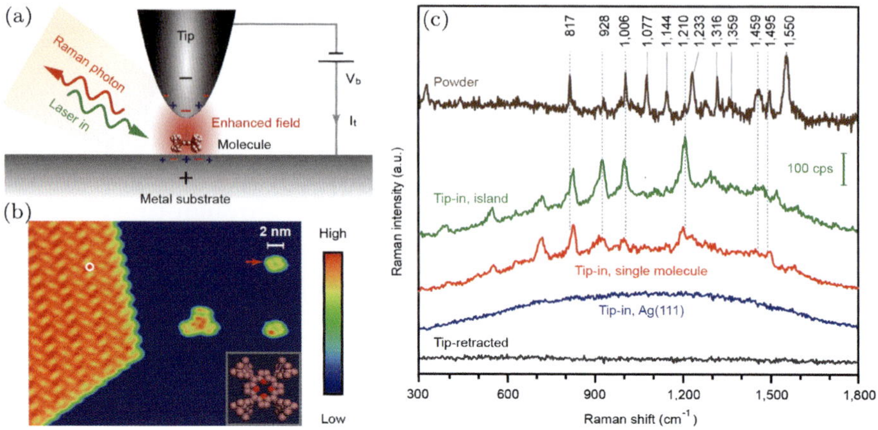

Figure 1. (a) Schematic tunneling-controlled TERS in a confocal-type side-illumination configuration, in which V_b is the sample bias and I_t is the tunneling current. (b) STM topograph of sub-monolayered H_2TBPP molecules on Ag(111) (1.5 V, 30 pA, 35 nm × 27 nm). The inset shows the chemical structure of H_2TBPP and the white circle indicates one representative site for TERS measurements on the molecular island. (c) TERS spectra for different conditions. The tip-in spectra were acquired at 120 mV, 0.5 nA and 3 s. The green spectrum was taken on top of the molecular island (the green scale bar shows the signal level detected by CCD). The red spectrum was taken on top of a single molecule (marked by the red arrow in b). The blue spectrum was taken on the bare Ag(111). The black spectrum was taken on top of the molecular island but with the tip retracted 5 nm from the surface (120 mV, 3 s). For comparison, a standard Raman spectrum (brown) is shown on the top for a powder sample of H_2TBPP molecules. [Reprinted with permission from Ref. [31]. Copyright (2013) Macmillan Publisher Limited].

The photon collection and detection systems were described previously [40, 41]. A single-longitude-mode diode-pumped laser at 532 nm was used to provide a linearly polarized laser beam (>100:1) for Raman excitation. The beam was fiber-coupled to a dark-box via a single-mode polarization-maintaining fiber with a collimated output beam of ~1.0 mm, passed through a half-wave plate to achieve desired *p*-polarization, and then introduced into UHV via a quartz viewport. The collimated beam was re-focused by an aspheric lens (f_b = 19.5 mm, NA = 0.51) on the tip–substrate junction at an angle of 60° from the surface normal. The diameter of the laser spot on the sample surface at optimum focus was ~30 μm. Raman scattered light was collected through the same lens, filtered to remove residual laser light, and then fiber-coupled to a spectrometer using a slit size of 100 μm. The Raman light was dispersed by a 600 grooves/mm grating and detected by a liquid-nitrogen-cooled charge-coupled device (CCD). The laser power used in our TERS experiments was ~0.8 mW, corresponding to a photon flux of ~110 W/cm^2. The spectral resolution was ~18 cm^{-1}, limited by the instrumental resolution that was selected in order to detect sufficiently strong Raman signals so that a reasonably good signal-to-noise ratio can be obtained for the spectral measurements. The STM operation was controlled by Nanonis-SPECS electronics. The Raman spectral mapping was carried out through a synchronization function between the STM controller and CCD camera at each pixel [40]. All Raman spectra presented here are not corrected for the wavelength-dependent sensitivity of photon detection systems.

We would like to note that a low-temperature UHV-STM is advantageous for single-molecule TERS measurements [42, 43]. Under UHV conditions, both the tip and sample surface can be cleaned up easily and maintained clean, minimizing serious contamination in ambient conditions and thus enhancing the stability of the tip–sample junction. In addition, clean molecular samples can be prepared via vacuum deposition onto surfaces with a controlled coverage. The latter is particularly important since the presence of isolated single molecules on the surfaces is desirable for well-defined single-molecule TERS studies. In this regard, a substrate cooled to low temperature is also very helpful for isolated single molecules to immobilize on the surface, thus suppressing the diffusion and desorption problem. Moreover, the thermal drift problem can also be suppressed at low temperature even under light illumination (~0.15 nm/min at ~80 K), and the UHV condition can also significantly decrease photodegradation of adsorbed molecules, thus allowing molecules to be measured with longer time and repeatedly.

Tip quality and system stability are two important factors to consider for performing TERS experiments. A low-temperature UHV-STM not only provides improved system stability in both lateral and vertical directions, but equally importantly, it also offers the capability to actively control the tip status that can be monitored *in situ* by STM induced luminescence. All these special conditions allow us to achieve exquisite tuning between the nanocavity plasmon (NCP) resonance and the molecular vibronic transitions. In particular, the spectral matching of the NCP resonance to the downward transitions associated with the emission of Raman photons is found to be crucial for enhancing and stabilizing the inelastic scattering signals. Remarkably, the very small incident photon flux used in our STM-controlled spectral-matching TERS enables stable single-molecule mapping with unambiguous chemical identification and unprecedented sub-molecular spatial resolution for a single H_2TBPP molecule on the Ag(111) surface, as described below.

2.2. *TERS spectral analysis*

Figures 1(b) and 1(c) illustrate the high quality and level of cleanness of the tip and sample under a low-temperature and ultrahigh-vacuum environment [17, 42, 43], which allows for reproducible chemical identifications through vibrational fingerprints. In the STM image of Figure 1(b) for H_2TBPP on Ag(111), we can identify two isolated single molecules on the right (featuring the characteristic four-lobe pattern [41, 44]), a three-molecule cluster at the center, and a densely packed monolayer island on the left. The bare Ag surface in blue serves as an *in situ* check-ground for monitoring the tip cleanness and the NCP resonance mode.

As shown in Figure 1(c), when a tip is positioned on the bare Ag surface, a featureless spectrum (blue) containing simply a broad continuum is obtained, consistent with the absence of molecules and also suggesting the operation of a contamination-free tip. On top of the molecular islands (marked with a white circle in Figure 1(b)), the TERS spectrum (green) shows clear vibrational fingerprints of H_2TBPP molecules over a broad continuum. The spectra do not change substantially when acquired at different positions of molecules on the island, due to the tightly packed structure of H_2TBPP molecules on Ag(111).

For comparison, the standard Raman signals (brown spectrum) from a H_2TBPP powder sample are also given in Figure 1(c), showing all Raman-active modes averaged over the full space of randomly oriented molecules. Many of the fingerprint peaks of the powder spectrum are in good correspondence with the vibrational fingerprints on the molecular island and single molecules (see dash lines), thus providing clear chemical identification of the molecules on the surface. However, the number and relative intensity of peaks reveal differences as well. These could be attributed to both the ordering of the H_2TBPP molecules on Ag(111) and the preferred axial polarization of the nanocavity plasmon, which selects particular Raman modes. In other words, the H_2TBPP molecules on Ag(111) are no longer randomly oriented, but rather packed in specific orientations with the porphyrin core plane almost parallel to the surface. In addition, the plasmonic excitations are confined within the nanocavity and the dominant plasmonic dipole prefers to align along the normal of the substrate.

When the tip is retracted about 5 nm from the island (black spectrum), the molecular fingerprints disappear, providing unambiguous evidence that the TERS signals observed in the green spectrum originate only from the molecular sample itself. The observation of tip-in TERS spectra with a relatively high signal-to-noise ratio (for example, for the 1,210 cm^{-1} peak) together with the absence of the far-field Raman signal associated with the tip-retracted mode implies a large enhancement in the present system [37, 42]. This is nicely illustrated by the TERS measurement on an *isolated single* H_2TBPP molecule (red spectrum), which exhibits unambiguous vibrational fingerprints similar to those on the molecular island and thus provides a clear demonstration of single-molecule TERS spectroscopy.

Of particular interest is the broad continuum that usually accompanies the TERS measurements on both the bare metal surface and the molecules (Figure 1(c)) [45–47]. This broad continuum turns out to correlate closely with the NCP resonance mode that is determined by the junction geometry of the STM nanocavity and dielectric properties of the tip and substrate [45] and can be monitored *in situ* by STM-induced luminescence through tunneling electron excitation [41]. The importance of this broad continuum in TERS experiments will be discussed in detail in Section 5 on the mechanism of ultrahigh spatial resolution.

2.3. *Orientation dependence of single-molecule TERS spectra*

The large enhancement assisted by the broadband NCP resonance not only allows for the non-invasive TERS measurements with single-molecule sensitivity but also provides an opportunity to explore the influence of molecular orientations on the vibrational spectral features. Figure 2(a) shows two representative TERS spectra acquired on two isolated molecules with different adsorption configurations (or molecular orientations). As shown in the STM images on the right, one spectrum was acquired on a single flat-lying H_2TBPP molecule adsorbed on the terrace of Ag(111) (blue spectrum), and the other was on a single tilted molecule adsorbed at the step edge (red spectrum). Both spectra reveal vibrational fingerprints characteristic of the H_2TBPP molecule in terms of Raman peak positions.

However, there also exist differences in the relative peak intensities between the two spectra in Figure 2(a), reflecting the changes in the adsorption configurations (or molecular orientations). With the help of the DFT calculations, the origin of such differences can be associated with the particular orientation of the single molecule that governs the Raman selection rules of the fingerprint. The simulated spectra in Figure 2(b) nicely reproduce the activation and deactivation of the experimental modes observed in Figure 2(a) when the tilting angle of the molecule is changed. (The details of the theoretical calculation of Raman spectra and the corresponding selection rules will be discussed in Section 4). Notably, as occasionally found in experiments, similar spectral variations can also occur for the "same" molecule on a surface during the TERS measurements, owing to configuration or orientation changes, particularly under the strong illumination condition. Previous single-molecule TERS studies under ambient conditions were usually based on statistical analysis of such spectral fluctuations within a timescale of several seconds [16, 18, 19, 42, 48]. However, a single molecule can also remain stable

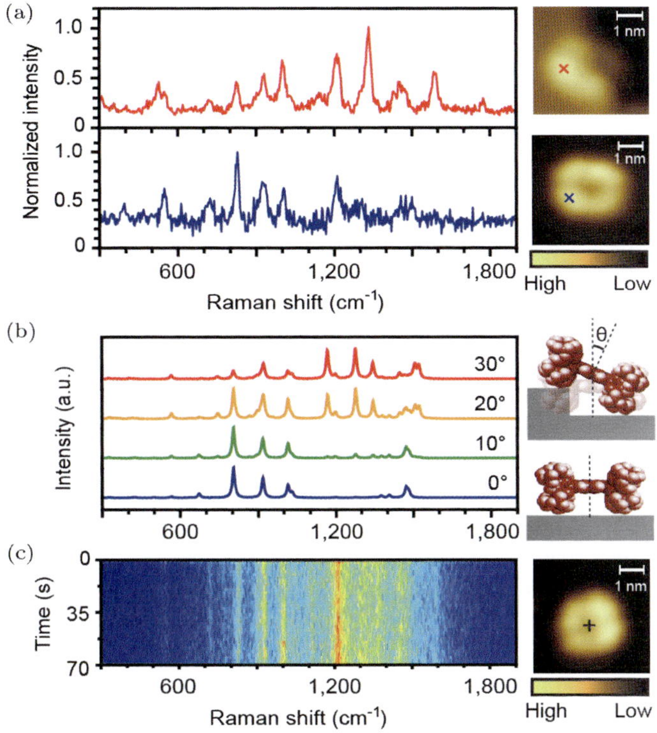

Figure 2. Single-molecule TERS spectra and their dependency on molecular orientations. (a) Single-molecule TERS spectra (100 mV, 1 nA, 3 s) for an isolated H_2TBPP molecule adsorbed on the terrace (bottom, blue) or at the step edge (top, red) of Ag(111). Both spectra were acquired on the molecular lobes marked with crosses in the STM images on the right (the broad continuum is subtracted for clarity). (b) Calculated TERS spectra of a molecule for different tilt angles θ. Shown on the right are the schematics of flat-lying and tilted molecules, respectively. (c) 35 sequential TERS spectra (120 mV, 1 nA, 2 s) acquired on the center of a single flat-lying H_2TBPP molecule adsorbed on the Ag(111) terrace (the terrace is the flat area between the step edges); the center position is marked by a cross in the corresponding STM image on the right. [Reprinted with permission from Ref. [31]. Copyright (2013) Macmillan Publisher Limited.]

for a certain period of time during the TERS measurements without revealing any substantial spectral changes.

Remarkably, Figure 2(c) shows one example of such stability over 70 s when 35 sequential TERS measurements were made on a single H_2TBPP molecule on the Ag terrace. The STM image on the right was taken before the sequential TERS measurements, but remains the same after the measurements. Such molecular stability over a sufficiently long period of time is crucial to perform TERS imaging experiments that can yield a complete and meaningful spectral map.

2.4. *Sub-nanometer resolved TERS mapping of a single H_2TBPP molecule on Ag(111)*

The realization of efficient nanocavity-plasmon-enhanced Raman scattering in our STM-controlled TERS technique provides a means to perform Raman mapping with unprecedented spatial resolution. Figure 3(a) indicates that, for an isolated single H_2TBPP molecule on Ag(111), the TERS spectra acquired on the molecular lobe are stronger than those on the center, whereas the nearby Ag gives a broad continuum, highlighting the highly localized nature of the TERS signals. Such contrast is best illustrated in the panoramic TERS mapping image in Figure 3(b) (top row), in which we plot experimental TERS mapping results of a flat-lying molecule on Ag for five selected Raman peaks. The characteristic four-lobe pattern of a H_2TBPP molecule is discernible in the TERS mapping, at least for low-wavenumber vibrational modes. The molecular lobes appear bright but the center appears dark. However, for relatively large wavenumbers of 1,210 cm^{-1} or above, the contrast and central dark area become smaller.

Such Raman image contrast and frequency dependence can be qualitatively understood by the relative locality of the vibrational modes of the H_2TBPP molecule with respect to the axial polarization of the highly confined local plasmonic fields [16]. In brief, the low-wavenumber vibrational modes are more localized in the lobe while the high-wavenumber modes above about 1,200 cm^{-1} contain more contribution from the porphyrin core. To help explain the contrast between these Raman images and their evolution, the bottom of Figure 3(b) shows simulated TERS mapping by assuming that the enhanced local electric field follows a Gaussian beam distribution (the details will be discussed in Section 4). These simulated images are consistent with the experimental TERS mapping results, although the consistency is better at low-wavenumber vibrational

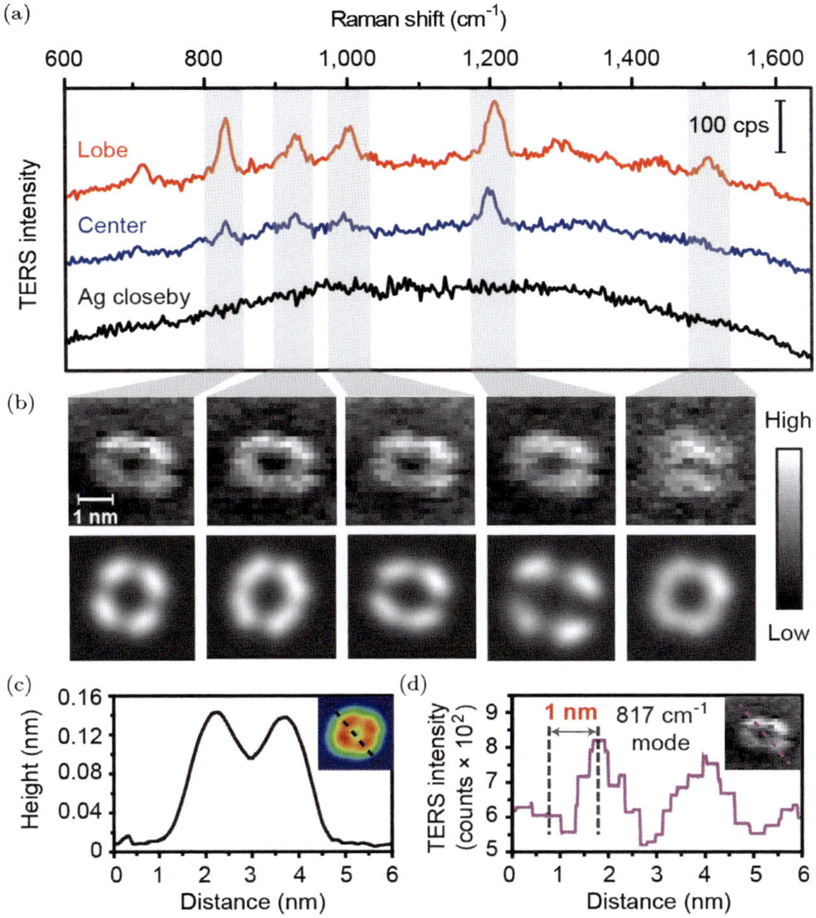

Figure 3. (a) Representative single-molecule TERS spectra on the lobe (red) and center (blue) of a flat-lying molecule on Ag(111). The TERS spectrum on the bare Ag surface about 1 nm away from the molecule is also shown, in black (120 mV, 1 nA, 3 s). (b) The top panels show experimental TERS mapping of a single H_2TBPP molecule for different Raman peaks (23 × 23, ~0.16 nm per pixel), processed from all individual TERS spectra acquired at each pixel (120 mV, 1 nA, 0.3 s; image size: 3.6 × 3.6 nm^2). The bottom panels show the theoretical simulation of the TERS mapping. (c) Height profile of a line trace in the inset STM topograph (1 V, 20 pA). (d) TERS profile of the same line trace for the inset Raman map associated with the 817 cm^{-1} Raman peak, integrated over 800–852 cm^{-1}. [Reprinted with permission from Ref. [31]. Copyright (2013) Macmillan Publisher Limited.]

modes. Strikingly, in comparison with the spatial resolution of the STM topograph (Figure 3(c)), the profile of the Raman spectral mapping shown in Figure 3(d) not only exhibits a comparable spatial resolution below 1 nm (about 0.5 nm estimated within a 10% to 90% contrast), but more importantly, also provides chemically resolved information revealing intramolecular features.

We also carried out TERS spectral and mapping experiments for a different H_2TBPP molecule that appears asymmetric in terms of its STM topograph on the surface (Figure 4(a)). Such asymmetry

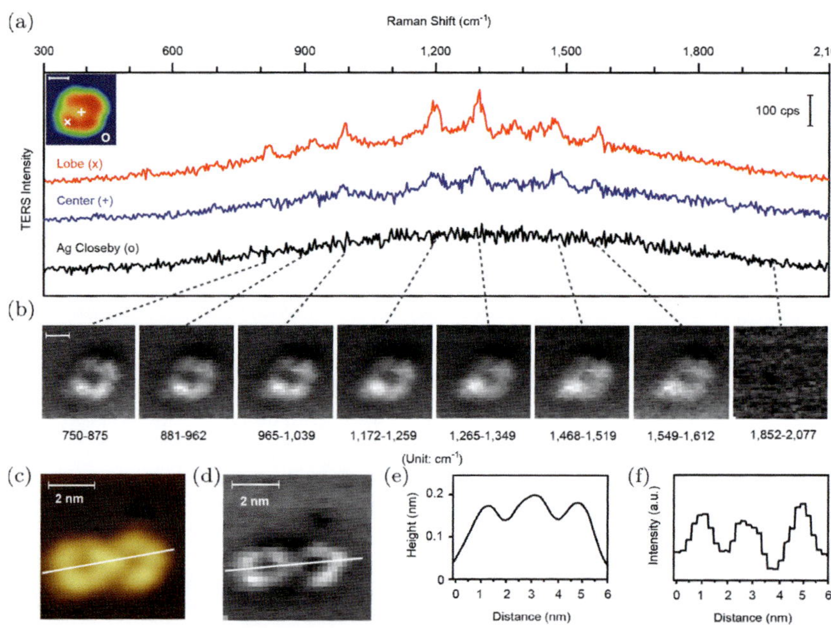

Figure 4. (a) TERS spectra (100 mV, 1 nA, 0.3 s) with the STM tip scanning over the lobe (×) or center (+) of a single asymmetric H_2TBPP molecule on the terrace of the Ag surface as well as over the nearby Ag substrate (o). The inset shows the STM topograph of the measured molecule with the extracted spectral positions marked (1 V, 20 pA, scale bar: 1 nm). (b) TERS mapping images of the asymmetric molecule at different wavenumbers corresponding to the dash lines (100 mV, 1 nA, 0.3 s; 32 × 32); scale bar: 1 nm). The integrated wavenumber ranges for each mapping are given below the images. (c) STM image of two adjacent H_2TBPP molecules on Ag(111). (d) Corresponding TERS mapping of the dimer integrated over 300–1900 cm^{-1} (6.5 × 6.5 nm^2, 200 mV, 0.5 nA, 0.3 s, 30 × 30). (e) and (f). The line profile for the STM height and TERS mapping corresponding to the white solid line in (c) and (d), respectively.

might arise from the tilting of one lobe and resultant deformation of the molecule. The TERS spectra acquired on the molecular lobe, center and nearby silver are overall similar to the flat-lying molecule in Figure 3(a), but the peak intensities at large wavenumbers are stronger, showing a similar feature to the tilted molecule at the step edge (Figure 2(a)) and thus suggesting a non-flat adsorption configuration for this asymmetric molecule. The mapping images for different Raman shift regions are shown in Figure 4(b), in which the four-lobe feature of the single H_2TBPP is also visible, with the central dark region becoming more and more blurred as the wavenumber increases. For the integrated region without any Raman spectral peaks, the mapping image shows no molecular features, ruling out the influence of background signals.

In order to examine the spatial distinction ability of the STM-controlled TERS technique, we also performed TERS mapping for a molecular dimer on Ag(111), as shown in Figures 4(c) and 4(d). The Raman mapping contrast shows nice similarities to the corresponding STM topographs, but now contains Raman spectral information at each pixel. Further analyses of line profiles in Figures 4(e) and 4(f) exhibit a TERS spatial resolution less than 1 nm, thus providing the possibility to distinguish adjacent different molecules on a surface.

3. Distinguishing Adjacent Different Molecules on the Ag(111) Surface

In the above, we have demonstrated single-molecule Raman mapping for a single type of porphyrin molecules on Ag(111), with an ultrahigh spatial resolution down to ~0.5 nm. However, we are still facing a challenging but practically important issue: Is such sub-nanometer optical contrast really capable of chemically resolving *adjacent different* molecules on surfaces where a molecule is surrounded by different Raman-active molecules rather than Raman-silent bare metal surface? How close and how similar can two adjacent different molecules be distinguished by TERS? In other words, the Raman spectral assignment of individual species becomes non-trivial in an interactive molecular system with different molecules positioned onto a surface next to each other. More complicatedly, for two close packed molecules with similar chemical units, their Raman signals may mix up. In this section, we report a proof-of-principle study to address these scientifically and technically important issues for two different adjacent molecules that are of similar chemical structure (a metal-centered porphyrin and a free-base prophyrin) and within the van der Waals contact.

3.1. *Distinguishing between ZnTPP and H_2TBPP molecular domains on Ag(111)*

In order to chemically distinguish two different chemical species on a surface, the samples of sub-monolayer coverage were prepared separately for two different porphyrin molecules, metal-centered zinc-5,10,15,20-tetraphenyl-porphyrin (ZnTPP) and free-base H_2TBPP, on Ag(111), respectively. By measuring the TERS spectra under the same procedure as described in Section 2, the TERS spectrum acquired above the ZnTPP island on Ag(111) is found to be different from that on the H_2TBPP island, although the two molecules belong to the same porphyrin family and have similar structural skeletons [32]. Such spectral distinction suggests the possibility to distinguish them from each other on a surface.

Having separately established the reference TERS database for ZnTPP and H_2TBPP molecules on Ag(111), we then examined the TERS spectra with both types of molecules co-deposited on the metal surface. To make a clear chemical identification we started with a simple system of molecular domain identification, distinguishing two different molecular islands separated by ~2.5 nm by a bare Ag surface (Figure 5(a)). Figure 5(b) plots TERS spectra that were measured consecutively along the line trace shown in Figure 5(a). All the Raman spectra on the top molecular island (positions 1–4) resemble the TERS spectrum of ZnTPP on Ag(111), while those on the bottom island (positions 8–10) all reproduce the TERS spectrum of H_2TBPP on Ag(111). In contrast, the spectra over the gap region of the bare Ag surface (positions 5–7) present a broad featureless continuum, the profile of which is associated with the nanocavity plasmon resonance under laser illumination [31, 45]. All these observations indicate that

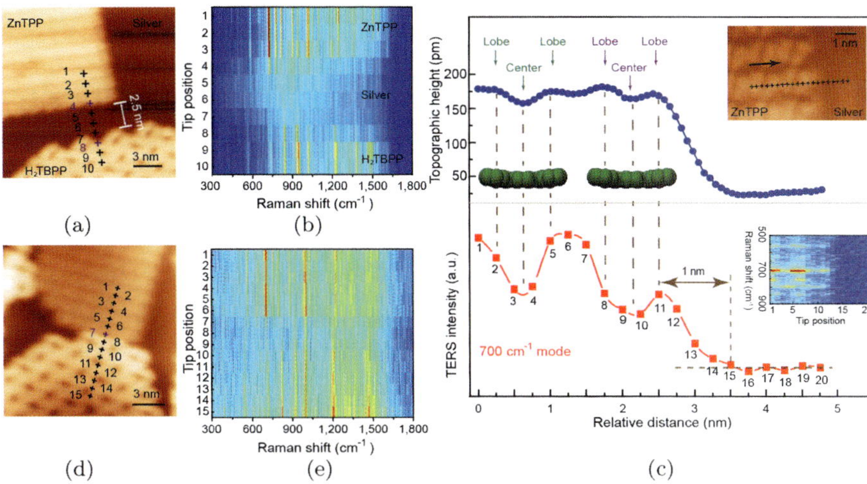

Figure 5. (a) STM topograph of two separated molecular island (1 V, 100 pA). (b) Corresponding spectral evolution for ten sequential TERS measurements along the trace marked in (a) (0.1 V, 1 nA, 30 s) with a step size of 1 nm. (c) Fine details of spectral evolution across the molecular island edge for ZnTPP molecules with a step size of 0.25 nm. Top: height profile for the line trace marked in the right inset (0.75 V, 10 pA). Bottom: corresponding TERS intensity profile for the 700 cm^{-1} mode (integrated over 670–730 cm^{-1}) shown in the right inset (0.1 V, 1 nA, 10 s). (d) STM topograph of two adjacent molecular islands (0.6 V, 10 pA). (e) Corresponding spectral evolution for 15 sequential TERS measurements along the line trace marked in **d** (0.1 V, 1 nA, 10 s) with a step size of 0.75 nm. [Reprinted with permission from Ref. [32]. Copyright (2015) Macmillan Publisher Limited.]

the molecules detected on the top island are all ZnTPP, while those on the bottom island are all H_2TBPP.

It should also be pointed out that the edge molecules, particularly the constituents facing the island edge (positions 4 and 8 in Figure 5(a)), appear to show weaker TERS intensities than those inside an island, as illustrated in Figure 5(b) for ZnTPP (position 4) and H_2TBPP (position 8). To find out exactly how TERS spectra evolve across the island edge, we ran 20 sequential TERS measurements along the line trace shown in the inset to Figure 5(c) with a small step size of 0.25 nm on the ZnTPP molecular island. Interestingly, in addition to the reproduction of weaker TERS intensities at the edges, the line profile for the Raman peak at ~700 cm^{-1} also exhibits Raman intensity undulations inside the molecular island (Figure 5(c)), strongest at the intermolecular positions and weakest at the molecular centers. The line profile analysis also suggests a spatial resolution of $\sim$0.6 nm, estimated within a 10–90% contrast.

Such a high spatial resolution should allow ZnTPP to be resolved chemically from the neighboring H_2TBPP molecules, as each molecule has a dimension of about 2 nm. Indeed, as shown in Figures 5(d) and 5(e), when two molecular islands are packed close to each other, within van der Waals contact ($\sim$0.3 nm), the analysis of the TERS spectral pattern along the line trace can still tell us that the edge molecule on the top island (position 6) is ZnTPP, while the edge molecule on the bottom island (position 8) is H_2TBPP. Position 7 appears to be located between these two molecules, and shows weak mixed TERS signals from both molecules.

3.2. *Distinguishing adjacent different molecules at step edges*

In the above demonstration for ZnTPP and H_2TBPP domains adsorbed on the terrace (Figure 5), the chemical distinction of different porphyrin species is quite clear according to the molecule-specific vibrational fingerprints. However, the situation becomes more complicated when these molecules are adsorbed at the step edges. This is because the distribution of relative peak intensities in the TERS spectral profile is quite sensitive to the adsorption configuration and orientation of the molecule on a surface due to the surface selection rule of TERS (Figure 6(a)), resulting in the change of vibrational fingerprints. Figure 6(b) shows how the simulated TERS spectra of ZnTPP change considerably when the structural parameters are changed, with three representative hypothetical configurations of ZnTPP on Ag schematically shown on the right. The fair agreement of the simulated TERS spectrum (bottom pink trace in Figure 6(b)) with the experimental data suggests an adsorption conformation being slightly tilted and bent ($\theta = 5°$, $\alpha = 5°$), which is consistent with the observation of slightly non-uniform brightness for the four-lobe feature shown on the right of Figure 6(b). In fact, even a small change in either the bending (α) or twisting (β) angle can lead to considerable

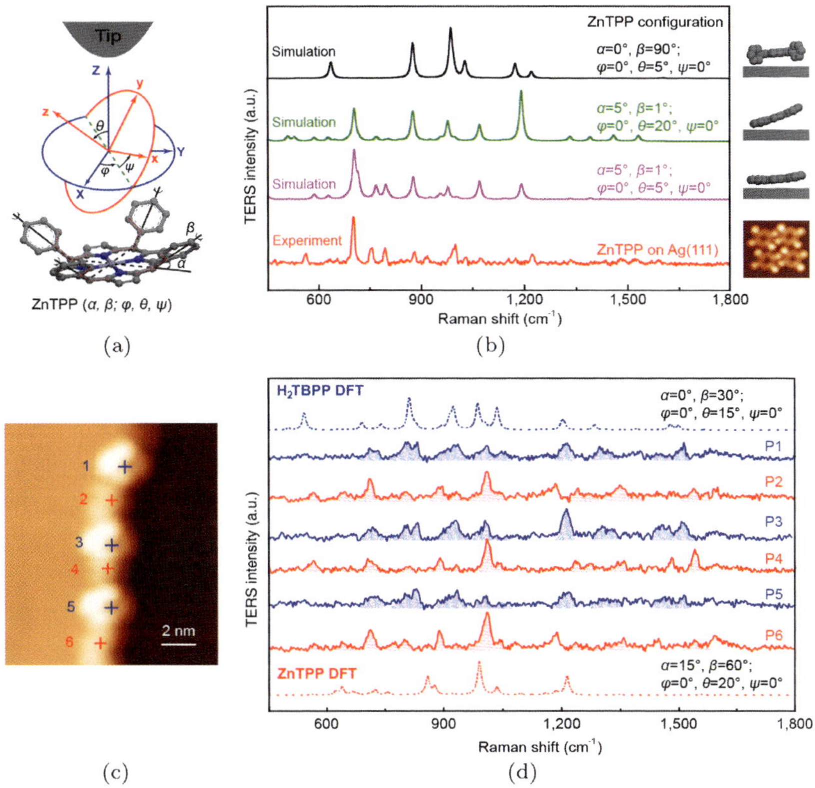

Figure 6. (a) ZnTPP configuration defined in terms of two angles (α, β). Bending angle α quantifies that out-of-plane bending of the phenyl group and twist angle β describe the rotation of the phenyl group about the connecting σ-bond to the porphyrin core. The molecular orientation is defined by the three Euler angles (φ, θ, ψ) according to the molecular coordinates (x, y, z) relative to the STM coordinate (X, Y, Z), with the tilting angle θ being most critical. (b) Calculated TERS spectra of ZnTPP for three representative hypothetical configurations show dramatic spectral vibrations. Also shown at the bottom is the experimental data measured above the ZnTPP molecule on the terrace (-1 V, 1 nA, 30 s, background subtracted). (c) STM topograph of a mixed molecular chain self-assembled at the step edge of Ag(111), with six molecules arranged adjacent to each other alternately (-0.6 V, 3 pA). (d) Corresponding TERS spectra acquired at the positions marked in (c) (0.2 V, 0.8 nA, 10 s, background subtracted). Calculated TERS spectra for tilted molecules are also shown at the top for H_2TBPP (blue: $\alpha = 0°$, $\beta = 30°$, $\varphi = 0°$, $\theta = 15°$, $\psi = 0°$) and at the bottom for ZnTPP (red: $\alpha = 15°$, $\beta = 60°$, $\varphi = 0°$, $\theta = 20°$, $\psi = 0°$) to help resolve the chemical identity of molecules via Raman fingerprints. [Reprinted with permission from Ref. [32]. Copyright (2015) Macmillan Publisher Limited.]

changes in the relative peak intensities (for more details see Section 4). We would like to note that the simulation procedure used in the present work is based on a simplified model without considering molecule–substrate and intermolecular interactions. More sophisticated calculations are needed to improve the agreement with experimental spectra.

Evidently, if a molecule is adsorbed at the step edge that has more active sites associated with dangling bonds, its TERS spectrum is expected to be different to some extent from the one on the terrace because the adsorption configuration there could be quite different. In particular, molecules on step edges are often more tilted than molecules on terraces. As shown in Figure 6(c), both ZnTPP and H_2TBPP molecules co-adsorbed at the step edge, in total six in number, appear to arrange alternately next to one another within the contact regime of van der Waals interactions. Their spectral variations in TERS are nicely demonstrated in Figure 6(d). Remarkably, the TERS spectra measured approximately above the center position of each molecule, despite some similarities, can still be classified into two types — those at positions 1, 3 and 5 (blue curves) forming one type and those at positions 2, 4 and 6 (red curves) representing the other type — with each type itself featuring essentially the same vibrational fingerprints in terms of overall spectral profiles.

In combination with the TERS spectral simulations for both H_2TBPP (top blue, Figure 6(d)) and ZnTPP (bottom red, Figure 6(d)), the molecules at positions 1, 3 and 5 can be clearly distinguished as H_2TBPP while those at positions 2, 4 and 6 are identified as ZnTPP molecules. As expected, the simulated spectra suggest a much more tilted orientation for both molecules, with a much larger tilting angle θ at step edges than flat terraces. Note that the simulation also suggests variations in the adsorption configuration, particularly for the twisting angle β of the phenyl ring. Together with the differences in configuration and orientation, the relative intensities of the Raman peaks are also different. For example, the dominant Raman peak around $700\,\text{cm}^{-1}$ for the nearly flat-lying ZnTPP on the terrace is weakened upon tilting, because this peak is mainly associated with an out-of-plane vibrational mode of the porphyrin core and phenyl ring, whose coupling with the axial plasmonic field is weakened with increased tilting. By contrast, the pronounced Raman peaks in the range of ~880–1,200 cm^{-1} are more related to the in-plane breathing vibrations of both the porphyrin core and phenyl rings. Their coupling with the axial plasmonic field is strengthened with increased molecular tilting, leading to enhanced Raman scattering. A similar principle also applies to the situation for a tilted H_2TBPP molecule adsorbed at the step edge. Note that there is a small mismatch in some spectral positions, especially for ZnTPP molecules, which may be related to the neglecting of molecule–substrate interactions in our simulations. Such interactions could lead to charge-transfer between the molecule and metal substrate, accompanied by a change in the strength of the associated chemical bonds.

4. Theoretical Consideration of Raman Selection Rules and Simulation of Raman Mapping

4.1. *Calculation of spontaneous Raman spectra for H_2TBPP and ZnTPP molecules*

In order to understand the Raman vibrational properties of porphyrin molecules under investigation, we firstly studied the Raman spectra of a single molecule in free space and analyze the vibrational modes corresponding to individual Raman peaks by using DFT calculations via the Gaussian 09 software with the B3LYP/6-31G(d) basis. The H_2TBPP molecule containing 174 atoms is optimized within a C_{2v} point-group symmetry, which consists of 516 vibrational modes that are classified into four irreducible representations, A_1, A_2, B_1 and B_2, namely, $\Gamma_{vib} = 131A_1 + 128A_2 + 128B_1 + 129B_2$. All these modes are Raman-active. For a ZnTPP molecule with 77 atoms and D_{4h} point-group symmetry, it consists of 225 vibrational modes which can be classified into ten irreducible representations: $\Gamma_{vib} = 19A_{1g} + 14A_{2g} + 19B_{1g} + 15B_{2g} + 22E_g + 7A_{1u} + 16A_{2u} + 8B_{1u} + 15B_{2u} + 34E_u$, where only the A_{1g}, B_{1g}, B_{2g} and E_g modes are Raman-active. The calculated Raman spectra for the two free-space porphyrin molecules at the optimized geometry agree quite well with the Raman spectra measured on a molecular powder sample.

4.2. *Calculation of TERS spectra for H_2TBPP and ZnTPP molecules adsorbed on Ag(111)*

When a molecule is adsorbed on the surface, the Raman signals will be modified due to the ordered molecular orientation and the axial polarization of the local field along the tip shaft. In order to understand the TERS spectra measured under the STM configuration, we made DFT calculations for the conventional Raman scattering process, taking into account the influence of the local electric field. The general expression of conventional surface-enhanced Raman scattering (SERS) cross-section for the kth mode of a molecular vibration can be written as

$$I_{\text{SERS}}(\omega_k) \propto \frac{d\sigma_k}{d\Omega} = \frac{\omega_R^4}{16\pi^2\varepsilon_0^2 c^4}|\mathbf{E}_{\text{loc}}(\omega_k) \cdot \hat{R}^{(k)} \cdot \mathbf{E}_{\text{loc}}(\omega_L)|^2, \tag{1}$$

in which $\omega_R = \omega_L - \omega_k$ denotes the frequency of the Stokes–Raman photons, $\hat{R}^{(k)}$ is the Raman tensor, ε_0 is the vacuum permittivity and c is the light speed. The local field $\mathbf{E}_{\text{loc}}(\omega_L)$ [or $\mathbf{E}_{\text{loc}}(\omega_k)$] is enhanced by a factor of $G(\omega_L)$ [or $G(\omega_k)$], which is usually viewed as an orientation-independent quantity in the SERS that involves many hot spots and many molecules with varieties of orientations. Consequently, in the description of conventional SERS signals the configuration and orientation of molecules are averaged and the orientation of the local field is ruled out. However, for a single molecule adsorbed on the surface and positioned into the nanojunction between the metallic tip and substrate, the local field direction and molecular configuration or orientation could influence the final Raman signals dramatically [49]. Therefore, the Raman tensor, instead of the Raman activity, of a molecule is adopted and an enhancement matrix is introduced to describe the orientation-dependent property. The schematic of molecular coordinate (x, y, z) relative to the STM coordinate (X, Y, Z) is shown in Figure 6(a). The molecular orientation can be expressed by Euler angles with three independent parameters: θ, ϕ and ψ. The incident angle and polarization of the incident laser are also considered in the calculation. The TERS intensity can be written as

$$I_{\text{TERS}}(\omega_k) \propto |[\hat{P}\hat{G}(\omega_k)](\hat{T}^{-1}\hat{R}^{(k)}\hat{T})[\hat{G}(\omega_L)\hat{P}]\mathbf{E}_L(\omega_L)|^2, \tag{2}$$

in which $\hat{T}$ is the transformation matrix and $\hat{P}$ is the polarization matrix. The Raman tensor $\hat{R}^{(k)}$ is calculated by the DFT method (Gaussian 09) mentioned above, and the enhancement factor is replaced by a matrix $\hat{G}$. By assuming the local field is enhanced along the Z-direction, the selection rule of TERS can be expressed as

$$I_{\text{TERS}}(\omega_k) \propto G_{II} T_{Ii}^{-1} R_{ij}^{(k)} T_{jJ} G_{JJ} \underset{\hat{T}=\hat{I}}{\overset{I=J=Z}{\Rightarrow}} I_{\text{TERS}} \propto |\alpha_{ZZ}|^2, \quad (3)$$

in which α_{ZZ} is the ZZ component of the Raman tensor, I and J as well as i and j denote the components of STM and molecular coordinates, respectively. According to Equation (3), one can see that the TERS signals would be only concerned with α_{ZZ} if there is no rotation of the molecule (i.e. $\hat{T} = \hat{I}$). During the calculation, the incident angle is chosen as 60° with p-polarization, and the rotation angle θ of the molecule is varied from 0° to 30° while keeping ϕ and ψ unchanged (set to zero in the present calculations). For the four types of vibration modes (A_1, A_2, B_1 and B_2) of a single H_2TBPP molecule that lies horizontally on the surface, only the total symmetric A_1 modes can survive to be Raman-active because only A_1 modes contain α_{zz} components. However, when a molecule is tilted on the surface (like the one on the top of Figure 2(a)), the Raman tensor of other modes (A_2, B_1 and B_2) could also contribute to a new non-zero α_{ZZ}' component that is related to the rotation angles. As a result, more modes could also become visible (see Figure 2(b)). To be more specific, when a H_2TBPP molecule is tilted, the vibrational modes associated with the porphyrin core–plane vibrations will become Raman-active and contribute to the Raman intensity. As shown in Figure 2(b), the more the porphyrin molecule is tilted, the more the core-plane modes will contribute, and thus the larger the Raman intensity will become. Since the low-lying vibrational modes are more localized in the lobe while the high-lying modes above ~1,210 cm^{-1} contain more contributions from the porphyrin core, the intensity of the high-lying Raman peaks above ~1.210 cm^{-1} will be considerably increased, as confirmed experimentally by TERS spectra shown in Figure 2(a).

For ZnTPP molecules, we can obtain similar results. The TERS spectra are very sensitive to different adsorption configurations and orientations of a ZnTPP molecule. Even a small change in the structural parameter of α, β, or θ can result in noticeable changes in the relative peak intensities, sometimes even making the spectral profile look different although the peak positions remain essentially the same. According to our simulations for the observed TERS spectra (the pink spectrum in Figure 6(b)), the ZnTPP molecule appears to adsorb on the terrace of Ag(111) in a relatively flat conformation. Such adsorption structure is quite different from the saddle-shape conformation of the macrocycle with much larger dihedral angles between the peripheral phenyl rings and the porphyrin planes, previously reported by Buchner *et al.* for similar porphyrin molecules (namely H_2TPP, FeTPP and CoTPP) on Ag(111) [50]. We performed TERS simulations based on the saddle-shape configurations reported by Buchner *et al.* (with the central atoms simply replaced by Zn atoms). But the simulated TERS spectra are quite similar to the spectrum for a hypothetical configuration with a large dihedral angle (the black spectrum in Figure 6(b)), which are evidently quite different from the TERS spectra experimentally observed in our system.

Nevertheless, the molecules Buchner *et al.* used are different from ours and the situation there could also be different. In addition, we would like to note that the simulation procedure used in the present work is based on a simplified model without considering molecule-substrate and intermolecular interactions. All the simulated TERS spectra given here are still far from perfect in light of its agreement with the experimental spectrum. But, in principle, one can write a sophisticated iteration program to fit the simulated Raman spectrum to the experimental data by fine tuning each parameter in the adsorption configuration and orientation of molecules, preferably starting from an optimized configuration that is obtained via sophisticated theoretical calculations taking into account the interaction of molecules with substrates. In this sense, the simulated TERS spectra presented here can merely serve as an example to highlight such fitting idea. More sophisticated theories are needed to correlate the observed TERS spectra with precise adsorption conformations of molecules on a surface. An advanced program package for TERS spectral fitting is highly demanding but desirable and remains to be developed.

4.3. *Simulation of TERS spectral mapping for a single H_2TBPP molecule on Ag(111)*

In the calculation described above for the TERS spectra, the molecule is excited by a plane wave and the molecule is treated as a whole piece (or a point dipole) to contribute to the Raman signals. In addition, the local field is also treated as a uniform beam by adding the direction and enhancement factor to the exciting field. However, in order to simulate the TERS mapping results, this plane wave should be replaced by a Gaussian-shape field with a spatial distribution in order to resolve the sub-molecular details. In other words, both the enhancement factor matrix G and the Raman tensor of the molecule should be position-dependent. This is fulfilled by replacing the normal coordinate by Cartesian coordinates. Upon the consideration above, the TERS intensity for Equation (3) can be rewritten as

$$I_{\text{TERS}}(\omega_k, \mathbf{X}) \propto \left| \sum_{n=1}^{3N} \left[\hat{P}\hat{G}\left(\omega_k, \mathbf{X}, \hat{T}\mathbf{x}^{(n)}\right) \right] \left(\hat{T}^{-1} \left[\frac{\varphi_k^{(n)}}{\sqrt{\mu_k}} \left(\frac{\partial \hat{\alpha}}{\partial \mathbf{x}^{(n)}} \right) \hat{T} \right] \left[\hat{G}\left(\omega_L, \mathbf{X}, \hat{T}\mathbf{x}^{(n)}\right) \hat{P}\mathbf{E}_L(\omega_L) \right] \right) \right|^2, \tag{4}$$

in which $\mathbf{X}$ is the position of the tip under the STM coordinate and $\mathbf{x}^{(n)}$ is the coordinate of the nth atom under the molecular coordinate. By tuning the full-width-at-half-maximum (FWHM) of the Gaussian distribution, i.e. the spatial resolution of the local field, different Raman mapping images corresponding to different Raman modes can be obtained. For the FWHM value varying from 0.6 nm to 2.0 nm, the four-lobe features of the single molecule become more and more blurred and eventually become a featureless bright spot. The simulated Raman image shows a reasonably good agreement with the experimental TERS mapping (Figure 3(b)) when the FWHM value is chosen as 1 nm (in other words, when the local field is highly confined). The requirement of such a highly confined local field for achieving an ultrahigh spatial resolution is also reported by other researchers [51–54], which suggests the possible involvement of additional mechanisms to further confine the local field. In fact, by probing the dependence of TERS spectra on the tunneling current and laser power, certain nonlinear optical processes appear to be involved in the TERS experiments demonstrated here, which also implies a highly localized response when the tip is brought very close to the substrate. The description of such nonlinear properties and their influence on the ultrahigh spatial resolution of TERS will be presented in the next section in detail.

5. Mechanism of High Spatial Resolution and Analysis of Nonlinear Processes

5.1. *Spectral matching in STM-controlled TERS*

The mechanism behind the sub-nanometer spatial resolution of the STM-controlled TERS is certainly related to the highly localized plasmonic field at the tip apex [51–54], but the exact connection and physical picture are still not completely clear. Experimentally, it is found that the spectral matching between the nanocavity plasmon resonance and the downward transitions associated with emission of Raman photons is crucial for enhancing and stabilizing the inelastic scattering signals.

In Figure 7(a), we show two TERS spectra (red) of H_2TBPP molecule obtained from the "on-resonance" and "off-resonance" conditions defined below. "On-resonance" here refers to good spectral matching between the nanocavity plasmon resonance (blue curves) and the downward molecular vibronic transition of $Q_y(1,1)$. The latter has an energy similar to that of the $Q_y(0,0)$ band measured by the photoluminescence excitation (PLE) technique (see brown curve at the bottom of Figure 7(a) and the schematic of molecular vibronic transitions in Figure 7(b)). In contrast, "off-resonance" refers to the situation in which the nanocavity plasmon resonance does not match well with the $Q_y(1,1)$ band, but does match with the laser line. We

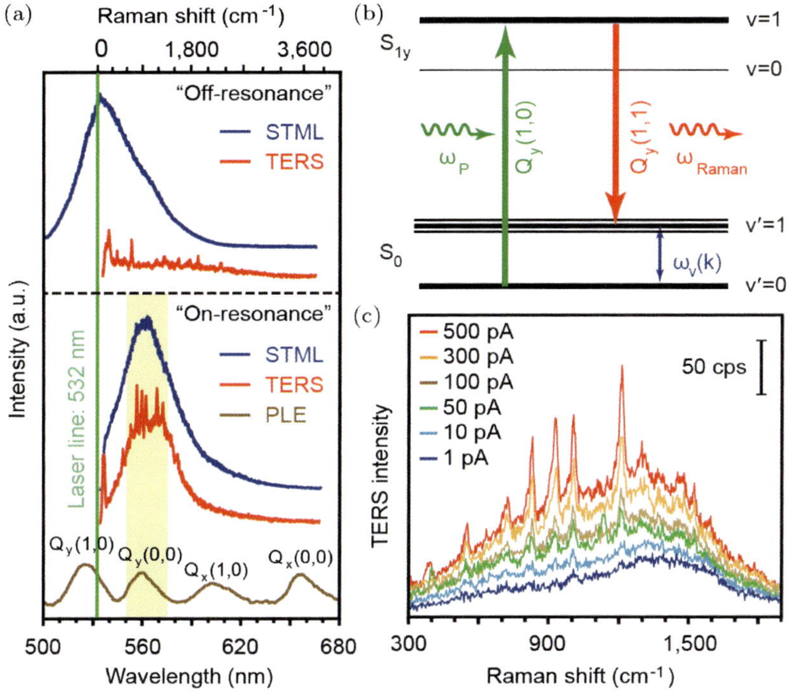

Figure 7. (a) Dependence of TERS spectra (red, 200 mV, 1 nA, 3 s) on the spectral matching among the laser line (green), nanocavity plasmon resonance (blue), and molecular vibronic transitions (brown). The lower panel shows the "on-resonance" situation with the nanocavity plasmon resonance matching the downward molecular vibronic transition $Q_y(0,0)$. The nanocavity plasmon profile (blue) is determined by STM-induced luminescence (STML) for the Ag tip on Ag(111) (2.8 V, 0.1 nA, 10 s), whereas the adsorption spectrum of H_2TBPP (brown) is measured by photoluminescence excitation. The upper panel shows the "off-resonance" situation in which the nanocavity plasmon resonance does not match the downward $Q_y(0,0)$ transition. The scale on the top axis shows the Raman shifts for the TERS measurements. (b) Schematic representation of the optical transitions involved in the Raman process between the ground state S_0 and excited state S_1, in which ω_P is the frequency of the incident pumping laser, ω_{Raman} is the frequency of Raman photons, and $\omega_v(k)$ is the vibrational frequency of the k_{th} mode of the molecule; v and v' stand for the vibrational levels of excited states and ground states, respectively. (c) TERS spectra on the molecular island as a function of tunneling currents (120 mV, 3 s). [Reprinted with permission from Ref. [31]. Copyright (2013) Macmillan Publisher Limited.]

note that for H_2TBPP molecules, the laser line at 532 nm is always resonant with the upward molecular vibronic $Q_y(1,0)$ transition, but under the "off-resonance" condition, the TERS spectrum shows very few observable spectral features. In contrast, a dramatic enhancement of the Raman signals can be observed under the "on-resonance" condition even under a small incident laser flux of about 10^2 W/cm^2. The Raman vibrational modes appear as sharp spectral features on top of a smooth continuum, whose profile matches very well with the nanocavity plasmon profile.

In other words, under the "on-resonance" condition, the broadband nature of the nanocavity plasmon field allows us to realize an efficient doubly resonant TERS process: not only producing sufficiently strong resonant excitation through the spectral overlap between the nanocavity plasmon shoulder, the laser line, and the upward molecular vibronic transitions, but, more importantly, also generating large resonant emission enhancement owing to the good spectral matching of the nanocavity plasmon resonance to the downward molecular vibronic transitions. It is worth mentioning that thanks to the large enhancement, the photon flux used for the "on-resonance" TERS experiments is about one to two order of magnitude smaller than what has been reported to date ($\sim 10^3$–10^4 W/cm^2) [16, 34, 42, 43], thus guaranteeing the stability and sustainability of the single molecule during the TERS spectral and mapping measurements.

Moreover, the broad continuum and associated TERS signals are found to be enhanced dramatically when tunneling currents are increased (Figure 7(c)) [42], which suggests a highly sensitive dependency of both signals on the gap distance (that is, on the local-field enhancement) [45]. We would like to note that in our spectral-matching TERS, the nanocavity plasmon resonance mode is tuned mainly by modifying the tip status [41] while the strength of the nanocavity plasmon mode is controlled by the gap distance set by the tunneling condition, particularly by the tunneling current. A sufficiently intense nanocavity plasmon field associated with a relatively short gap distance is important to yield pronounced TERS signals.

5.2. *An "analogue" to the broadband femtosecond stimulated Raman scattering*

However, as shown in Figure 7(a), the proper setting of the nanocavity plasmon resonance to satisfy the "on-resonance" condition appears much more critical to achieve large enhancement and high signal level. These observations correlate directly with how the nanocavity plasmon field is involved in the Raman scattering process. The overall spectral profile of the "on-resonance" TERS spectrum strikingly resembles that in the broadband femtosecond stimulated Raman scattering (SRS) process (Figures 8(a) and 8(b)) [55–59]. In addition, we also observed a nonlinear relationship between the TERS response and the incident laser power (see Figures 8(c) and 8(d)). Such nonlinear power dependency, together with the spectral profile similarity to the broadband stimulated Raman scattering process, suggests that our spectral-matching TERS can be viewed as an analogue to the third-order nonlinear stimulated Raman scattering process [55], providing both enhanced signals and improved spatial resolution. In other words, although our TERS experiments were performed using a single beam of continuous-wave laser as the excitation source, the laser beam shining into the STM nanocavity can generate not only a highly enhanced local excitation field that has the same frequency as the incident light (i.e. via elastic scattering), but also a pronounced broadband emission from the nanocavity whose spectral profile reflects the nanocavity plasmon (NCP) resonance mode (via inelastic scattering). The latter can also be confirmed and monitored *in situ* by STM-induced luminescence (STML) technique [41]. The local field associated with such broadband plasmonic emission (or Stokes photons) is highly confined in space (and also femtoseconds in lifetime) and can serve as a microscopic analogue to the Raman probe source used in the conventional stimulated Raman scattering process, thus creating a plasmon-stimulated nonlinear Raman scattering scenario. Due to the additional higher-order dependence of the TERS signals on the strength and gradient of the highly localized plasmonic field, the signals obtained from such nonlinear Raman scattering are expected to be dramatically enhanced with improved spatial resolution as well.

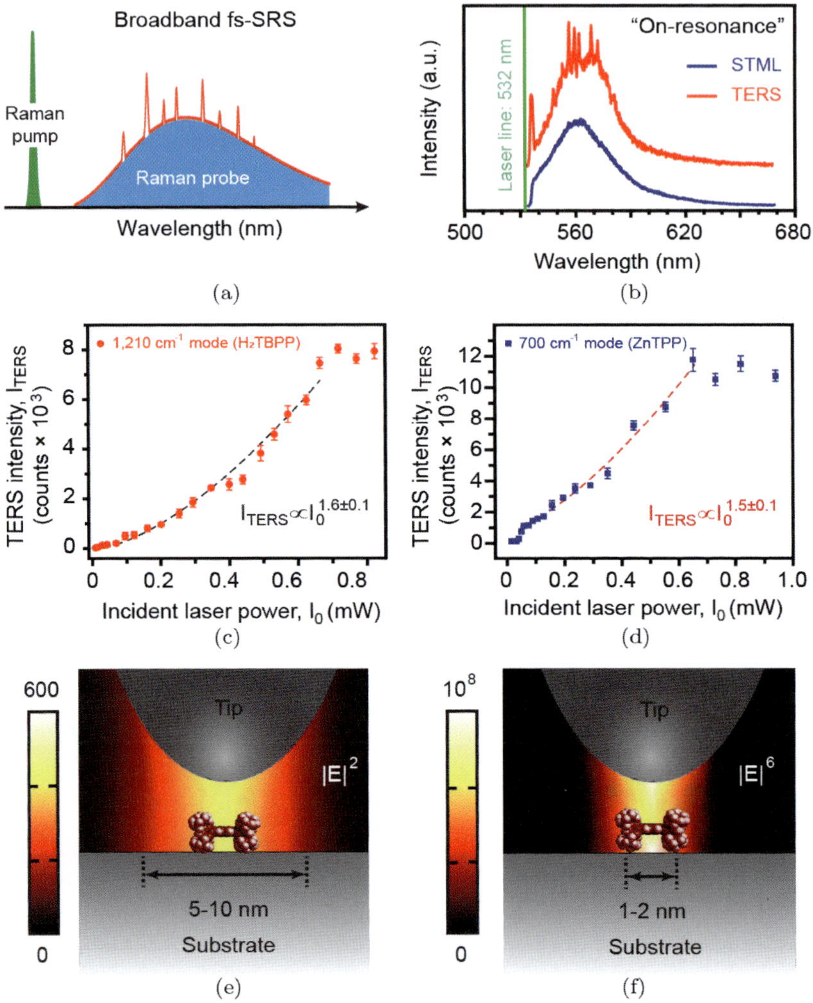

Figure 8. (a) A schematic draw of the broadband femtosecond stimulated Raman scattering (fs-SRS) [55–59]. (b) Typical spectral profiles under the "on-resonance" condition. The TERS spectrum shows a striking similarity to that of the fs-SRS process. (c) and (d) The dependence of the TERS signals (I) on the incident laser power (I_0) for one of the dominant Raman peaks at $1{,}210\,\mathrm{cm}^{-1}$ (red circles) for H_2TBPP (c) and $700\,\mathrm{cm}^{-1}$ (blue squares) for ZnTPP (d). The fitted curve by a power-law dependency is also shown as dashed curves. (e) and (f) Schematic drawing of simulated electric-field distributions for $|\mathbf{E}|^2$ and $|\mathbf{E}|^6$, respectively. The molecule is simply stacked over the simulation result, which was obtained without considering the role of the molecule. The tip radius in the plot is not scaled with the size of the molecule and the gap separation for the sake of

Figure 8(c) shows the dependence of the TERS signals (I) of H_2TBPP molecule on the incident laser power (I_0) for one of the dominant Raman peaks at 1,210 cm^{-1} (red circles). The data can be nicely fitted by an overall power-law dependency (black dotted curve: $I = kI_0^n$, $k = 1.3(1) \times 10^4$, $n = 1.6(1)$). For the ZnTPP molecule, the dependence of TERS signals on the power of incident laser also indicates a similar nonlinear behavior. A power-law fitting gives a nonlinear exponent of ~1.5(1) for the net TERS signals of the dominant 700 cm^{-1} Raman peak (as shown in Figure 8(d)). Note that the power dependency of the Raman signals has been studied in SERS previously by the Kneipp group [60] and Itoh group [61]. In Figures 8(c) and 8(d), the TERS signals appear to approach saturation at high incident power above ~0.7 mW, which suggests the quantum saturation of the level population of the single molecule. This is probably unique to the single-molecule study since the strong local field generated by the pumping laser can easily excite the single molecule to populate at excited vibronic states. The schematic drawing of simulated electric-field distributions for $|\mathbf{E}|^2$ and $|\mathbf{E}|^6$ is plotted in Figures 8(e) and 8(f), respectively, to show the evolution of the spatial extent of the confined plasmonic fields, which are calculated with the finite element analysis method.

As illustrated in the fitting curve of Figures 8(c) and 8(d), the occurrence of a power-law exponent around 1.5–1.6 strongly suggests a nonlinear behavior and the involvement of high-order Raman processes. In the Raman scattering process, such nonlinear relation can be approximately treated as a combination of the linear spontaneous Raman emission and a third-order stimulated Raman process. In other words, in the context of our experimental setup with only a single continuous-wave laser pumped into

Figure 8. (*continued*) "artistic" presentation. The actual calculations were carried out for a model system with the following parameters: the tip radius is 25 nm, the gap distance is 2 nm, silver is selected as the material for both tip and substrate with the dielectric constant from the literature. A plane wave with a wavelength of 532 nm is incident at an angle of 60°. [Reprinted with permission from Ref. [31]. Copyright (2013) Macmillan Publisher Limited.]

the STM junction, stimulated scattering processes can occur only after sufficiently strong spontaneous Raman photons are generated. Indeed, a fitting according to such a proposition (i.e. $I = aI_0 + bI_0^3$) would give a set of fitting coefficients of $a = 0.58(4)$ and $b = 1.7(1)$ for H_2TBPP as well as $a = 1.1(1)$ and $b = 1.0(2)$ for ZnTPP molecules, suggesting considerable contributions of the third-order term to the TERS signals (particularly at higher incident power). Therefore, in terms of the nonlinear power-dependence relationship observed, it appears safe to describe our spectral-matching TERS as an "analogue" to the broadband femtosecond stimulated Raman scattering process, in which the incident laser acts as a Raman pump source and the NCP field generated *in situ* by the laser as an "analogue" to the Raman probe source. Of course, what we observed is not a conventional stimulated Raman scattering that uses two laser sources in the standard nonlinear optics [55–59, 62]. In our case, these two sources are always correlated, and it is the NCP field generated by the pumping laser that acts as an analogue to the Stokes field for the possible stimulated Raman emission. Since the broadband NCP field is naturally highly confined in space, one can expect greatly improved spatial resolution, apart from the large enhancement offered by the nonlinear optical process.

It is worth noting that an exponential dependence of Raman gain signals on the pump power was previously reported by McCamant *et al.* in conventional broadband stimulated Raman scattering process [63]. However, unlike the conventional stimulated Raman scattering process where two independent laser sources are used, in our TERS setup, as described above, the plasmonic probe source is generated *in situ* by the laser pump source and thus these two fields are always closed related to each other, namely, $E_{\text{probe}} \propto E_{\text{pump}}$. In other words, no Raman probe signals can be detected if there is no Raman pump source. Therefore, we can only measure directly the TERS signals rather than Raman gain signals. In this case, if we view the third-order response of the polarization dipole of a single molecule as $p \propto \chi_R^3 E_{\text{probe}} E_{\text{pump}}^2$ [57, 64], the intensity of emitted Raman photons could be expressed as $I \propto |p|^2 \propto E_{\text{probe}}^2 E_{\text{pump}}^4 = I_{\text{probe}} I_{\text{pump}}^2$. Taking into account the relation between E_{probe} and E_{pump}, we can have $I_R \propto I^3_{\text{pump}}$. As a result, the final Raman signals could show a nonlinear dependence with the pump light intensity. In this sense, an observed power exponent of 1.5–1.6 suggests that the process involved is not strictly nonlinear.

5.3. *Role of the tunneling current in the STM-controlled TERS and gap-distance dependency of TERS signals*

We note that inelastic electron tunneling spectroscopy in cryogenic STM can also provide information about certain vibrational modes of a single molecule through inelastic electron excitation [65]. By contrast, the STM-controlled TERS provides full vibrational fingerprints of molecules through the excitation of optical fields alone. Localized tunneling electrons do not contribute to the TERS signals because the vibrational fingerprints of the H_2TBPP molecule show up in the spectrum even when the bias voltage is well below the excitation threshold of vibrational modes.

However, the increase of TERS signals with increased currents shown in Figure 7(c) indicates that there exists certain relation between the TERS signal intensity and the current. In order to examine the role of tunneling electrons in the TERS process, we performed feedback-controlled TERS experiments for a fixed gap distance while reducing the current from a certain value (e.g. 1 nA) to almost zero. This is realized by suddenly changing the bias (e.g. 120 mV) to a very small value (e.g. 1 mV) while the feedback loop is switched off simultaneously. As shown in Figure 9, the current is changed by about 300 times (namely, from 1 nA to ~3 pA), but there are no noticeable changes in both the overall TERS spectral profile and the intensities of individual peaks. In other words, the tunneling current itself (or electrons) does not contribute directly to the generation of TERS signals. In addition, the bias voltage of 1 mV applied under the feedback-off condition is well below the excitation threshold of any molecular vibrational modes. As a result, the tunneling electrons set by the bias cannot vibrationally excite any molecules. All these observations strongly suggest that the tunneling electrons only play a role of controlling the gap distance and resultant

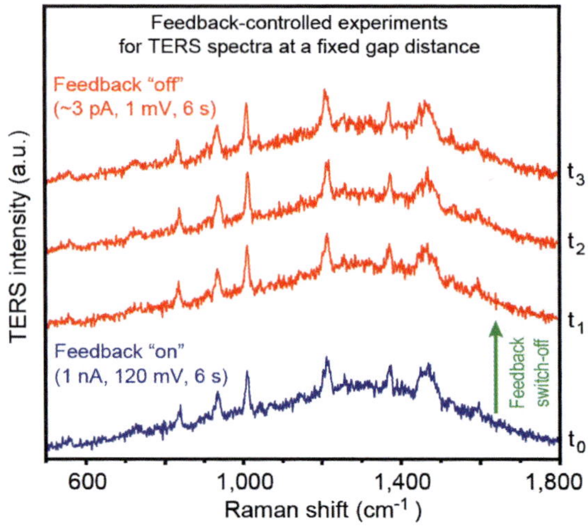

Figure 9. Feedback-controlled TERS spectra with and "without" tunneling currents. The spectrum in blue was acquired with the feedback "on" for a predetermined gap distance at 1 nA and 120 mV. The spectra in red were acquired at the same gap distance right after switching off the feedback loop and changing the bias to 1 mV simultaneously. The spectral feature remains essentially the same before and after feedback-off, and even stays unchanged from t_1 to t_3 for more than ten seconds, which is a good indication of the tip and junction stability in our experiments. [Reprinted with permission from Ref. [31]. Copyright (2013) Macmillan Publisher Limited.]

local-field strength and have no direct contributions to the TERS signals.

We would also like to note that the NCP field excited by electrons in the STM junction also appears not capable of generating TERS signals. In our STM-controlled TERS experiments, the NCP field can be produced in two ways: one is directly from the incident optical laser, and the other is from the inelastic excitation by tunneling electrons if the bias voltage is sufficiently high to excite the plasmon resonance [66, 67]. The quantum cutoff relation has to be satisfied for the plasmonic excitation. Since the plasmon resonance is peaked around 2.2 eV in the present system, a bias voltage above 2.2 is required in order to electrically generate the plasmonic resonance around 2.2 eV. Most of the TERS spectra given in this work were acquired at a very small bias voltage (e.g. around 120 mV), which means that the molecules are excited by the optically generated plasmonic fields. However, even when the bias voltage is abruptly increased from a low bias (e.g. 120 mV) to a high value (e.g. 2.5 V) but with the gap distance kept fixed by switching off the feedback loop simultaneously, the additional NCP field generated by the large tunneling current (more than 10 nA in this case) does not help to enhance the TERS signals at all despite an evident increase in the intensity of the NCP "background". Therefore, it is the NCP field generated by the pumping laser, rather than that by the tunneling electrons, that enhances significantly the TERS signals.

As stated above, when the STM feedback is "on", the tunneling current (I) can be used to control the gap distance (d), allowing us to measure the distance dependence of TERS signals, as shown in Figures 10(a) and 10(b) for ZnTPP molecules on Ag(111). An exponential fitting for the net peak intensity at the 700 cm^{-1} Raman mode gives a small decay length of ~110(8) pm for the TERS signals. Note that this number is only about three times of the decay length for the gap-distance dependency of the tunneling current itself (~42 pm, as shown in Figure 10(c)). In other words, the nonlinear TERS signals

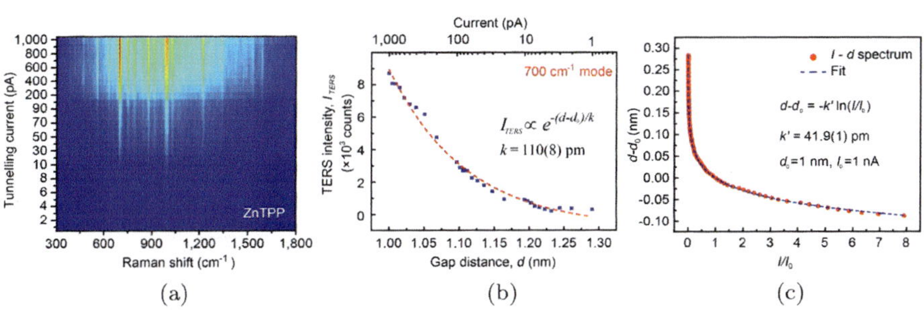

Figure 10. Dependence of TERS signals on the gap distance. (a) Tunneling current dependency of TERS spectra over 1–1000 pA on the ZnTPP molecular island (0.1 V, 10 s, 0.45 mW). (b) Dependence of the net TERS signals (I_{TERS}) on the gap distance (d) for the dominant Raman mode at ~700-cm^{-1} shown in (a). The distance d is estimated according to the measured I–d tunneling spectrum assuming an initial reference value of 1 nm (d_0) for the set-point of 0.1 V and 1 nA. (c) I–d tunneling spectrum on the ZnTPP island for a fixed bias voltage of 0.1 V during laser illumination. [Reprinted with permission from Ref. [32]. Copyright (2015) Macmillan Publisher Limited.]

are highly localized too, almost comparable to the localization of tunneling currents. In a phenomenological sense, such an ultrasensitive nature of the TERS signals to the gap distance is likely the key factor for the sub-nanometer spatial resolution experimentally observed.

6. Summary and Outlook

By using STM-controlled nonlinear TERS techniques that exploit the broadband, confinement, and enhancement features of the nanocavity plasmonic field, we have demonstrated sub-nanometer resolved single-molecule Raman mapping and chemical distinguishing of two adjacent different molecules on surfaces in real space. In combination with theoretical simulations, TERS data can also provide information on molecular adsorption configurations and orientations. In this regard, more sophisticated calculations that are based on optimized configurations taking into account the molecule–substrate and inter-molecular interactions are still needed to improve the agreement of theoretical simulations with the experimental spectra. Furthermore, the exact mechanism of sub-nanometer spatial resolution and its relation to the nonlinear behavior observed still remain to be explored. The ability to access the structure and conformation of a single molecule with both chemical recognition and sub-nanometer resolution by optical means provides a new potential to explore the nanometer-scale world, offering new ways to design, control and engineer the functionality of molecules on demand. Such capability should substantially affect the fields of nanophotonics, molecular nanotechnology, surface science, bio-science, and materials science, in which identifying surface species with single-molecule resolution is important. The application of such technique is very broad: catalysis, photochemistry, DNA sequencing, and protein folding are just a few examples in a very long list.

References

1. Klar, T. A. and Hell, S. W. (1999). Subdiffraction resolution in far-field fluorescence microscopy. *Opt. Lett.* **24**(14), pp. 954–956.
2. Rittweger, E., Han, K. Y., Irvine, S. E., Eggeling, C. and Hell, S. W. (2009). STED microscopy reveals crystal colour centres with nanometric resolution. *Nat. Photonics* **3**(3), pp. 144–147.

3. Rust, M. J., Bates, M. and Zhuang, X. (2006). Sub-diffraction-limit imaging by stochastic optical reconstruction microscopy (STORM). *Nat. Methods* **3** (10), pp. 793–796.
4. Huang, B., Wang, W., Bates, M. and Zhuang, X. (2008). Three-dimensional super-resolution imaging by stochastic optical reconstruction microscopy. *Science* **319**(5864), pp. 810–813.
5. Betzig, E., Patterson, G. H., Sougrat, R., Lindwasser, O. W., Olenych, S., Bonifacino, J. S., Davidson, M. W., Lippincott-Schwartz, J. and Hess, H. F. (2006). Imaging intracellular fluorescent proteins at nanometer resolution. *Science* **313**(5793), pp. 1642–1645.
6. Dickson, R. M., Cubitt, A. B., Tsien, R. Y. and Moerner, W. (1997). On/off blinking and switching behaviour of single molecules of green fluorescent protein. *Nature* **388**(6640), pp. 355–358.
7. Huth, F., Govyadinov, A., Amarie, S., Nuansing, W., Keilmann, F. and Hillenbrand, R. (2012). Nano-FTIR absorption spectroscopy of molecular fingerprints at 20 nm spatial resolution. *Nano Lett.* **12**(8), pp. 3973–3978.
8. Pollard, B., Muller, E. A., Hinrichs, K. and Raschke, M. B. (2014). Vibrational nano-spectroscopic imaging correlating structure with intermolecular coupling and dynamics. *Nat. Commun.* **5**, p. 3587.
9. Anderson, M. S. (2000). Locally enhanced Raman spectroscopy with an atomic force microscope. *Appl. Phys. Lett.* **76**(21), pp. 3130–3132.
10. Hayazawa, N., Inouye, Y., Sekkat, Z. and Kawata, S. (2000). Metallized tip amplification of near-field Raman scattering. *Opt. Commun.* **183**(1), pp. 333–336.
11. Pettinger, B., Picardi, G., Schuster, R. and Ertl, G. (2000). Surface enhanced Raman spectroscopy: Towards single moleculer spectroscopy. *Electrochem.-Tokyo* **68**(12), pp. 942–949.
12. Stöckle, R. M., Suh, Y. D., Deckert, V. and Zenobi, R. (2000). Nanoscale chemical analysis by tip-enhanced Raman spectroscopy. *Chem. Phys. Lett.* **318**(1), pp. 131–136.
13. Bharadwaj, P., Deutsch, B. and Novotny, L. (2009). Optical antennas. *Adv. Opt. Photon.* **1**(3), pp. 438–483.
14. Mauser, N. and Hartschuh, A. (2014). Tip-enhanced near-field optical microscopy. *Chem. Soc. Rev.* **43**(4), pp. 1248–1262.
15. Domke, K. F., Zhang, D. and Pettinger, B. (2006). Toward Raman fingerprints of single dye molecules at atomically smooth Au (111). *J. Am. Chem. Soc.* **128**(45), pp. 14721–14727.
16. Neacsu, C. C., Dreyer, J., Behr, N. and Raschke, M. B. (2006). Scanning-probe Raman spectroscopy with single-molecule sensitivity. *Phys. Rev. B* **73**(19), p. 193406.
17. Steidtner, J. and Pettinger, B. (2008). Tip-enhanced Raman spectroscopy and microscopy on single dye molecules with 15 nm resolution. *Phys. Rev. Lett.* **100**(23), p. 236101.
18. Liu, Z., Ding, S.-Y., Chen, Z.-B., Wang, X., Tian, J.-H., Anema, J. R., Zhou, X.-S., Wu, D.-Y., Mao, B.-W. and Xu, X. (2011). Revealing the molecular structure of single-molecule junctions in different conductance

states by fishing-mode tip-enhanced Raman spectroscopy. *Nat. Commun.* **2**, p. 305.

19. Sonntag, M. D., Klingsporn, J. M., Garibay, L. K., Roberts, J. M., Dieringer, J. A., Seideman, T., Scheidt, K. A., Jensen, L., Schatz, G. C. and Van Duyne, R. P. (2011). Single-molecule tip-enhanced Raman spectroscopy. *J. Phys. Chem. C* **116**(1), pp. 478–483.
20. Klingsporn, J. M., Jiang, N., Pozzi, E. A., Sonntag, M. D., Chulhai, D., Seideman, T., Jensen, L., Hersam, M. C. and Duyne, R. P. V. (2014). Intramolecular insight into adsorbate–substrate interactions via low-temperature, ultrahigh-vacuum tip-enhanced Raman spectroscopy. *J. Am. Chem. Soc.* **136**(10), pp. 3881–3887.
21. Anderson, N., Hartschuh, A., Cronin, S. and Novotny, L. (2005). Nanoscale vibrational analysis of single-walled carbon nanotubes. *J. Am. Chem. Soc.* **127**(8), pp. 2533–2537.
22. Yano, T.-A., Verma, P., Saito, Y., Ichimura, T. and Kawata, S. (2009). Pressure-assisted tip-enhanced Raman imaging at a resolution of a few nanometres. *Nat. Photon.* **3**(8), pp. 473–477.
23. Chen, C., Hayazawa, N. and Kawata, S. (2014). A 1.7 nm resolution chemical analysis of carbon nanotubes by tip-enhanced Raman imaging in the ambient. *Nat. Commun.* **5**, p. 3312.
24. Domke, K. F., Zhang, D. and Pettinger, B. (2007). Tip-enhanced Raman spectra of picomole quantities of DNA nucleobases at Au (111). *J. Am. Chem. Soc.* **129**(21), pp. 6708–6709.
25. Ichimura, T., Watanabe, H., Morita, Y., Verma, P., Kawata, S. and Inouye, Y. (2007). Temporal fluctuation of tip-enhanced Raman spectra of adenine molecules. *J. Phys. Chem. C* **111**(26), pp. 9460–9464.
26. Bailo, E. and Deckert, V. (2008). Tip-enhanced Raman spectroscopy of single RNA strands: Towards a novel direct-sequencing method. *Angew. Chem. Int. Ed.* **47**(9), pp. 1658–1661.
27. Zhang, D., Domke, K. F. and Pettinger, B. (2010). Tip-enhanced Raman spectroscopic studies of the hydrogen bonding between adenine and thymine adsorbed on Au (111). *Chem. Phys. Chem.* **11**(8), pp. 1662–1665.
28. Treffer, R., Bhme, R., DeckertGaudig, T., Lau, K., Tiede, S., Lin, X. and Deckert, V. (2012). Advances in TERS (tip-enhanced Raman scattering) for biochemical applications. *Biochem. Soc. Trans.* **40**(4), p. 609–614.
29. Najjar, S., Talaga, D., Schué, L. O., Coffinier, Y., Szunerits, S., Boukherroub, R., Servant, L., Rodriguez, V. and Bonhommeau, S. B. (2014). Tip-enhanced Raman spectroscopy of combed double-stranded DNA bundles. *J. Phys. Chem. C* **118**(2), pp. 1174–1181.
30. Deckert-Gaudig, T., Kämmer, E. and Deckert, V. (2012). Tracking of nanoscale structural variations on a single amyloid fibril with tip - enhanced Raman scattering. *J. Biophoton.* **5**(3), pp. 215–219.
31. Zhang, R., Zhang, Y., Dong, Z., Jiang, S., Zhang, C., Chen, L., Zhang, L., Liao, Y., Aizpurua, J. and Luo, Y. E. (2013). Chemical mapping of a single molecule by plasmon-enhanced Raman scattering. *Nature* **498**(7452), pp. 82–86.

32. Jiang, S., Zhang, Y., Zhang, R., Hu, C., Liao, M., Luo, Y., Yang, J., Dong, Z. and Hou, J. (2015). Distinguishing adjacent molecules on a surface using plasmon-enhanced Raman scattering. *Nat. Nanotechnol.* **10**(10), p. 865–869.
33. Ichimura, T., Fujii, S., Verma, P., Yano, T., Inouye, Y. and Kawata, S. (2009). Subnanometric near-field Raman investigation in the vicinity of a metallic nanostructure. *Phys. Rev. Lett.* **102**(18), p. 186101.
34. Stadler, J., Schmid, T. and Zenobi, R. (2010). Nanoscale chemical imaging using top-illumination tip-enhanced Raman spectroscopy. *Nano Lett.* **10**(11), pp. 4514–4520.
35. Treffer, R., Lin, X., Bailo, E., Deckert-Gaudig, T. and Deckert, V. (2011). Distinction of nucleobases–a tip-enhanced Raman approach. *Beilstein J. Nanotechnol.* **2**(1), pp. 628–637.
36. Berweger, S., Neacsu, C. C., Mao, Y., Zhou, H., Wong, S. S. and Raschke, M. B. (2009). Optical nanocrystallography with tip-enhanced phonon Raman spectroscopy. *Nat. Nanotechnol.* **4**(8), pp. 496–499.
37. Berweger, S. and Raschke, M. B. (2010). Signal limitations in tip-enhanced Raman scattering: the challenge to become a routine analytical technique. *Anal. Bioanal. Chem.* **396**(1), pp. 115–123.
38. Aizpurua, J., Hoffmann, G., Apell, S. and Berndt, R. (2002). Electromagnetic coupling on an atomic scale. *Phys. Rev. Lett.* **89**(15), p. 156803.
39. Wang, X., Liu, Z., Zhuang, M.-D., Zhang, H.-M., Wang, X., Xie, Z.-X., Wu, D.-Y., Ren, B. and Tian, Z.-Q. (2007). Tip-enhanced Raman spectroscopy for investigating adsorbed species on a single-crystal surface using electrochemically prepared Au tips. *Appl. Phys. Lett.* **91**(10), p. 101105.
40. Zhang, C., Gao, B., Chen, L., Meng, Q., Yang, H., Zhang, R., Tao, X., Gao, H., Liao, Y. and Dong, Z. (2011). Fabrication of silver tips for scanning tunneling microscope induced luminescence. *Rev. Sci. Instrum.* **82**(8), p. 083101.
41. Dong, Z., Zhang, X., Gao, H., Luo, Y., Zhang, C., Chen, L., Zhang, R., Tao, X., Zhang, Y. and Yang, J. (2010). Generation of molecular hot electroluminescence by resonant nanocavity plasmons. *Nat. Photon.* **4**(1), pp. 50–54.
42. Pettinger, B., Schambach, P., Villagómez, C. J. and Scott, N. (2012). Tip-enhanced Raman spectroscopy: near-fields acting on a few molecules. *Annu. Rev. Phys. Chem.* **63**, pp. 379–399.
43. Jiang, N., Foley, E., Klingsporn, J., Sonntag, M., Valley, N., Dieringer, J., Seideman, T., Schatz, G., Hersam, M. and Van Duyne, R. (2012). Observation of multiple vibrational modes in ultrahigh vacuum tip-enhanced Raman spectroscopy combined with molecular-resolution scanning tunneling microscopy. *Nano Lett.* **12**(10), pp. 5061–5067.
44. Dong, Z.-C., Guo, X.-L., Trifonov, A., Dorozhkin, P., Miki, K., Kimura, K., Yokoyama, S. and Mashiko, S. (2004). Vibrationally resolved fluorescence from organic molecules near metal surfaces in a scanning tunneling microscope. *Phys. Rev. Lett.* **92**(8), p. 086801.
45. Pettinger, B., Domke, K. F., Zhang, D., Picardi, G. and Schuster, R. (2009). Tip-enhanced Raman scattering: Influence of the tip-surface geometry on optical resonance and enhancement. *Surf. Sci.* **603**(10), pp. 1335–1341.

46. Itoh, T., Biju, V., Ishikawa, M., Kikkawa, Y., Hashimoto, K., Ikehata, A. and Ozaki, Y. (2006). Surface-enhanced resonance Raman scattering and background light emission coupled with plasmon of single Ag nanoaggregates. *J. Chem. Phys.* **124**(13), p. 134708.
47. Yorulmaz, M., Khatua, S., Zijlstra, P., Gaiduk, A. and Orrit, M. (2012). Luminescence quantum yield of single gold nanorods. *Nano Lett.* **12**(8), pp. 4385–4391.
48. van Schrojenstein Lantman, E. M., Deckert-Gaudig, T., Mank, A. J., Deckert, V. and Weckhuysen, B. M. (2012). Catalytic processes monitored at the nanoscale with tip-enhanced Raman spectroscopy. *Nat. Nanotechnol.* **7**(9), pp. 583–586.
49. Le Ru, E. and Etchegoin, P. (2008). *Principles of surface-enhanced Raman spectroscopy: And Related Plasmonic Effects.* Elsevier.
50. Buchner, F., Warnick, K.-G., Wölfle, T., Görling, A., Steinrück, H.-P., Hieringer, W. and Marbach, H. (2009). Chemical fingerprints of large organic molecules in scanning tunneling microscopy: Imaging adsorbate–substrate coupling of metalloporphyrins. *J. Phys. Chem. C* **113**(37), pp. 16450–16457.
51. Meng, L., Yang, Z., Chen, J. and Sun, M. (2015). Effect of electric field gradient on sub-nanometer spatial resolution of tip-enhanced Raman spectroscopy. *Sci. Rep.* **5**, p. 9240.
52. Zhang, C., Chen, B.-Q. and Li, Z.-Y. (2015). Optical origin of subnanometer resolution in tip-enhanced Raman mapping. *J. Phys. Chem. C* **119**(21), pp. 11858–11871.
53. Duan, S., Tian, G., Ji, Y., Shao, J., Dong, Z. and Luo, Y. (2015). Theoretical modeling of plasmon-enhanced Raman images of a single molecule with subnanometer resolution. *J. Am. Chem. Soc.* **137**(30), pp. 9515–9518.
54. Barbry, M., Koval, P., Marchesin, F., Esteban, R., Borisov, A., Aizpurua, J. and Sánchez-Portal, D. (2015). Atomistic near-field nanoplasmonics: Reaching atomic-scale resolution in nanooptics. *Nano Lett.* **15**(5), pp. 3410–3419.
55. Kukura, P., McCamant, D. W. and Mathies, R. A. (2007). Femtosecond stimulated Raman spectroscopy. *Annu. Rev. Phys. Chem.* **58**, pp. 461–488.
56. Frontiera, R. R., Henry, A.-I., Gruenke, N. L. and Van Duyne, R. P. (2011). Surface-enhanced femtosecond stimulated Raman spectroscopy. *J. Phys. Chem. Lett.* **2**(10), pp. 1199–1203.
57. Min, W., Freudiger, C. W., Lu, S. and Xie, X. S. (2011). Coherent nonlinear optical imaging: beyond fluorescence microscopy. *Annu. Rev. Phys. Chem.* **62**, p. 507–530.
58. Wessel, J. (1985). Surface-enhanced optical microscopy. *J. Opt. Soc. Am. B* **2**(9), pp. 1538–1541.
59. Lin, H.-B. and Campillo, A. (1994). CW nonlinear optics in droplet microcavities displaying enhanced gain. *Phys. Rev. Lett.* **73**(18), p. 2440.
60. Kneipp, K., Wang, Y., Kneipp, H., Itzkan, I., Dasari, R. R. and Feld, M. S. (1996). Population pumping of excited vibrational states by spontaneous surface-enhanced Raman scattering. *Phys. Rev. Lett.* **76**(14), p. 2444.
61. Itoh, T., Yoshida, K., Biju, V., Kikkawa, Y., Ishikawa, M. and Ozaki, Y. (2007). Second enhancement in surface-enhanced resonance Raman

scattering revealed by an analysis of anti-Stokes and Stokes Raman spectra. *Phys. Rev. B* **76**(8), p. 085405.

62. Ichimura, T., Hayazawa, N., Hashimoto, M., Inouye, Y. and Kawata, S. (2004). Tip-enhanced coherent anti-Stokes Raman scattering for vibrational nanoimaging. *Phys. Rev. Lett.* **92**(22), p. 220801.
63. McCamant, D. W., Kukura, P., Yoon, S. and Mathies, R. A. (2004). Femtosecond broadband stimulated Raman spectroscopy: Apparatus and methods. *Rev. Sci. Instrum.* **75**(11), pp. 4971–4980.
64. Shen, Y. R. (2003). *The Principles of Nonlinear Optics.* Wilkey-Interscience.
65. Stipe, B., Rezaei, M. and Ho, W. (1998). Single-molecule vibrational spectroscopy and microscopy. *Science* **280**(5370), pp. 1732–1735.
66. Berndt, R., Gimzewski, J. and Johansson, P. (1993). Electromagnetic interactions of metallic objects in nanometer proximity. *Phys. Rev. Lett.* **71**(21), p. 3493.
67. Ren, B., Picardi, G. and Pettinger, B. (2004). Preparation of gold tips suitable for tip-enhanced Raman spectroscopy and light emission by electrochemical etching. *Rev. Sci. Instrum.* **75**(4), pp. 837–841.

Chapter 14

SERS Theory: The Chemical Effect of Rhodamine 6G Adsorption on Silver Surfaces on Its Raman Spectrum

Lindsey R. Madison, Mark A. Ratner and George C. Schatz*

Department of Chemistry,
Northwestern University,
Evanston, IL 60208-3113, USA
**g-schatz@northwestern.edu*

1. Introduction

The SERS spectra of small organic molecules can be very sensitive to the physisorption or chemisorption of the molecules onto metallic surfaces. However, it is challenging to model the results, as typical SERS substrates have a broad distribution of possible surface sites, and the populations of these sites and the SERS enhancement factors associated with them are very difficult to determine, especially when electromagnetic hot spots dominate the observed spectra. However, it is important to account for these effects if one is to make a quantitative interpretation of measured SERS spectra. Indeed, the influence of different sites is important to both the electromagnetic and chemical mechanisms in SERS, where the former mechanism is concerned with local field enhancement effects arising from plasmon excitation, while the latter arises from the effect of chemisorption-induced charged transfer on the Raman intensities, as can be modeled using static limit Raman intensity calculations [1]. Extensive studies have been performed on simple organics like pyridine bound to gold and silver clusters, and it was found that both the chemical and electrodynamic enhancement factors associated with binding pyridine in various orientations result in noticeable changes in the SERS spectra compared to the gas phase spectrum of pyridine [2].

One of the most important benchmark molecules for SERS and TERS studies is Rhodamine 6G (R6G). This is a strongly fluorescent dye molecule in solution, but when adsorbed onto SERS substrates, its fluorescence is quenched, and the resulting SERS intensities are among the largest observed for any molecule. As a result, R6G has played a crucial role in the development of single-molecule SERS, and in the development of SERS substrates [3, 4].

Previously, the SERS spectrum of R6G was modeled as if it were a gas phase molecule. Jensen and Schatz [5] incorporated resonant effects as well as vibronic effects and found that resonance excitation to the S_1 excited state of R6G resulted in good agreement with spectra obtained from SERS experiments. In another study, intermolecular charge-transfer was studied with R6G bound to an Ag_2 cluster [6]. It is unclear if a cluster model as small as two silver atoms can accurately capture the surface effects of a nanoparticle with thousands of atoms.

Recent single-molecule, ultra-high vacuum, tip-enhanced Raman spectroscopy (SM-UHV-TERS) measurements for R6G [3] were moderately reproduced by gas-phase resonance Raman calculations. Of particular importance in this work was the comparison of TERS spectra at room temperature with those at a low temperature of 10 K. It was found that lowering the temperature led to both sharpening of the vibrational spectra and shifts in the spectra (up to $20\,\text{cm}^{-1}$ with some lines shifting to red, some to blue and some not at all), suggesting molecule transitions from a largely gas-phase molecule at room temperature to being strongly chemisorbed at 10 K. A potential energy distribution analysis performed by Chulhai and Jensen [3] showed that the modes that were shifted from the gas phase had vibrational activity associated with the ethylamine component of R6G, which is the likely moiety associated with binding of R6G to the metal surface. However, the metal surface was not incorporated into the Raman calculations.

What remains unresolved concerning the SM-UHV-TERS studies of R6G is direct evidence that the chemical binding of R6G to a silver nanoparticle surfaces causes frequency shifts and/or intensity changes in the Raman scattering intensities of specific modes, especially those involving vibrations of the ethylamine moiety. The theoretical work described in this chapter directly addresses this issue by calculating Raman spectra for R6G interacting with a 20-atom silver cluster, and with a flat Ag(111) surface. Tetrahedral Ag_{20} is chosen for part of these studies as it provides a model nanoparticle structure that has a well-defined fragment of a (111) surface, as well as having edges and vertices that can be used to model coordinately unsaturated sites on a particle. In addition, Ag_{20} has been found to provide a reasonable reference structure for determining the chemical contribution to SERS enhancements [1]. Two different geometries of the Ag_{20}/R6G system are considered in the present study, corresponding to two different binding configurations of R6G to a pristine Ag(111) surface. In addition, we study adsorption of R6G on a flat Ag(111) surface of using plane wave DFT calculations to determine optimized geometries and normal mode properties.

2. Methods and Computational Approaches

For R6G bound to an Ag_{20} cluster, simple molecular mechanics calculations using the Universal Force field [7] implemented in the software Avogadro [8] were performed to roughly optimize the binding geometry in two configurations: a configuration with the xanthene plane perpendicular to the (111) face of the Ag_{20} cluster, and geometry with the xanthene plane parallel to the face of the Ag_{20} cluster. These two configurations are represented in Figures 1(a) and 1(b). Subsequently, a Hessian calculation was performed using DFT, a triple ζ with polarization (TZP) basis set, and the Becke–Perdew (BP86) [9,10] exchange-correlation functional. The BP86 exchange-correlation functional was chosen because it has been shown to result in harmonic vibrational frequencies that more closely match experimental results without incorporating a scaling factor [11]. Given the

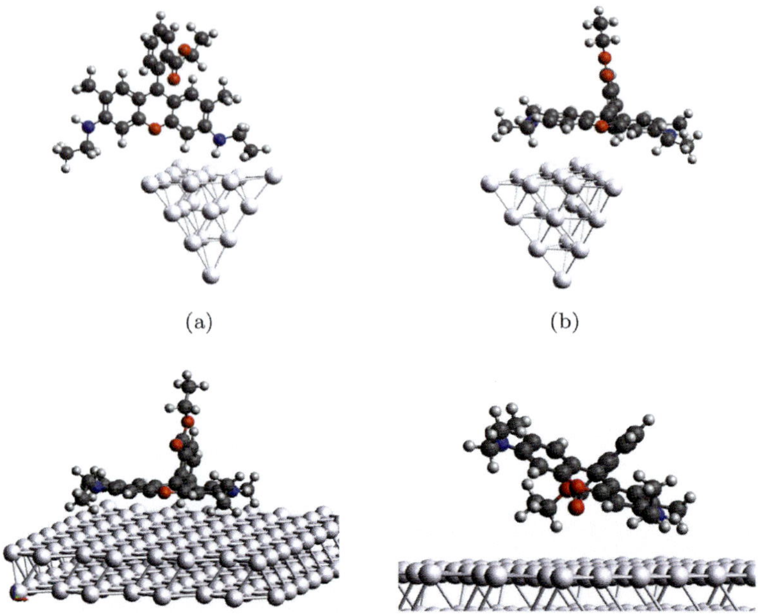

(a) (b)

(c) (d)

Figure 1. The four different configurations of R6G bound to a silver cluster or silver surface. (a) The xanthene plane is oriented perpendicular to the Ag_{20} cluster (configuration $C_{\perp}$). (b) The xanthene plane is oriented parallel to the Ag_{20} cluster ($C_{||}$). (c) The xanthene plane is oriented parallel to a Ag(111) surface (D1). (d) The normal vector from the xanthene plane is oriented 28° from the normal of the Ag(111) surface (D2).

very simplistic model of geometry optimization, some imaginary frequencies were found in the system, indicating that the structure is not at a local minimum. Since the imaginary frequencies refer to intermolecular modes and modes of the silver cluster that have not been studied in the TERS measurements, we have ignored this problem in the present work. To determine Raman spectra, normal Raman spectroscopy (NRS) calculations were performed using the Amsterdam density functional (ADF) 2013.01 program and the AOResponse module implemented in ADF [12–14], following the method described by Jensen *et al.* [15]. The wavelength of the perturbation used was the standard excitation wavelength used in the measurements, 785 nm. This wavelength is such that the R6G cannot be resonantly excited, which means that the Raman spectrum is in the low frequency limit where intensities are primarily a measure of the static chemical contribution to the SERS enhancement [1]. A damping parameter $\gamma = 0.004$ au (approximately 0.1 eV) was used, consistent with previous work [15–18]. The zero-order regular approximation (ZORA) [19,20] was applied to the calculation of the polarizability tensors and the energy convergence values for the AOResponse module were set to 10^{-8} Hartree.

In addition to using ADF with a finite cluster model for the silver surfaces, geometry optimization calculations of R6G binding to a flat silver surface using a periodic DFT approach have been performed [21] and are depicted in Figures 1(c) and 1(d). In these calculations, a two-layer deep (111) silver surface slab was used. The Vienna *Ab initio* Simulation Package (VASP) [22] and DFT with the PBE exchange-correlation functional [23] and a triple zeta basis set was used to determine likely binding geometries of R6G to the silver surface and to calculate the vibrational Hessian matrix. These vibrational modes were then used to calculate the Raman response from the polarizability derivatives using the AOResponse module in ADF. The level of theory was the same as was used for calculating the Raman response for the system consisting of R6G and a Ag_{20} cluster, with the BP86 exchange correlation functional and a TZP basis set. This alternative approach of calculating the vibrational Hessian with the plane-wave VASP software and using the normal modes as input to the polarizability tensor calculations accomplished with ADF will be referred to as "Hessian: VASP/ Polarizability: ADF" approach and thus distinguished from the traditional methodology of computing both the Hessian and the polarizability tensors with ADF, referred to as "Hessian: ADF/Polarizability: ADF".

Also investigated was the approximation that the chemical effect on the Raman scattering due to binding to the silver surface predominantly affects the frequencies of the normal modes and minimally affects the polarizability. This approximation is accomplished by calculating the vibrational modes with the silver present and eliminating the silver atoms, leaving only the R6G molecule for the polarizability calculation. This second method for evaluating polarizability derivatives will be referred to as the "polarizability sans-metal" technique. This technique was applied with both ADF and VASP.

3. Results and Discussion

To establish that the VASP/ADF approach compares well with the ADF/ADF approach for computing Raman scattering spectra, the normal Raman scattering of gas-phase R6G was computed and is compared in Figure 2. The two spectra are in excellent agreement with a small exception in two regions. There is evidence of peak splitting at 1250 cm^{-1} and a slight red-shift at 1480 cm^{-1} from the VASP calculations. These slight changes are very minor and overall, the two spectra compare favorably enough to proceed with this

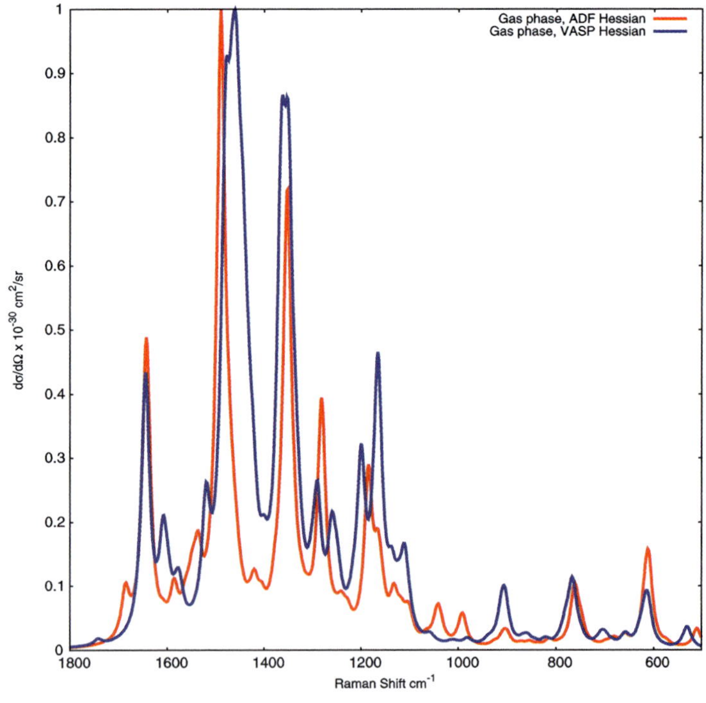

Figure 2. The normalized resonance Raman scattering spectra of gas-phase R6G with a 785 nm excitation with the vibrational frequencies calculated by ADF (blue) and VASP (red).

methodology for the R6G-Ag systems. The gas-phase calculations also compare favorably with the previous resonance Raman modeling of R6G performed by Jensen and experimental spectra collected by Klingsporn [3,5].

Geometry optimization studies (with VASP) of R6G binding to a pristine Ag (111) surface resulted in numerous optimized structures. Two representative geometries were used to perform normal Raman scattering calculations for the normal modes of R6G. For points of comparison, the Raman scattering from R6G in two different binding geometries to an Ag_{20} cluster were also considered (with ADF), one with the xanthene plane in a parallel configuration to the cluster's face and another geometry with the xanthene plane perpendicular to the face.

Figure 3 summarizes the resonance Raman calculations of the systems referenced in Table 1. The first two panels show the ADF versus VASP comparison for gas-phase spectra described above in Figure 2. The 3rd and 4th panels show spectra for the two $C_{||}$ calculations in Table 1, while the 5th and 6th panels refer to the $C_{\perp}$ calculations. The 7th and 8th panels refer to the two VASP conformations D1 and D2 in Table 1. The binding configuration that shows the greatest Raman sensitivity is the R6G bound in a parallel configuration to the Ag_{20} face ($C_{||}$). These spectra are very different from the predicted spectra for the perpendicular binding configuration ($C_{\perp}$). Spectral features that change the most are associated with the in-plane and out-of-plane xanthene ring distortions. This may indicate that the resonance Raman scattering signal will be sensitive to small clusters or defect sites on the silver surface and not as sensitive to changes in the binding geometry on pristine silver surfaces. However, we note that the experimental spectra are most like the $C_{\perp}$ results, so the $C_{||}$ structures must not be present in significant amounts when R6G adsorbs onto silver particles.

The "polarizability sans-metal" approximation simplifies the calculation of the Raman scattering by incorporating the silver cluster in only the vibrational frequency calculations and eliminating the cluster in the polarizability calculations. By making this approximation, we assume that the chemical effect of an organic molecule binding to a silver cluster predominantly affects the frequency of the vibrational modes but has minimal effect on the derivative of the polarizability with respect to the displacement of the normal modes. The 3rd/4th, and 5th/6th panels of Figure 3 compare these two methods. The results are qualitatively very similar regardless of whether the silver is present during the polarizability calculations or not. Only the intensity of a few select modes seems to be sensitive to the presence of the silver cluster, notably the vibrations at 1650, 1120 and 810 cm^{-1} which all have some ethyl amine vibrational characters. The consistency of the spectra after the "sans-metal" approximation indicates that the majority of the chemical effect in the normal Raman response is due to structural changes in the molecule that lead to shifts in the frequencies of the vibrational modes.

The seventh and eighth panels of Figure 3 have been calculated using VASP to identify local energy minimum for R6G binding to a pristine silver surface as well as the Hessian of that system. The Raman response was subsequently calculated at the same level of theory as the previous six panels with ADF, but using the "sans metal" technique. Both the D1 and D2 configurations yield similar Raman response spectra with the most marked difference being a small peak at 1450 cm^{-1} for D1 configuration. There are subtle differences between the Raman response spectra of the D1 and D2 configurations on the silver surface when compared to the Raman response of gas-phase R6G, most notably the region between 1300 and 1350 cm^{-1} which corresponds to xanthene ring stretching coupled to hydrogen bending modes on the ethylamine. We also note that panels 7 and 8 show roughly similar spectra to panels 5 and 6, so these VASP-based structures are consistent with the perpendicular conformation.

Table 2 summarizes the peak assignments for the vibrational modes, here comparing the gas-phase results, the $C_{||}$ and $C_{\perp}$ results from panels 3 and 5, and the experimental results (mostly RT TERS, but LT TERS included for two modes with significant ethylamine character) from Klingsporn [3]. The comparison of gas-phase results with RT experiment is similar to that presented previously [3], so the important new results in the work are the frequency shifts in going

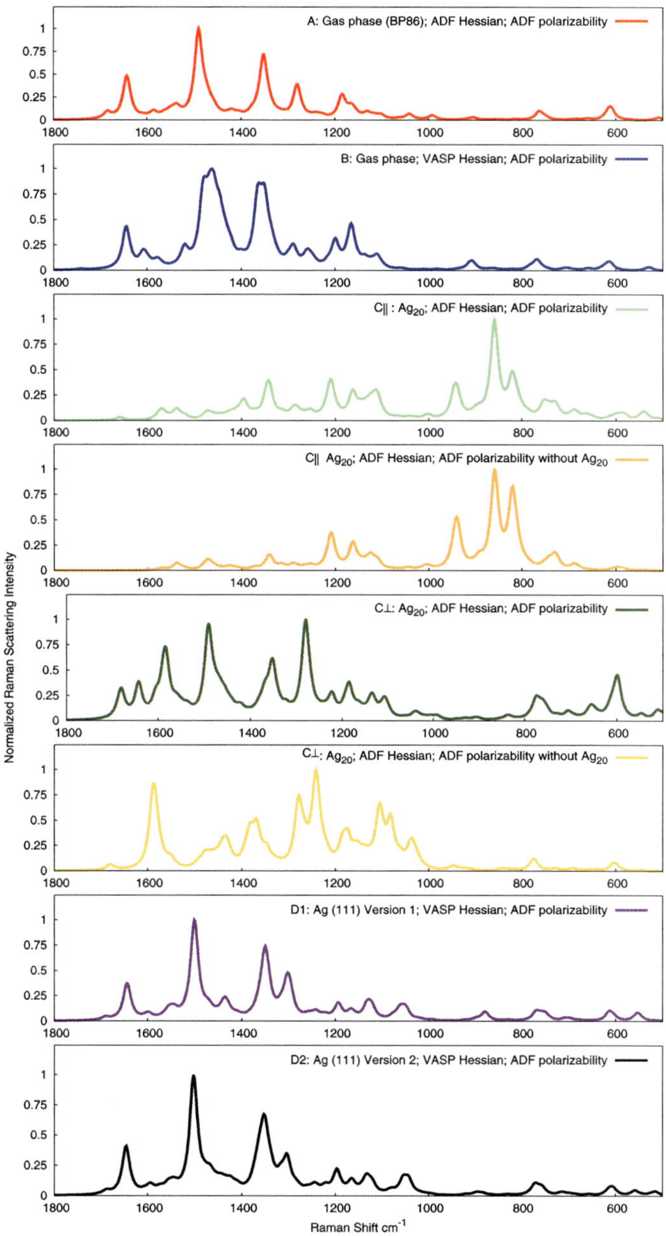

Figure 3. The normal Raman scattering spectra of R6G species listed in Table 1. Note that the panels are ordered differently than in Table 1 as indicated in the legends.

Table 1. Rhodamine 6G and Ag systems of varying binding geometries evaluated with indicated software packages and the sans-metal approximation.

Species ID	Binding configuration to Ag surface	Hessian calculations	Raman response
A	Gas phase (no cluster)	ADF	ADF
B	Gas phase (no cluster)	VASP	ADF
$C_{\perp}$	Perpendicular binding to Ag_{20} face	ADF	ADF with Ag_{20} cluster
$C_{\|\|}$	Parallel binding to Ag_{20} face	ADF	ADF with Ag_{20} cluster
$C_{\perp}$	Perpendicular binding to Ag_{20} face	ADF	ADF without Ag_{20} cluster
$C_{\|\|}$	Parallel binding to Ag_{20} face	ADF	ADF without Ag_{20} cluster
D1	Conformation 1 on Ag (111) surface	VASP	ADF without Ag(111) surface
D2	Conformation 2 on Ag (111) surface	VASP	ADF without Ag(111) surface

Table 2. NRS Calculations of the Raman response of R6G (gas phase) and R6G-Ag20 and Experiment [3].

Stretch	R6G	R6G-Ag_{20} parallel	R6G-Ag_{20} perpendicular	Experiment [3]
ip XRD		600	605	625
op XRD	616	538		625
ip XRD	762		775	775
w H-XR	1184	1112–1163	1106–1184	1175–1200
XRD, w H-XR	1280	1283	1279	1310–1325
XRS, w H-ethane	1352		1368	1350–1375 (LT shifts red by 14)
w H-N, b CH on EA	1490	1395	1493–1441	
XRD	1508	1522		
XRD ip, b. HN, BRS	1578 (v.w.)	1538–1569	1588	1538 (LT-TERS shifts blue by 9)
XRS	1643	1569		1640–1660
CO st.	1648	1659	1680	1660

Notes: ip = in plane motion, op = out of plane, st = stretch, b = bend, w = wag (symmetric bending), XR = xanthene ring; XRD = xanthene ring distortions, XRS = xanthene ring stretching; EA = ethyl amine; BRS = benzoate ring stretch. All numbers are in cm^{-1} units.

from the gas-phase to Ag_{20} results. Also, we have already argued that the $C_{||}$ results are not representative of what is found in the experiment, so this column can be ignored. Looking at the $C_{\perp}$ results, we see that there are generally small frequency shifts in going from the gas-phase to Ag_{20} frequencies, however, the mode at $1352\,cm^{-1}$ (gas phase) becomes blue-shift by $16\,cm^{-1}$ upon adsorption on Ag_{20}, which is a similar shift but in the opposite direction to what is seen in the experiment for that mode. Another comparison is with the mode at $1578\,cm^{-1}$ (gas phase). This shifts blue by $10\,cm^{-1}$ in the calculations, which is similar to a $9\,cm^{-1}$ blue shift seen in the experiment. While these comparisons at least show shifts that are similar in magnitude, a more general comparison between theory and experiment is at best qualitative, and not particularly better for the Ag_{20} results than for the gas-phase comparison.

4. Conclusion

The chemical effect of binding R6G to a silver surface and a silver nanoparticle model have been investigated. Since most of the bright modes in the gas-phase R6G spectrum involve vibrations in the xanthene ring, and it is predicted through simple geometry optimizations that R6G binds to a silver surface through that ring, it is not surprising that there are noticeable spectral changes upon binding to the metal surface. Indeed, the most significant changes are for a structure where the xanthene ring is parallel to the surface, however, the spectra we obtain for this structure are quite different from the experiment so we assume this is not significantly populated in the experiment. Qualitative comparisons with experiment are found when the xanthene is perpendicular to a Ag_{20} cluster. In this case, we found shifts in vibrational frequencies due to adsorption that are similar in magnitude to those seen experimentally. However, the overall appearance of the spectrum is not in better agreement with experiment than the gas-phase calculated spectrum.

Another result of this paper is that we have explored two new methodologies for calculating spectra for adsorbed molecules that provide useful simplifications for what is generally a very complicated calculation. In the first, we considered an approach that couples a vibrational frequency calculation performed with the plane-wave DFT-based software package, VASP, to polarizability calculations performed with the AOResponse module implemented in ADF. The second methodology was the "sans-metal" approach and it was confirmed that the spectral changes upon elimination of the metal during the polarizability calculations usually have a minimal effect on the off resonant Raman spectra. From this study, it seems that the majority of the effect of binding R6G to a silver substrate can be captured by optimizing the R6G geometry and performing the Hessian calculation in the presence of a silver cluster or surface. However, the lack of agreement between details of the calculated spectrum and experiment indicates that the structural models we have used are still inadequate.

Acknowledgment

This research was supported by the NSF GRFP under Grant DGE-1324585, Department of Energy grant DE-FG02-10ER16153 (for methods development), and by AFOSR MURI grant FA9550-14-1-0003 (applications).

References

1. Valley, N., Greeneltch, N., Van Duyne, R. P. and Schatz, G. C. (2013). A look at the origin and magnitude of the chemical contribution to the enhancement mechanism of surface-enhanced Raman spectroscopy (SERS): Theory and experiment. *J. Phys. Chem. Lett.* **4**, pp. 2599–2604.
2. Aikens, C. M. and Schatz, G. C. (2006). TDDFT studies of absorption and SERS spectra of pyridine interacting with Au_{20}. *J. Phys. Chem. A* **110**, pp. 13317–13324.
3. Klingsporn, J. M., Jiang, N., Pozzi, E. A., Sonntag, M. D., Chulhai, D., Seideman, T., Jensen, L., Hersam, M. C. and Van Duyne, R. P. (2014). Intramolecular insight into adsorbate-substrate interactions via low-temperature, ultrahigh-vacuum tip-enhanced Raman spectroscopy. *J. Am. Chem. Soc.* **136**, pp. 3881–3887.
4. Camden, J. P., Dieringer, J. A., Wang, Y., Masiello, D. J., Marks, L. D., Schatz, G. C. and Van Duyne, R. P. (2008). Probing the structure of single molecule surface-enhanced Raman scattering hot spots. *J. Am. Chem. Soc.* **130**, pp. 12616–12617.
5. Jensen, L. and Schatz, G. C. (2006). Resonance Raman scattering of Rhodamine 6G as calculated using time-dependent density functional theory. *J. Phys. Chem. A* **110**, pp. 5973–5977.

6. Liu, S., Wan, S., Chen, M. and Sun, M. (2008). Theoretical study on SERRS of rhodamine 6G adsorbed on Ag_2 cluster: Chemical mechanism via intermolecular or intramolecular charge transfer. *J. Raman Spectrosc.* **39**, pp. 1170–1177.
7. Rappe, A. K., Casewit, C. J., Coldwell, W. A. and Skiff, W. M. (1992). UFF, a full periodic table force field for molecular mechanics and molecular dynamics simulations. *J. Am. Chem. Soc.* **114**, pp. 10024–10035.
8. Hanwell, M. D., Curtis, D. E., Lonie, D. C., Vandermeersch, T., Zurek, E. and Hutchison, G. R. (2012). Avogadro: An advanced semantic chemical editor, visualization, and analysis platform. *J. Chem. Inf.* **4**, p. 17.
9. Becke, A. D. (1988). Density-functional exchange-energy approximation with correct asymptotic behavior. *Phys. Rev A* **38**, pp. 3098–3100.
10. Perdew, J. (1986). Density-functional approximation for the correlation energy of the inhomogeneous electron gas. *Phys. Rev. B* **33**, pp. 8822–8824.
11. Neugebauer, J. and Hess, B. A. (2003). Fundamental vibrational frequencies of small polyatomic molecules from density-functional calculations and vibrational perturbation theory. *J. Chem. Phys.* **118**, p. 7215.
12. ADF2013, SCM, Theoretical Chemistry, Vrije Universiteit, Amsterdam, The Netherlands, http://www.scm.com/.
13. Guerra, C. F., Snijders, J. G., Te Velde, G. and Baerends, E. J. (1998). Towards an order-N DFT method. *Theor. Chem. Acc.* **99**, pp. 391–403.
14. Te Velde, G., Bickelhaupt, F. M., Baerends, E. J., Fonseca Guerra, C., van Gisbergen, S. J., Snijders, J. G. and Ziegler, T. (2001). Chemistry with ADF. *J. Comput. Chem.* **22**, pp. 931–967.
15. Jensen, L., Autschbach, J. and Schatz, G. C. (2005). Finite lifetime effects on the polarizability within time-dependent density-functional theory. *J. Chem. Phys.* **122**, pp. 224115–224115.
16. Jensen, L., Zhao, L. L., Autschbach, J. and Schatz, G. C. (2005). Theory and method for calculating resonance Raman scattering from resonance polarizability derivatives. *J. Chem. Phys.* **123**, pp. 174110–174110.
17. Greeneltch, N. G., Davis, A. S., Valley, N. A., Casadio, F., Schatz, G. C., Van Duyne, R. P. and Shah, N. C. (2012). Near-Infrared surface-enhanced Raman spectroscopy (NIR-SERS) for the identification of Eosin Y: Theoretical calculations and evaluation of two different nanoplasmonic substrates. *J. Phys. Chem. C* **116**, pp. 11863–11869.
18. Aquino, F. W. and Schatz, G. C. (2014). Time-dependent density functional methods for Raman spectra in open-shell systems. *J. Phys. Chem. A* **118**, pp. 517–525.
19. van Lenthe, E., Baerends, E. J. and Snijders, J. G. (1993). Relativistic regular two-component Hamiltonians. *J. Chem. Phys.* **99**, p. 4597.
20. van Lenthe, E., Aikens, C. M., Baerends, E. J., Li, S., Snijders, J. G. and Schatz, G. C. (1994). Relativistic total energy using regular approximations. *J. Chem. Phys.* **101**, pp. 9783–9792.
21. Fu, B., Van Dyck, C., Zaleski, S. Van Duyne, R. P. and Ratner, M. A. (2016). Single Molecule Electrochemistry: Impact of Surface Site Heterogeneity. *J. Phys. Chem. C* **120**, pp. 27241–27249.

22. Kresse, G. and Furthmüller, (1996). Efficiency of ab-initio total energy calculations for metals and semiconductors using a plane-wave basis set. *Comput. Mater. Sci.* **6**, pp. 15–50.
23. Perdew, J. P., Burke, K. and Ernzerhof, M. (1996). Generalized gradient approximation made simple. *Phys. Rev. Lett.* **77**, pp. 3865–3868.

Chapter 15

Graphene-Enhanced Raman Scattering (GERS): Chemical Effect

Xi Ling[*,†,§], Shengxi Huang[‡], Jing Kong[*] and Mildred Dresselhaus[*]

[*]*Department of Electrical Engineering and Computer Science, Massachusetts Institute of Technology, Cambridge, MA 02139, USA*

[†]*Department of Chemistry, Division of Materials Science and Engineering, Boston University, Boston, MA 02215, USA*

[‡]*Department of Electrical Engineering, The Pennsylvania State University, University Park, PA 16802, USA*

[§]*xiling@bu.edu*

1. Introduction

Although there are many books and much literature on the topic of surface-enhanced Raman scattering (SERS) [1–8], most of the publications focus on the metal-based SERS, where the rough surface of some noble metals can generate a plasmonic excitation and lead to giant magnification of the electromagnetic field at certain hot spots. However, it is also widely accepted that besides the electromagnetic mechanism (EM), a chemical enhancement mechanism (CM) can also exist in a conventional SERS system, with the major contribution from EM and a minor contribution from CM. The theory behind the electromagnetic mechanism has been well developed to describe the main feature of SERS as early as the 1980s. In contrast, the

CM effect has historically been neglected as being less important because of the low practical impact, and therefore, a poor understanding about the CM mechanism remains. Because EM is studied and widely discussed by the scientific community, gradually, it has become increasingly likely that when talking about SERS, people usually refer to the plasmonic effect coming from metals with rough surfaces. One reason for the delayed development of CM may be that there is no ideal system that can be used to separate CM from EM for study.

The current understanding about CM is that it originates from the charge-transfer between the adsorbate and the substrate [9–13]. To learn what the previous researchers did to contribute to the CM study, readers are advised to refer to the experimental work done by Otto *et al.* [14], and the theoretical work done by Jensen *et al.* [12, 15], Lombardi *et al.* [16–18], and Arenas *et al.* [19, 20]. It should be mentioned that those experimental and theoretical works were all done on metal systems.

Several factors may lead to the unilateral understanding about CM: (i) CM may not only exist in the metal system, as there are some reports on the Raman enhancement effect on semiconducting nanoparticles [21–25]; (ii) metals are normally chemically active, which usually induces chemical bonding (e.g. Au–S, Au–N) with the adsorbate, while this may not always be the case for the other systems; (iii) for some experimental studies, it may still be difficult to remove the EM contribution completely. Nevertheless, some conclusions about the characteristics of the CM mechanism are obtained from previous studies [14]. For example, it is widely accepted that CM is a short-range effect relevant to the charge-transfer effect; the enhancement factors are normally $\sim 10^2$, while the EM effect is many orders of magnitude larger. At the same time, there is scientific interest in gaining a deeper understanding about CM, as well as developing a more detailed theory for CM.

From an experimental perspective, a system with the following characteristics is preferred for the CM study: the preferred system (i) has only CM contribution and no EM disturbance nor an abundance of other noisy effects; (ii) has a uniform and defect-free surface to produce uniform enhancement; (iii) is chemically inert to avoid ongoing chemical reactions; (iv) enables a controllable adsorbate configuration to develop on the substrate; (v) can produce stable and repeatable signals for a quantifiable study. However, metal surfaces are not qualified for many of the standards. Therefore, although noble metals are satisfying substrates for providing giant enhancement, they are probably not a good standard for CM study.

Two-dimensional (2D) materials [26–28], such as graphene, hexagonal boron nitride (h-BN), molybdenum sulfide (MoS_2), and black phosphorus (BP), are all layered materials with flat surfaces and strong chemical interactions with many organic molecules. In 2010, we reported that graphene can be used to enhance the Raman signals of some dye molecules, and this process was then named "graphene-enhanced Raman scattering (GERS) effect" [29]. Systematic studies were done on the surfaces after this discovery, and similar effects were observed on the surfaces of other 2D materials [30]. It was found that this effect provided an effective system for the CM study with stable and quantifiable enhancement signals.

In this chapter, we provide an overview on the use of 2D materials, particularly graphene, as a substrate for Raman signal enhancement. Experiments designed to consider a variety of factors, including molecule parameters, substrate parameters and laser parameters, will be described to gain an understanding of the GERS effect. Based on these experiments, the characteristics of the chemical mechanism are discussed for the GERS system. Toward the development of practical applications, an advanced system combining 2D materials and a metal substrate will be introduced to enable both large enhancement and clean, stable, reproducible signals to be observed.

2. Observation of GERS

2.1. *Photoluminescence suppression of dye molecules on a graphene surface*

Molecular photoluminescence (PL) is the main disturbing effect for the measurement of the resonant Raman spectra (RRS) of dye molecules [31, 32]. Different from the normal spontaneous Raman

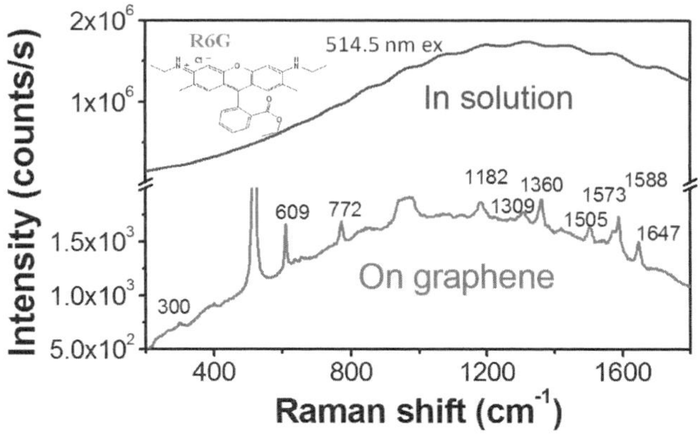

Figure 1. Typical result of the photoluminescence suppression effect of R6G on graphene. The top spectra is from the R6G water solution, and the bottom spectra is from the R6G molecules on graphene [29].

scattering, RRS involves the electron transition between real electronic states, which share similar processes as PL [33]. As the PL emission cross-section ($\sim 10^{-16}$ cm^2) is much larger than the resonant Raman scattering cross-section ($\sim 10^{-27}$ cm^2) process, the PL intensity is much stronger than the signal in RRS. In addition, the full width at half-maximum intensity (FWHM) of PL is about 100 times larger than that of RRS. Therefore, the RRS signals are usually masked under the PL background. To obtain the RRS of dye molecules, a method to suppress or quench the PL is therefore required.

Before the discovery of GERS, we noticed that graphene can suppress the PL intensity of the dye molecules effectively [34], which enables us to see the RRS of the dye molecules. As shown in Figure 1, for the Rhodanmine 6G (R6G) water solution, a strong PL background was observed when trying to measure the RRS. However, when depositing the R6G molecules on a graphene surface (graphene is supported by a 300 nm SiO_2/Si substrate) by solution soaking, the PL background got suppressed by about three orders of magnitude, and the RRS was then observed. Similarly, for other dye molecules (such as protoporphyrin IX, PPP), the PL suppression phenomena was also observed. The mechanism for the PL suppression by graphene was attributed to the charge-transfer and energy transfer between the dye molecules and graphene [34].

2.2. *Evidence of GERS*

In the above PL suppression experiment, we notice that even though we decrease the concentration of R6G solution to a relatively low value (e.g. around 10^{-9} mol/L), obvious Raman signals from R6G can still be observed [34]. This indicates that besides the PL suppression effect, graphene probably plays another role in the enhancement process of the Raman signals of R6G. In this case, since we do not know how strong the Raman signals of R6G are for the case without graphene due to the PL disturbance, we are not able to compare the Raman intensity of R6G on the graphene area and the same area without graphene. Thus, we cannot confirm independently if there is a Raman enhancement effect contributing to the Raman signals.

To clarify whether the Raman enhancement does exist when using graphene as a substrate, instead of R6G, a molecule without showing any PL disturbing effect was used as a probe molecule (e.g. phthalocyanine, Pc). The Pc molecules were deposited on the areas with and without graphene by vacuum thermal evaporation in a batch; thus, the number of the molecules deposited on both areas are the same. It should be mentioned that with the interference enhancement coming from the 300 nm SiO_2/Si substrate, even though there is no graphene present, the Raman signals from 2 Å thick Pc are still visible. By comparing the Raman intensity on the region with and without graphene, we found that stronger Raman intensity can be obtained on the region with graphene [29]. As shown in Figure 2, the Raman enhancement is observed not only for the resonant excitation (634.8 nm) but also for 514.5 and 457.5 nm. For the case under 632.8 nm excitation, we calculated the enhancement factor to be up to 17, depending on the vibrational modes.

It should be mentioned that to exclude the adsorption difference due to the different surfaces (i.e. the graphene surface and SiO_2 surface in this case), an additional experiment was performed by depositing the molecules on a blank SiO_2/Si substrate first and then transferring graphene on part of the substrate [35]. In this way, as molecules are deposited on a uniform surface, the amount of the molecules is considered to be uniform everywhere. After graphene was transferred, the Raman signals from the molecules on

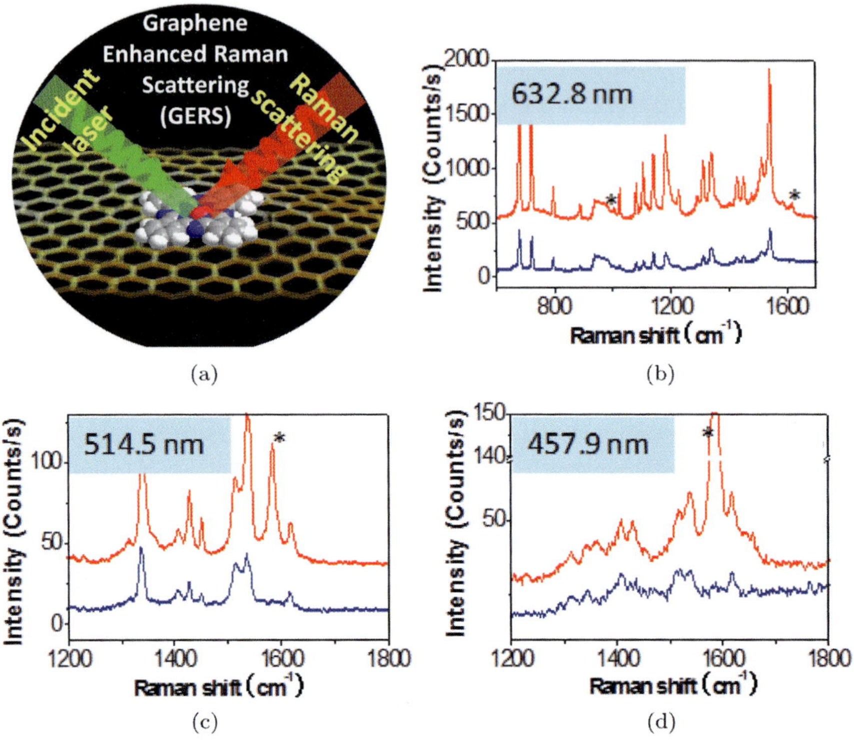

Figure 2. (a) Schematic illustration of the GERS effect. (b)–(c) the comparison of the Raman spectra of Pc molecules on the regions of graphene (red trace) and the blank SiO2/Si substrate (blue trace) under the laser excitation of 632.8 nm (b), 514.5 nm (c) and 457.9 nm (d). The signals marked with "*" are from the G-band of graphene. Other signals are all from the Pc molecules [29].

the graphene region still show a stronger intensity than those from the region without graphene. This procedure further proves that the Raman enhancement effect is associated with the graphene substrate. Here, a benefit from graphene is that the laser can pass through graphene and excite the molecules underneath with almost no signal decay due to the transparent character of graphene in the visible range (transmittance of graphene is about 98% in the visible range).

Therefore, as we directly compare the Raman intensity of the probe molecules for the regions with and without graphene, this experiment provides strong evidence that graphene can be used as a substrate for Raman enhancement. We call this effect graphene-enhanced Raman scattering (GERS). Regarding this effect, systematic studies were done to try to answer the questions readers may have, for example, concerning the enhancement mechanism-related questions, how the various parameters in this systems play a role in the enhancement process, the characteristics and advantages of this effect compared to conventional SERS in practical applications. The following discussion may help the reader find such answers.

2.3. *Analysis of the enhancement mechanism*

Does the GERS enhancement originate from the EM or CM process? First, a theoretical analysis is given. According to the requirements for the EM and CM processes, to be able to support the EM processes, it is very important to have a substrate which can absorb the incident light and generate the plasmon (which is what some noble metals can do). However, graphene has a very high transmittance up to 98% in the visible range [36], which means that most of the incident light will not be absorbed by the graphene substrate. Also, the surface plasmon wavelength of graphene is in the terahertz range rather than in the visible range [37]. On the basis of these considerations, graphene does not support EM. While to support the CM, the distance between the adsorbate and the substrate is required to be very close and enable strong coupling between them. Here, the strong coupling mainly means the charge-transfer effect according to a previous study on CM where the details were not clarified [14]. Considering a large planar conjugated structure of graphene, it is easy for graphene to adsorb the aromatic molecules on top of its surface [38, 39]. Due to the strong π–π interaction between graphene and the aromatic molecules, graphene may support the CM mechanism.

Besides, from the experimental data, the enhanced signals match the important characteristics of the CM very well. (i) One of the most important evidence to show a strong influence of the CM process is that the enhancement factor shows a vibrational mode dependence. In Figure 3(a), we show that the enhancement factor of Pc molecules on graphene depends on the vibrational modes. For example, the signal is enhanced 17 times for the 1140 cm^{-1} mode, and less than 2

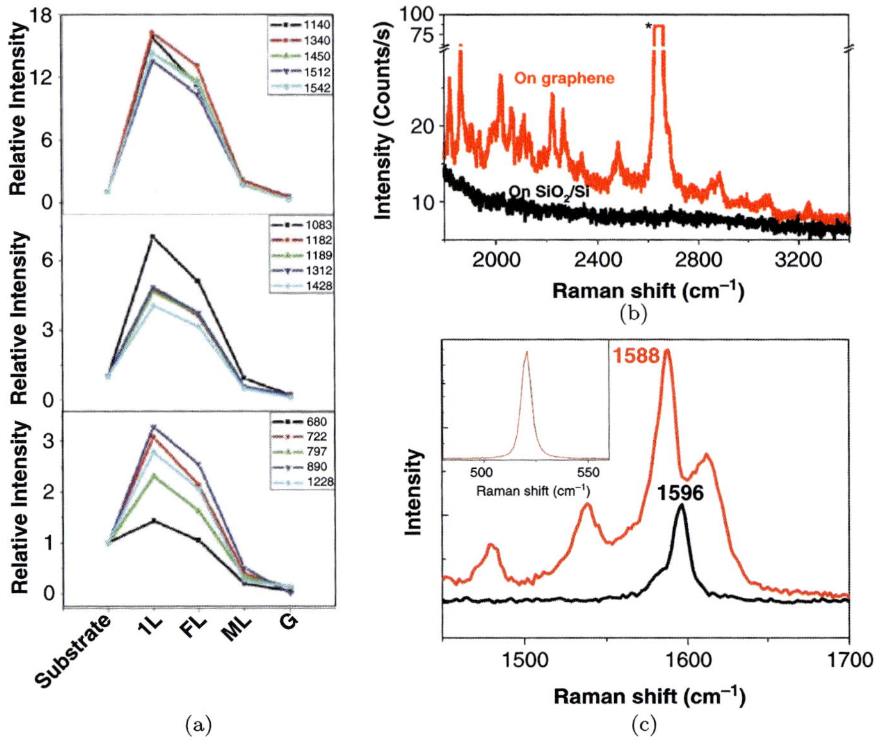

Figure 3. (a) Relative Raman intensity of the vibrational modes (labeled on the right corner) of Pc molecules on different thicknesses of graphene [29]. (b) The overtone and combination modes of Pc molecules on graphene (red trace) and the blank SiO_2/Si substrate (black trace) [29]. (c) G-band shift before (black trace) and after (red trace) the deposition of PPP molecules on graphene [40]. Inset shows the absence of frequency shift of the Raman signal of Si 520 cm^{-1} as reference.

times for 680 cm^{-1} mode. The enhancement factors calculated here are less than the reality, which will be explained in Section 3.2. (ii) Overtone and combinational modes are more remarkable with GERS (Figure 3(b)). This is attributed to the strong coupling-induced breaking of the symmetry. As we can obtain richer Raman spectra through GERS, it provides more information for the structure characterization. (iii) The charge-transfer effect is considered as a necessary condition for CM. The G-band shift of graphene is very sensitive to the charge-transfer, which can be used as an internal standard for

GERS. For example, we observed that the G-band red-shifted after depositing the PPP molecules on graphene (Figure 3(c)). (iv) The enhancement factor for the CM is normally low compared to that for the EM process. Although the GERS effect shows strong enhancement, the enhancement factor is usually less than 100. This finding matched with the previous understanding of CM well also. Besides, in the following sections, the characteristics of GERS that we will introduce also indicate the importance of CM in controlling GERS.

3. How Does the Molecule Influence the Effect?

To investigate the GERS effect in more detail, some experiments are designed by looking at each part of the system which has three components: the molecule, the graphene substrate, and the laser. Sections 3–5 cover each component sequentially. Here, in this section, we mainly see the effect from the point of view of the molecule, such as the distance between the molecule and graphene, the configuration of molecules on the graphene surface, and the type of molecules. These three factors lead us to discuss the first layer effect of molecules, the molecular configuration and the molecular selectivity, respectively.

3.1. *First-layer effect*

Let us first consider the effect of the molecule–graphene distance. To study this effect, we constructed Langmuir–Blodgett (LB) films of molecules PPP with a controllable number of layers [40]. It should be mentioned that the intensity of the Raman signals of PPP should increase with the increase of the number of layers (or the number of molecules) if the enhancement effect is equal for every layer of molecules on graphene. However, the Raman signals do not change much after the first layer (Figure 4). This indicates that the molecule layer adjacent to the graphene substrate dominates the GERS enhancement, which we call the "first layer effect". This suggests that only the molecules at a very small distance from the graphene surface can interact with graphene and show GERS enhancement. Moreover, even for the functional groups, stronger Raman signals can be obtained for molecular groups that are closer to the graphene substrate. Further evidence for the first layer effect

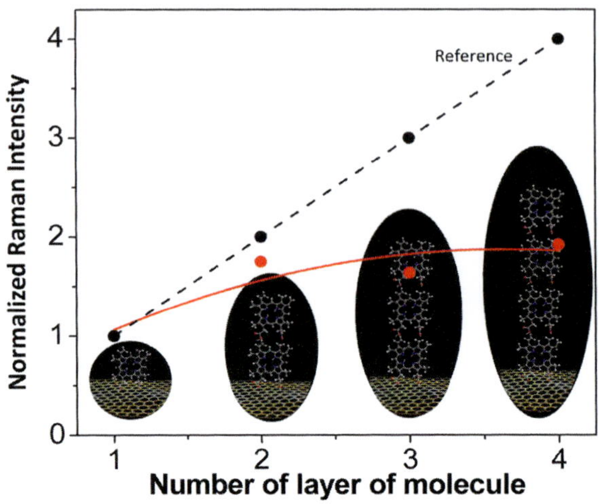

Figure 4. Relative intensities of the 1360 cm^{-1} Raman mode from PPP as a function of the number of layers of the LB film of PPP (red dots) [40]. The red line is the corresponding polynominal fit. The black dotted line is a reference considering the different numbers of layers of PPP that contribute equally. Inset: Schematic diagrams of a 1–4 layer PPP LB film on a monolayer graphene substrate.

is that if a spacing layer, even though only a few nanometers thick, is inserted between graphene and the molecule, the enhancement effect is eliminated. The "first layer effect" has been demonstrated as an important feature of the chemical mechanism, referring to the short-range effect [41–45]. The intensive embodiment of this effect in the GERS observations further indicates that the GERS effect is promising for the study of the CM effect.

3.2. *Molecular configuration*

From the first layer effect, it seems that the interaction between graphene and the molecules plays a really important role in the GERS effect. The molecular configuration on the graphene substrate will also affect the interaction. How will it apply to the GERS effect? Here, we use the planar molecule, CuPc, as a probe. With the help of the LB technique, an LB film of CuPc molecules is prepared on graphene. The as-prepared film contains molecules of the upstanding configuration on graphene. After annealing the molecule film at

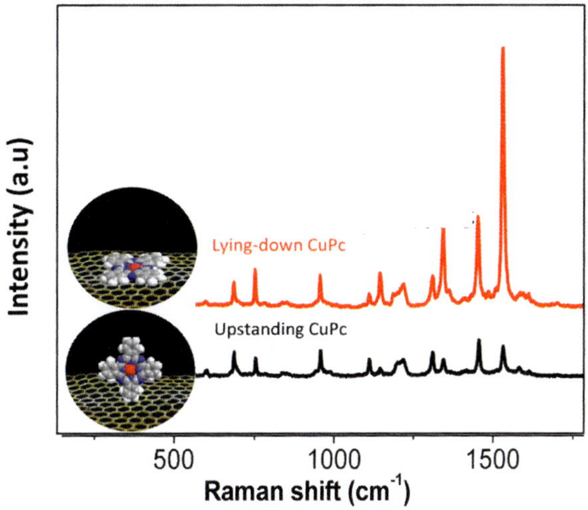

Figure 5. Comparison of the Raman spectra of an as-prepared CuPc LB film (upstanding, black line) and that after annealing at 300°C (lying-down, red line) with graphene on the SiO_2/Si substrate [46]. Inset: Schematic diagrams of upstanding and lying-down CuPc molecules on graphene.

300°C in an Ar atmosphere, the molecule orientation changes from upstanding to lying-down, with a decrease of the number of molecules due to desorption. The lying-down configuration after annealing was confirmed by AFM measurement, which shows a reduced thickness for the LB film after annealing. However, the measurement of GERS reveals a GERS enhancement several times larger after annealing than the as-prepared molecule film (Figure 5), which suggests that the molecules lying down have a larger GERS enhancement effect [46]. This is due to the larger molecule–graphene interaction areas and the smaller distance between the molecule center and the graphene surface for the CuPc molecule lying down compared to upstanding, and therefore stronger molecule–graphene interactions for the CuPc molecule lying down.

The controllable molecular configuration on graphene allows us to perform a more quantitative analysis. For example, by considering the change of both the number of the molecules and the signal intensity during the transformation of the molecular configuration, we can calculate the precise enhancement factors for the Raman

Table 1. Contribution of the CM to the GERS effect of lying-down CuPc molecules.

Raman frequency (cm^{-1})	I_2/I_1	EF_{CM}
233.0	0.8	3.5
255.3	0.9	3.9
483.8	0.4	1.7
679.0	1.4	6.0
747.2	2.2	9.5
951.9	1.1	4.7
1106.1	0.7	3.0
1140.4	6.0	25.9
1307.0	1.2	5.2
1341.8	5.6	24.1
1453.8	2.5	10.8
1530.8	10.3	43.2

Note: I_2/I_1 is the Raman intensity ratio after and before annealing at 300°C. EF_{CM} is the enhancement factor after considering the decrease of the number of molecules during annealing [46].

modes (Table 1). It is found that the 1530 cm^{-1} mode is enhanced the most with an enhancement factor of around 45. This value is several times larger than what we mentioned in Section 2.2. That is because the result in Section 2.2 is an average effect from the molecules with random configurations, while the result here is from the full contribution of the lying-down molecules.

3.3. *Molecule selectivity*

In addition to the above two factors, another important factor that can significantly change the molecule–graphene interaction is the molecules themselves. With the same molecular configuration, some molecules can experience a strong GERS enhancement, while some others cannot, as seen in Table 2. This is due to the nature of the molecules themselves. Based on our study [47], two properties of the molecules are closely connected to the GERS enhancement effect quantitatively: molecular HOMO–LUMO level and molecular structure.

Table 2. List of molecules with different structures and HOMO–LUMO levels and their GERS enhancement. Molecules above (below) the red horizontal lines are those with considerable (weak) enhancement.

Molecule	Structure	Symm.	HOMO/ LUMO (eV)	ΔE_{HF}/ ΔE_{LF} (eV)	d_z (Å)	E^{ads} (eV/atom)	ω_{ph} (cm^{-1})	EF
CuPc	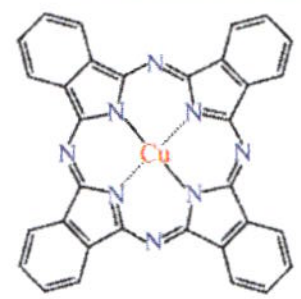	D_{4h}	−5.2/−3.5	−0.6/1.1	3.24	−0.06	1048 (532)[a]	37.5
							1538	11.6
							1451	24.3
							1530 (633)	47.3
							1451	35.6
ZnPc	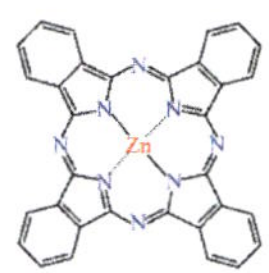	D_{4h}	−5.2/−3.8	−0.6/0.8	3.22	−0.06	1448 (532)	4.1
							1508	11.9
							1508 (633)	12.3
							1448	2.2
F_{16}CuPc	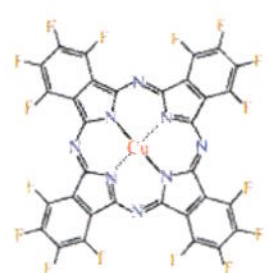	D_{4h}	−6.3/−4.8	−1.7/−0.2	3.19	−0.07	1540 (532)	4.2
							1461	2.7
							1540 (633)	6.2
							1461	3.1
PTCDA	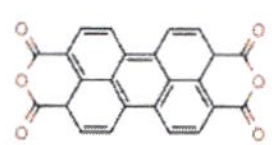	D_{2h}	−6.8/−4.7	−2.2/−0.1	3.17	−0.07	1532 (532)	5.8
							1455	3.9
							1531 (633)	8.3
							1454	3.9

(*Continued*)

Table 2. (*Continued*)

Molecule	Structure	Symm.	HOMO/ LUMO (eV)	ΔE_{HF}/ ΔE_{LF} (eV)	d_z (Å)	E^{ads} (eV/atom)	ω_{ph} (cm^{-1})	EF
TTP	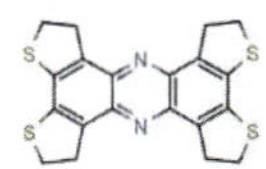	D_{2h}	−5.6/−2.3	−1.0/2.3	3.24	−0.07	1450 (633)	23.3
TCTA	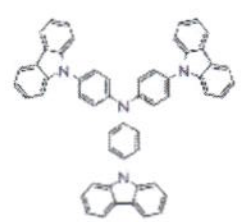	C_3	−5.7/−2.4	−1.1/2.2	4.34	−0.04	1340 (532) 1455 1449 (633)	1.3 0.9 6.9
Sp2-NPB	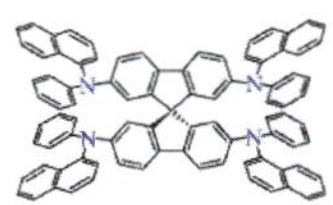	S_4	−5.5/−2.4	−0.9/2.2	4.36	−0.04	1503 (532) 1450 1447 (633)	5.3 1.7 4.3
Sp2-TPD	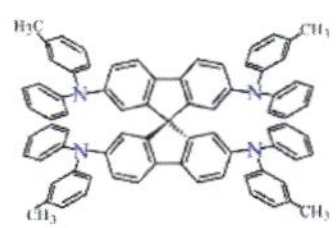	S_4	−5.5/−2.3	−0.9/2.3	4.27	−0.03	1450 (532) 1290 1446 (633) 1290	0.9 2.3 0.9 1.0
TPBi	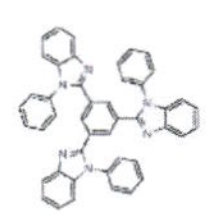	C_3	−6.7/−2.7	−2.1/1.9			1450 (532) 1280 1449 (633)	1 1.5 2.9

$Ir(ppy)_3$	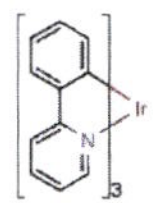	C_3	−5.6/−3.0	−1.0/1.6	none	
BCP	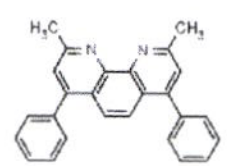	C_2	−7.0/−3.5	−2.4/1.1	1446 (633) (532 none)	1
Alq_3	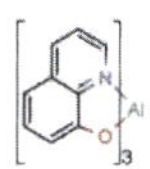	C_3	−5.8/−3.1	−1.2/1.5	none	

Note: [a]Excitation wavelength in nanometer.
Source:The table is from Ref. [47].

The molecular HOMO–LUMO level affects the density of states which in turn affects the Raman process. According to the quantum theory of Raman scattering, the HOMO and LUMO energy levels can strongly affect the GERS enhancement, and the proper energy levels can induce a resonant Raman effect with an extremely high intensity. Barros *et al.* [48] constructed a GERS model based on the quantum theory of Raman scattering by considering the graphene and molecule as one coupled system. It is concluded that there are four resonance conditions necessary to achieve a strong Raman intensity, and the one with the strongest enhancement occurs when graphene's Fermi level differs from the HOMO or LUMO level by the phonon energy.

The experiment performed using molecules CuPc, ZnPc and $F_{16}CuPc$ in the GERS enhancement confirms the importance of this HOMO–LUMO level factor [47]. These three molecules have very similar molecular structures, but different HOMO–LUMO levels. This difference in the HOMO–LUMO level results in the different GERS enhancement of the three molecules. CuPc can render an enhancement factor of over 40 under 532 nm laser excitation for the 1530 cm^{-1} mode, while the same mode has enhancement factors of only $\sim$12 for ZnPc and $\sim$6 for $F_{16}CuPc$. This is because CuPc has a LUMO level closer to graphene's Fermi level than the other two molecules.

Another factor of molecular selectivity is the molecular structure. Huang *et al.* [47] selected the model molecules TTP, TCTA and sp2-NPB. These three molecules have very similar HOMO–LUMO levels, but distinct molecular structures. Thus, TTP's GERS enhancement factors are one order of magnitude larger than for the other two molecular structures. Digging more into their structures, it can be noted that TTP has a highly symmetric structure with a D_{2h} symmetry, which is similar to graphene's D_{6h} symmetry. The D_{2h} symmetry also dictates TTP to be planar. These features enable TTP to attach well on the graphene surface. The other two molecules have symmetries of C_3 and S_4, which are not compatible with graphene's D_{6h} symmetry, and are non-planar. Therefore, the coupling between these molecules and the graphene surface cannot be as strong as for TTP–graphene. From the DFT calculations [47], when molecules are adsorbed on graphene, the molecule–graphene distance and adsorption energy in TTP is the smallest among the three, confirming that the TTP–graphene interaction is strong, and therefore its GERS enhancement is strong.

The above analyses show that molecular selectivity mainly resides in the HOMO–LUMO levels and in the structures of the molecules. The HOMO–LUMO levels that bring the system close to the resonance condition of the Raman spectra, and the molecular structures compatible with graphene's D_{6h} structure, can help boost the GERS enhancement factor. These factors can also be used to explain why some molecules have very small or even no GERS enhancement, as listed in Table 2.

4. Influence of the 2D Materials on the Enhancement

4.1. *Graphene thickness*

One question that now arises is whether this effect only exists on monolayer graphene. The answer is "no". Indeed, we found that not only monolayer graphene but also few-layer graphene shows remarkable Raman enhancement for the adsorbate [49]. Furthermore, some graphene derivatives (such as chemical vapor deposition CVD, graphene, graphene oxide, doped graphene, graphene quantum dots, etc.), as well as some other layered materials (which are introduced in Section 4.3), also show more or less Raman enhancement effects [50–56]. Some reports show a stronger enhancement on some of the graphene derivatives, as the energy level of the adsorbate and the graphene Fermi level may match better for those derivatives than for the pristine graphene, or that the absorption ability is better for some of the graphene derivatives.

Here, we will focus on the dependence of the GERS enhancement on the thickness of the pristine graphene. The PPP molecules are deposited on a clean 300 SiO_2/Si substrate through vacuum thermal evaporation, and then graphene flakes with different numbers of layers are mechanically exfoliated on the substrate. Figure 6(a) shows the Raman spectra of the PPP molecules on graphene flakes with the number of layers from 1 to 6. It is found that the intensities of the Raman signals of PPP do not change much for those six

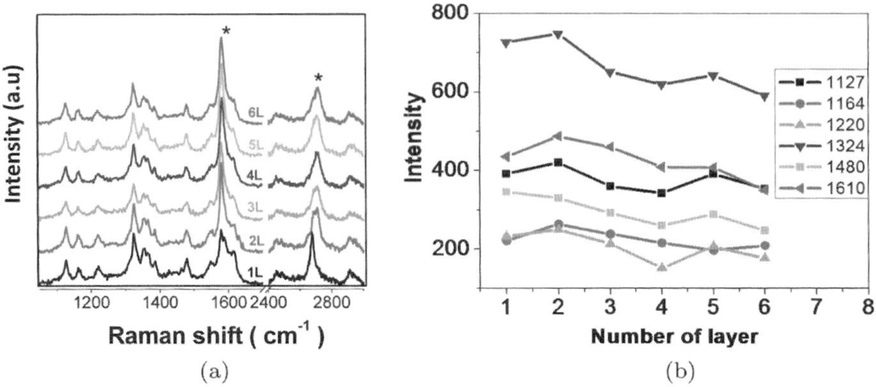

Figure 6. (a) Raman spectra of the PPP molecules collected on different numbers of graphene layers. The peaks labeled by "*" are from graphene. (b) Raman intensity of the Raman modes of PPP on graphene as a function of the number of PPP LB layers [49].

layers. From the plot of the intensities of the PPP Raman modes as a function of the number of graphene layers (Figure 6(b)), the intensity of each Raman mode decreases slightly, which is due to the 2.3% absorption of each layer of graphene. However, when the thickness of graphene increases to a certain thickness, the intensities of the Raman modes from the molecules will show a remarkably large decrease compared to the few-layer graphene. It should be mentioned that for the case where we deposit the molecules on top of graphene with different thicknesses, it shows a similar trend that the enhancement effect does not change much over a few-layer range, while it decreases significantly on multi-layer graphene, as well as graphite and highly oriented pyrolytic graphite (HOPG). Therefore, the question is: since the surface structures of graphene and HOPG are the same, how come the GERS effect shows such a big difference? Comparing the Raman spectra of CuPc molecules on graphene and on HOPG, we found that although the Raman intensity on graphene is much stronger than that on HOPG, the relative intensity of each mode is the same [35]. Since we already mentioned that the enhancement strongly depends on the vibrational mode in a CM system, the same relative intensity of the Raman modes on graphene and on HOPG indicates that similar CM also exists on HOPG. However, due to the different dielectric properties of graphene and HOPG, the GERS effect is suppressed in HOPG.

In conclusion, because of the high transparency of graphene, the GERS effect can be observed remarkably well in both monolayer graphene and few-layer graphene, but the effect is strongly suppressed in graphite and HOPG due to the dielectric loss. Therefore, when applying the GERS effect to a detection of practical interest, we have a large flexibility in the selection of the number of graphene layers. This flexibility offers a great convenience for applications.

4.2. *Graphene Fermi level*

Energy match is an important condition for observing the CM effect, as energy match affects the interaction strength between the substrate and the adsorbate. In the conventional metal-based SERS system, the electrochemistry method is widely used to study the CM effect as a function of the Fermi level of the metal by adding a potential to the electrode (which acts as the SERS substrate) [57]. Since the Fermi level of graphene is very sensitive to the doping, the variation of the Raman intensities of molecules on graphene with the shift of the Fermi level of graphene has been studied using an electrical field modification method [58]. For a series of metal phthalocyanine (M-Pc) molecules, the Raman intensities become weaker when the graphene Fermi level is up-shifted, while the intensities become stronger when the graphene Fermi level is down-shifted, as shown in Figure 7. The modulation of the GERS effect with the Fermi level of graphene is the result of the change of the interaction between graphene and the molecule, consistent with the discussion of the molecule selection rules in terms of the energy level rule.

Another interesting phenomenon we observed is that the Raman intensity enhancement on bilayer graphene can be stronger than that on monolayer graphene in some cases when depositing the molecules by solution soaking [49]. For example, we use CuPc molecules in a mixed solvent ($CF_3COOH{\cdot}CH_2Cl_2$, 1:10, v/v) as a probe to deposit CuPc molecules on a bilayer and monolayer graphene by solution soaking. It is found that when the solution concentration is low (e.g. 4×10^{-6} mol/L), the Raman intensities of CuPc molecules are significantly stronger on bilayer graphene than that on monolayer graphene, while when the solution concentration is high (e.g. 4×10^{-4} mol/L), the Raman intensities of CuPc molecules are slightly stronger on monolayer graphene than that on bilayer graphene. The reason for this abnormal dependence of the GERS effect on graphene thickness is due to the p-doping effect of CF_3COOH and n-doping effect of CuPc on graphene. (i) Under the low concentration, the p-doping effect from the solvent molecules is dominant, leading to a large shift of the Fermi level of monolayer graphene and a relatively small shift of the Fermi level of bilayer graphene to a lower energy. In this case, the Fermi level of bilayer graphene matches better with the energy level of CuPc than that of monolayer graphene. (ii) Under the high concentration, the n-doping effect from CuPc also contributes to the doping, leading to a decrease in p-doping in both monolayer and bilayer gaphene. In this case, both the Fermi levels of monolayer and bilayer graphenes match well with the energy level of the CuPc molecule. Therefore, the thickness dependence of the GERS effect under the high molecule concentration follows the usual pattern, whereas the GERS effect is abnormal under the low concentration situation. The abnormal thickness dependence of the GERS effect further indicates that the CM effect strongly depends

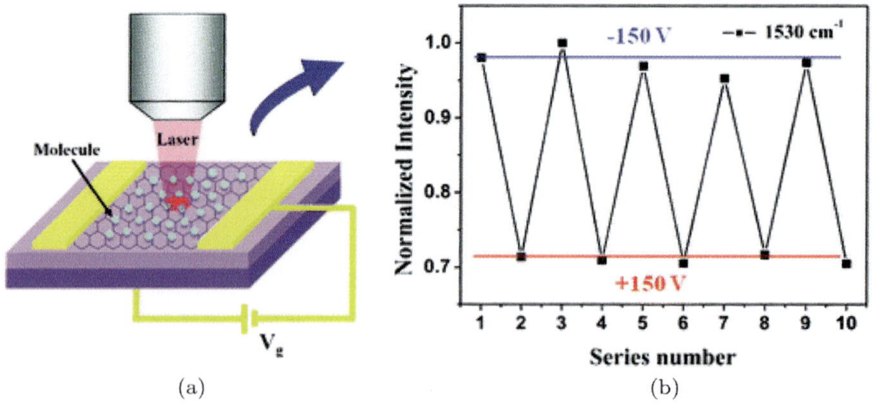

Figure 7. (a) Schematic illustration of the electrical field modification of the GERS effect. (b) Raman intensity variations of the 1530 cm^{-1} mode of CuPc molecules with the gate voltage change between −150 and +150 V [58].

on the energy match of the electron states. This is why the chemical enhancement is also called "electronic enhancement" [14].

4.3. *Graphene analogs*

There are many advantages in using 2D graphene as a Raman enhancement substrate, since we can thereby have a uniform molecule distribution and a controllable molecular configuration on the substrate. Graphene is just one typical member of the 2D materials family. There are many other 2D members in the family, and they share the same characteristics of flatness, transparency and flexibility. Moreover, the study of the Raman enhancement effect using other 2D materials will help in understanding the CM effect, as well as extending the number and properties of available substrates.

Similar experiments have already been implemented on other 2D materials, when used as GERS-type substrates, such as h-BN and monolayer MoS_2 [59]. In Figure 8(a), results obtained for the Raman enhancement effect as observed on three 2D materials, are compared to the use of the blank SiO_2/Si substrate. A stronger Raman enhancement effect was observed on the graphene substrate than on h-BN

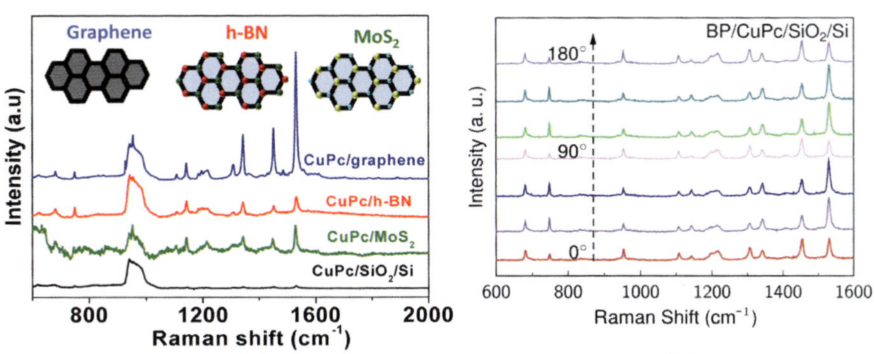

(a) (b)

Figure 8. (a) Typical results for the Raman spectra of the CuPc molecule on the 2D graphene, h-BN, MoS_2 and blank SiO_2/Si substrates [59]. (b) Corresponding polarized Raman spectra of the CuPc molecules on BP/SiO_2/Si substrate [60]. All the Raman spectra are collected under parallel polarization configurations with 633 nm laser excitation.

and MoS_2 substrates. In addition, the relative intensity of the vibrational modes of the CuPc molecule is different on these substrates, which indicates differences in the enhancement mechanism. In this connection, it is interesting to observe the enhancement effect on a h-BN substrate. Since h-BN is an insulator, it is difficult for the charge-transfer interaction to occur between h-BN and the adsorbate. However, h-BN has a strong polar B–N bond, leading to a strong dipole–dipole coupling between h-BN and the adsorbate. According to Fermi's golden rule, the electron transition probability rate can be expressed as

$$w_{lk} = \frac{2\pi}{\hbar} g(E_k)|H'_{kl}|^2. \tag{1}$$

The interface dipole interaction in h-BN will result in a local symmetry-related perturbation, which can increase the matrix element H'_{kl} in Equation (1), and therefore the enhancement of the Raman signals. For the case of MoS_2, the charge-transfer and dipole–dipole coupling strength with the adsorbate are both relatively weak compared to that of graphene and h-BN, leading to a relatively weak signal enhancement. Therefore, the origins of the Raman enhancement for graphene and h-BN are different from the charge-transfer effect for graphene and the interface dipole interaction for h-BN. This conclusion is further supported by the UV–visible absorption spectra of the CuPc molecule on these 2D materials, where the absorption peak of CuPc red-shifts on graphene, and blue-shifts on h-BN [59]. The different mechanisms are also consistent with the fact that the enhancement factors on a graphene substrate become larger and larger with the increasing frequency of the vibrational mode, since the larger the frequency of the vibrational mode, the easier the occurrence of charge-transfer. However, for h-BN, the vibration with a larger dipole will be enhanced the most ($1143\,\text{cm}^{-1}$ vibrational mode).

The Raman enhancement has also been investigated on 2D materials with lower symmetry than MoS_2, such as black BP and rhenium disulfide (ReS_2) [60]. Because of the anisotropic electronic properties of those 2D materials, the charge interactions become anisotropic on such materials. It is observed that the Raman enhancement effect shows anisotropic effects on such 2D materials due to the unequal contributions from the different orientations of the molecules on the surface. Taking BP as an example, the CuPc molecules with their primary axis along the armchair direction of BP contribute much more to the Raman enhancement than the CuPc molecules orientated in other directions, since there is a maximum charge overlap and thus charge interaction between CuPc molecules and BP in this configuration. This orientational effect results in the observed anisotropic Raman enhancement for some of the vibrational modes (such as the 1530 cm^{-1} mode) of CuPc molecules (Figure 8(b)). This experiment embodies the "electronic" enhancement character of CM, which offers a new insight into the CM process in SERS.

5. Influence of the Excitation Laser on GERS

5.1. *Wavelength scanned GERS effects*

Concerning the CM effect, at the beginning of the 1980s, Adrian [61] and Lippitsch [62] presented two distinct charge-transfer models, the excited state charge-transfer model and the ground state charge-transfer model, respectively. One of the important differences between the two models is the excitation laser wavelength dependence. For the former, the excited state mechanism is usually attributed to the formation of the charge-transfer complex by chemical bonding. The maximum enhancement is usually associated with the transition energy to the charge-transfer resonance state which depends on both the electronic structures of the molecule and the substrate. For the ground state mechanism, the charge-transfer complex is not necessary for the Raman enhancement. Charge-transfer can occur when the substrate and the molecule are both in the ground state, since the charge-transfer process does not need the help of the light radiation. However, the maximum enhancement wavelength is dependent on the electronic structure of the molecule itself. Therefore, by measuring the Raman spectra using several wavelengths, we can then obtain the Raman excitation profile, and the type of charge-transfer mechanisms that are most important. The technique, used for studying the mechanism of SERS by the Raman excitation profile, is called "wavelength-scanned Raman excitation spectroscopy". Van Duyne's group and Moskovits's group have investigated the electromagnetic mechanism in SERS using this technique successfully, and they have established this technique as a useful tool [63–66].

To investigate the CM in the GERS mechanism acting through the system being either in the excited state charge-transfer or the ground state charge-transfer mechanism, we carried out the wavelength-scanned Raman excitation spectroscopy measurement in the CuPc/graphene system [67]. Figure 9 shows the Raman excitation profiles of the different vibrational modes of the CuPc molecule in the GERS system, as well as the fitting lines. It was found that the profiles can be fitted well by the functional expression function for the ordinary resonant Raman intensity:

$$E_L = K \left| \frac{\langle f|E_{e-r}|b\rangle\langle b|E_{e-\text{ph}}|a\rangle\langle a|E_{e-r}|i\rangle}{(E_L - E_g - i\Gamma_a)(E_L - E_{\text{ph}} - E_g - i\Gamma_a)} \right|^2. \tag{2}$$

Instead of seeing an additional charge-transfer resonant state, we only observe the regular incident resonant peak (left peak in the dash line fitted peaks) and the scattered resonant peak (right peak in the dash line fitted peaks) for each of the vibrational modes. The energy difference between the two red peaks corresponds to the phonon energy of each vibrational mode. This indicates that the CM in the GERS system originates from the ground state charge-transfer rather than from the excited state charge-transfer.

It should be noted that most previous works on the chemical mechanism have been discussed using the excited state charge-transfer model because these studies were usually in a traditional SERS system based on a metal substrate to which the molecules were usually chemically bonded to the surface of the metal, where the electronic structure of the system was strongly changed [14, 68–71]. In such a situation, the resonant condition changed after the molecule was bonded on the surface of the metal, and the incident light could drive the charge-transfer transition between the metal and the molecule, thereby inducing the enhancement of the Raman signal. The strongest vibrational mode enhancement was usually related with the vibration involving those atoms that were

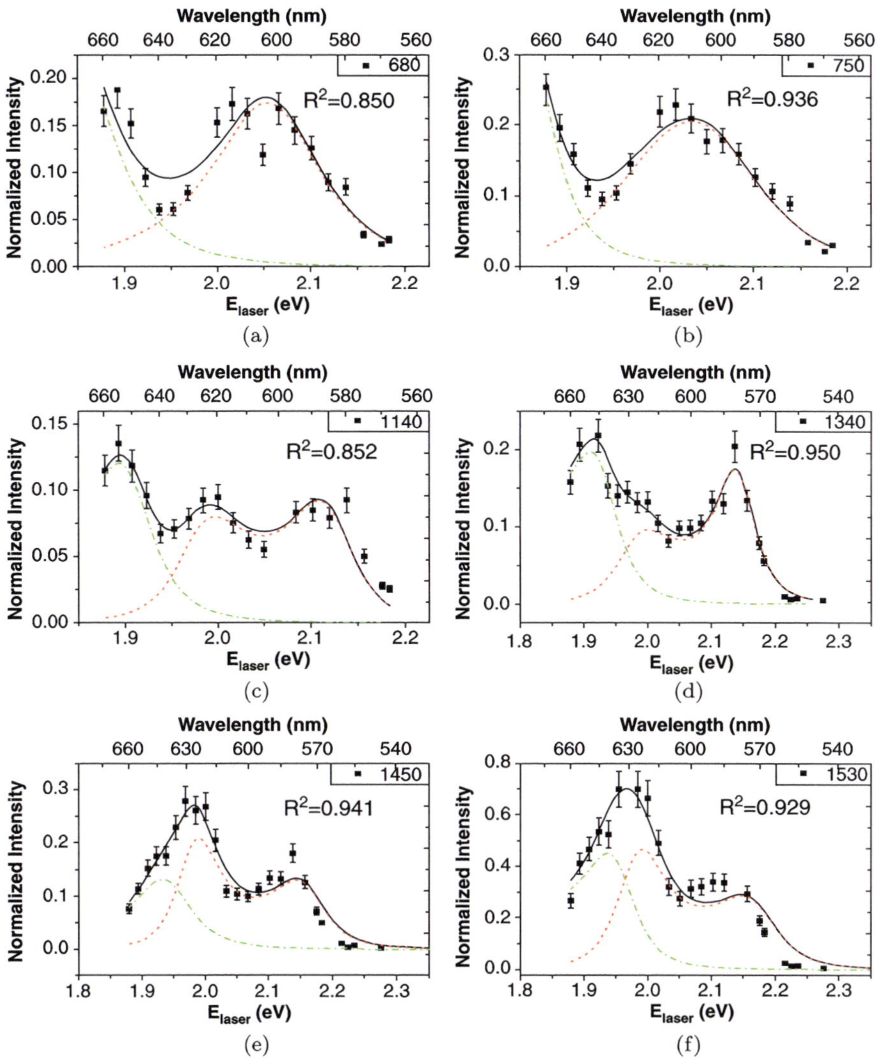

Figure 9. The Raman excitation profiles of different vibrational modes of the CuPc molecules in the GERS system [66]. The dash and dash dot lines are the two curves used to fit the experimental data. The black lines are the results of the fitting. The corresponding peak positions and the fitting related coefficients (R^2) are shown on the upper right corner. The error bars are from the uncertainty of the laser power. Since some Raman vibrational modes are too weak to resolve, for excitation by some laser lines, the data are not shown in this figure.

close to the metal–molecule bond. Here, in the GERS system, it is the physical interaction that occurs between graphene and the molecules that is important, and there is no chemical bonding between the molecules and the graphene surface. The electronic structure of the molecule will not change too much (this is also supported by the UV–visible absorption spectra). In addition, due to the abundance of π electrons on graphene and the continuity of the energy band of graphene dispersion relation, the ground state charge-transfer can easily occur between graphene and the conjugated organic molecules. Therefore, the origins of the chemical enhancement in the GERS system and of the metal-based SERS system are different in terms of the different manner of the charge-transfer.

5.2. *Laser polarization-difference technique to study the GERS effect of fluorescence molecules*

In Section 2.1, although we observe the PL suppression and the RRS of R6G on graphene, there is no evidence for a GERS effect and if there is Raman enhancement for R6G or not [72], since we cannot obtain the RRS of R6G without graphene in a normal Raman measurement (Figure 10(a)). In this section, we introduce the polarization-difference resonant Raman spectroscopy (PD-RRS) technique which enables us to observe the RRS of the fluorescence molecules (such as R6G) without graphene [73]. In the PD-RRS technique, a half-wave plate and a polarizer are added to the standard system. Two spectra (collected using parallel and perpendicular excitation laser polarization) are subtracted to cancel out the fluorescence background, leaving only the Raman signal. The parallel incoming polarization laser serves as a reference light source, and perpendicular incoming polarization laser serves as the measurement light source (the reverse use of the polarization beams can also be used). Thanks to the different depolarization ratios of the fluorescence (ρ_F) and the Raman signals (ρ_R), in this way, we can get the Raman spectra without the "fixed-structure noise" when the two spectra are subtracted from each other to cancel out the fluorescence background, leaving only the Raman signal. Then the absolute RR

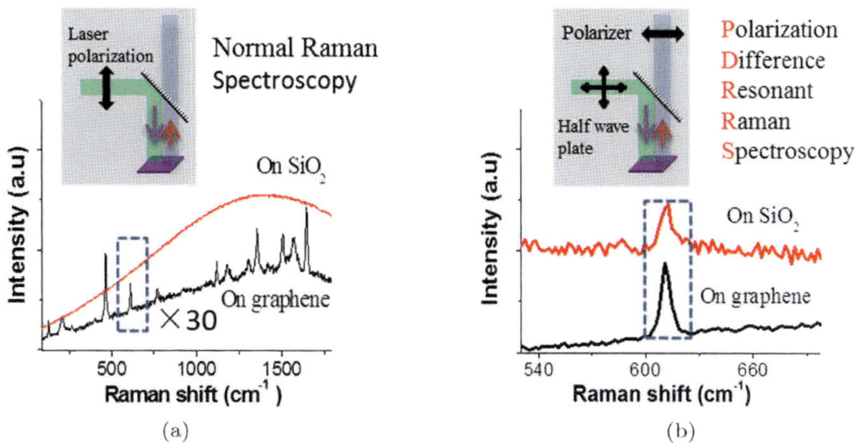

Figure 10. (a) The resonance Raman spectra of R6G on SiO_2 (red trace) and R6G on graphene (black trace) collected by the normal setup (see inset). (b) The resonance Raman spectra of R6G on SiO_2 (red trace) and R6G on graphene (black trace) collected by the polarization-difference Raman setup (see inset) [72].

intensity I^R is given by

$$I^R = \frac{1+\rho_R}{1-\rho_R/\rho_F} I^{\text{PDRRS}}. \tag{3}$$

Taking the 610 cm^{-1} mode of R6G as an example, Figure 10(b) shows the experimental setup of the PD-RRS, as well as the RRS signal of R6G on SiO_2 (red line) and on graphene (black line) from the PD-RRS. In this way, we can see that the RRS of R6G on graphene is about 1.6 times stronger than that on the blank SiO_2 substrate. Through the PD-RRS technique, we give the Raman enhancement evidence for the fluorescence molecules on graphene. Since the PD-RRS method allows us to study a large class of GERS related situations involving high fluorescence quantum yield dyes, it provides us with more capabilities to further explore the GERS effect in detail.

6. Combined System of Graphene and Metal

The GERS technique offers many advantages, especially in obtaining clean background, repeatable, and quantitative spectra. However, toward use for some practical applications, larger enhancement

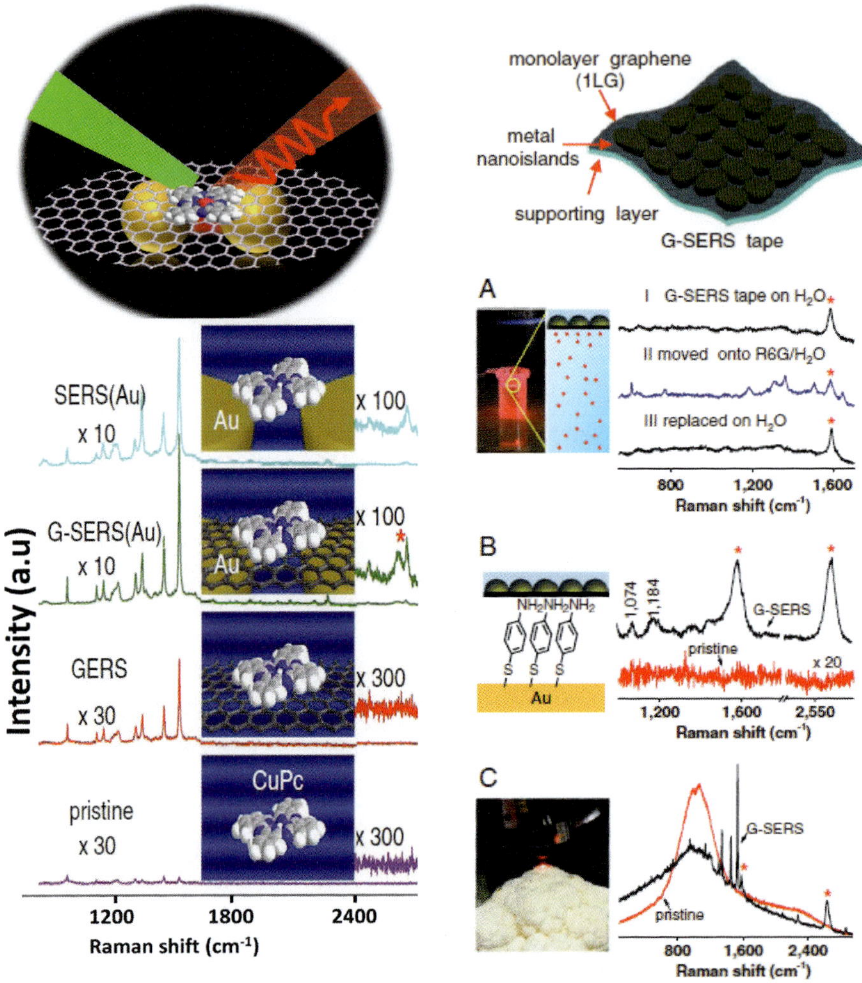

(a) (b)

Figure 11. (a) Raman spectra of CuPc in different systems, including pristine CuPc, the GERS system, the G-SERS system and the SERS system. Inset: Schematic illustration of Raman scattering with SERS substrates combining graphene and metal nanostructures. (b) Schematic illustration of the G-SERS tape and its application in detecting the chemicals in an aqueous solution (A), a flat self-assembly system (B), and a cauliflower vegetable surface (C) [73].

factors are expected. Combining GERS and conventional SERS is expected to maintain the advantages from both. A graphene-mediated SERS (G-SERS) system has been developed [74, 75], where graphene is placed over a metal to separate the adsorbate and the metal, as shown in Figure 11(a). A typical comparison of the Raman spectra in different systems is also shown in Figure 11(a). There are several significant advantages of combining graphene/graphene derivatives with conventional metal SERS substrates. First, graphene/graphene derivatives separate the metal and the molecules from each other, which avoids the side reaction of the molecules. Second, a graphene/graphene derivative protects the surface of the metal, which reduces the laser-induced carbonization of the metal surface. In addition, the flat and inert surface of graphene makes the density and the orientation of the molecules more controllable. All these conditions help obtain clean, reproducible and strong SERS signals. In addition, the results show that the G-SERS setup preserves the advantages from the GERS effect and obtains an even higher enhancement factor than the SERS case [74, 75].

This graphene/metal enhancement effect depends on several factors, such as the type of metal, the metal surface morphology, graphene thickness [76, 77], etc. For example, graphene has been integrated with various shapes of gold nanoparticles for SERS enhancement, such as arrays of gold nanodisks [78], photonic-crystal nanocavities [79], gold nanorods [80], and silver/copper/gold nanospheres [81]. The effectiveness of graphene in stabilizing the SERS system of thermally grown Au and Ag nanoparticles and achieving strong enhancement has been demonstrated [74, 81]. These studies showed that the detection limit is ~100 fold larger with graphene coupled with the SERS system [82, 83]. In future research, additional promising structures tailored for specific applications are expected.

Based on the concept of G-SERS, a flexible and transparent G-SERS tape was designed with a PMMA/metal/graphene structure [74], which is expected to have promising applications in analyzing micro/trace amounts of chemicals in different systems. Figure 11(b) shows illustrations of the G-SERS tape used for three typical applications by taping the G-SERS tape on to the surfaces of an aqueous solution (A), a flat self-assembled film (B), and a cauliflower vegetable surface (C). This platform has both the advantages from GERS and SERS and could lead to the effective detection of both chemicals in the environment and medical applications such as detection of possible life-threatening diseases.

7. Summary

In summary, we introduced a newly developed technique for Raman enhancement, named GERS. It is a novel branch of SERS, which uses graphene, as well as other 2D materials, as an active substrate. Different from the conventional SERS substrate, which is based on the rough surface of a metal and is dominated by the EM, GERS shows its superiority in lighting up molecules uniformly, repeatably and quantifiably, due to the flat and chemically inert surface of 2D materials. In particular, it is an available model and substrate to study the CM, since GERS has no disturbance from the EM effect, which opens a window to study the CM effect in more detail and offers possibilities to realize the additional picture for the SERS mechanism. In addition, great progress has been made to extend this effect to other graphene derivatives, and to enhance the effect of more diverse molecules with particular functions. It offers potential stages to study this effect more deeply and to reveal new applications for the CM effect. In the future, from the fundamental study side, more detailed theoretical models are expected based on this system to understand the principles of the CM effect. From the application side, more promising applications are expected in the process monitoring of chemical and biological systems, utilizing the ability of graphene in "lighting up" the molecule as it is excited.

References

1. Kneipp, K., Moskovits, M. and Kneipp, H. (2006). *Surface-Enhanced Raman Scattering: Physics and Applications* (Springer, Berlin, New York).
2. Schlücker, S. (2011). *Surface Enhanced Raman Spectroscopy: Analytical, Biophysical and Life Science Applications*, (Wiley-VCH Verlag, Weinheim, Germany).

3. Ozaki, Y., Kneipp, K. and Aroca, R. (2014). *Frontiers of Surface-Enhanced Raman Scattering: Single-Nanoparticles and Single Cells* (Wiley, Chichester, West Sussex).
4. Schlücker, S. (2014). Surface-enhanced Raman spectroscopy: Concepts and chemical applications. *Angew Chem. Int. Ed.* **53**, pp. 4756–4795.
5. Fan, M., Andrade, G. F. S. and Brolo, A. G. (2011). A review on the fabrication of substrates for surface enhanced Raman spectroscopy and their applications in analytical chemistry. *Anal. Chim. Acta.* **693**, pp. 7–25.
6. Otto, A., Mrozek, I., Grabhorn, H. and Akemann, W. (1992). Surface-Enhanced Raman Scattering. *J. Phys. Condens. Matter* **4**, pp. 1143–1212.
7. Tian, Z. Q., Ren, B. and Wu, D. Y. (2002). Surface-enhanced Raman scattering: From noble to transition metals and from rough surfaces to ordered nanostructures. *J. Phys. Chem. B* **106**, pp. 9463–9483.
8. Kneipp, K., Wang, Y., Kneipp, H. *et al.* (1997). Single molecule detection using surface-enhanced Raman scattering (SERS). *Phys. Rev. Lett.* **78**, pp. 1667–1670.
9. Campion, A., Ivanecky, J. E., Child, C. M. and Foster, M. (1995). On the mechanism of chemical enhancement in surface-enhanced Raman scattering. *J. Am. Chem. Soc.* **117**, pp. 11807–11808.
10. Wu, D. Y., Liu, X. M., Duan, S. *et al.* (2008). Chemical enhancement effects in SERS spectra: A quantum chemical study of pyridine interacting with copper, silver, gold and platinum metals. *J. Phys. Chem. C* **112**, pp. 4195–4204.
11. Liu, L., Chen, D., Ma, H. and Liang, W. (2015). Spectral characteristics of chemical enhancement on SERS of benzene-like derivatives: Franck–Condon and Herzberg–Teller contributions. *J. Phys. Chem. C* **119**, pp. 27609–27619.
12. Jensen, L., Zhao, L. L. and Schatz, G. C. (2007). Size-dependence of the enhanced Raman scattering of pyridine adsorbed on Ag_n ($n = 2 - 8$, 20) Clusters. *J. Phys. Chem. C* **111**, pp. 4756–4764.
13. Morton, S. M. and Jensen, L. (2009). Understanding the molecule–surface chemical coupling in SERS. *J. Am. Chem. Soc.* **131**, pp. 4090–4098.
14. Otto, A. and Futamata, M. (2006). *Surface-Enhancemnt Raman Scattering,* eds. Kneipp, K., Moskovits, M. and Kneipp, H., Electronic Mechanisms of SERS (Springer, Berlin, Heidelberg), pp. 147–182.
15. Morton, S. M. and Jensen, L. (2009). Understanding the molecule–surface chemical coupling in SERS. *J. Am. Chem. Soc.* **131**, pp. 4090–4098.
16. Lombardi, J. R. and Birke, R. L. (2008). A unified approach to surface-enhanced raman spectroscopy. *J. Phys. Chem. C* **112**, pp. 5605–5617.
17. Lombardi, J. R., Birke, R. L., Lu, T. and Xu, J. (1986). Charge-transfer theory of surface enhanced Raman spectroscopy: Herzberg–Teller contributions. *J. Chem. Phys.* **84**, p. 4174.
18. Lombardi, J. R. and Birke, R. L. (2007). Time-dependent picture of the charge-transfer contributions to surface enhanced Raman spectroscopy. *J. Chem. Phys.* **126**, p. 244709.
19. Arenas, J. F., Soto, J., Pelaez, D. *et al.* (2005). Understanding complex surface-enhanced Raman scattering, using quantum chemical calculations. *Int. J. Quantum Chem.* **104**, pp. 681–694.

20. Arenas, J. F., Woolley, M. S., Otero, J. C. and Marcos, J. I. (1996). Charge-transfer processes in surface-enhanced Raman scattering. Franck–Condon active vibrations of pyrazine. *J. Phys. Chem.* **100**, pp. 3199–3206.
21. Yang, L., Jiang, X., Ruan, W. *et al.* (2008). Observation of enhanced Raman scattering for molecules adsorbed on TiO_2 nanoparticles: Charge-transfer contribution. *J. Phys. Chem. C* 112, pp. 20095–20098.
22. Musumeci, A., Gosztola, D., Schiller, T. *et al.* (2009). SERS of Semiconducting Nanoparticles (TiO_2 Hybrid Composites). *J. Am. Chem. Soc.* **131**, pp. 6040–6041.
23. Tarakeshwar, P., Finkelstein-Shapiro, D., Hurst, S. J. *et al.* (2011). Surface-enhanced Raman scattering on semiconducting oxide nanoparticles: Oxide nature, size, solvent, and pH effects. *J. Phys. Chem. C* **115**, pp. 8994–9004.
24. Hayashi, S., Koh, R., Ichiyama, Y. and Yamamoto, K. (1988). Evidence for surface-enhanced Raman scattering on nonmetallic surfaces: Copper phthalocyanine molecules on GaP small particles. *Phys. Rev. Lett.* **60**, pp. 1085–1088.
25. Wang, Y., Ruan, W., Zhang, J. *et al.* (2009). Direct observation of surface-enhanced Raman scattering in ZnO nanocrystals. *J. Raman. Spectrosc.* **40**, pp. 1072–1077.
26. Gupta, A., Sakthivel, T. and Seal, S. (2015). Recent development in 2D materials beyond graphene. *Prog. Mater. Sci.* **73**, pp. 44–126.
27. Geim, A. K. and Novoselov, K. S. (2007). The rise of graphene. *Nat. Mater.* **6**, pp. 183–191.
28. Ling, X., Wang, H., Huang, S. *et al.* (2015). The renaissance of black phosphorus. *Proc. Natl. Acad. Sci.* **112**, pp. 4523–4530.
29. Ling, X., Xie, L., Fang, Y. *et al.* (2010). Can graphene be used as a substrate for Raman enhancement?. *Nano Lett.* **10**, pp. 553–561.
30. Ling, X., Huang, S., Deng, S. *et al.* (2015). Lighting up the Raman signal of molecules in the vicinity of graphene related materials. *Acc. Chem. Res.* **48**, pp. 1862–1870.
31. Matousek, P., Towrie, M., Ma, C. *et al.* (2001). Fluorescence suppression in resonance Raman spectroscopy using a high-performance picosecond Kerr gate. *J. Raman. Spectrosc.* **32**, pp. 983–988.
32. Martyshkin, D. V., Ahuja, R. C., Kudriavtsev, A. and Mirov, S. B. (2004). Effective suppression of fluorescence light in Raman measurements using ultrafast time gated charge coupled device camera. *Rev. Sci. Instrum.* **75**, p. 630.
33. Lakowicz, J. R. (2006). *Principles of Fluorescence Spectroscopy* 3rd edn. (Springer, New York).
34. Xie, L., Ling, X., Fang, Y. *et al.* (2009). Graphene as a substrate to suppress fluorescence in resonance Raman spectroscopy. *J. Am. Chem. Soc.* **131**, pp. 9890–9891.
35. Ling, X. (2012). Graphene enhanced Raman scattering effect. Ph.D. thesis, Peking University.
36. Bruna, M. and Borini, S. (2009). Optical constants of graphene layers in the visible range. *Appl. Phys. Lett.* **94**, p. 31901.

37. Ju, L., Geng, B., Horng, J. *et al.* (2011). Graphene plasmonics for tunable terahertz metamaterials. *Nat. Nanotechnol.* **6**, pp. 630–634.
38. Dantham, V. R., Bisht, P. B., Kalanoor, B. S. *et al.* (2012). Restricting charge transfer in dye-graphene system. *Chem. Phys. Lett.* **521**, pp. 130–133.
39. Lu, Y. H., Chen, W., Feng, Y. P. and He, P. M. (2009). Tuning the electronic structure of graphene by an organic molecule. *J. Phys. Chem. B* **113**, pp. 2–5.
40. Ling, X. and Zhang. J. (2010). First-layer effect in graphene-enhanced Raman scattering. *Small* **6**, pp. 2020–2025.
41. Mrozek, I. and Otto, A. (1990). Quantitative separation of the "classical" electromagnetic and the "chemical" contribution to surface enhanced Raman scattering. *J. Electron Spectrosc. Relat. Phenom.* **54–55**, pp. 895–911.
42. Mrozek, I. and Otto, A. (1990). Long and short-range effects in SERS from silver. *Europhys. Lett.* **11**, pp. 243–248.
43. Sanda, P. N., Warlaumont, J. M., Demuth, J. E. *et al.* (1980). Surface-enchanced Raman scattering from pyridine on Ag(111). *Phys. Rev. Lett.* **45**, pp. 1519–1523.
44. Eesley, G. L. (1981). Coverage dependence of enhanced adsorbate Raman scattering. *Phys. Lett. A* **81**, pp. 193–196.
45. Abe, H. (1981). Surface-enhanced Raman spectroscopy of CO adsorbed on colloidal silver particles. *J. Chem. Phys.* **74**, p. 792.
46. Ling, X., Wu, J., Xu, W. and Zhang, J. (2012). Probing the effect of molecular orientation on the intensity of chemical enhancement using graphene-enhanced Raman spectroscopy. *Small* **8**, pp. 1365–1372.
47. Huang, S., Ling, X., Liang, L. *et al.* (2015). Molecular selectivity of graphene-enhanced Raman scattering. *Nano Lett.* **15**, pp. 2892–2901.
48. Barros, E. B. and Dresselhaus, M. S. (2014). Theory of Raman enhancement by two-dimensional materials: Applications for graphene-enhanced Raman spectroscopy. *Phys. Rev. B.* **90**, p. 035443.
49. Ling, X., Wu, J., Xie, L. and Zhang, J. (2013). Graphene-thickness-dependent graphene-enhanced Raman scattering. *J. Phys. Chem. C* **117**, pp. 2369–2376.
50. Huh, S., Park, J., Kim, Y. S. *et al.* (2011). UV/Ozone-oxidized large-scale graphene platform with large chemical enhancement in surface-enhanced Raman scattering. *ACS Nano* **5**, pp. 9799–9806.
51. Yang, H., Hu, H., Ni, Z. *et al.* (2013). Comparison of surface-enhanced Raman scattering on graphene oxide, reduced graphene oxide and graphene surfaces. *Carbon* **62**, pp. 422–429.
52. Cheng, H., Zhao, Y., Fan, Y. *et al.* (2012). Graphene-quantum-dot assembled nanotubes: A new platform for efficient Raman enhancement. *ACS Nano* **6**, pp. 2237–2244.
53. Lv, R., Li, Q., Botello-Méndez, A. R. *et al.* (2012). Nitrogen-doped graphene: Beyond single substitution and enhanced molecular sensing. *Sci. Rep.* **2**, p. 586.
54. Yu, X., Cai, H., Zhang, W. *et al.* (2011). Tuning chemical enhancement of SERS by controlling the chemical reduction of graphene oxide nanosheets. *ACS Nano* **5**, pp. 952–958.

55. Sun, S., Zhang, Z. and Wu, P. (2013). Exploring graphene nanocolloids as potential substrates for the enhancement of Raman scattering. *ACS Appl. Mater. Interfaces.* **5**, pp. 5085–5090.
56. Liang, W., Chen, X., Sa, Y. *et al.* (2012). Graphene oxide as a substrate for Raman enhancement. *Appl. Phys. A* **109**, pp. 81–85.
57. Wu, D. Y., Li, J. F., Ren, B. and Tian, Z. Q. (2008). Electrochemical surface-enhanced Raman spectroscopy of nanostructures. *Chem. Soc. Rev.* **37**, p. 1025.
58. Xu, H., Xie, L., Zhang, H. and Zhang, J. (2011). Effect of Graphene Fermi level on the Raman scattering intensity of molecules on graphene. *ACS Nano* **5**, pp. 5338–5344.
59. Ling, X., Fang, W., Lee, Y. H. *et al.* (2014). Raman enhancement effect on two-dimensional layered materials: Graphene, h-BN and MoS_2. *Nano Lett.* **14**, pp. 3033–3040.
60. Lin, J., Liang, L., Ling, X. *et al.* (2015). Enhanced Raman scattering on in-plane anisotropic layered materials. *J. Am. Chem. Soc.* **137**, pp. 15511–15517.
61. Adrian, F. J. (1982). Charge transfer effects in surface-enhanced Raman scatteringa. *J. Chem. Phys.* **77**, p. 5302.
62. Lippitsch, M. E. (1984). Ground-state charge transfer as a mechanism for surface-enhanced Raman scattering. *Phys. Rev. B* **29**, pp. 3101–3110.
63. Zhao, J., Dieringer, J. A., Zhang, X. *et al.* (2008). Wavelength-scanned surface-enhanced resonance Raman excitation spectroscopy. *J. Phys. Chem. C* **112**, pp. 19302–19310.
64. Vlčková, B., Gu, X. J. and Moskovits, M. (1997). SERS excitation profiles of phthalazine adsorbed on single colloidal silver aggregates as a function of cluster size. *J. Phys. Chem. B* **101**, pp. 1588–1593.
65. Ye, J., Hutchison, J. A., Uji-i, H. *et al.* (2012). Excitation wavelength dependent surface enhanced Raman scattering of 4-aminothiophenol on gold nanorings. *Nanoscale* **4**, p. 1606.
66. McFarland, A. D., Young, M. A., Dieringer, J. A. and Van Duyne, R. P. (2005). Wavelength-scanned surface-enhanced Raman excitation spectroscopy. *J. Phys. Chem. B* **109**, pp. 11279–11285.
67. Ling, X., Moura, L. G., Pimenta, M. A. and Zhang, J. (2012). Charge-transfer mechanism in graphene-enhanced Raman scattering. *J. Phys. Chem. C* **116**, pp. 25112–25118.
68. Nakai, H. and Nakatsuji, H. (1995). Electronic mechanism of the surface enhanced Raman scattering. *J. Chem. Phys.* **103**, p. 2286.
69. Sun, M., Liu, S., Chen, M. and Xu, H. (2009). Direct visual evidence for the chemical mechanism of surface-enhanced resonance Raman scattering via charge transfer. *J. Raman Spectrosc.* **40**, pp. 137–143.
70. Sun, M., Liu, S., Li, Z. *et al.* (2009). Direct visual evidence for the chemical mechanism of surface-enhanced resonance Raman scattering via charge transfer: (II) Binding-site and quantum-size effects. *J. Raman Spectrosc.* **40**, pp. 1172–1177.

71. Sun, M., Wan, S., Liu, Y. *et al.* (2008). Chemical mechanism of surface-enhanced resonance Raman scattering via charge transfer in pyridine–Ag_2 complex. *J. Raman Spectrosc.* **39**, pp. 402–408.
72. Thrall, E. S., Crowther, A. C., Yu, Z. and Brus, L. E. (2012). R6G on graphene: High Raman detection sensitivity, yet decreased Raman cross-section. *Nano Lett.* **12**, pp. 1571–1577.
73. Deng, S., Xu, W., Wang, J. *et al.* (2014). Direct measurement of the Raman enhancement factor of rhodamine 6G on graphene under resonant excitation. *Nano Res.* **7**, pp. 1271–1279.
74. Xu, W., Ling, X., Xiao, J. *et al.* (2012). Surface enhanced Raman spectroscopy on a flat graphene surface. *Proc. Natl. Acad. Sci.* **109**, pp. 9281–9286.
75. Xu, W., Mao, N. and Zhang, J. (2013). Graphene: A platform for surface-enhanced Raman spectroscopy. *Small* **9**, pp. 1206–1224.
76. Lee, J., Novoselov, K. S. and Shin, H. S. (2011). Interaction between metal and graphene: Dependence on the layer number of graphene. *ACS Nano* **5**, pp. 608–612.
77. Kuo, C. C. and Chen, C. H. (2014). Graphene thickness-controlled photocatalysis and surface enhanced Raman scattering. *Nanoscale* **6**, pp. 12805–12813.
78. Schedin, F., Lidorikis, E., Lombardo, A. *et al.* (2010). Surface-enhanced Raman spectroscopy of graphene. *ACS Nano* **4**, pp. 5617–5626.
79. Gan, X., Mak, K. F., Gao, Y. *et al.* (2012). Strong enhancement of light–matter interaction in graphene coupled to a photonic crystal nanocavity. *Nano Lett.* **12**, pp. 5626–5631.
80. Nguyen, T. H. D., Zhang, Z., Mustapha, A. *et al.* (2014). Use of graphene and gold nanorods as substrates for the detection of pesticides by surface enhanced Raman spectroscopy. *J. Agric. Food Chem.* **62**, pp. 10445–10451.
81. Liu, Y., Hu, Y. and Zhang, J. (2014). Few-layer graphene-encapsulated metal nanoparticles for surface-enhanced Raman spectroscopy. *J. Phys. Chem. C* **118**, pp. 8993–8998.
82. Zhao, Y., Li, X., Du, Y. *et al.* (2014). Strong light–matter interactions in sub-nanometer gaps defined by monolayer graphene: Toward highly sensitive SERS substrates. *Nanoscale* **6**, pp. 11112–11120.
83. Zhao, Y., Chen, G., Du, Y. *et al.* (2014). Plasmonic-enhanced Raman scattering of graphene on growth substrates and its application in SERS. *Nanoscale* **6**, pp. 13754–13760.

Chapter 16

Charge-Transfer-Induced Enhancement of Raman Scattering Based on Semiconductors

Wei Ji*, Xiao Xia Han† and Bing Zhao†,‡

*School of Chemistry,
Dalian University of Technology,
Dalian 116023, P. R. China

†State Key Laboratory of Supramolecular Structure and Materials,
Jilin University, Changchun
130012, P. R. China
‡zhaob@mail.jlu.edu.cn

1. Introduction

Surface-enhanced Raman scattering (SERS) has been a particularly fast growing area of research over the last 40 years, which has been widely used as a powerful tool for fundamental research as well as for ultrasensitive analysis [1–4]. As a surface phenomenon, the advance in SERS technique is closely associated with the evolution of SERS-active substrates. Much attention has been paid to the design and construction of SERS substrates with strong Raman enhancement ever since the discovery of SERS [5]. Despite the promising results that have been achieved, the selection of substrate materials is rather limited. So far, the large enhancements were observed in a handful of noble metal surfaces (mainly Au and Ag). Unfortunately, noble metal materials are not ideal for all purposes and perform rather poorly with regard to stability. The difficulty in generating the desired active SERS substrate with high reproducibility, high stability, and excellent signal uniformity continues to press the need for new substrate materials. Furthermore, the generation of a substrate with excellent biocompatibility to explore biological system also requires new substrate materials. Thus, the synthesis of non-metal substrate becomes an active area of research as new and synthetic techniques are developed. Recently, SERS-active semiconductor materials have been extensively studied due to their high photoactivity, low cost, low toxicity, and good chemical and thermal stability, as well as possessing intriguing application potential in a variety of fields. Intense research has been carried out on enhancement of Raman scattering based on semiconductors, which has been focused on understanding the fundamental principles, developing the active substrate materials, and expanding the scope of applications. Considering the increasing number of publications, herein, we would like to introduce the recent advances in charge-transfer-induced enhancement of Raman scattering based on semiconductors.

2. SERS-Active Semiconductor Materials

With the explosive growth in nanoscience and nanotechnology, enhancement of Raman scattering based on semiconductors has undergone a significant development. A large number of semiconductor materials have been demonstrated to be SERS-active substrates. Table 1 shows the summary of the reported SERS-active semiconductor materials and possible mechanism proposed in the corresponding references. The substrate materials may be roughly grouped under the following headings.

2.1. *Metal oxides*

The study of SERS on semiconductor materials can be traced back to the 1980s, Yamada *et al.* observed enhanced Raman signal of pyridine adsorbed on the surface of single-crystal NiO (110) [6]. The Raman frequencies for pyridine adsorbed on NiO are identical to that observed on Ni. Similar observations have also been made on

Table 1. Summary of the reported SERS-active semiconductor materials.

Semi-conductor	Form or size	Probe	EF	Mechanism	References
NiO	Crystal surface (100)	Pyridine	Low	Resonance Raman	[6]
TiO_2	Crystal surface(001)	Pyridine	Low	Resonance Raman	[7]
TiO_2	Mesoporous film	Ru-bpy	Low	Resonance Raman	[8]
TiO_2	10 nm particle	Mercaptobenzoic acid	10^3	CT	[9–10]
TiO_2	2–5 nm colloid	Enediol molecules	High	CT	[11–12]
TiO_2	5 nm colloid	Nitrothiophenol isomers	10^2–10^3	CT	[13]
ZnO	50 nm colloid	D266	50	CT	[14–15]
ZnO	15–40 nm particle	Mercaptopyridine	10^3	CT	[16]
Fe_2O_3	70–80 nm colloid	Pyridine	10^3	EM	[17]
Fe_2O_3	70-80 nm colloid	Pyridine	10^3	CT	[10, 18]
Fe_3O_4	120 nm colloid	TSPP	30	CT	[19]
Al_2O_3	80 nm AAO film	Cytochrome C	30	EM	[20]
Ag_2O	28 nm colloid	Pyridine	10^5	EM	[21]
Cu_2O	Roughness	Pyridine	Low	CT	[22]
Cu_2O	150 nm particle	Mercaptopyridine	High	CT, EM	[23]
CuO	20–60 nm particle	Mercaptopyridine	10^2	CT	[24–25]
SnO_2	40 nm particle	Mercaptobenzoic acid	High	CT	[26]
Pb_3O_4	100 nm particle	Mercaptopyridine	10^2	CT	[27]
MoO_3	nanobelt	Mercaptobenzoic acid	10^3	CT	[28]
$W_{18}O_{49}$	nanowire	Rodamine 6G	10^5	CT	[29]
CdS	8 nm particle	Mercaptopyridine	10^2	CT	[30]
CdSe	Quantum dots	Pyridine	10^5	CT	[31]
CdTe	3 nm particle	Mercaptopyridine	10^4	CT	[32]
PbS	Quantum dots	Mercaptopyridine	10^3	CT	[33]

(Continued)

Table 1. (*Continued*)

Semi-conductor	Form or size	Probe	EF	Mechanism	References
ZnS	3 nm particle	Mercaptopyridine	10^3	CT	[34]
ZnSe	Thin film	Mercaptopyridine	10^6	CT, EM	[35–36]
AgX	Colloid	Dyes	50–100	Resonance Raman	[37–40]
PbS	5–10 nm particle	Mercaptopyridine	High	CT	[41]
InAs/GaAs	Quantum dots film	Pyridine	10^3	CT	[42]
Doped TiO_2	10 nm particle	Mercaptobenzoic acid	10^3	CT	[43–45]
GaP	40–100 nm particle	CuPc	700	EM	[46]
Si	200–370 nm particle	Mercaptobenzoic acid	High	Mie resonance	[47]
Si	nanowire	Rodamine 6G	Low	CT	[48]
Ge	nanowire	N719	Low	CT	[48]
Graphene	layer	Dye molecules	2–17	CT	[49]
SiO2@TiO_2	Particle	Methylene blue	High	Mie resonance	[50]
TiO_2	Photonic microarray	Methylene blue	10^4	Matter-light	[51]

the surface of Ti and single-crystal TiO_2 (001) [7]. The plasmon resonance frequency of metal oxides such as NiO and TiO_2 lies in the infrared region, which is far away from the excitation wavelengths used in the SERS experiments. Meanwhile, the surface plasmon of Ni or Ti is damped due to the absorption of the metal themselves even though the surfaces might have the appropriate roughness. Therefore, the electromagnetic mechanism cannot be the dominant contribution to SERS in those systems. The chemical bonding between nitrogen of pyridine and the atomic sites of semiconductor could form a new chemisorbed species, which gives rise to the SERS effect via charge-transfer excitation from the adsorption site to the pyridine. These observations are similar to the mechanism of resonance Raman scattering effect. However, the enhancement factor is just 10 orders of magnitude.

In comparison with single-crystal semiconductors, nanoscale TiO_2 shows a stronger SERS activity and their enhancement factors can reach as high as 10^3 [9, 13]. For example, the anatase crystal TiO_2 nanoparticles (NPs) with the diameter of 10 nm were fabricated by calcining the sol-hydrothermal production at a certain temperature. The enhancement mechanism is regard as the charge-transfer mechanism corresponding to electron transitions from surface states of TiO_2 to molecule. Furthermore, SERS spectra were performed for a family of mercapto-compounds, including 4-mercaptobenzoin acid (4-MBA), 4-mercaptopyridine (4-MPY), and 4-aminothiophenol (PATP). While these three molecules have same Raman scattering cross-section, the enhancement effects for them adsorbed on TiO_2 nanoparticles are different with each other. The order of enhancements is 4-MBA $>$ 4-MPY $>$ PATP, which agrees with the attracting electron ability of functional group at the *para*-position of mercapto group. Thus, we can conclude that the SERS enhancements increase with the increase of the electron-attracting ability at the *para*-position of mercapto group, which further suggests that the SERS could be attributed to TiO_2-to-molecule charge-transfer transitions.

Using enediol molecules as probe, similar SERS experiments were also performed on colloidal TiO_2 nanoparticles [11]. However, the SERS is attributed to the molecule-to-TiO_2 charger-transfer transitions. In this work, the strong electronic coupling of enediol molecule to colloidal TiO_2 nanoparticles results in the formation of a charge-transfer complex. The charge-transfer complex is regard as the key role in the observation of SERS. On the basis of these facts, it has been clarified that the enhancement of Raman scattering based on semiconductors largely depends not only on the surface properties of semiconductor, but also on the intrinsic nature of the adsorbed molecules. Besides, the particle size of semiconductor is another important factor that influenced both the SERS and the phonon modes of TiO_2 nanoparticles due to the quantum confinement [52]. The highest Raman enhancement of 4-MBA adsorbed on TiO_2 nanoparticles was observed at the particle size 10.9 nm with the 514.4 nm excitation.

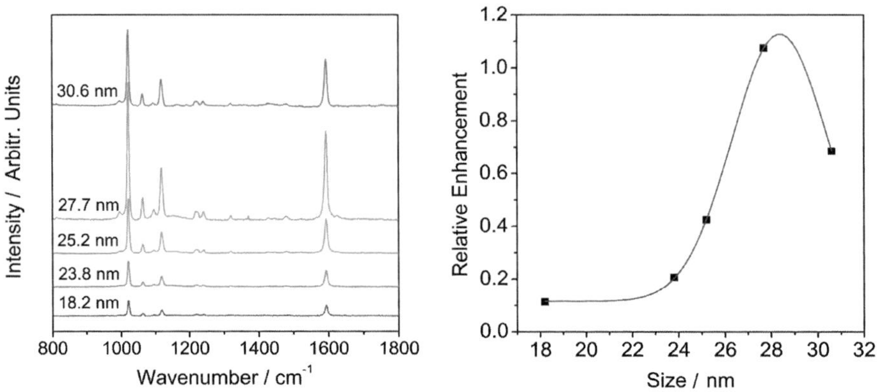

Figure 1. (a) SERS spectrum of 4-MPY adsorbed on ZnO nanoparticles with various size. (b) Relative Intensity of the 1021 cm^{-1} mode of 4-MPY adsorbed on ZnO nanoparticles as a function of particle size. Reproduced with permission from Ref. [16] copyright 2007, American Institute of Physics.

ZnO has been an extensively researched semiconductor for various optoelectronic applications, and the SERS of cyanine dye D266 molecules adsorbed on ZnO colloids was observed by using 476.5 nm excitation [14]. The Raman enhancement factor is estimated to be more than 50. It should be noted that some vibrational modes of D226 were selectively enhanced due the charge-transfer contribution, which is similar with the chemical enhancement mechanism existing in metal-based SERS. Further studies were performed on ZnO nanoparticles by using 4-Mpy and 1,4-bis[2-(4-pyridyl)ethenyl]-benzene as probe molecules [15, 16]. Based on the UV–vis spectrum of Zn^{2+}/4-Mpy complex, ZnO nanoparticle, 4-Mpy-modified ZnO nanoparticle, and 4-Mpy aqueous solution, the resonance effect originating from Zn^{2+}/4-Mpy complex was clearly ruled out. The charge-transfer transitions between valence band of the ZnO and the empty π^* levels of the 4-Mpy are considered to be the dominant contribution to SERS (see Figure 1). More importantly, a particle size-dependent SERS phenomenon was observed on the surface of ZnO nanoparticles with the size in the range of 18–31 nm diameter [16]. This result is very similar to the observation of the size-dependent resonance effect in 4-MBA-adsorbed TiO_2 system.

Other oxide semiconductor materials that have been chosen as the substrates, including Ag_2O, α-Fe_2O_3, Fe_2O_4, Cu_2O, $W_{18}O_{49}$. The enhancement mechanism of the probe molecules adsorbed on these oxide semiconductor materials is also considered as the charge-transfer contribution. It should be noted that the excitation wavelength dependence of Raman enhancement has been demonstrated on CuO nanocrystals by using 4-Mpy as a probe [25]. The relative peak intensities vary with the excitation wavelength and it exhibits a maximum enhancement under 532 nm excitation. This result is not surprising since the charge-transfer mechanism is a resonance Raman-like process that can be turned into and out of resonance by changing excitation energy. Besides, nanoparticle size and excitation wavelength-dependent SERS of 4-Mpy on PbS quantum dots were studies by Lombardi *et al.* [41]

2.2. *Metal chalcogenides*

In addition to oxide semiconductors, chalcogenide semiconductors, including CdS [30], CdSe [31], CdTe [32], PbS [41], and ZnSe [35], have also been well studies, the dominant contribution for Raman enhancement of molecules adsorbed on their surfaces is derived from charge-transfer contribution. Interestingly, the enhancement factor on the order of 10^5 has been observed for adsorbed pyridine on CdSe quantum dots and ZnSe quantum dots, respectively. As such, it was found that a new phonon mode appears at 248 cm^{-1}, which is attributed to the forbidden phonon mode in colloidal ZnSe quantum dots [36]. Because the excitation wavelength is near the exciton resonance of ZnSe quantum dots, the additional enhancement was ascribed to the exciton resonance contribution. Besides, the CdTe quantum dot is described to provide an enhancement factor on the order of 10^4, which is also associated with the exciton resonance [32]. Further observations show that the photoluminescence of CdTe quantum dots is suppressed after molecule adsorption to its surface, indicating the strongly interaction between exciton and adsorbed molecule. This result further proved that the exciton resonance of semiconductor could contribute to the SERS enhancement.

2.3. *Doped semiconductor*

Nanoscale semiconductors possess a large amount of intrinsic defects, that is, oxygen vacancies and metal interstitials. These defects give rise to surface states in the bandgap. As demonstrated in the previous works, the surface states play an important role in semiconductor-to-molecule charge-transfer enhancement [9]. The SERS enhancement capacity of broad bandgap semiconductors can be enhanced obviously by the appropriate doping of transition metals [43–45, 53]. It was clearly found that TiO_2 doped with transition metals such as Fe, Co, Ni, Cu, Mn and Zn has extended the spectral response of TiO_2well into the visible region. The XPS measurements show that the electronegativity of doping ions has certain influence on electron cloud densities of Ti and O ions in the crystal lattice [44]. Accodingly, the doping concentration and intrinsic nature of doping ions could significantly influence the interfacial charge-transfer, thus leading to the variation in charge-transfer contribution of SERS.

The enhancement in Raman intensities is depended on the doping amount of the transition metals, because the dopant can act as an effective trap for interfacial charge-transfer or charge-carries recombination. For example, Co^{2+} acted as the traps to capture the photoinduced electrons when the dopant content was less than 5% mol, which prohibited the recombination of photoinduced carriers and improved the TiO_2-to-moleucle charge-transfer transition [45]. When the dopant content exceeded 5% mol, CoO formed and became the recombination centers of photoinduced carriers, which prohibited the TiO_2-to-moleucle charge-transfer transition (see Figure 2). Besides, the ion radius of transition metals also affects the SERS enhancement.

2.4. *Other semiconductor materials*

The SERS from other semiconductor materials has also been reported, most experimental investigations indicated that the origin of SERS in these systems is mainly due to charge-transfer contribution. Thus, we will ignore that for this section and focus on the special enchantment contribution. Graphene has been demonstrated as an active SERS substrate which can quench the fluorescence of dye molecules, including rhodamine 6G, phthalocyanine,

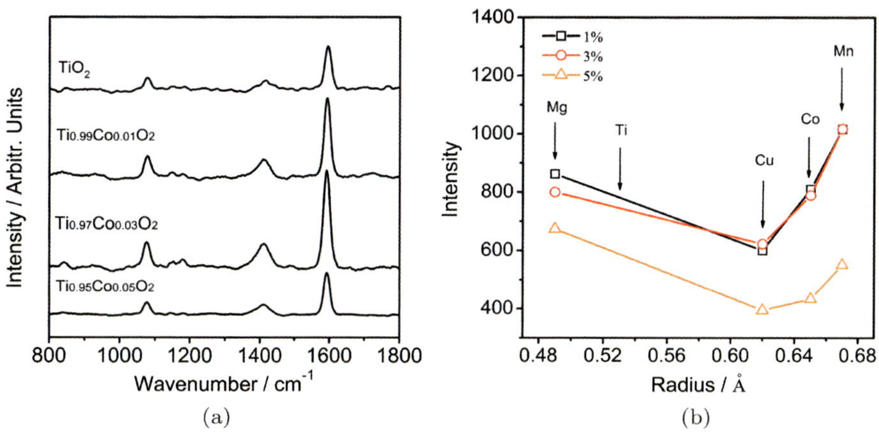

Figure 2. (a) SERS spectra of 4-MBA adsorbed on the $Co_xTi_{1-x}O_2$ Nanoparticles. Reproduced with permission from Xue *et al.*, copyright 2013, Chinese Chemical Society. (b) SERS intensity of the $1594\,cm^{-1}$ mode of 4-MBA adsorbed on doped TiO_2 nanoparticle as function of ionic radius of dopant.

protoporphyrin IX, and crystal violet, and enhance the Raman signal [49, 54–56]. It was found that single-layer graphene provided the highest enhancement effect and the Raman intensities decreased as increase in the number of graphene layers [49]. The surface plasmon of graphene is in the range of terahertz, which is far away from the visible excitation. Thus, the enhancement in Raman signals observed in graphene has been attributed to charge-transfer mechanism. It should be pointed out that the charge-transfer mode is considered as the ground state charge-transfer contribution and chemical bonding is unnecessary for such enhancement [54]. Similar observations have also been found on Si and Ge nanowire, respectively [57].

Besides, the SERS from copper phthalocyanine adsorbed on GaP particles was observed and the enhancement was ascribed to the electromagnetic contribution originated from the Mie near-field theory [46]. Combined with the electric-field simulation, the SERS of *p*-aminobenzoic acid-adsorbed Si microparticle was well studied, and the origin of SERS is regarded as the Mie resonance contribution [47]. It has been found the Mie resonance of Si particles results in the same or even large near-field scattering intensity than those for gold plasmons at a certain particle size. Moreover, the semiconductor micro/nanostructures have the similar electromagnetic contribution to Raman scattering enhancement based on semiconductors. Alessandri *et al.* have reported so-called all-oxide-based core–shell colloidal crystals (CSCCs) formed by atomic layer deposition of a 100 nm thick TiO_2 film onto SiO_2 microsphere assemblies [50]. The CSCCs exhibit a good enhancement of Raman scattering due to the multiple-reflection effects between the core and shell. Moreover, TiO_2 inverse opal photonic microarray was also applied as a SERS substrate, the enhancement factor of methylene blue adsorbed on TiO_2 invers opal is as high as 10^4 due to the light-matter coupling effect [51].

3. Important Factors for Enhanced Raman Scattering Based on Semiconductors

According to the previous experimental analysis, several important factors, including the size, surface defects, and surface morphology, are found to have remarkable impact on the enhancement of Raman scattering based on semiconductors.

(i) The enhanced Raman scattering is just observed on the semiconductor with the particle size in a particular range. In theory, decreasing the particle size increase the specific surface area and subsequently increase the number of molecules adsorbed on the particle's surface. Thus, the SERS intensity will be enhanced as the particle size decrease. However, for a semiconductor nanoparticle small enough whose size approaches that of the Bohr radius of bulk exciton, the electron energy levels are discrete rather than continuous due to the quantization effects. In this case, the allowed transition levels in semiconductor would plummet, thus resulting in the decrease in charge-transfer contribution to SERS. In some case, the size-dependent SERS for semiconductors has no obvious relationship to the Bohr radius of bulk exciton. For example, the largest enhancements were observed on 29 nm ZnO and 8 nm CdS, respectively [16]. The sizes do not match with their exciton Bohr radius (the exciton Bohr radius is 2.4 for ZnO and 3.0 for CdS). This may be due to the variation in exciton Bohr radius of semiconductor after the adsorption of molecules on its surface.

(ii) The contribution from chemical mechanism is generally believed to be less than 10^2 in metal-based SERS. However, the enhancement factor achieved on semiconductor surfaces ranges from 10 to 10^6, which far exceeds the expectation of chemical enhancement contribution. These results indicate that chemical mechanism for metal-based SERS does not fully apply to that of enhancement of Raman scattering based on semiconductors.

(iii) In most experiments, it has been found that the profile of Raman spectrum observed on semiconductor surface is totally different from that observed on metal surface. The Raman enhancement varies with the excitation wavelength. As such, the relative intensities between different vibrational modes are associated with the excitation wavelength as well. Furthermore, the half-peak width of Raman band becomes narrow as the particle size decrease, which is probably due to the narrow size distribution of semiconductor. These results all indicate that the enhancement of Raman scattering based on semiconductors is a quasi-resonance phenomenon.

(iv) Most semiconductors generally have good photostability under the visible excitation. However, silver halide nanoparticles are susceptible to be reduced to form Ag cluster under the laser illumination during the Raman spectroscopic measurement process, and thus it may be difficult to estimate the real enhancement contribution on silver halide surfaces. Similarly, the molecule adsorbed on semiconductor is susceptible to decomposition under laser irradiation. For the study of SERS on semiconductor substrates, the photostability of the molecule should be carefully considered as well.

(v) The surface defects of semiconductor and its interaction with molecule has crucial contribution to the SERS effect. This is mainly due the fact that the surface defects can bind electron and form the so-called surface states energy levels, providing more energy levels for charge-transfer transitions.

Thus, the surface defects of semiconductor can increase the charge-transfer contribution. It should be pointed out that the surface defects are close to or at a semiconductor surface, and thus the charge-transfer contribution originating from surface defects might be offset by the interaction between adsorbed molecule and surface defects.

(vi) Transition metal doping is a double-edged sword for SERS enhancement, the role of doping depends on the dopant concentration as well as the ion radius. The dopant serves as a mediator of interfacial charge-transfer. Appropriate dopant can increase the Raman intensities of molecule owing to the increased surface states in semiconductor. In contrast, excessive dopant will suppress the charge-transfer due to the electron recombination, resulting the decrease in the Raman intensities of molecule. However, it is not yet clear if there is any distinct relationship between the enhancement contribution of transition metals doping and ion radius of transition metals.

(vii) While the chemical bonding is not essential for the charge-transfer transition in some reports, especially for the dye molecule–semiconductor system, it plays an important role in molecular adsorption, especially in the real sample analysis. Moreover, the chemical bonding may keep the stability of the adsorbed molecule, which benefits to the observation of SERS effects.

(viii) Semiconductor enhanced Raman scattering strongly depends on the excitation wavelength. In most cases, SERS on semiconductor materials is mainly due to the electronic excitation of charge-transfer transition between semiconductor and adsorbed molecules, and thus it generally requires enough energy to excite the charge-transfer transition. Besides, the Mie resonance existing in semiconductor micro/nanostructure is associated with the excitation wavelength as well.

4. Possible Enhancement Mechanism

In general, electromagnetic (EM) and chemical enhancements are the two primary mechanisms operating in metal-based SERS [58–61].

However, the mechanism involved in the enhancement of Raman scattering based on semiconductors are not yet known in detail, because it is difficult to compare the results of different studies owing to the wide variation in material preparation and properties. Moreover, many of new SERS-active semiconductors possess previously unseen or record-breaking enhancement factors, leading to a reconsideration of the intrinsic nature of the enhancement of Raman scattering based on semiconductors. So far, a large amount of research indicates that the enhancement of Raman scattering based on semiconductors is dependent on the size, surface defects, and shape of nanoparticles as well as the morphology of nanostructures. Accordingly, the enhancement can be considered to be derived from various resonances in semiconductor system [62].

4.1. *Plasmon resonance*

The EM effect results from the local electric-field generated by the collective oscillations of surface plasmons in metallic particles [61]. When a molecule is placed in the close vicinity of a metal surface, its Raman enhancement factor is proportional to the 4th power of the local electric field. Theoretically, the EM enhancement is associated with the surface geometry, the size, shape, and aggregation forms of the nanoparticles. The strongest EM enhancement factor is up to 10^{10} when a molecule resides in the narrow gap of two closely approaching nanoparticles (forming the so-called "hotspot"). It should be noted that the surface plasmons of metal nanoparticles or nanostructures (Au, Ag and Cu) generally fall in the visible region, which can be resonantly excited by the corresponding laser used in the Raman experiment, and thus the surface plasmon resonance is dominant in metal-based SERS experiment.

For semiconductor materials, the surface plasmon frequency could be determined by the formula:

$$\omega_{\rm sp} = \left[\frac{4\pi ne^2}{m_e\varepsilon_\infty}\right]^{1/2}. \tag{1}$$

In this equation, n is the electron carrier density and m_e is the effective electron mass. ε_∞ is the high frequency dielectric constant. Since the electron density (n) is low in the conduction band, the conduction band plasmon is considered to fall in the infrared region. The electron density (n) of valence electrons in solids is typically around 10^{22} cm^{-3} to 10^{24} cm^{-3}, and thus the valence band plasmon is expected to fall in the ultraviolet region, rather than in the visible range. The plasmon resonance cannot be excited by visible excitation and could not be responsible for the SERS enhancement in most semiconductor systems.

4.2. *Mie resonance*

High electric field can generate in semiconductor micro- and nanostructures as well, which originates from the optical resonances of the whispering gallery modes (WGMs), also known as Mie resonances. In contrast to plasmons inside the metallic particles, Mie resonances typically are supported by a dielectric particle, and the modes are determined by the size of the particle and its refractive index [63, 64]. The strength of the electric field also amplifies the Raman scattering for a molecule either insider the cavities or attached to them. In general, this resonance becomes more prominent for cavities or particles at larger diameters.

Take the spherical semiconductor for example, the near-field scattering efficiency (Q_{NF}) can be presented as follows:

$$Q_{\mathrm{NF}} = 2\sum_{n=1}^{\infty}\{|a_n|^2[(n+1)|h_{n-1}^{(2)}(ka)|^2 + n|h_{n+1}^{(2)}(ka)|^2] + (2n+1)|b_n|^2|h_n^{(2)}(ka)|^2\}, \tag{2}$$

where a_n and b_n are the scattering coefficients, $h_n^{(2)}$ is the Hankel function of the second kind. The parameter a is the radius of the sphere and $k = 2\pi/\lambda$ is wavenumber of the incident light. Accordingly, the Mie scattering near-field intensity display resonances as a function of the size parameter $x = 2\pi a/\lambda$. Theoretical studies revealed that the Raman enhancement factor is approximately proportional to the second power of the near-field scattering efficiency. The near-field scattering efficiency can be calculated by MATLAB program and the input parameters are the particle size and the refractive index of semiconductor at the calculated wavelength. The Mie scattering near-field intensity display resonances as a function of the size parameter, which shows narrow peaks at a large particle size corresponding to the high-order Mie resonances and a broad structure at low values of the particle size corresponding to the low-order modes resonances. It should be noted that the refractive index is expressed in a complex form ($m = n + ik$), where the real part n is the refractive index and the imaginary part k is extinction coefficient. In principle, the semiconductor materials with high refractive index and low extinction coefficient will show a high near-field scattering efficiency.

There have been only a few examples to discuss the Mie resonance contribution to SERS. Hayashi *et al.* demonstrated firstly the enhanced Raman scattering from copper phthalocyanine on GaP particle using the Mie resonance theory [46]. Lombardi *et al.* studied the Mie resonance contribution on the surface of chemically etched ZnSe. In the recently demonstrated Mie resonance contribution, aspherical particle Si is employed as SERS-active substrate the due to its high refractive index. It has been found the Mie resonance of Si particles results in the same or even large near-field scattering intensity than those for gold plasmons at certain particle size. Beside those described earlier, other micro- and nanostructures with more complex morphologies, such as TiO_2-based core–shell colloidal crystals [50], 2D SiO_2 photonic crystal [65], and TiO_2 photonic microarray [51], have also been designed for the purpose of obtaining desired Mie resonance effect for SERS.

4.3. *Charge resonance*

Charge-transfer resonance is a widely accepted explanation for the mechanism that generates charge-transfer-induced enhancement of Raman scattering based on semiconductors. The charge-transfer model for metal-based SERS is well known using the vibronic theory, which is involved in the vibronic coupling of the molecular states to the metal states. Semiconductors have an energy gap between a full valence band and an empty conduction band. Accordingly, the charge-transfer process would depend on the vibronic coupling

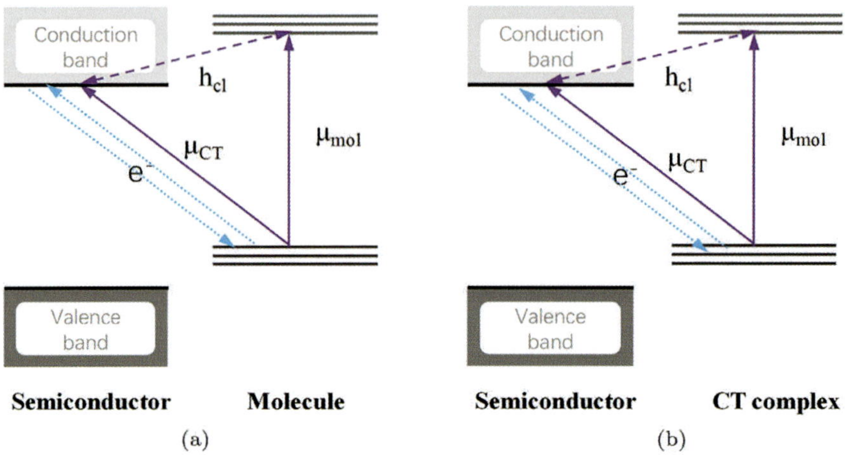

Figure 3. Mechanism of photoexcitation electrons transfer process (blue line) and coupling scheme (purple line) for (a) electron transfer from the HOMO to conduction band of semiconductor and (b) electron transfer from charge-transfer complex to conduction band of semiconductor.

between the conduction band and valence with the molecular excited state and ground state. The charge-transfer pathway is dependent on the band structure compatibility between the semiconductor and the molecule. The charge-transfer in the semiconductor-molecule system could take place in five different pathways.

(i) Electron transfer from the highest occupied molecular orbital (HOMO) of molecule to the conduction band of the semiconductor (see Figure 3(a)). The electron occupying the ground state of the molecule is directly excited from the HOMO to an energy level in the semiconductor corresponding to the conduction band by the incident light. Then the excited electron immediately transits back to some ground vibrational energy level of the molecule and subsequently releases a Raman photon. The SERS intensity is borrowed from some allowed molecular transition μ_{mol} by the molecule to semiconductor transition μ_{CT} through the Herzberg–Teller coupling term h_{cl}.

(ii) Electron transfer from charge-transfer complex to the conduction band of the semiconductor (see Figure 3(b)). The charger-transfer complex is formed by the strong chemical bonding between the molecule and semiconductor. It provides a new electron transfer pathway for the charge-transfer process. The molecular polarizability is enhanced by such a charge-transfer transition, leading to the increase in SERS signals. The SERS intensity is borrowed from the allowed molecular transition μ_{mol} by the charge-transfer complex to semiconductor transition μ_{CT} through the Herzberg–Teller coupling term h_{cl}.

(iii) Electron transfer from the valence band of the semiconductor to the lowest unoccupied molecular orbital (LUMO) of molecule (see Figure 4(a)). The electron in valence band is first excited from the semiconductor to the high energy level in the molecule corresponding to the HOMO, and then quickly transits back to the valence band of the semiconductor, releasing a Raman phonon with the molecule at some vibrational ground state. The SERS intensity is borrowed from some allowed molecular transition μ_{mol} by the semiconductor to molecule transition μ_{CT} through the Herzberg–Teller coupling term h_{vh}.

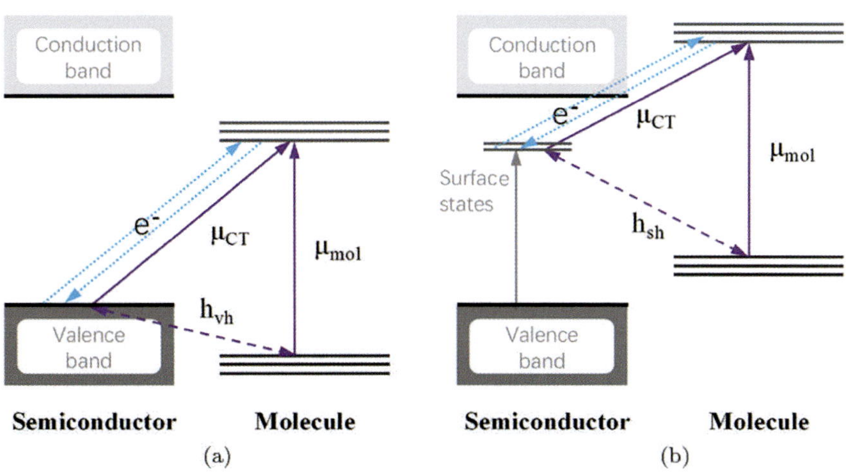

Figure 4. Mechanism of photoexcitation electrons transfer process (blue line) and coupling scheme (purple line) for (a) electron transfer from the valance band of semiconductor to LUMO of molecule and (b) electron transfer from surface state to LUMO of molecule.

(iv) Electron transfer from the surface state of semiconductor to the LUMO of molecule (see Figure 4(b)). The electron is first excited from the valence band to the surface defect, resulting in the formation of a surface state at the surface of the semiconductor. Then, the electron is further excited from the surface state to the HOMO of molecule. The SERS intensity is borrowed from the allowed molecular transition μ_{mol} by the surface state to semiconductor transition μ_{CT} through the Herzberg–Teller coupling term h_{sh}.

(v) Electron injection from the photo-excited molecule to the semiconductor (see Figure 5). Such a change-transfer model usually occurs in the dye molecule-adsorbed semiconductor system. Generally, the dye molecules can be facilely excited by visible light and have more negative potential than the conduction band of the semiconductor, and thus the electrons are initially excited from the HOMO to a higher energy level in the dye corresponding to the LUMO by the incident light energy. Then, the electron is injected into a matching energy level in the semiconductor corresponding to the conduction band through

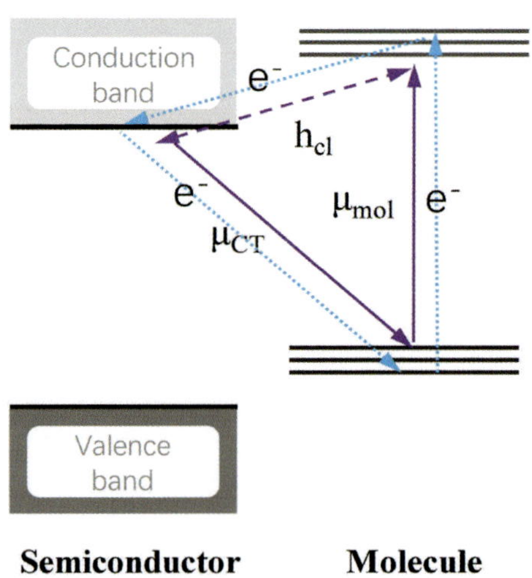

Figure 5. Mechanism of photoexcitation electron transfer process (blue line) and coupling scheme (purple line) for electron transfer from the conduction band of semiconductor to the HOMO of molecule.

resonant tunneling. Subsequently, the electron would transit back to some ground vibrational energy level of the molecule and release a Raman photon with the molecule at some vibrational ground state. Such charge-transfer process gives rise to Raman enhancement, which is similar with the resonance Raman scattering mechanism.

4.4. *Exciton resonance*

As stated above, the energy bands in semiconductor are separated into valence band and conduction band by an energy known as the bandgap. The electron is capable of being excited from the valence band into the conduction band, resulting in strongly correlated electron–hole state in the form of exciton. The exciton Bohr radius is regarded as the distance in an electron–hole pair. Recent studies revealed that the quantum effect of semiconductors is associated with the exciton Bohr radius. As the size of semiconductor nanoparticle is smaller in comparison to the Bohr radius of bulk exciton, quantum confinement effect becomes apparent and gives rise to unique optical properties.

For an exciton resonance process, a vibrational quantum of energy would transfer to the vibrational level of the molecule, while a Raman photon would be radiated from the molecule at some vibrational state. Extensive research suggests that the Raman enhancement factor depends on the size of semiconductor nanoparticles. For example, 5 nm CdSe quantum dots show an enhancement factor as high as 10^5 under 488 nm excitation, which is significantly more than the contribution originated from charge-transfer resonance. The additional enhancement is considered to arise from the exciton resonance contribution because the excitation wavelength is in the vicinity of the exciton resonance. It is still difficult to estimate the exciton resonance contribution to SERS due to the lack of appropriate experimental system.

5. Applications

Semiconductor materials inject new vitality into traditional SERS and may lead to promising functions based on their unique optical, chemical, electrical, and catalytic properties. The applications of enhancement of Raman scattering based on semiconductors widely extend from fundamental studies to real practical systems. For example, the chemical and thermal stability of semiconductor materials enhance the SERS sensing performance including spectral reproducibility, increasing the reliability of SERS-based analytical experiments. Meanwhile, their superior bio-compatibility allows them to have great potential application in bioscience. Besides, the enhancement of Raman scattering based on semiconductors opens the possibility of analyzing the structural and photoelectronic properties of the adsorbents on the semiconductor surfaces, such as dye-sensitized solar cells, lithium ion batteries, and semiconductor-supported photocatalytic process.

5.1. *Merits of semiconductor materials as SERS-active substrates*

The enhancement of Raman scattering based on semiconductors inherits the advantages of SERS including molecular specificity and *in situ* analysis capability. More importantly, semiconductor materials have some inherent physicochemical features that are attractive for theoretical studies and application research. The main features of the enhancement of Raman scattering based on semiconductors technique are summarized as follows:

(i) *Novel surface properties*: The surfaces of semiconductors have better surface bonds to capture analytes, extending the possible systems and functionalities that can be studied. It is worth noting that the superior biocompatibility of semiconductors has assured them an inherent advantage in biochemical and biomedical applications.

(ii) *High chemical stability*: Difficulty with the reproducibility of Raman signals related to substrate degradation has been a major constraint in developing further SERS applications. Semiconductor materials and their structures have superior thermodynamic stability compared with conventional metals. Furthermore, the robustness of semiconductor materials makes them highly stable in various detection environments.

(iii) *Diversity of active substrates*: A variety of semiconductor materials have been examined for SERS and good enhancement effects have been demonstrated in each case. These semiconductor substrates display intriguing structural diversity, including quantum dots, nanowires, nanosheets, 3D nanostructures, and multi-dimensional nanofilms, which could meet the different criteria for a particular purpose.

(iv) *Controllable photoelectrical properties*: The SERS intensity can be borrowed from some allowed molecular transitions by the molecule-to-substrate transition or substrate-to-molecule transition through the Herzberg–Teller coupling term. The band structure of semiconductor materials is classified by the fully occupied valence band and unoccupied conduction band, which can be easily turned by varying the nanoparticle size, shape, material, and doping. Consequently, the charge-transfer transition can be tailored on the basis of analyte, providing more possible resonant contributions to the enhancement.

(v) *Other*: The low cost and diverse synthetic techniques for semiconductor substrates are also particularly important for their applications in SERS. Moreover, semiconductor materials have applications in a wide variety of fields, including (opto)electronics, energy conversion and storage, and (photo)catalysis, which can further broaden the application area of SERS.

5.2. *Biological application*

It is important to study the temporal evolution of a biochemical reaction *in vitro* for further understanding the mechanisms and dynamic regulation *in vivo*. Most biological reactions take place in a complex aqueous environment. Thus, the high selectively of SERS makes this an ideal tool to monitor reaction processes at the molecular level, in particular for the chemistry of enzyme-catalyzed reactions [66]. However, a common problem is the extreme difficulty in retaining biomolecular native structures after adsorption on an undecorated metal (e.g. protein denaturation), which may inhibit bioactivity and function. Moreover, the chemical stability of metals also restricts the operating conditions for SERS application. Semiconductor materials are basically biocompatible and have been frequently proposed as a prospective interface for the immobilization of biomolecules [67]. In addition, semiconductors form coordination bonds with the amine and carboxyl groups of biomolecules, allowing their bioactivity to be maintained. Thus, the enhancement of Raman scattering based on semiconductors is a good candidate for studies and simulations of biological reactions.

Using a combination of SERS and electrochemical experiments, Han *et al.* recently studied the redox process of a heme protein on nanostructured TiO_2 electrodes (see Figure 6(a)) [68] Here, the ellipsoidal TiO_2 nanorods were fabricated by anodization of titanium foils, and their SERS performance was optimized by tuning the anodization voltage for the excitation wavelength of 413 nm, which corresponds to the resonance frequency of the heme structure. The SERS effect is ascribed to local electric-field enhancement

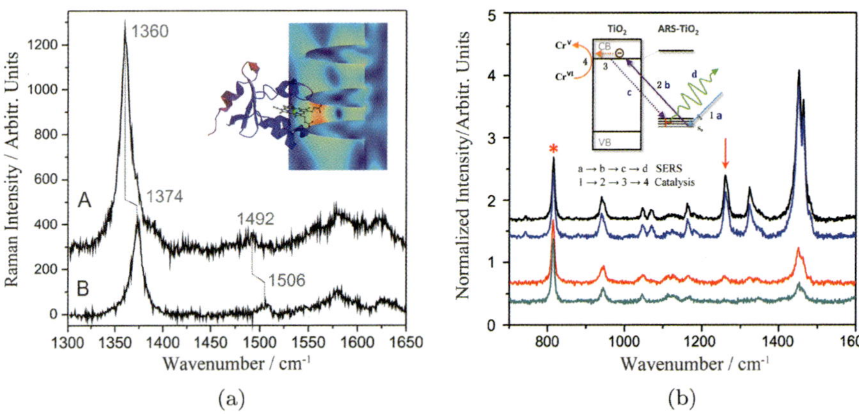

Figure 6. (a) Surface-enhanced resonance Raman spectra of cyt b_5 on the TiO_2 electrodes measured with 413 nm excitation at (A) −0.45 V and (B) 0.0 V. Inset is the calculated electric-field enhancement of TiO_2 nanostructures using half ellipsoids with an aspect ratio of 6. Reproduced with permission from Han *et al.* Copyright 2013, John Wiley and Sons, Ltd. (b) SERS spectra obtained from the ARS–TiO_2 mixtures with the various Cr(VI) concentrations. The concentrations from top to bottom are 0, 10^{-4}, 100 and 140 mM, respectively. Inset: suggested mechanism of photoexcitation charge-transfer process and catalysis process of ARS–TiO_2 complexes. Reproduced with permission from Ref. [12], Copyright 2015, The Royal Society of Chemistry.

originating from the synergistic effect of the adjacent TiO_2 nanorods. Based on the analysis of the obtained results, it was clearly found that the adsorption of the heme protein on the nanostructured TiO_2 electrodes did not induce any change in its enzymatic activity. Compared with the electrochemical SERS experiments on metal electrodes, the attractive feature of this approach is the ability to conduct a SERS study of the electrochemical redox process of a heme protein without complex surface modification and/or the presence of mediators [69].

In addition to the redox process of enzyme proteins, another important type of redox reaction is the thiol-disulfide exchange reaction, which plays a major role in protein folding and heavy metal detoxification [61, 70]. Apparently, conventional SERS approaches are not suitable for investigating this reaction because metals have a strong affinity toward thiol groups. More recently, semiconductor-based SERS has been exploited by Alessandri *et al.* for the detection of a model glutathione-based thiol-disulfide exchange reaction [72]. The all-oxide-based core–shell colloidal crystals (CSCCs) are formed by atomic layer deposition of a 100-nm thick TiO_2 film onto SiO_2 microsphere assemblies. This new substrate was successfully used to analyze disulfide-containing biomolecules without introducing any interfering elements, which may be very useful for monitoring *in situ* the redox reactions of disulfide-containing proteins in biological environments [71].

5.3. *Metal ion and small inorganic molecule sensing*

An entirely different approach toward metal ion and small inorganic molecule sensing was recently developed based on a semiconductor-driven "turn-off" SERS strategy [12, 73]. Several research groups have been focusing on studies of SERS on colloidal semiconductor nanoparticles, extremely small-sized particles wherein quantization effects are expected [9, 11, 13]. The near-field enhancement is negligible for small-sized semiconductor-based SERS because the efficiency of adsorption dominates over the scattering efficiency, as the particle size is less than 10 nm. In fact, mostly, SERS with visible or NIR laser excitation on small-sized semiconductor nanoparticles is induced by charge-transfer transitions between the semiconductor and the molecule, which can produce an enhancement factor on the order of 10^3–10^4. Although the sensitivity is lower than that of noble metal substrates, the unique charge-transfer characteristic in such SERS systems assures their potential application for novel electronic and biological sensing.

Ji *et al.* recently proposed a method for ion detection via monitoring the "turn-off" SERS signal that can efficiently improve the sensitivity of charge-transfer-induced enhancement of Raman scattering based on semiconductors sensing [12]. They employed a dye-sensitized colloidal TiO_2 nanoparticle system to demonstrate this possibility (see Figure 6(b)). In this system, the strong interaction between alizarin red S (ARS) and TiO_2 nanoparticles is conducive to electron transfer and recombination, and thus provides a stable SERS intensity borrowed from the charge-transfer process resulting from the vibronic coupling between the excited molecular level and the conduction band of TiO_2. A decrease in the SERS intensity was observed upon increasing the concentration of Cr(VI) from 0.6 to 140 μM. The most interesting feature of this approach lies in the SERS "turn-off" mechanism, which is a result of the catalysis properties of TiO_2 nanoparticles. Selectivity originates from the energy level matching between the redox potential of Cr(VI)/Cr(V) and the energy level of the semiconductor. This implies that other metal ions could also be sensed by choosing a different semiconductor enhancement system in which the energy level is matched with the redox potential of the targeted metal ion.

In addition to utilizing the properties of semiconductor materials for the sensor design, Ji *et al.* proposed a sensing system for phosphate anion detection based on the first-layer effect of SERS [73]. In acidic conditions (pH 2.3), the association constant of phosphate anions on colloidal TiO_2 surface is similar to that of pyrocatechol. Thus, competitive adsorption between phosphate anions and pyrocatechol leads to a decrease in the SERS intensity of pyrocatechol, which can be used as an index for the detection of phosphate anions. More importantly, this simple competitive active site system provides an excellent experimental system to evaluate the stability and reproducibility of semiconductor-based SERS sensing systems. The results showed that SERS intensities were reduced by only 2% after storage of colloidal TiO_2 nanoparticles at low temperature for one month, and the corresponding sensing abilities were still very robust. Charge-transfer-induced enhancement of Raman scattering based on semiconductors provides a new platform to achieve highly robust SERS sensing with excellent stability and reproducibility.

5.4. *Other applications*

As mentioned in the last section, charge-transfer transitions play a key role in the Raman enhancement of molecules adsorbed on semiconductor surfaces. The charge-transfer intensity can be borrowed from some allowed molecular transitions by the molecule-to-semiconductor transition or semiconductor-to-molecule transition through the Herzberg–Teller coupling. One of the notable observations in these systems is that the intensities of non-totally symmetric modes are more sensitive to such charge-transfer transitions than the totally symmetric modes. Thus, the analysis of the vibrational modes and their associated changes could be efficiently employed to study the nature of the charge-transfer transitions in molecule–semiconductor hybrids, such as dye-sensitized solar cells, photocatalysts, and semiconductor photoelectronic devices [74–77] Besides, remarkable enhancements in phonon modes of semiconductor nanoparticles have been observed as molecules adsorbed on semiconductor surfaces [78]. This is probably attributable to the same mechanism as that responsible for the molecular enhancement, that is, charge-transfer transitions between the molecule and the semiconductor. Thus, the variations in Raman phonon modes of semiconductor nanoparticles also can be used to study the charge-transfer transitions in molecule–semiconductor hybrids.

The doping of semiconductors, which results in surface defects, leads to dramatic changes in their photocatalysis properties, as well as their SERS performance [44, 53, 79]. Carbon-doped TiO_2 has been synthesized and used as a self-cleaning SERS substrate [80]. Moreover, SERS has also been used to study adsorption problems on semiconductor surfaces [81]. With the aid of semiconductor materials, the new SERS phenomenon in these systems would be advantageous to basic research on SERS. We believe that an increasing number of applications in different fields will emerge in the near future with the rapid development of charge-transfer-induced enhancement of Raman scattering based on semiconductors.

6. Summary

This chapter provides an overview of the enhancement of Raman scattering based on semiconductors, including the active materials, important influence factors, possible enhancement mechanism, and their developments in applications. It has been clearly shown that the enhancement of Raman scattering based on semiconductors has been increasingly advanced with the rapid development of nanoscience, and most semiconductor materials have been demonstrated as the active SERS substrates, which are capable of amplifying the Raman signals of adsorbed molecules by 1–6 orders of magnitude. The enhancement is considered to be derived from various resonances, such as Mie resonance, charge-transfer resonance, molecular resonance, and exciton resonance. Charge-transfer-induced enhancement of Raman scattering based on semiconductors has already demonstrated some promising applications. However, more work will be required to fully understand the enhancement mechanism and to fully exploit the potential of this technique.

Acknowledgments

The authors thank their coworkers in the studies reported in this article, particularly Prof. John R Lombardi, Dr. Yanfei Wang, Libin Yang, Zhihua Sun, and Xiangxin Xue for their significant contributions to the studies. The related research was supported by the National Natural Science Foundation (20473029, 20873050, 21273091 and 21603021) of P. R. China and the Fundamental Research Funds for the Central Universities (DUT15RC(3)115).

References

1. Acroca, R. (2006). *Surface-Enhanced Vibrational Spectroscopy*, 1st edn. John Wiley & Sons Ltd, United Kingdom.

2. Kneipp, K., Moskovits, M. and Kneipp, H. (2006). *Surface-Enhanced Raman Scattering: Physics and Applications*, 1st edn. Springer, Germany.
3. Schlücker, S. (2010). *Surface Enhanced Raman Spectroscopy: Analytical, Biophysical and Life Science Applications*, 1st edn. Wiley-VCH Verlag & Co. KGaA, Germany.
4. Y. Ozaki, K. Kneipp and Aroca, R. (2014). *Frontiers of Surface-Enhanced Raman Scattering: Single Nanoparticles and Single Cells*, 1st edn. John Wiley & Sons Ltd, United Kingdom.
5. Li, J. F., Huang, Y. F., Ding, Y., Yang, Z. L., Li, S. B., Zhou, X. S., Fan, F. R., Zhang, W., Zhou, Z. Y., Wu, D. Y., Ren, B., Wang, Z. L. and Tian, Z. Q. (2010). Shell-isolated nanoparticle-enhanced Raman spectroscopy. *Nature* **464**, pp. 392–395.
6. Yamada, H., Yamamoto, Y. and Tani, N. (1982). Surface-enhanced Raman scattering (SERS) of adsorbed molecules on smooth surfaces of metals and a metal oxide. *Chem. Phys. Lett.* **86**, pp. 397–400.
7. Yamada, H. and Yamamoto, Y. (1983). Surface enhanced Raman scattering (SERS) of chemisorbed species on various kinds of metals and semiconductors. *Surf. Sci.* **134**, pp. 71–90.
8. Pérez León, C., Kador, L., Peng, B. and Thelakkat, M. (2006). Characterization of the adsorption of Ru-bpy dyes on mesoporous TiO_2 films with UV–vis, Raman, and FTIR spectroscopies. *J. Phys. Chem. B* **110**, pp. 8723–8730.
9. Yang, L. B., Jiang, X., Ruan, W. D., Zhao, B., Xu, W. Q. and Lombardi, J. R. (2008). Observation of enhanced Raman scattering for molecules adsorbed on TiO_2 nanoparticles: Charge-transfer contribution. *J. Phys. Chem. C* **112**, pp. 20095–20098.
10. Fu, X., Bei, F., Wang, X., Yang, X. and Lu, L. (2009). Surface-enhanced Raman scattering of 4-mercaptopyridine on sub-monolayers of α-Fe_2O_3 nanocrystals (sphere, spindle, cube). *J. Raman Spectrosc.* **40**, pp. 1290–1295.
11. Musumeci, A., Gosztola, D., Schiller, T., Dimitrijevic, N. M., Mujica, V., Martin, D. and Rajh, T. (2009). SERS of semiconducting nanoparticies (TiO_2 hybrid composites). *J. Am. Chem. Soc.* **131**, pp. 6040–6041.
12. Ji, W., Wang, Y., Tanabe, I., Han, X., Zhao, B. and Ozaki, Y. (2015). Semiconductor-driven "turn-off" surface-enhanced Raman scattering spectroscopy: Application in selective determination of chromium(vi) in water. *Chem. Sci.* **6**, pp. 342–348.
13. Teguh, J. S., Liu, F., Xing, B. and Yeow, E. K. L. (2012). Surface-enhanced Raman scattering (SERS) of nitrothiophenol isomers chemisorbed on TiO_2. *Chem. Asian J.* **7**, pp. 975–981.
14. Wen, H. A. O. (1996). Surface enhancement of Raman and absorption spectra from cyanine dye D266 adsorbed on ZnO colloids. *Mol. Phys.* **88**, pp. 281–290.
15. Wang, Y. F., Ruan, W. D., Zhang, J. H., Yang, B., Xu, W. Q., Zhao, B. and Lombardi, J. R. (2009). Direct observation of surface-enhanced Raman scattering in ZnO nanocrystals. *J. Raman Spectrosc.* **40**, pp. 1072–1077.

16. Sun, Z., Zhao, B. and Lombardi, J. R. (2007). ZnO nanoparticle size-dependent excitation of surface Raman signal from adsorbed molecules: Observation of a charge-transfer resonance. *App. Phys. Lett.* **91**, p. 221106.
17. Zhang, P., Wang, Y., He, T., Zhang, B., Wang, X., Xin, H. and Liu, F. C. (1988). SERS of pyridine, 1,4-dioxane and 1-ethyl-3′-methyl-2-thiacyanine iodide adsorbed on α-Fe_2O_3 colloids. *Chem. Phys. Lett.* **153**, pp. 215–222.
18. Fu, X., Bei, F., Wang, X., Yang, X. and Lu, L. (2009). Two-dimensional monolayers of single-crystalline α-Fe_2O_3 nanospheres: Preparation, characterization and SERS effect. *Mater. Lett.* **63**, pp. 185–187.
19. Mou, C., He, T., Wang, X., Liu, F., Jiang, J. and Chen, L. (1996). Surface enhanced Raman scattering (SERS) from H_2TSPP and Ag(II)TSPP adsorbed on uniform Fe_3O_4 colloids. *Acta. Phys. Chim.* **12**, pp. 841–844.
20. Zhang, C., Smirnov, A. I., Hahn, D. and Grebel, H. (2007). Surface enhanced Raman scattering of biospecies on anodized aluminum oxide films. *Chem. Phys. Lett.* **440**, pp. 239–243.
21. Wang, X., He, T., Wen, H., Xu, C., Zuo, J. and Liu, F. C. (1997). Surface-enhanced Raman scattering from citrate, azobenzene, pyridine and cyanine dyes adsorbed on Ag_2O colloids. *Spectrochim. Acta A* **53**, pp. 1411–1417.
22. Kudelski, A., Grochala, W., Janik-Czachor, M., Bukowska, J., Szummer, A. and Dolata, M. (1998). Surface-enhanced Raman scattering (SERS) at copper(i) oxide. *J. Raman Spectrosc.* **29**, pp. 431–435.
23. Jiang, L., You, T., Yin, P., Shang, Y., Zhang, D., Guo, L. and Yang, S. (2013). Surface-enhanced Raman scattering spectra of adsorbates on Cu(2)O nanospheres: Charge-transfer and electromagnetic enhancement. *Nanoscale* **5**, pp. 2784–2789.
24. Blajiev, O., Terryn, H., Hubin, A., Soukupova, L. and Geerlings, P. (2006). Adsorption of some benzohydroxamic acid derivatives on copper oxide: Assignment and interpretation of SERS spectra. *J. Raman Spectrosc.* **37**, pp. 777–788.
25. Wang, Y. F., Hu, H. L., Jing, S. Y., Wang, Y. X., Sun, Z. H., Zhao, B., Zhao, C. and Lombardi, J. R. (2007). Enhanced Raman scattering as a probe for 4-mercaptopyridine surface-modified copper oxide nanocrystals. *Anal. Sci.* **23**, pp. 787–791.
26. Jiang, L., Yin, P., You, T., Wang, H., Lang, X., Guo, L. and Yang, S. (2012). Highly reproducible surface-enhanced Raman spectra on semiconductor SnO_2 octahedral nanoparticles. *Chem. Phys. Chem.* **13**, pp. 3932–3936.
27. Wang, Y. X., Wang, Y. F., Gao, Y., Sun, Z. H., Zhao, C., Hu, H. I., Xu, W. Q., Wang, Z. C. and Zhao, B. (2006). Surface-enhanced Raman spectroscopy of 4-mercaptopyridine molecules on Pb_3O_4 nanoparticles. *Chem. Res. Chinese Univ.* **22**, pp. 388–389.
28. Dong, B., Huang, Y., Yu, N., Fang, Y., Cao, B., Li, Y., Xu, H. and Sun, M. (2010). Local and remote charge-transfer-enhanced Raman scattering on one-dimensional transition-metal oxides. *Chem. Asian J.* **5**, pp. 1824–1829.
29. Cong, S., Yuan, Y., Chen, Z., Hou, J., Yang, M., Su, Y., Zhang, Y., Li, L., Li, Q., Geng, F. and Zhao, Z. (2015). Noble metal-comparable SERS

enhancement from semiconducting metal oxides by making oxygen vacancies. *Nat. Commun.* **6**, p. 7800.

30. Wang, Y., Sun, Z., Wang, Y., Hu, H., Zhao, B., Xu, W. and Lombardi, J. R. (2007). Surface-enhanced Raman scattering on mercaptopyridine-capped cds microclusters. *Spectrochim. Acta A* **66**, pp. 1199–1203.
31. Livingstone, R., Zhou, X. C., Tamargo, M. C., Lombardi, J. R., Quagliano, L. C. and Jean-Mary, F. (2010). Surface enhanced Raman spectroscopy of pyridine on CdSe/ZnBeSe quantum dots crown by molecular beam epitaxy. *J. Phys. Chem. C* **114**, pp. 17460–17464.
32. Wang, Y. F., Zhang, J. H., Jia, H. Y., Li, M. J., Zeng, J. B., Yang, B., Zhao, B., Xu, W. Q. and Lombardi, J. R. (2008). Mercaptopyridine surface-functionalized cdte quantum dots with enhanced Raman scattering properties. *J. Phys. Chem. C* **112**, pp. 996–1000.
33. Fu, X., Pan, Y., Wang, X. and Lombardi, J. R. (2011). Quantum confinement effects on charge-transfer between PbS quantum dots and 4-mercaptopyridine. *J. Chem. Phys.* **134**, p. 024707.
34. Wang, Y. F., Sun, Z. H., Hu, H. L., Jing, S. Y., Zhao, B., Xu, W. Q., Zhao, C. and Lombardi, J. R. (2007). Raman scattering study of molecules adsorbed on ZnS nanocrystals. *J. Raman Spectrosc.* **38**, pp. 34–38.
35. Islam, S. K., Tamargo, M., Moug, R. and Lombardi, J. R. (2013). Surface-enhanced Raman scattering on a chemically etched ZnSe surface. *J. Phys. Chem. C* **117**, pp. 23372–23377.
36. Islam, S. K. and Lombardi, J. R. (2014). Enhancement of surface phonon modes in the Raman spectrum of ZnSe nanoparticles on adsorption of 4-mercaptopyridine. *J. Chem. Phys.* **140**, p. 074701.
37. Li, D. W., Wang, J., Xin, H., Xu, S. and Liu, F. (1987). Enhancement origin of SERS from pyridine adsorbed on AgCl colloids. *Spectrochim. Acta A* **43**, pp. 379–382.
38. Wang, X., Wen, H., He, T., Zuo, J., Xu, C. and Liu, F. C. (1997). Enhancement mechanism of SERS from cyanine dyes adsorbed on Ag_2O colloids. *Spectrochim. Acta A* **53**, pp. 2495–2504.
39. Zhang, H., Xin, H., He, T. and Liu, F. (1991). Time dependent surface enhanced Raman spectroscopy of pyridine in AgBr sol. *Spectrochim. Acta A* **47**, pp. 927–932.
40. Mou, C., Chen, D., Wang, X., Zhang, B., He, T., Xin, H. and Liu, F. C. (1991). Surface-enhanced Raman scattering of TSPP, Ag(ii)TSPP, and Pb(ii)TSPP adsorbed on AgI and AgCl colloids. *Spectrochim. Acta A* **47**, pp. 1575–1581.
41. Fu, X. Q., Pan, Y., Wang, X. and Lombardi, J. R. (2011). Quantum confinement effects on charge-transfer between PbS quantum dots and 4-mercaptopyridine. *J. Chem. Phys.* **134**, p. 024707.
42. Quagliano, L. G. (2004). Observation of molecules adsorbed on III–V semiconductor quantum dots by surface-enhanced Raman scattering. *J. Am. Chem. Soc.* **126**, pp. 7393–7398.

43. Yang, L., Zhang, Y., Ruan, W., Zhao, B., Xu, W. and Lombardi, J. R. (2009). Improved surface-enhanced Raman scattering properties of TiO_2 nanoparticles by Zn dopant. *J. Raman Spectrosc.* **41**, pp. 721–726.
44. Yang, L. B., Jiang, X. and Yang, M. (2011). Improvement of surface-enhanced Raman scattering performance for broad band gap semiconductor nanomaterial (TiO(2)): Strategy of metal doping. *Appl. Phys. Lett.* **99**, p. 111114.
45. Xue, X. X., Ji, W., Mao, Z., Li, Z. S., Guo, Z. N., Zhao, B. and Zhao, C. (2013). SERS study of co-doped TiO_2 nanoparticles. *Chem. Res. Chin. Univ.* **29**, pp. 751–754.
46. Hayashi, S., Koh, R., Ichiyama, Y. and Yamamoto, K. (1988). Evidence for surface-enhanced Raman scattering on nonmetallic surfaces: Copper phthalocyanine molecules on gap small particles. *Phys. Rev. Lett.* **60**, pp. 1085–1088.
47. Rodriguez, I., Shi, L., Lu, X., Korgel, B. A., Alvarez-Puebla, R. A. and Meseguer, F. (2014). Silicon nanoparticles as Raman scattering enhancers. *Nanoscale* **6**, pp. 5666–5670.
48. Wang, X., Shi, W., She, G. and Mu, L. (2011). Using Si and Ge nanostructures as substrates for surface-enhanced Raman scattering based on photoinduced charge-transfer mechanism. *J. Am. Chem. Soc.* **133**, pp. 16518–16523.
49. Ling, X., Xie, L. M., Fang, Y., Xu, H., Zhang, H. L., Kong, J., Dresselhaus, M. S., Zhang, J. and Liu, Z. F. (2010). Can graphene be used as a substrate for Raman enhancement. *Nano Lett.* **10**, pp. 553–561.
50. Alessandri, I. (2013). Enhancing Raman scattering without plasmons: Unprecedented sensitivity achieved by TiO_2 shell-based resonators. *J. Am. Chem. Soc.* **135**, pp. 5541–5544.
51. Qi, D., Lu, L., Wang, L. and Zhang, J. (2014). Improved SERS sensitivity on plasmon-free TiO_2 photonic microarray by enhancing light-matter coupling. *J. Am. Chem. Soc.* **136**, pp. 9886–9889.
52. Xue, X. X., Ji, W., Mao, Z., Mao, H. J., Wang, Y., Wang, X., Ruan, W. D., Zhao, B. and Lombardi, J. R. (2012). Raman investigation of nanosized TiO_2: Effect of crystallite size and quantum confinement. *J. Phys. Chem. C* **116**, pp. 8792–8797.
53. Xue, X. X., Ji, W., Mao, Z., Zhao, C., Zhao, B. and Lombardi, J. R. (2013). Simultaneous enhancement of phonons modes with molecular vibrations due to Mg doping of a TiO_2 substrate. *Rsc Adv.* **3**, pp. 20891–20895.
54. Ling, X., Moura, L. G., Pimenta, M. A. and Zhang, J. (2012). Charge-transfer mechanism in graphene-enhanced Raman scattering. *J. Phys. Chem. C* **116**, pp. 25112–25118.
55. Ling, X. and Zhang, J. (2010). First-layer effect in graphene-enhanced Raman scattering. *Small* **6**, pp. 2020–2025.
56. Xie, L., Ling, X., Fang, Y., Zhang, J. and Liu, Z. (2009). Graphene as a substrate to suppress fluorescence in resonance Raman spectroscopy. *J. Am. Chem. Soc.* **131**, pp. 9890–9891.
57. Liu, Q., Jiang, L. and Guo, L. (2014). Precursor-directed self-assembly of porous ZnO nanosheets as high-performance surface-enhanced Raman scattering substrate. *Small* **10**, pp. 48–51.

58. Lombardi, J. R. and Birke, R. L. (2009). A unified view of surface-enhanced Raman scattering. *Acc. Chem. Res.* **42**, pp. 734–742.
59. Moskovits, M. (2005). Surface-enhanced Raman spectroscopy: A brief retrospective. *J. Raman Spectrosc.* **36**, pp. 485–496.
60. Otto, A. (2005). The 'chemical' (electronic) contribution to surface-enhanced Raman scattering. *J. Raman Spectrosc.* **36**, pp. 497–509.
61. Brown, R. J. C., Wang, J., Tantra, R., Yardley, R. E. and Milton, M. J. T. (2006). Electromagnetic modelling of Raman enhancement from nanoscale substrates: A route to estimation of the magnitude of the chemical enhancement mechanism in SERS. *Faraday Discuss.* **132**, pp. 201–213.
62. Lombardi, J. R. and Birke, R. L. (2014). Theory of surface-enhanced Raman scattering in semiconductors. *J. Phys. Chem. C* **118**, pp. 11120–11130.
63. Eversole, J. D., Lin, H. B., Huston, A. L. and Campillo, A. J. (1990). Spherical-cavity-mode assignment of optical resonances in microdroplets using elastic scattering. *J. Opt. Soc. Am. A* **7**, pp. 2159–2168.
64. Shi, L., Tuzer, T. U., Fenollosa, R. and Meseguer, F. (2012). A new dielectric metamaterial building block with a strong magnetic response in the sub-1,5-micrometer region: Silicon colloid nanocavities. *Adv. Mater.* **24**, pp. 5934–5938.
65. Christie, D., Lombardi, J. and Kretzschmar, I. (2014). Two-dimensional array of silica particles as a SERS substrate. *J. Phys. Chem. C* **118**, pp. 9114–9118.
66. Murgida, D. H. and Hildebrandt, P. (2008). Disentangling interfacial redox processes of proteins by SERS spectroscopy. *Chem. Soc. Rev.* **37**, pp. 937–945.
67. Tao, F. and Bernasek, S. L., *Functionalization of Semiconductor Surfaces*, 1st edn. John Wiley & Sons Ltd, United Kingdom.
68. Han, X. X., Köhler, C., Kozuch, J., Kuhlmann, U., Paasche, L., Sivanesan, A., Weidinger, I. M. and Hildebrandt, P. (2013). Potential-dependent surface-enhanced resonance Raman spectroscopy at nanostructured TiO_2: A case study on cytochrome b5. *Small* **9**, pp. 4175–4181.
69. Murgida, D. H. and Hildebrandt, P. (2004). Electron-transfer processes of cytochrome c at interfaces. New insights by surface-enhanced resonance Raman spectroscopy. *Acc. Chem. Res.* **37**, pp. 854–861.
70. Schafer, F. Q. and Buettner, G. R. (2001). Redox environment of the cell as viewed through the redox state of the glutathione disulfide/glutathione couple. *Free Radical Bio. Me.* **30**, pp. 1191–1212.
71. Tu, B. P. and Weissman, J. S. (2004). Oxidative protein folding in eukaryotes: Mechanisms and consequences. *J. Cell Bio.* **164**, pp. 341–346.
72. Alessandri, I. and Depero, L. E. (2014). All-oxide Raman-active traps for light and matter: Probing redox homeostasis model reactions in aqueous environment. *Small* **10**, pp. 1294–1298.
73. Ji, W., Song, W., Tanabe, I., Wang, Y., Zhao, B. and Ozaki, Y. (2015). Semiconductor-enhanced Raman scattering for highly robust SERS sensing: The case of phosphate analysis. *Chem. Commun.* **51**, pp. 7641–7644.

74. Finkelstein-Shapiro, D., Petrosko, S. H., Dimitrijevic, N. M., Gosztola, D., Gray, K. A., Rajh, T., Tarakeshwar, P. and Mujica, V. (2013). CO_2 pre-activation in photoinduced reduction via surface functionalization of TiO_2 nanoparticles. *J. Phys. Chem. Lett.* **4**, pp. 475–479.
75. Tarakeshwar, P., Palma, J. L., Finkelstein-Shapiro, D., Keller, A., Urdaneta, I., Calatayud, M., Atabek, O. and Mujica, V. (2014). SERS as a probe of charge-transfer pathways in hybrid dye/molecule–metal oxide complexes. *J. Phys. Chem. C* **118**, pp. 3774–3782.
76. Finkelstein-Shapiro, D., Tarakeshwar, P., Rajh, T. and Mujica, V. (2010). Photoinduced kinetics of SERS in bioinorganic hybrid systems. A case study: Dopamine$-TiO_2$. *J. Phys. Chem. B* **114**, pp. 14642–14645.
77. Mao, Z., Song, W., Xue, X. X., Ji, W., Chen, L., Lombardi, J. R. and Zhao, B. (2012). Multiphonon resonant Raman scattering and photoinduced charge-transfer effects at ZnO-molecule interfaces. *J. Phys. Chem. C* **116**, pp. 26908–26918.
78. Ma, S., Livingstone, R., Zhao, B. and Lombardi, J. R. (2011). Enhanced Raman spectroscopy of nanostructured semiconductor phonon modes. *J. Phys. Chem. Lett.* **2**, pp. 671–674.
79. Xue, X., Ruan, W., Yang, L., Ji, W., Xie, Y., Chen, L., Song, W., Zhao, B. and Lombardi, J. R. (2012). Surface-enhanced Raman scattering of molecules adsorbed on Co-doped ZnO nanoparticles. *J. Raman Spectrosc.* **43**, pp. 61–64.
80. Kiran, V. and Sampath, S. (2012). Enhanced Raman spectroscopy of molecules adsorbed on carbon-doped TiO_2 obtained from titanium carbide: A visible-light-assisted renewable substrate. *ACS Appl. Mater. Inter.* **4**, pp. 3818–3828.
81. Mao, Z., Song, W., Chen, L., Ji, W., Xue, X., Ruan, W., Li, Z., Mao, H., Ma, S., Lombardi, J. R. and Zhao, B. (2011). Metal–semiconductor contacts induce the charge-transfer mechanism of surface-enhanced Raman scattering. *J. Phys. Chem. C* **115**, pp. 18378–18383.

Index

D

O

P